The SAGES Manual of Groin Pain

The SAGES Manual of Groin Pain

Brian P. Jacob, MD, FACS
Mount Sinai School of Medicine, NY, USA

David C. Chen, MD
Lichtenstein Amid Hernia Clinic at UCLA,
Santa Monica, CA, USA

Bruce Ramshaw, MD, FACS
Transformative Care Institute, Daytona Beach, FL, USA

Shirin Towfigh, MD, FACS
Beverly Hills Hernia Center, Beverly Hills, CA, USA

Editors

Springer

Editors
Brian P. Jacob, MD, FACS
Associate Clinical Professor of Surgery, Icahn School of Medicine at Mount Sinai
Partner, Laparoscopic Surgical Center of New York
Regional Medical Director
New York, NY, USA

Bruce Ramshaw, MD, FACS
Co-Director, Advanced Hernia Solutions at Transformative Care Institute
Chief Medical Officer, Surgical Momentum
Chairman, the Bruce Kennedy General Surgery Residency Program at Halifax Health
Associate Clinical Professor, Florida State University
Daytona Beach, FL, USA

David C. Chen, MD
Associate Professor of Clinical Surgery
Clinical Director, Lichtenstein Amid Hernia Clinic, Department of Surgery, David Geffen School of Medicine at UCLA
Santa Monica, CA, USA

Shirin Towfigh, MD, FACS
President, Beverly Hills Hernia Center
Beverly Hills, CA, USA

Videos can also be accessed at http://link.springer.com/book/10.1007/978-3-319-21587-7

ISBN 978-3-319-21586-0 ISBN 978-3-319-21587-7 (eBook)
DOI 10.1007/978-3-319-21587-7

Library of Congress Control Number: 2015953840

Springer Cham Heidelberg New York Dordrecht London

Printed on acid-free paper

Springer International Publishing AG Switzerland is part of Springer Science+Business Media (www.springer.com)

Foreword

Inguinal hernia repair is one of the most common operations done by general surgeons in practice today. In the early 1990s, the surgical approach to inguinal hernias underwent a major transformation from primary sutured tissue repairs to the widespread use of tension-free mesh repairs using both open and laparoscopic techniques. The primary reason for this shift was twofold: (1) there was an unacceptably high recurrence rate with primary tissue repairs, and (2) the tension-free approach allowed an earlier return to full and unrestricted activity. However, this move toward mesh-based repairs has come with a price—that is, a higher rate of chronic or lingering groin pain post-herniorrhaphy.

The SAGES Manual of Groin Pain provides a comprehensive look at all aspects of groin pain that might be encountered by a surgeon in practice. The book is organized into sections according to primary or secondary groin pain with chapters on definitions, the various etiologies, and the approach to diagnosis and management across the spectrum of disorders. It is important to note that not all groin pain is due to a hernia or occurs post-hernia repair. Of particular interest to the reader in this regard will be the section on athletic pubalgia or the so-called "sports hernia," which has gained increasing attention in the sports and general surgery community in recent years and which continues to confound clinicians who are asked to see these individuals.

The book also takes on several contemporary areas of debate in the prevention of groin pain post-hernia repair, including the role of biologic mesh, technical tips and tricks to minimize postoperative pain, and the role of prophylactic neurectomy. It concludes with a unique section of case reports that cover the gamut of difficult groin pain scenarios likely to be encountered as well as patient perspectives on these topics.

Like *The SAGES Manual of Hernia Repair* that preceded it, *The SAGES Manual of Groin Pain* fills an important need in contemporary hernia practice that will be a valued reference for any surgeon who manages these patients. Drs. Jacob, Chen, Ramshaw, and Towfigh are to be commended for bringing this work together into one compendium that should become a mainstay of any hernia surgeon's library.

St. Louis, MO, USA L. Michael Brunt, MD

Contents

Part I Primary Groin Pain

Part IV Case Reports and Patients' Perspectives

Editors and Contributors

Editors

Brian P. Jacob, MD, FACS
Icahn School of Medicine at Mount Sinai, New York, NY, USA
Laparoscopic Surgical Center of New York, New York, NY, USA

David C. Chen, MD
Clinical Surgery, Lichtenstein Amid Hernia Clinic, Santa Monica, CA, USA
Department of Surgery, David Geffen School of Medicine at UCLA, Santa Monica, CA, USA

Bruce Ramshaw, MD, FACS
Advanced Hernia Solutions at Transformative Care Institute, Daytona Beach, FL, USA
Surgical Momentum, Daytona Beach, FL, USA
The Bruce Kennedy General Surgery Residency Program at Halifax Health, Daytona Beach, FL, USA
Florida State University, Daytona Beach, FL, USA

Shirin Towfigh, MD, FACS
Beverly Hills Hernia Center, Beverly Hills, CA, USA

Contributors

Naif A. Al-Enazi, MD, MBA
Clinical Fellow of Laparoscopic and Bariatric Surgery, Mount Sinai Medical Center, New York, NY, USA

Rigoberto Álvarez, MD
Hernia Surgery, Proben Centro Especializado en Biocontención, Guadalajara, Jalisco, Mexico

Parviz K. Amid, MD, FACS
Clinical Surgery, Lichtenstein Amid Hernia Clinic, Santa Monica, CA, USA
Department of Surgery, David Geffen School of Medicine at UCLA, Santa Monica, CA, USA

Jeffrey A. Blatnik, MD
Minimally Invasive Surgery and Abdominal Wall Reconstruction, Department of Surgery, University Hospitals Case Medical Center, Cleveland, OH, USA

Grazia Bombini, MD
General and Day Surgery Unit, Center of Research and High Specialization for the Pathologies of Abdominal Wall and Surgical Treatment and Repair of Abdominal Hernia, Istituto Clinico Sant'Ambrogio, Milan, Italy

Jamin V. Brahmbhatt, MD
The PUR Clinic, Personalized Urology and Robotics, Clermont, FL, USA
Department of Urology, South Lake Hospital in Partnership with Orlando Health, Clermont, FL, USA

Frederick J. Brody, MD, MBA
Department of Surgery, George Washington University Medical Center, Washington, DC, USA

Piero Bruni, MD
General and Day Surgery Unit, Center of Research and High Specialization for the Pathologies of Abdominal Wall and Surgical Treatment and Repair of Abdominal Hernia, Istituto Clinico Sant'Ambrogio, Milan, Italy

Giampiero Campanelli, MD
University of Insubria di Varese, Milan, Italy
General and Day Surgery Unit, Center of Research and High Specialization for the Pathologies of Abdominal Wall and Surgical Treatment and Repair of Abdominal Hernia, Istituto Clinico Sant'Ambrogio, Milan, Italy

Joshua C. Campbell, MD
Orthopedic Surgery, Cedars-Sinai Medical Center, Los Angeles, CA, USA

Marta Cavalli, MD
University of Catania, Milan, Italy
General and Day Surgery Unit, Center of Research and High Specialization for the Pathologies of Abdominal Wall and Surgical Treatment and Repair of Abdominal Hernia, Istituto Clinico Sant'Ambrogio, Milan, Italy

Victor W. Chang, MD
Department of Neurology and Neurosurgery, Henry Ford Hospital, Detroit, MI, USA

Catherine Coyne, BA
University of Arizona College of Medicine, Phoenix, AZ, USA

Lisa A. Cunningham, MD
General Surgery Resident, PGY2, Daytona Beach, FL, USA
Department of General Surgery, Halifax Health Medical Center, Daytona Beach, FL, USA

Jorge Daes Daccarett, MD, FACS
Minimally Invasive Surgery, Clinica Bautista, Barranquilla, Atlántico, Columbia

Ibrahim M. Daoud, MD, FACS
Management Information System, St. Francis Hospital and Medical Center, Hartford, CT, USA
University of Connecticut School of Medicine, Hartford, CT, USA

Valentina De Berardinis, MD
Resident in General Surgery, University of Insubria di Varese, Milan, Italy
General and Day Surgery Unit, Center of Research and High Specialization for the Pathologies of Abdominal Wall and Surgical Treatment and Repair of Abdominal Hernia, Istituto Clinico Sant'Ambrogio, Milan, Italy

Christopher G. DuCoin, MD, MPH
Division of Minimally Invasive Surgery, Minimally Invasive and Advanced Gastrointestinal Surgery, La Jolla, CA, USA
Department of Surgery, University of California San Diego, La Jolla, CA, USA

Brian J. Dunkin, MD, FACS
Clinical Surgery, Weil Cornell College of Medicine, New York, NY, USA
Methodist Institute for Technology, Innovation, and Education, Houston Methodist Hospital, Houston, TX, USA

Katherine Dunn, MD
Department of Surgery, Whidden Memorial Hospital, Cambridge Health Alliance, Boston, MA, USA

David S. Edelman, MD, FACS
The Gallbladder and Laparoscopic Surgery Center of Miami, Miami, FL, USA
Department of SurgeryDoctor's Hospital, Coral Gables, Miami, FL, USA

Edward L. Felix, MD, FACS
Department of Surgery, Marian Regional Medical Center, Santa Maria, CA, USA

Brandie Forman
Advanced Hernia Solutions, Transformative Care Institute, Daytona Beach, FL, USA

Irmina Anna Gawlas, MD
Department of Surgery, David Geffen School of Medicine at UCLA, Los Angeles, CA, USA

Matthew I. Goldblatt, MD, FACS
Department of Surgery, Medical College of Wisconsin, Milwaukee, WI, USA

Jacob A. Greenberg, MD, EdM, FACS
Department of Surgery, University of Wisconsin, Madison, WI, USA

Ahmet Gudeloglu, MD
Memorial Ankara Hospital, Ankara, Turkey

Jeffrey Harr, MD
Department of Surgery, George Washington University Medical Center, Washington, DC, USA

Michael Hibner, MD, PhD, FACOG, FACS
Division of Surgery and Pelvic Pain, Department of Obstetrics and Gynecology, St. Joseph's Hospital and Medical Center, Phoenix, AZ, USA
Creighton University School of Medicine, Omaha, NE, USA
University of Arizona College of Medicine, Phoenix, AZ, USA

David N. Ishimitsu, MD
Department of Radiology, VA West Los Angeles Medical Center, Los Angeles, CA, USA

Garth R. Jacobsen, MD, FACS
Center for the Future of Surgery, San Diego, CA, USA
UCSD Hernia Center, San Diego, CA, USA
Department of Surgery, University of California at San Diego, San Diego, CA, USA

Juzar Jamnagerwalla, MD
Department of Urology, Cedars-Sinai Medical Center, Los Angeles, CA, USA

Evish Kamrava, MD
Nuvo Spine and Sports Institute and Ortho Regenerative Center, Encino, CA, USA

Kent W. Kercher, MD, FACS
Department of General Surgery, Carolinas Medical Center, Charlotte, NC, USA

Howard H. Kim, MD
Male Reproductive Medicine and Microsurgery, Los Angeles, CA, USA
Department of SurgeryCedars-Sinai Medical Center, Los Angeles, CA, USA

Stephanie A. Kingman, MD
Department of Surgery, David Geffen School of Medicine at UCLA, Ronald Reagan UCLA Medical Center, Los Angeles, CA, USA

Johan F.M. Lange Jr., MD
Department of Surgery, The University Medical Center Groningen, Groningen, The Netherlands

Charles H. Li, BS
Department of Neurosurgery, David Geffen School of Medicine at UCLA, Los Angeles, CA, USA

Daniel Lu, MD, PhD
Department of Neurosurgery, David Geffen School of Medicine at UCLA, Los Angeles, CA, USA

Charles Ma, MD
Department of Surgery, University of Minnesota, Minneapolis, MN, USA

Ian T. MacQueen, MD
Department of Surgery, David Geffen School of Medicine at UCLA, Los Angeles, CA, USA

Anuj Malhotra, MD
Pain Management Division, Department of Anesthesiology, Icahn School of Medicine at Mount Sinai, New York, NY, USA

Gregory J. Mancini, MD, FACS
Department of Surgery, University of Tennessee, Knoxville, TN, USA

Dean J. Mikami, MD, FACS
Dept of Gastrointestinal Surgery, Center for Minimally Invasive Surgery, Columbus, OH, USA
The Ohio State University Wexner Medical Center, Columbus, OH, USA

Joseph M. Miller, MD, MS
S. Mark Taper Foundation Imaging Center, Cedars-Sinai Medical Center, Los Angeles, CA, USA

Calin Stefan Moucha, MD
Division of Adult Reconstruction & Joint Replacement,
Leni and Peter W. May Department of Orthopaedic Surgery, Icahn School of Medicine at Mount Sinai, New York, NY, USA

Joseph S. Pachman, MD, PhD, MPH
Adjunct Faculty, University of Connecticut, Storrs, CT, USA

Guy D. Paiement, MD
Orthopedic Surgery, Cedars-Sinai Medical Center, Los Angeles, CA, USA

Sijo J. Parekattil, MD
The PUR Clinic, Personalized Urology and Robotics, Clermont, FL, USA
Department of Urology, South Lake Hospital in Partnership with Orlando Health, Clermont, FL, USA

Warwick J. Peacock, MD, FRCS
David Geffen School of Medicine at UCLA, Los Angeles, CA, USA

Ajita S. Prabhu, MD
Department of Surgery, Case Comprehensive Hernia Center, University Hospitals Case Medical Center, Cleveland, OH, USA

Archana Ramaswamy, MD, FACS, FRCSC
Department of Surgery, Minneapolis VA Health Care System, University of Minnesota, Minneapolis, MN, USA

Wolfgang M.J. Reinpold, MD
Department of Surgery and Hernia Center, Gross Sand Hospital Hamburg, Teaching Hospital of the University of Hamburg, Hamburg, Germany

David B. Renton, MD, MSPH, FACS
Division of General and Gastrointestinal Surgery, Department of Surgery, The Ohio State University Wexner Medical Center, Columbus, OH, USA

James A. Rydlewicz, MD
Department of General Surgery, Aurora Medical Center in Grafton, Grafton, WI, USA

Rola Saouaf, MD
Body MRI and Congenital Cardiac MRI Program, S. Mark Taper Foundation Imaging Center, Cedars-Sinai Medical Center, Los Angeles, CA, USA

Alexander D. Schroeder, MD
Department of Surgery and Hernia Center, Gross Sand Hospital Hamburg, Teaching Hospital of the University of Hamburg, Hamburg, Germany

Jennifer S. Schwartz, MD
Division of General and Gastrointestinal Surgery, Department of Surgery, The Ohio State University Wexner Medical Center, Columbus, OH, USA

Keri A. Seymour, DO
Department of Surgery, Division of Metabolic and Weight Loss Surgery, Duke University, Durham, NC, USA

Matthew Thomas Siedhoff, MD, MSCR
Department of Obstetrics and Gynecology, Division of Advanced Laparoscopy and Pelvic Pain, University of North Carolina at Chapel Hill, Chapel Hill, NC, USA

Shane D. Smith, MD
Columbus Radiology, Columbus, OH, USA

M. Jonathon Solnik, MD, FACS
Department of Obstetrics and Gynecology, Division of Urogynecology and Center for Minimally Invasive Gynecologic Surgery, Cedars-Sinai Medical Center, Los Angeles, CA, USA

Nathaniel F. Stoikes, MD, FACS
Department of Surgery, University of Tennessee Health Science Center, Memphis, TN, USA

David S. Strosberg, MD
Department of Surgery, The Ohio State University Wexner Medical Center, Columbus, OH, USA

Payam Vahedifar, MD
Physical Medicine and Rehabilitation and Pain, Nuvo Spine and Sports Institute and Ortho Regenerative Center, Encino, CA, USA

Guy R. Voeller, MD, FACS
Department of Surgery, University of Tennessee Health Science Center, Memphis, TN, USA

David L. Webb, MD, FACS
Division of Minimally Invasive Surgery, Department of Surgery, University of Tennessee Health Science Center, Memphis, TN, USA

Jin S. Yoo, MD, FACS
Division of Metabolic and Weight Loss Surgery, Duke University, Durham, NC, USA

Part I
Primary Groin Pain

1. Introduction to Primary and Secondary Groin Pain: What Is Inguinodynia?

Brian P. Jacob, David C. Chen, Bruce Ramshaw, and Shirin Towfigh

If dedicated inguinodynologists already existed and were easily accessed by patients with groin pain, we would have much less of a need for this type of manual. However, as of today, inguinodynology is not yet a specialty, nor do specific inguinodynologists readily exist. In fact, patients with groin pain and chronic groin pain either primarily existing or secondarily existing following a hip or hernia surgical procedure are often lost, mainly because they have no place or physician to turn to with this chief complaint. The Internet is filled with an equal number of myths and facts and is often not helpful with finding patients a specialist. With such an extensive differential diagnosis, the optimal treating physician may be a pain specialist, a physical therapist, a psychologist, a radiologist, a gastroenterologist, a general surgeon, an orthopedic surgeon, a urologist, a neurosurgeon, a neurologist, a rehabilitation specialist, a chiropractor, an acupuncturist, a gynecologist, or even a plastic surgeon. The differential diagnosis for groin pain crosses into 15 different specialties, so no wonder patients are lost.

Inguinodynia is the technical term for *groin pain*, and chronic groin pain is a complex topic. In an attempt to organize this complex disease entity, we set out to gather chapters that covered the entire differential diagnosis of a patient with groin pain. In doing so, we quickly realized that patients could be divided into two broad categories: *primary groin pain*, or groin pain not related to a prior surgery (this would include pain after sports, a sprain, or overuse during work), and *secondary groin pain*, or groin pain that began after a surgical procedure (including hernia repairs and orthopedic surgeries). Our chapters are thus divided up

B.P. Jacob et al. (eds.), *The SAGES Manual of Groin Pain*,
DOI 10.1007/978-3-319-21587-7_1

as such. That being said, complex systems science tells us that there will not be a single pathway to work up and cure each groin pain patient, and that each patient should expect an individualized outcome.

If a reader takes away only one message from this entire manual, it is that the single most important initial steps in helping a patient with groin pain, even if they also present with a hernia bulge, is to *take a full and detailed pain history* that focuses on that pain complaint and includes information on the patient's back, hip, groin, pubis, and legs. Never assume that the pain is from the hernia alone. A full and detailed groin pain exam should then follow, which would include documenting any obvious hernias.

Document, document, and document some more. The specific history and exam will often help dictate which approach is optimal for each patient. For primary groin pain, starting your approach with the patient's *back*—evidence for entities that cause groin pain like sacroiliac joint dysfunction, thoracolumbar syndrome, and degenerative disc disease—should be sought. The *hip* pathologies causing groin pain should then be discussed, and include intra- and extra-articular diseases, with femoral acetabular impingement (FAI) and acetabular labral tears being among the more common intra-articular etiologies causing groin pain. Extra-articular hip causes are extensive and include iliopsoas bursitis, trochanteric bursitis, snapping hip syndrome, pelvic stress fractures, obturator nerve (and other nerve) entrapment syndromes, and lumbar radiculopathies. The *pubic bone* itself can be to blame with either osteitis pubis or pubic rami stress fractures. In addition, each *muscle and tendon* that inserts on the pubis can have a tendonopathy, tendonitis, a sprain, or an avulsion injury. Finally, the muscles and tendons of the buttock and leg that insert on the pubic bone can also be sprained or torn, causing groin pain. An adductor sprain is the most common etiology of the leg tendons to blame. The concept of a sports hernia, now accepted as a misnomer, is really just a weak transversalis fascia bulging through a widened internal ring, and is a diagnosis of exclusion when all other disruption injuries have been excluded by exam and MRI.

Nerve compression or entrapments may be to blame and should be considered in the differential. These nerves, which can be compressed or entrapped, include the T12 nerve, the iliohypogastric, the ilioinguinal, the genitofemoral, the lateral femoral cutaneous, the pudendal, and the obturator nerves. *True inguinal hernias* and difficult-to-palpate "occult" hernias are included in the broad differential. To add complexity, there is an additional long list of *GI, GU, and GYN* etiologies for groin pain should the history and exam suggest these. Some etiologies in this list include chronic relapsing appendicitis or diverticulitis, IBD, adhesions, orchitis, prostatitis, and in women, round ligament pain, endometrioma, and endometriosis—to name a few.

When it comes to discussing secondary groin pain, a few definitions are helpful. *Nociceptive pain* is caused by activation of nociceptors by nociceptive molecules due to tissue injury or inflammatory reaction. These signals are then transmitted to the brain via A-delta and C-fibers. *Neuropathic pain* is caused by direct nerve injury due to direct contact of nerves with mesh and/or nerve entrapment by sutures, staples, tacks, folded mesh, or meshoma. For complex pain histories lasting for more than 6 months, or years, where no real etiology can be found, a referral to a pain specialist and/or a neurologist or neurosurgeon can sometimes be helpful in differentiating between *chronic regional pain syndromes type 1 and 2,* and can provide useful information for surgeon and patient alike.

Neuropathic pain complaints following hernia surgery can be subdivided into either a *chronic regional pain syndrome type 1 or type 2*, depending on (a) when the pain began after surgery: after a time lag (type 1) or immediately (type 2), (b) whether or not the pain follows a specific nerve distribution: no (type 1) or yes (type 2), and, finally, (c) whether the pain is alleviated with local anesthetic blocks: no response (type 1) or immediate but temporary response (type 2). It is believed that neurectomy or removal of noxious material has a better chance of resolving pain if type 2 exists while type 1 carries a worse prognosis. Patients with type 1 often need referral to physicians specializing in coping mechanisms and alternative therapeutic remedies. Nonetheless, it is very important to ascertain whether the pain began before or after the hernia repair, and if after the repair, how long after the repair.

Generally (and more practically), pain after an inguinal hernia repair is caused by (a) the material inserted (mesh, tacks, or sutures), (b) an inadequately reduced hernia, or (c) a missed lipoma or hernia. Unfortunately, many recurrent hernias are actually just inadequately dissected hernia fields the first time. The pain resulting from inserted mesh, fixation tacks, technique, or sutures can be caused by direct irritation from the material or by adjacent nerve damage. In an open repair, the nerves that may be involved include the iliohypogastric, ilioinguinal, and the genital branch of the genital femoral nerve. In a laparoscopic repair, those at risk are the lateral femoral cutaneous nerve, the entire genital femoral nerve and its distal branches, and—if tacking—the ilioinguinal and iliohypogastric nerves. A careful history and physical should be able to identify the affected nerve.

The history and workup for secondary groin pain are equally important. An operative report should be obtained and reviewed, and a CT scan and/or a MRI performed. Attention should be directed toward looking for surgical material and recurrences on the CT, and nerve pathology as well as musculoskeletal damage on the MRI. It is important to remember that hip pathologies such as osteoarthrosis, labral tears, and femoral-acetabular impingement syndromes present as groin pain near the internal ring over 70 % of the time, and can be the etiology even if a patient has had hernia repair in the past.

Office-based local nerve blocks can be used for diagnostic purposes, paying careful attention to the nerve that is blocked, whether or not immediate relief is obtained, and when after the block the pain returns. Pain that responds immediately to a local block and remains alleviated for a short time tends to respond better to neurectomy or foreign body removal than does pain that is not decreased immediately. Neurectomy should be used as a last resort, as even a triple neurectomy carries a 10 % failure rate in terms of mitigating the pain.

The choice of operation, if needed for secondary groin pain, will depend on the previous surgery, as well as the patient's response to local and regional nerve blocks, which can be performed for diagnostic purposes, as indicated above. Open surgical procedures can be very useful in treating many of the primary groin pain etiologies that are neuro-muscular-skeletal in origin. Tendonotomies, hernia repairs, neurolysis, fascial strengthening and reappoximation, and neurectomies are just some techniques employed during open surgeries, but the list is extensive. On the other hand, the minimally invasive (laparoscopic or robotic) transabdominal preperitoneal (TAPP) approach is very useful as a diagnostic, and possibly therapeutic, tool for patients presenting with groin pain either primarily or after hernia repairs. However, patients should be warned that all of *these operations carry significant risk* of side effects or injury to vessels, nerves, and surrounding viscera, and these risks must be weighed against the significance of the patient's pain complaint before embarking on remedial surgery. Patients should be educated that there is a chance the surgery will not resolve their pain, but can still contribute greatly to the workup, with the goal being an eventual diagnosis and resolution.

If performing a TAPP for groin pain, the surgeon should mark the patient's pain spot with a marker before the surgery. During the TAPP procedure, potential pain-inducing tacks and mesh can be removed, adhesions can be identified and lysed, viscera can be examined, and the femoral, direct, and indirect spaces can be carefully examined for missed, new, or recurrent hernias.

Neurectomy and laparoscopic neurectomy are emerging once again as a possible last resort surgery that, if performed by an experienced surgeon, can be life-improving for sure. That being said, if a surgeon is not familiar with revisional hernia surgery or neurectomy, the patient should be referred elsewhere.

In conclusion, inguinodynia, or groin pain, can be acute onset or chronic in nature. Patients suffering from either deserve to have easier access to find a physician specializing in groin pain, and we hope a manual like this one will at least inspire more surgeons to master the anatomy, differential diagnosis, history and physical exam, workup, and nonoperative and operative techniques involved with inguinodynia that are required to assist some of these patients. While a groin pain team will always be required to fully treat inguinodynia patients, for a general surgeon, sometimes just being able to facilitate the diagnosis and workup can be as rewarding to these groin pain patients as would be an entire surgery. For just that reason alone, every general surgeon should read this manual.

2. Groin Pain: A Neurologic and Musculoskeletal Anatomic Review

Irmina Anna Gawlas and Warwick J. Peacock

Boundaries of the Inguinal Canal

As shown in Fig. 2.1, the anterior wall of the inguinal canal is the external oblique aponeurosis, reinforced laterally by the internal oblique muscle [1]. The posterior wall is formed by the transversalis fascia laterally and the conjoint tendon medially. The inguinal and lacunar ligaments are the floor. The roof is formed by the fusion of the lowest fibers of the transversus abdominus and internal oblique muscles, which become the conjoint tendon, and insert onto the pubic crest.

Fascial Layers and the Inguinal Rings

In order to understand the course of the various structures that enter and exit the inguinal canal, listed in Table 2.1, the fascial layers and their openings must first be conceptualized. In the male, the *internal ring* is an opening in the transversalis fascia through which the vas deferens enters on its course from the pelvis. The vas meets the gonadal vessels and the genital branch of the genitofemoral nerve as they approach from an inferolateral direction and picks up the cremasteric artery as it branches off the inferior epigastric artery. These comprise the *cord structures* that are covered by the internal spermatic fascia, formed from the connective tissue of the transversalis fascia. Fibers from the internal oblique form a second covering of the cord, the cremaster muscle, and cremasteric fascia. In the female, the internal ring transmits the round ligament, which suspends the uterus anteriorly, and the genital branch of the genitofemoral nerve. The cord structures that approach and pass

B.P. Jacob et al. (eds.), *The SAGES Manual of Groin Pain*,
DOI 10.1007/978-3-319-21587-7_2

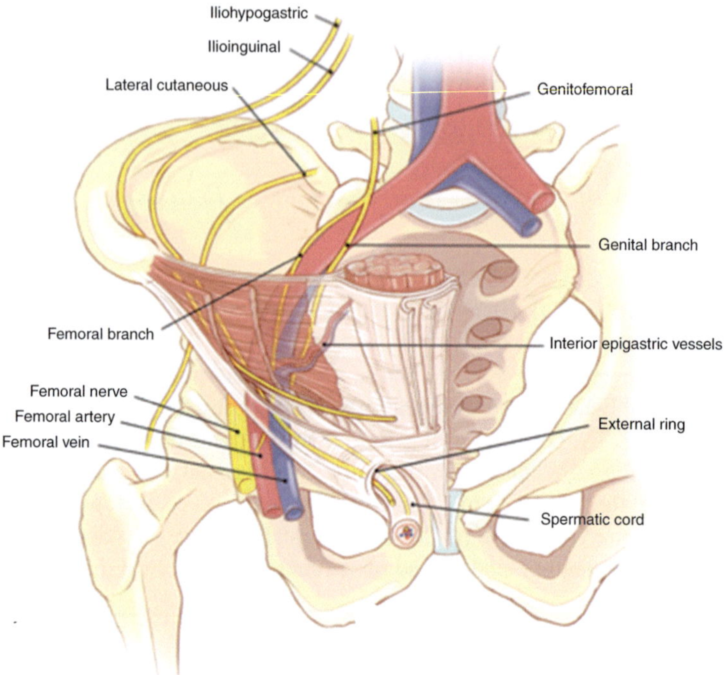

Fig. 2.1. Borders of the inguinal canal (from Wagner et al. [1], with kind permission © McGraw-Hill Education).

Table 2.1. Structures of the inguinal canal.

Vas deferens (or round ligament)
Artery to the vas
Gonadal artery
Cremasteric artery
Cremasteric vein
Gonadal vein (pampiniform plexus)
Genital branch of genitofemoral nerve
Ilioinguinal nerve
Sympathetic nerves
Lymphatics
Internal spermatic fascia
Cremasteric fascia and muscle
External spermatic fascia

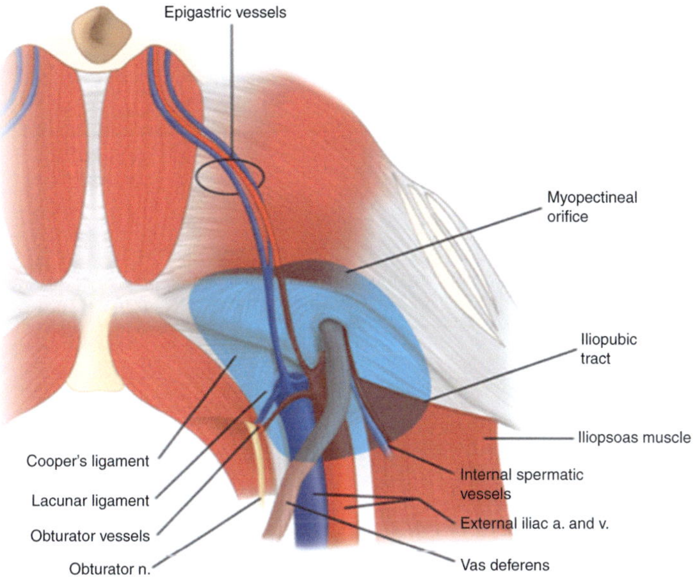

Fig. 2.2. Laparoscopic view of the internal ring (from Wagner et al. [1], with kind permission © McGraw-Hill Education).

through the internal ring are best appreciated from the laparoscopic view, which is illustrated in Fig. 2.2 [1].

The *external ring* is a triangular opening in the aponeurosis of the external oblique muscle. To form the external ring, the aponeurosis splits into a lateral crus, which attaches to the pubic tubercle, and a medial crus attached to the pubic crest. The two crura are held together by intercrural fibers at the apex of the triangular opening. The ilioinguinal nerve and the cord structures destined for the scrotum (or the round ligament into the labium majus in the female) pass through the external ring where fibers from the external oblique continue downward to form the third covering of the cord, the external spermatic fascia.

Some authors refer to a "*third ring*" which is deep and slightly lateral to the internal ring, and is formed by an opening in the preperitoneal fascia [2, 3]. This ring transmits the gonadal vessels and the vas from the visceral space of the preperitoneum into the parietal space of the preperitoneum, where they are joined by the genital branch of the genitofemoral nerve. These fascial planes and spaces, and the structures they contain, are illustrated from an axial view in Fig. 2.3. Mesh placement during laparoscopic repair should be in the visceral compartment;

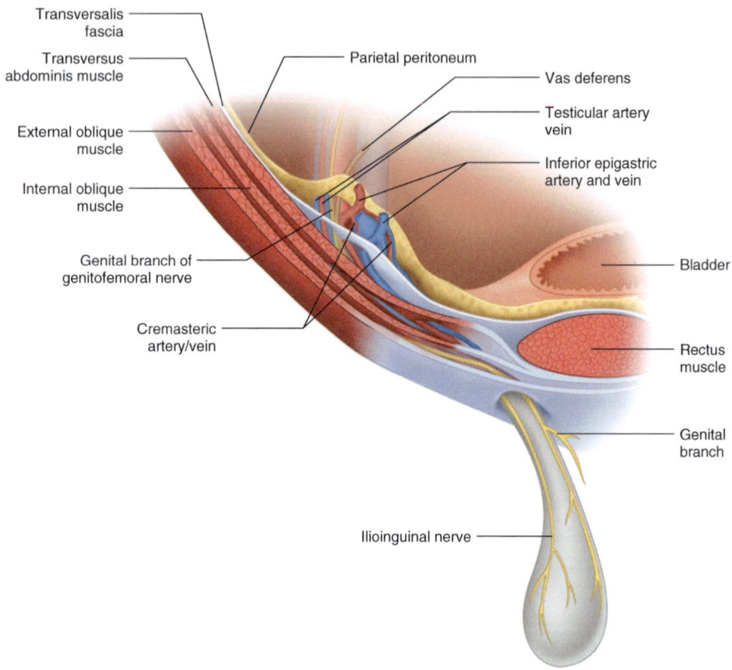

Fig. 2.3. Axial illustration of preperitoneal planes.

incorrect placement in the parietal preperitoneal space is a potential cause of postoperative pain because the mesh may directly contact the genital and/or femoral branches of the genitofemoral nerve and the lateral femoral cutaneous nerve [1].

Anatomical Course and Innervation of the Three Inguinal Nerves

The anterior ramus of L1, with perhaps a contribution from T12, emerges lateral to the psoas muscle and crosses quadratus lumborum in the retroperitoneum. The L1 nerve branches into the upper iliohypogastric and lower ilioinguinal nerves. The *iliohypogastric nerve* pierces the transversus abdominus muscle above the iliac crest, and then travels in the neurovascular plane between the transversus abdominus and the internal oblique muscles, to both of which it provides motor innervation. It then pierces internal oblique at a variable point along the anterior abdominal wall, eventually passing through external oblique about 2.5 cm superior to the external ring, providing sensory innervation to the suprapubic skin [3].

The lower branch of the L1 nerve is the *ilioinguinal nerve*, which similarly pierces the transversus abdominus muscle to travel in the neurovascular plane and then passes through the internal oblique to enter the inguinal canal, providing motor innervation to the lower fibers of both muscles. It runs with the cord structures and exits the canal via the external ring to provide sensory innervation to the overlying skin of the upper medial thigh, anterior scrotum, and base of the penis (or labium majus and mons pubis).

The *genitofemoral nerve* arises from L1 and L2 within the substance of psoas major and then courses downward on its anterior surface. It divides into a genital and a femoral branch. The genital branch travels along the external iliac artery and then ascends to meet the vas deferens at the internal ring. Entering the inguinal canal, it becomes part of the spermatic cord lying on its inferior surface with a companion vein. In the male, it passes into the scrotum via the external ring and provides motor innervation to the cremaster muscle (and therefore is responsible for the cremasteric reflex), and sensory innervation to a small part of the scrotum. In the female, it provides sensory innervation to the mons pubis. The femoral branch passes deep to the inguinal ligament, lateral to the femoral artery, supplying the skin of the upper anterior thigh immediately inferior to the ligament.

The sympathetic nerves of the spermatic cord originate in the T10-11 spinal cord segments and travel with the greater and lesser splanchnic nerves to the pre-aortic plexus. These autonomic fibers then run with the gonadal artery to the testis. The sympathetic efferent fibers are vasomotor, and the afferent fibers conduct testicular pain, which is thus referred to the umbilical region, innervated by the tenth intercostal nerve.

Anatomy of the Femoral Ring

Immediately posterior and inferior to the inguinal ligament is the *femoral ring*. It is bounded posteriorly by the pectineal ligament, medially by the lacunar ligament, and laterally by the femoral vein. Immediately below the inguinal ligament from lateral to medial lie the femoral nerve, femoral artery and vein, both contained within the femoral sheath, and most medially, the femoral canal with its contained fat and lymphatic tissue.

Variations in Neuroanatomy and Intraoperative Considerations

It is estimated that, during open repair, all three inguinal nerves can be distinctly identified in 70–90 % of patients [3]. The anatomy of the iliohypogastric and ilioinguinal nerves is highly variable. The L1 nerve trunks may divide into its two nerves early or late as it crosses quadratus lumborum, or even may continue as one nerve as far as the anterior abdominal wall before dividing. They are generally inversely proportional in size.

Within the inguinal canal, the ilioinguinal nerve may be found within the cremasteric sheath (as opposed to its usual position where it lies on the anterior surface of the cord) [4]. Occasionally, it may not pass through the external ring but may pierce the external oblique aponeurosis more proximally [5].

During open repair, the genital branch of the genitofemoral nerve is prone to injury as it enters the canal at the internal ring when the cord is encircled with a Penrose drain while the floor of the inguinal ligament is being exposed [6].

The iliohypogastric nerve may be injured in open repairs while securing the superior edge of mesh to the aponeurosis of the transversus and internal oblique muscles. During laparoscopy, it can also be injured while the superior edge of the mesh is tacked into place, as this nerve will not be seen where it runs in the neurovascular plane between the transversus and internal oblique [6]. As they approach the internal ring, the genital branch of the genitofemoral nerve will be inferior, and the ilioinguinal nerve lateral, and both also may be inadvertently tacked during laparoscopic repair.

Clearly, an awareness of the classic anatomy, as well as its possible variations, is extremely important to avoid nerve damage.

References

1. Wagner JP, Brunicardi FC, Amid PK, Chen DC. Inguinal hernias. In: Brunicardi FC, Andersen DK, Billiar TR, Dunn DL, Hunter JG, Matthews JB, Pollock RE, editors. Schwartz's principles of surgery. 10th ed. New York: McGraw Hill Medical; 2014. p. 1495–521. Ch. 37.
2. Mirilas P, Mentessidou A, Skandalakis JE. Secondary internal inguinal ring and associated surgical planes: surgical anatomy, embryology, applications. J Am Coll Surg. 2008;206(3):561–70.

3. Amid PK, Hiatt JR. Surgical anatomy of the preperitoneal space. J Am Coll Surg. 2008;207(2):295. Author reply 295–6.
4. Alfieri S, Amid PK, Campanelli G, Izard G, Kehlet H, Wijsmuller AR, Di Miceli D, Doglietto GB. International guidelines for prevention and management of post-operative chronic pain following inguinal hernia surgery. Hernia. 2011;15(3):239–49.
5. Amid PK. Causes, prevention, and surgical treatment of postherniorrhaphy neuropathic inguinodynia: triple neurectomy with proximal end implantation. Hernia. 2004;8(4):343–9.
6. Lichtenstein IL, Shulman AG, Amid PK, Montllor MM. Cause and prevention of postherniorrhaphy neuralgia: a proposed protocol for treatment. Am J Surg. 1988;155(6):786–90.

3. Complex Regional Pain Syndrome Types I and II

Payam Vahedifar and Evish Kamrava

Introduction

Chronic pain is described by the International Association for the Study of Pain as pain lasting greater than 3 months. Some etiologies of chronic groin pain are radicular in nature, such as from lumbar impingement of the L1–3 lumbar nerves, muscle strains and sports hernias, pain from pubic structures and enthesopathy, osteoarthritis of the hip, and labral tears of the hip. These pain problems are often treatable, and despite their chronicity, follow an anatomical and pain pattern that is consistent with the underlying injury.

Additionally, chronic groin pain has become a common problem associated with hernia operations. Pain after inguinal hernia repair can be classified as acute postoperative pain, hernia recurrence, nerve injury, foreign body reaction, and injury due to surgical technique. In the past, hernia repair had complication and recurrence rates of up to 67 %; however, newer techniques using mesh reinforcement with lighter-weight mesh have shown reduction in post-procedure chronic pain. Despite technical improvements, chronic groin pain continues to be a frequent complaint after hernia repair, with incidence of at least 10 %. In the majority of patients, pain can be the result of localized disruptions that are directly the consequence of the operation, such as from the sutures, clips, or scar tissue. It can also be due to entrapped nerves. Pain can also be due to recurrence of the hernia. Revisional procedures and appropriate directed care can resolve the symptoms in a good proportion of patients.

In some patients, however, pain continues or even intensifies despite treatment and revision surgery and, in a subset of patients, may have a faster course. These patients should be further evaluated for the

B.P. Jacob et al. (eds.), *The SAGES Manual of Groin Pain*,
DOI 10.1007/978-3-319-21587-7_3

possibility of complex regional pain syndrome (CRPS). In these circumstances, patients may have pain that is clearly more severe than expected based on their injury. The pain in this subgroup does not usually respond to typical treatments. The pain is often accompanied with allodynia, with hyperesthesia, and with autonomic changes. This subgroup of patients may have complex regional pain syndrome (CRPS).

Complex regional pain syndrome (CRPS) defines a range of painful conditions following an insult, together with abnormal changes in autonomic function. Pain in CRPS is disproportionate to any instigating physical cause or injury. The goal of this chapter is to define CRPS and to help identify the clinical presentations, treatments, and plausible therapy as related to chronic groin pain.

Complex Regional Pain Syndrome Type I and Type II

The terms "CRPS type I and II" have recently been used as a diagnostic and descriptive tool for the previously known syndromes of reflex sympathetic dystrophy and causalgia, respectively. This change in typology occurred to help with diagnosis and to establish specific criteria for the pain syndromes. The hallmarks of CRPS as defined by the International Association for the Study of Pain (IASP) include the following: (a) specific injury or noxious stimuli, which may include surgery; (b) continued pain that is disproportionate to the noxious stimuli or injury, including allodynia and hyperalgesia; (c) changes in localized skin, including edema and changes in blood flow and coloration of the skin; and (d) no specific dermatomal or nerve pattern. This diagnosis is excluded by the existence of other conditions that would otherwise account for the degree of pain and dysfunction [1]. Furthermore, type I designation is used to describe conditions that are not associated with nerve injury, whereas type II is used to describe those that are associated with nerve injury.

CRPS classification can also be clustered into four distinct groups. Group 1 has pain-processing abnormalities such as allodynia and hyperesthesia. Group 2 has skin color changes and vasomotor dysfunction. Group 3 shows edema and pseudo-motor changes such as sweating and piloerection. Group 4 has motor trophic signs such as localized atrophy, spasticity, dystonia, and tremors [2].

Recently proposed clinical criteria for the diagnosis of CRPS include painful conditions that are characterized by continuous pain disproportionate to the degree of the usual course and symptoms of the underlying injury. The pain is regional by definition, and it does not follow a dermatomal or nerve pattern. The patient also must report at least three of the following changes: (a) sensory hyperesthesia, (b) vasomotor changes such as temperature asymmetry, (c) skin color changes, sudomotor changes, and edema such as sweating, piloerection, and swelling, and (d) motor trophic changes such as decreased range of motion, weakness, dystonia, tremors, and changes in hair, nail, and skin. Physical findings on examination at the time of evaluation should contain at least two of the following: (a) evidence of hyperalgesia, (b) evidence of vasomotor changes, (c) evidence of edema, (d) sweating and changes in sweating asymmetry, and (e) motor trophic changes that are noted on examination. Furthermore, there should be no other diagnosis that better explains the underlying conditions symptoms (Table 3.1) [3].

Clinical Management of CRPS

The most important factor in the management of CRPS is early diagnosis and early treatment. CRPS can affect a person at any age but has a peak onset at approximately 37–50 years [4]. Treatment of CRPS should be directed at restoring function and decreasing pain as soon as possible. Therapeutic guidelines include the multidisciplinary approach to treatment of pain, including physical therapy, medical intervention, and behavioral and psychological interventions. These have been proven most effective in the overall treatment of patients and their return to function with effective decrease in pain. Stanton-Hicks et al. reported specific guidelines in therapeutic intervention and algorithm for functional improvement, with the use of physical therapy at a measured pace and time contingent, to alleviate symptoms in a timely manner [5]. The following are essential to the treatment: the process of desensitization of the involved area, mobilization of the area, electrical stimulation if tolerated to reduce the pain secondary to myofascial changes, and isometric strengthening. Furthermore, a range of motion, stress loading, aerobic conditioning, and movement therapies, including vocational and functional rehabilitation, have been proposed in treatment algorithms for patients with CRPS. Adjunctive psychological intervention has been used in parallel with physical therapy to manage expectations, to improve motivation, and to use with behavioral, biofeedback, relaxation, imagery, and hypnosis techniques to help improve overall outcome (Fig. 3.1) [6].

Table 3.1. Criteria for diagnosis of CRPS, per International Association for the Study of Pain.

1. Continued pain that is disproportionate to the inciting event
2. No other diagnosis better explains the signs and symptoms
3. Signs and symptoms below

	Symptom (complaint)	**Sign (physical exam)**
	At least 1 symptom in at least 3 of the following categories	**At least 1 sign at the time of evaluation in at least 2 of the following categories**
Sensory	– Hyperesthesia – Allodynia	– Hyperalgesia to pin prick – Allodynia to light touch – Allodynia to temperature sensation – Allodynia to deep somatic pressure – Allodynia to joint movement
Vasomotor	– Temperature asymmetry – Skin color changes – Skin color asymmetry	– Temperature asymmetry greater than 1 °C – Skin color changes – Skin color asymmetry
Sudomotor/ edema	– Edema – Sweating changes – Sweating asymmetry	– Edema – Sweating changes – Sweating asymmetry
Motor/trophic	– Decreased range of motion – Motor dysfunction such as weakness, tremor, dystonia – Trophic changes such as hair, nail, or skin changes	– Decreased range of motion – Motor dysfunction such as weakness, tremor, dystonia – Trophic changes such as hair, nail, or skin changes

Adapted from Harden et al. [3], with kind permission John Wiley & Sons

Pharmacologic Treatment

Oral pharmacological treatment of CRPS has shown beneficial treatment with early-onset CRPS. The use of corticosteroids in early stages of CRPS has proven effective in some patients. Kozin et al. demonstrated that the pulsed use of steroids in patients with chronic regional pain syndrome showed improvement in 60–80 % of patients after 2 weeks [7]. Similarly, Christensen et al. confirmed decrease in pain in the first 3–4 months [8]. Farah et al. have also shown effectiveness of NSAIDS in some forms of CRPS in early stages of disease [9].

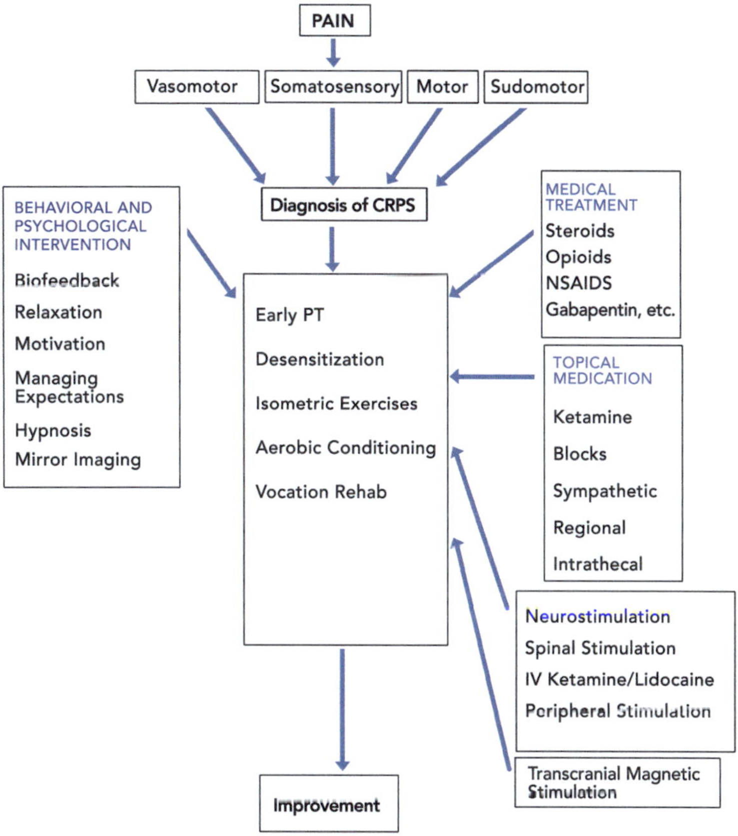

Fig. 3.1. Treatment algorithm for multimodality therapy of patients with CRPS.

Opioid use has not been studied in controlled trials for the treatment of CRPS, although it is frequently used early in the treatment of CRPS. Intravenous use of opioids has shown some response in patients with neuropathic pain and may have a role in the treatment of severe CRPS [10, 11]. Other oral medication treatments such as gabapentin, carbamazepine, valproic acid, phenytoin, and tricyclic antidepressants have also shown beneficial effects in a subgroup of patients with CRPS [12–15].

Topical Medication

Topical medications have shown efficacy in the treatment of neuropathic pain in those with diabetic and postherpetic neuralgia. This treatment may be an intuitive approach in treating allodynia and hyperalgesia at the affected area. Robbins et al. reported significant efficacy in patients with CRPS using large doses of topical capsaicin [16]. Davis et al. studied the topical use of clonidine to relieve the localized hyperalgesia in patients with sympathetically mediated pain, with favorable results [17]. Similarly, topical ketamine use as described by Finch et al. has been effective in reducing symptoms of allodynia among patients with CRPS [18].

Intravenous Medications

Several forms of intravenous medication have shown promise in the treatment of CRPS. There are a number of controlled studies on bisphosphonate such as pamidronate, alendronate, and clodronate, all of which have shown considerable improvement in pain with patients with CRPS. Also, intravenous use of ketamine in the treatment of CRPS has shown a significant reduction in allodynia [19, 20]. Other intravenous medications such as lidocaine as demonstrated by Wallace et al to have shown to help decrease pain in patients with CRPS types I and II [21].

Interventional Therapy

Interventional therapies have been used in conjunction with manual and physical therapy as well as behavioral therapy in order to reduce pain and increase the likelihood of positive outcomes. These treatments should not be used in the absence of multimodality treatment and should not be started if there is no improvement with physical therapy.

Interventional therapies are best used as an adjunctive treatment to decrease pain and to allow faster improvement in symptoms. One such therapy, sympathetic nerve blockade, has been historically used among those with CRPS as a diagnostic and therapeutic intervention to alleviate pain. Several studies have shown a reduction in sympathetically mediated pain with this blockade [22]. These nerve blocks should be continued as long as they provide improvement. If the effect of the nerve blockade plateaus, other treatments should be considered. Intrathecal use of morphine and baclofen has also been studied in the treatment of CRPS [23].

Intravenous regional anesthesia refers to the use of clonidine and lidocaine, as described by Reuben and Sklar. They showed complete pain relief after 4–6 sessions of such treatment [24].

Spinal cord stimulation has been shown in a randomized control trial to alleviate pain among patients with refractory CRPS [25]. Furthermore, Harke et al. found improvement in pain among those who had previously responded well to sympathetic block, with reduction in their level of pain. This was a prerequisite to spinal cord stimulation [26].

Peripheral nerve stimulation and peripheral field stimulation are also viable options when conventional treatment protocols have not provided adequate relief [27]. The use of ultrasound-guided intervention to help directly visualize the affected nerve has allowed for the use of direct visualization techniques, such as imaging, to implant peripheral electrodes without need for surgical dissection. Others have used peripheral field stimulation for localized extremity pain or regional pain rather than directly stimulating the nerve [28]. Finally, the stimulation of the dorsal root ganglion has also been considered [29]. This offers more specific targeting of a regional area and may have value in treatment of refractory conditions.

Other more novel interventional techniques for treatment of CRPS have promising outcomes but have not been studied extensively. One such treatment is transcranial magnetic stimulation. In one study, continued stimulation of the motor cortex by TMS of the affected side showed a decrease in pain as compared to sham stimulation [30].

Although interventional treatment has been a hallmark of decreasing pain and improving symptoms, the use of these interventions has longer-term success when used with conservative treatment including physical and behavioral therapies (see Fig. 3.1).

Summary

The current diagnostic criteria for CRPS are delineated in Table 3.1. However, it is important to understand that the vast majority of patients who have chronic and/or acute pain that is disproportionate to the inciting injury do not fulfill all the diagnostic criteria for CRPS. And yet, at the same time, care must be taken to appropriately treat these patients with a structured plan of early pharmacological therapy, interventional therapy, and restorative therapy in order to avoid a full-blown case of CRPS (see Fig. 3.1).

Interventional therapies, such as nerve blocks, should be initiated to not only help in the diagnosis of the patient's underlying condition but also to help in the differential diagnosis of the underlying cause of their pain. Using topical medications over the affected area that triggers the pain, such as the area of the surgical scar after a hernia repair, should be initiated early in the course of treatment. Also, appropriate use of opioids should be considered.

Psychological factors should always be taken into consideration for treatment of patients. Guided interventional techniques can help with appropriate diagnosis and to help avoid overdiagnosis of CRPS, as may be in the case of certain neuropathic pain, for example. Furthermore, as CRPS is a disabling and devastating disease for both patients and their loved ones, consideration of interventional and novel techniques to help alleviate pain, such as transcranial magnetic stimulation, sympathectomy, spinal cord stimulation and peripheral nerve stimulation, should not be abandoned.

References

1. Merskey H, Bogduk N. Classification of chronic pain: descriptions of chronic pain syndromes and definitions of pain terms. Seattle, WA: IASP Press; 1994.
2. Harden RN, Bruehl S, Galer BS, Saltz S, Bertram M, Backonja M, et al. Complex regional pain syndrome: are the IASP diagnostic criteria valid and sufficiently comprehensive? Pain. 1999;83(2):211–9.
3. Harden RN, Bruehl S, Stanton-Hicks M, Wilson PR. Proposed new diagnostic criteria for complex regional pain syndrome. Pain Med. 2007;8(4):326–31.
4. Sandroni P, Benrud-Larson LM, McClelland RL, Low PA. Complex regional pain syndrome type I: incidence and prevalence in Olmsted county, a population-based study. Pain. 2003;103(1-2):199–207.
5. Stanton-Hicks M, Baron R, Boas R. Complex regional pain syndromes: guidelines for therapy. Clin J Pain. 1998;14(2):155–66.
6. Mugge W, van der Helm FC, Schouten AC. Integration of sensory force feedback is disturbed in CRPS-related dystonia. PLoS One. 2013;8(3):e60293.
7. Kozin F, McCarty DJ, Sims J, Genant H. The reflex sympathetic dystrophy syndrome: I. Clinical and histological studies: evidence for bilaterality, response to corticosteroids and articular involvement. Am J Med. 1976;60(3):321–31.
8. Christensen K, Jensen EM, Noer I. The reflex dystrophy syndrome response to treatment with systemic corticosteroids. Acta Chir Scand. 1982;148(8):653–5.

9. Farah BA. Ketorolac in reflex sympathetic dystrophy. Clin Neuropharmacol. 1993; 16(1):88–9.
10. Portenoy RK, Foley KM, Inturrisi CE. The nature of opioid responsiveness and its implications for neuropathic pain: new hypotheses derived from studies of opioid infusions. Pain. 1990;43(3):273–86.
11. Rowbotham MC, Reisner-Keller LA, Fields HL. Both intravenous lidocaine and morphine reduce the pain of postherpetic neuralgia. Neurology. 1991;41(7):1024–8.
12. Mellick GA, Mellicy LB, Mellick LB. Gabapentin in the management of reflex sympathetic dystrophy. J Pain Symptom Manage. 1995;10(4):265–6.
13. Tan AK, Duman I, Taskaynatan MA, Hazneci B, Kalyon TA. The effect of gabapentin in earlier stage of reflex sympathetic dystrophy. Clin Rehumatol. 2007;26(4):561–5.
14. Perez RS, Zuurmond WW, Bezemer PD, Kuik DJ, van Loenen AC, de Lange JJ, et al. The treatment of complex regional pain syndrome type I with free radical scavengers: a randomized controlled study. Pain. 2003;102(3):297–307.
15. Zollinger PE, Tuinebreijer WE, Kreis RW, Breederveld RS. Effect of vitamin C on frequency of reflex sympathetic dystrophy in wrist fractures: a randomised trial. Lancet. 1999;354(9195):2025–8.
16. Robbins WR, Staats PS, Levine J, Fields HL, Allen RW, Campbell JN, et al. Treatment of intractable pain with topical large-dose capsaicin: preliminary report. Anesth Analg. 1998;86(3):579–83.
17. Davis KD, Treede RD, Raja SN, Meyer RA, Campbell JN. Topical application of clonidine relieves hyperalgesia in patients with sympathetically maintained pain. Pain. 1991;47(3):309–17.
18. Finch PM, Knudsen L, Drummond PD. Reduction of allodynia in patients with complex regional pain syndrome: a double-blind placebo-controlled trial of topical ketamine. Pain. 2009;146(1-2):18–25.
19. Everett A, Mclean B, Plunkett A, Buckenmaier C. A unique presentation of complex regional pain syndrome type I treated with a continuous sciatic peripheral nerve block and parenteral ketamine infusion: a case report. Pain Med. 2009;10(6):1136–9.
20. Koffler SP, Hampstead BM, Irani F, Tinker J, Kiefer RT, Rohr P, et al. The neurocognitive effects of 5 day anesthetic ketamine for the treatment of refractory complex regional pain syndrome. Arch Clin Neuropsychol. 2007;22(6):719–29.
21. Wallace MS, Ridgeway BM, Leung AY, Gerayli A, Yaksh TL. Concentration-effect relationship of intravenous lidocaine on the allodynia of complex regional pain syndrome types I and II. Anesthesiology. 2000;92(1):75–83.
22. Cepeda MS, Lau J, Carr DB. Defining the therapeutic role of local anesthetic sympathetic blockade in complex regional pain syndrome: a narrative and systematic review. Clin J Pain. 2002;18(4):216–33.
23. Zuniga RE, Perera S, Abram SE. Intrathecal baclofen: a useful agent in the treatment of well-established complex regional pain syndrome. Reg Anesth Pain Med. 2002; 27(1):90–3.
24. Reuben SS, Sklar J. Intravenous regional anesthesia with clonidine in the management of complex regional pain syndrome of the knee. J Clin Anesth. 2002;14(2):87–91.

25. Kemler MA, Barendse GA, van Kleef M, de Vet HC, Rijks CP, Furnée CA, et al. Spinal cord stimulation in patients with chronic reflex sympathetic dystrophy. N Engl J Med. 2000;343(9):618–24.
26. Harke H, Gretenkort P, Ladleif HU, Rahman S. Spinal cord stimulation in sympathetically maintained complex regional pain syndrome type I with severe disability. A prospective clinical study. Eur J Pain. 2005;9(4):63–73.
27. McRoberts WP, Roche M. Novel approach for peripheral subcutaneous field stimulation for the treatment of severe, chronic knee joint pain after total knee arthroplasty. Neuromodulation. 2010;13(2):131–6.
28. Huntoon M, Burgher A. Ultrasound-guided permanent implantation of peripheral nerve stimulation (PNS) system for neuropathic pain of the extremities: original cases and outcomes. Pain Med. 2009;10(8):1369–77.
29. Van Buyten JP, Smet I, Liem L, Russo M, Huygen F. Stimulation of dorsal root ganglia for the management of complex regional pain syndrome: a prospective case series. Pain Pract. 2015;15(3):208–16.
30. Pleger B, Janssen F, Schwenkreis P, Völker B, Maier C, Tegenhoff M. Repetitive transcranial magnetic stimulation of the motor cortex attenuates pain perception in complex regional pain syndrome type I. Neurosci Lett. 2004;356(2):87–90.

4. Chief Complaint of Groin Pain: How to Take and Document a Specific Groin Pain History, Exam, and What Studies to Order

Jacob A. Greenberg

Editor's Comment (BPJ)

The chief complaint of groin pain is serious. Surgeons who perform inguinal hernia repairs should develop a thorough understanding of the neuro and musculoskeletal anatomy of the inguinal region so that the history provided by a patient in pain can be expertly matched to a focused pain physical exam. A patient with a complaint of groin pain and a simultaneous inguinal hernia should not automatically be told that the pain is absolutely from the inguinal hernia until a thorough pain physical exam has been performed and documented. Some patients with chronic pain complaints after inguinal hernia repairs have undiagnosed sport injuries that were misdiagnosed as an inguinal hernia, and this situation is absolutely preventable by increasing awareness. Repairing a hernia in a patient with a sports injury risks ongoing pain complaints, excessive office visits and associated costs, and worst of all, an unsatisfied patient. This thorough chapter clearly explains how to take and document a proper history and physical on a patient whose chief complaint is groin pain.

Introduction

When the patient's chief complaint is "pain" in the groin, surgeons need to have an established routine to proceed with a proper evaluation. This routine is very different than if the complaint was for a simple

B.P. Jacob et al. (eds.), *The SAGES Manual of Groin Pain*,
DOI 10.1007/978-3-319-21587-7_4

inguinal hernia. In fact, most hernias present as asymptomatic, or as a bulge without accompanying pain, and so when the patient says they have pain in the region of the groin, the physician's focus must shift. Do not assume that the pain is simply because the patient has a hernia.

Groin pain is a common complaint seen in primary care clinics throughout the world. In sports medicine clinics across the United States, groin pain accounts for roughly 10 % of office visits [1]. While groin pain is a common complaint among elite athletes, there are multiple causes of groin pain, including inguinal and femoral hernias that affect a large percentage of nonathletes as well [2]. With a wide variety of pathologies as potential causes of hip or groin pain, it is imperative to begin all evaluations with a thorough history and physical exam. An accurate history and physical accompanied by appropriate ancillary imaging studies can help to determine the specific cause of groin pain in the majority of patients. In this chapter, we will review the pertinent initial workup for patients presenting with groin pain.

History

As with all other medical problems, the workup for patients with groin pain should begin with a thorough and accurate history of the present illness. The history should start with questions about the spine and back, then the hip, and then the abdominal wall, groin, and accompanying upper leg. This history will guide not only the physical exam maneuvers required for a thorough evaluation but will also lead to prompt diagnosis and treatment through the ordering of appropriate diagnostic tests and referrals if needed.

The pain should be characterized in terms of location, duration, sensation, onset, severity, aggravating or alleviating factors, and pattern of radiation. Additionally, in patients with chronic groin pain, these symptoms may change with time, and this should be elicited as part of the history. Patients with inguinal hernias will frequently note a bulge in the groin, while many other pathologies, such as hip sources of groin pain, musculoskeletal strain, and sports hernias, may not be associated with a bulge. Patients should be asked about pain with specific movements and activities as this may help to further narrow the differential diagnosis. Additionally, patients should be queried about the presence or absence of mechanical symptoms with their gait as this points to the presence of labral tears or loose bodies of the hip as the likely diagnoses [3, 4]. Table 4.1 shows many of the possible causes of groin pain.

Table 4.1. Possible causes of groin pain.

Category	Causes
Hernia	Indirect inguinal hernia
	Direct inguinal hernia
	Femoral hernia
Hip joint injury	Avulsion fracture
	Stress fracture
	Labral tear
	Loose bodies
	Degenerative joint disease
	Femoroacetabular impingement
	Legg–Calvé–Perthes disease
	Slipped capital femoral epiphysis
	Osteonecrosis
Athletic injuries	Sportsmans hernia (inguinal disruption)
	Osteitis pubis
	Adductor muscle strain
Genitourinary	Ectopic pregnancy
	Round ligament pain
	Endometriosis
	Ovarian cyst
	Ovarian torsion
	Varicoceles
	Prostatitis
	Orchialgia
	Urinary tract infection
Gastrointestinal	Appendicitis
	Diverticulitis
	Inflammatory bowel disease
	Irritable bowel syndrome
	Adhesions

Particular attention should be paid to the acuity of the injury. In athletes, many acute groin injuries arise from strains of the adductor muscles or hip flexors and will resolve with conservative measures [5]. Chronic causes of groin pain are less likely to resolve and generally require more definitive treatments [5–8]. Schilders and colleagues classified chronic sources of groin pain in athletes into four separate categories: hip joint injury, osteitis pubis, adductor dysfunction, and sports hernias [9]. Patients with hip joint injuries will commonly indicate their pain with the C sign (Fig. 4.1), where the hand is cupped over the hip in the shape of the letter C with the ipsilateral index finger positioned over the groin and

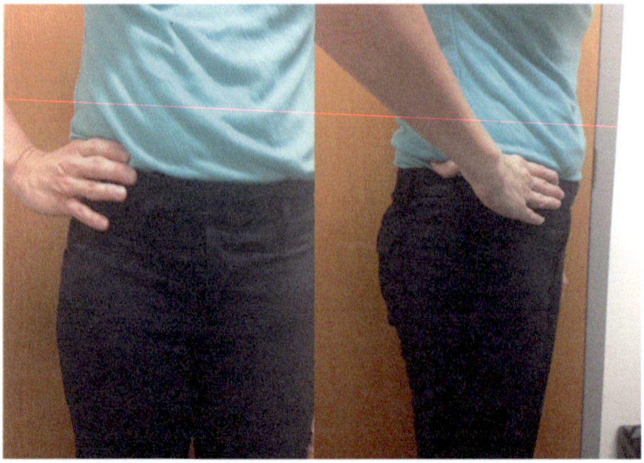

Fig. 4.1. In The "C Sign," the hand is cupped over the hip in the shape of the letter C with the ipsilateral index finger positioned over the groin and the thumb located proximal to the greater trochanter.

the thumb located proximal to the greater trochanter [10]. Patients with adductor injuries often complain of a pulling or tearing sensation in the groin with activity, while those with osteitis pubis note tenderness over the pubic symphysis. Patients with sports hernias typically complain of pain that is unilateral and burning or sharp in nature. The pain may radiate to a variety of locations, including the proximal thigh, lower back, lower abdomen, and downward to the scrotum as well [11]. Patients are generally able to sleep comfortably through the night, but upon awakening may experience extreme pain while attempting to get out of bed. Sudden movements, especially rotational or forceful activities such as sit-ups, cutting, and rapid acceleration or deceleration, will exacerbate these symptoms [5], while periods of rest will often relieve them, only to have them return upon resuming athletic activities [12].

Patients with inguinal hernias as their source of groin pain may experience a different set of symptomatology from those with sports hernias. The pain or discomfort associated with the inguinal hernia tends to be progressive over the course of the day and will be worse even in the evenings. Certain positions that increase intra-abdominal pressure, such as sitting, may exacerbate these symptoms, while lying supine may relieve them and return the hernia contents to their intra-abdominal location. Many patients will note increases in pain with forceful activities such as sneezing, coughing, and bowel movements, and some will reflexively hold their groin during these activities in order to mitigate these symptoms. Unlike patients with sports hernias, those with inguinal or femoral hernias will often notice a bulge and may associate their symptoms with a change in its size. All patients should be asked about any change in bowel or bladder habits to assess for any obstructive type symptoms and should also be asked about any history of incarceration of their hernias.

While traditional hernias, sports hernias, and other sport-related musculoskeletal injuries comprise the majority of causes for groin pain, there are many other congenital and acquired causes of groin pain that should be factored into one's differential diagnosis. Congenital disorders associated with osteonecrosis of the hip, including slipped capital femoral epiphysis, congenital hip dysplasia, and Legg–Calvé–Perthes disease, may all be a source of groin or hip pain in adulthood [13]. Other causes of osteonecrosis such as chronic steroid use or alcohol abuse should also be assessed. Many genitourinary conditions may present with groin pain as well, and these should be discussed in detail, especially with sexually active women and women of childbearing age where the differential diagnosis should include conditions such as pelvic inflammatory disease, ectopic pregnancy, ovarian cysts, endometriosis, ovarian torsion, and round ligament pain.

An accurate and thorough history is the key initial step in the workup of all patients with groin pain. While the history alone may not confirm the diagnosis, it will help determine which physical exam maneuvers and ancillary studies will be most appropriate and helpful in determining the cause of the patients' complaints.

Physical Exam

The physical examination should begin with vital signs, including accurate height and weight. Overweight and obese patients have a lower incidence of inguinal hernia formation compared to normal weight individuals [14, 15]. Routine physical examination of the thorax and abdomen should be performed, but the majority of the physical exam should be focused on the groin and the hip. The back, pelvis, groin, and upper thigh should be completely exposed in order to facilitate a thorough examination.

Examination should begin in the upright position with inspection. Palpation of the spine and paraspinal muscles should be done. Unilateral inguinal hernias will often be apparent as an asymmetric bulge in the inguinal area that may or may not extend into the scrotum in males. Most other pathologies will have no obvious findings on inspection alone. Palpation of the groin should also begin in the upright or standing position. Many inguinal hernias can be palpated simply by placing the hand over the inguinal canal and reducing any hernia contents into their intra-abdominal position. The patient is then asked to cough or to perform the Valsalva maneuver, and the hernia contents should slide past one's fingers. If a hernia cannot be appreciated, the index finger can be placed into the inguinal canal by invaginating the scrotum in male patients. With a finger placed deep in the canal, hernias can again be appreciated with Valsalva or cough. Additionally, the inguinal occlusion test can be performed to determine if the hernia is direct or indirect [16]. With this maneuver, the hernia contents are reduced and manual pressure is applied over the presumed site of the deep inguinal ring. The patient then performs a Valsalva maneuver, and one can observe if the hernia appears with continued compression (direct) or only after release of the internal ring (indirect). While this maneuver may help to differentiate the types of inguinal hernia, its accuracy is relatively low and is not likely to alter the surgical intervention [17, 18]. If no hernia is appreciated, then the groin is similarly examined with the patient lying supine. If hernias still cannot be recognized, then ancillary imaging may be necessary or an alternative diagnosis should be entertained.

Patients with sports injuries, often suspected by the patient's history, should undergo a sequential exam of the back, hip, and groin [19, 20]. For a general surgeon, the exams will in most cases be basic, but even a basic exam will help direct referral or image ordering. The spine and back should be palpated along the thoracic, lumbar, and sacral vertebrae. The paraspinal muscles should also be palpated, and the rare entity of thoracolumbar syndrome should be ruled out when this entity is suspected. The hip should then be examined with some simple maneuvers that examine hip rotation, extension, and flexion. The rest of the exam should focus on the groin, where a firm understanding of the musculoskeletal anatomy will help greatly in figuring out the precise cause of the athletic pubalgia. The rectus muscle insertion on the pubis should be examined with palpation and a sit-up or crunch maneuver while palpating the conjoint tendon Fig. 4.2 [11]. Reproducible pain in this area suggests rectus sheath or conjoined tendon pathology. The pubic tubercle should be palpated and pain with direct pressure can suggest osteitis pubis [21]. Finally, the leg muscles, specifically the adductors and abductors, can be examined by asking the patient to adduct and abduct against resistance and noting any reproducible symptoms [20]. The hip

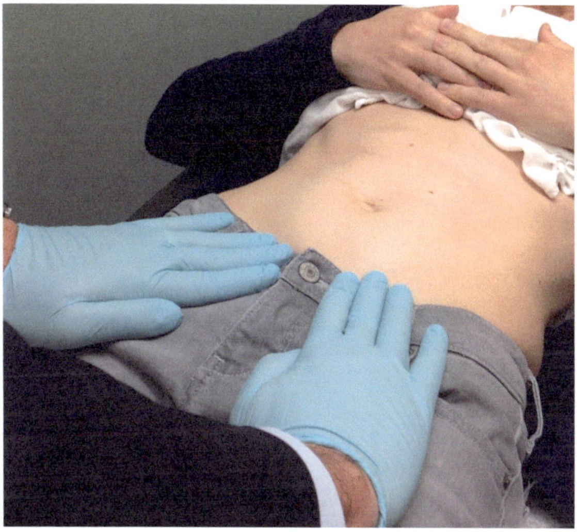

Fig. 4.2. The examiner places pressure on both groins, while the patient actively sits up. Pain indicates a possible inguinal disruption injury.

flexors (iliopsoas and rectus femoris) can also be tested at this point with leg flexion against resistance. If all is normal, yet the patient only has tenderness over the internal ring region of the canal without a palpable hernia, a "sports hernia" can then be suspected.

There are a wide variety of physical exam maneuvers that can be employed to assess for hip joint injury as the cause of groin pain (Table 4.2) [1, 10, 22–25]. Range of motion, strength, and provocative maneuvers may all be necessary; however, the majority of these clinical tests have not been found to be of substantial quality to dictate clinical decision making [26]. The majority of these maneuvers are outside the scope of practice for most general surgeons, and if hip pathology is suspected, then early referral to Sports Medicine or Orthopedic Surgery is warranted, as radiographic studies will likely be necessary to make a firm diagnosis.

Another step in a pain physical exam will depend on the history. For athletes, asking them to mimic the maneuver that was associated with the onset of the symptom can help pinpoint the cause. Having the patient mimic a basketball layup or a defensive tennis stance, for example, may help reproduce the pain of an adductor tear.

Finally, the iliohypogastric, ilioinguinal, and genitofemoral nerve distributions should also be examined and documented regarding the

Table 4.2. Clinical maneuvers for physical examination of the hip [1, 10, 22–25].

Maneuver	Examination procedure	Results	Possible diagnoses
Dynamic internal rotary impingement	Patient is laid supine with the contralateral leg flexed beyond 90° at the hip. The ipsilateral hip is flexed to 90° and passively taken through a wide arc of adduction and internal rotation	Pain	Femoroacetabular impingement Labral tears
Dynamic external rotary impingement	Patient is laid supine with the contralateral leg flexed beyond 90° at the hip. The ipsilateral hip is flexed to 90° and passively taken through a wide arc of abduction and external rotation	Pain	Femoroacetabular impingement Labral tears
Log roll	In the supine position, the leg is internally and externally rotated	Pain and guarding	Synovitis Septic joint
Heel strike	In supine position, the heel is struck firmly	Pain	Femoral neck stress fracture
Foveal distraction	With the patient supine, the leg is actively abducted to 30° and the leg is pulled away from the body	Relief of pain	Intra-articular hip injury
Patrick's test (FABER)	In the supine position, the hip is flexed, abducted, and externally rotated until the lateral heel is proximal to the contralateral knee. The ipsilateral leg is then lowered towards the table	Pain	Sacroiliac joint pathology
McCarthy	In the supine position, the hip is moved from maximal flexion, adduction, and internal rotation to full extension. This is repeated with the hip starting in full abduction as well	Pain in a specific position	Acetabular labral tear

Stinchfield	In the supine position, the leg is actively raised to 30° against downward force from the examiner	Pain or weakness	Arthritis Synovitis Femoral neck fracture Iliopsoas tendinitis
Ely	In the prone position, the knee is passively flexed	Involuntary compensatory hip flexion	Rectus femoris contracture
Gaenslen	In the supine position with both knees flexed, one thigh is extended over the edge of the table	Pain	Spondyloarthritis Arthritis Sciatica
FADIR impingement test	In the supine position, the hip is flexed, adducted, and internally rotated	Pain	Intra-articular hip pathology
Ober	In the lateral position with the knee flexed to 90°, the hip is brought into maximal extension and the leg is adducted	Unable to adduct past midline	IT band contractures Adductor dysfunction
Thomas	Both hips are flexed at the same time while one leg is brought back down to the table	Thigh cannot reach the table	Hip flexor contracture

presence or absence of pain or hypersensitivity along any of the distributions. Pain mapping, described in Chap. 21, is another exam tool to help isolate specific nerve dermatomes that are involved in the pathology.

Radiographic Studies

While radiographic studies certainly play a role in the diagnosis and potential management of groin pain, they are not needed for all patients. In patients with symptomatic inguinal hernias that are palpable on physical exam, radiographic studies are unnecessary, as they will add cost to the workup without significantly changing management. However, for the patient with symptoms suggestive of an inguinal hernia and a normal physical examination, imaging studies may be extremely beneficial. Ultrasonography, magnetic resonance imaging (MRI), computed tomography (CT), and herniography under real-time fluoroscopy can all be used to aid in the diagnosis of occult inguinal hernias. In a systematic review and meta-analysis of imaging techniques in the diagnosis of occult inguinal hernias, herniography was found to have a higher sensitivity and specificity than both ultrasound and computed tomography [27]. Herniography is an invasive procedure and not without its own complications; thus, ultrasound has become popular for the diagnosis of occult hernias in the United States. While ultrasound has a sensitivity of 86 % and a specificity of 77 % [27], it is operator dependent and requires some institutional expertise [28]. Despite being less sensitive and specific than other modalities, both CT and MRI of the groin can be performed to assess for inguinal hernias. Of these two tests, the MRI has less radiation exposure and a higher likelihood of discovering alternative causes of groin pain.

For patients with histories consistent with sports injuries, osteitis pubis, or hip joint injury, MRI of the groin and/or hip is likely to provide the most high-yield information. Sports injuries have a variety of findings seen on MRI. An MRI is dependent on the radiology read, and a specializing sports injury MRI radiologist is often needed to get a thorough enough dictation.

In skilled hands, patients with a true “sports hernia” can also be found to have a protrusion of the transversalis fascia on dynamic ultrasound [8]. Osteitis pubis can be diagnosed on MRI based on signs of inflammation at the pubic symphysis [29]. Hip joint injury such as stress fractures, labral tears, femoroacetabular impingement, and iliopsoas bursitis can all be visualized on MRI as well [29]. Plain films and bone scans may also be helpful in the detection of hip joint pathology such as regular or stress fractures and osteonecrosis.

There are no widely accepted imaging algorithms for the evaluation of groin pain; studies should be ordered as needed on a case-by-case basis. In my institution, herniography is not routinely performed, and thus I utilize ultrasound when needed to diagnose an occult hernia. For patients in whom the history and physical exam is more consistent with an alternative source of groin pain, MRI is my diagnostic test of choice. As expertise varies between institutions, it is prudent to discuss with radiologists which diagnostic test is most high yield in one's hospital or practice setting.

Documentation

Documentation of the history and physical exam should be extremely thorough, both for accuracy and for medicolegal purposes. Whether dictating, utilizing an electronic medical record, or handwriting notes, all of the above aspects of the patient's history should be incorporated into the medical record. Many electronic medical records allow for the creation of templates that may make it easier for practitioners to incorporate all of the pertinent aspects of the patient's history into the patient's visit as well as its documentation. When utilizing templates or copied notes, great care must be taken to amend all documentation as appropriate, as there is a high prevalence of errors, which may affect patient care and expose both the patient and the physician to unnecessary risk [30].

Summary

The treatment of most causes of primary and secondary groin pain will be thoroughly discussed throughout the remainder of this manual. However, the initial step to all evaluations still remains obtaining and documenting a thorough history and physical examination. When combined with appropriate imaging studies, these key initial steps can help to identify the cause of groin pain for the overwhelming majority of patients.

Disclosures **Dr. Greenberg is a paid consultant for Covidien and Bard-Davol.**

References

1. Quinn A. Hip and groin pain: physiotherapy and rehabilitation issues. Open Sports Med J. 2010;4:93–107.
2. Zendejas B, Ramirez T, Jones T, Kuchena A, Ali SM, Hernandez-Irizarry R, et al. Incidence of inguinal hernia repairs in Olmsted County, MN: a population-based study. Ann Surg. 2013;257(3):520–6.
3. Plante M, Wallace R, Busconi BD. Clinical diagnosis of hip pain. Clin Sports Med. 2011;30(2):225–38.
4. Safran M. Evaluation of the hip: history, physical examination, and imaging. Oper Tech Sports Med. 2005;13(1):2–12.
5. Lovell G. The diagnosis of chronic groin pain in athletes: a review of 189 cases. Aust J Sci Med Sport. 1995;27(3):76–9.
6. Hackney RG. The sports hernia: a cause of chronic groin pain. Br J Sports Med. 1993;27(1):58–62.
7. Lynch SA, Renstrom PA. Groin injuries in sport: treatment strategies. Sports Med. 1999;28(2):137–44.
8. Orchard JW, Read JW, Neophyton J, Garlick D. Groin pain associated with ultrasound finding of inguinal canal posterior wall deficiency in Australian Rules footballers. Br J Sports Med. 1998;32(2):134–9.
9. Schilders E, Bismil Q, Robinson P, O'Connor PJ, Gibbon WW, Talbot JC. Adductor-related groin pain in competitive athletes. Role of adductor enthesis, magnetic resonance imaging, and entheseal pubic cleft injections. J Bone Joint Surg Am. 2007;89(10):2173–8.
10. Suarez JC, Ely EE, Mutnal AB, Figueroa NM, Klika AK, Patel PD, et al. Comprehensive approach to the evaluation of groin pain. J Am Acad Orthop Surg. 2013;21(9):558–70.
11. Minnich JM, Hanks JB, Muschaweck U, Brunt LM, Diduch DR. Sports hernia: diagnosis and treatment highlighting a minimal repair surgical technique. Am J Sports Med. 2011;39(6):1341–9.
12. Meyers WC, Foley DP, Garrett WE, Lohnes JH, Mandlebaum BR. Management of severe lower abdominal or inguinal pain in high-performance athletes. PAIN (Performing Athletes with Abdominal or Inguinal Neuromuscular Pain Study Group). Am J Sports Med. 2000;28(1):2–8.
13. Zacher J, Gursche A. 'Hip' pain. Best Pract Res Clin Rheumatol. 2003;17(1):71–85.
14. Ruhl CE, Everhart JE. Risk factors for inguinal hernia among adults in the US population. Am J Epidemiol. 2007;165(10):1154–61.
15. Zendejas B, Hernandez-Irizarry R, Ramirez T, Lohse CM, Grossardt BR, Farley DR. Relationship between body mass index and the incidence of inguinal hernia repairs: a population-based study in Olmsted County, MN. Hernia. 2014;18(2):283–8.
16. Tromp WG, van den Heuvel B, Dwars BJ. A new accurate method of physical examination for differentiation of inguinal hernia types. Surg Endosc. 2014;28(5):1460–4.

17. Moreno-Egea A, Girela E, Canteras M, Martinez D, Aguayo J. Accuracy of clinical diagnosis of inguinal and femoral hernia and its usefulness for indicating laparoscopic surgery. Hernia. 2000;4(1):23–7.
18. Sanjay P, Fulke JL, Shaikh IA, Woodward A. Anatomical differentiation of direct and indirect inguinal hernias: is it worthwhile in the modern era? Clin Anat. 2010;23(7):848–50.
19. Caudill P, Nyland J, Smith C, Yerasimides J, Lach J. Sports hernias: a systematic literature review. Br J Sports Med. 2008;42(12):954–64.
20. Garvey JF, Read JW, Turner A. Sportsman hernia: what can we do? Hernia. 2010;14(1):17–25.
21. Choi H, McCartney M, Best TM. Treatment of osteitis pubis and osteomyelitis of the pubic symphysis in athletes: a systematic review. Br J Sports Med. 2011;45(1):57–64.
22. Evans RC, editor. Illustrated orthopedic physical assessment. 3rd ed. St. Louis, MO: Mosby; 2009.
23. Cleland J, editor. Netter's orthopaedic clinical examination: an evidence-based approach for physical therapists. Carlstadt, NJ: Icon Learning Systems; 2005.
24. Braly BA, Beall DP, Martin HD. Clinical examination of the athletic hip. Clin Sports Med. 2006;25(2):199–210.
25. Brown MD, Gomez-Marin O, Brookfield KF, Li PS. Differential diagnosis of hip diseases versus spine disease. Clin Orthop Relat Res. 2004;419:280–4.
26. Reiman MP, Goode AP, Hegedus EJ, Cook CE, Wright AA. Diagnostic accuracy of clinical tests of the hip: a systematic review with meta-analysis. Br J Sports Med. 2013;47(14):893–902.
27. Robinson A, Light D, Kasim A, Nice C. A systematic review and meta-analysis of the role of radiology in the diagnosis of occult inguinal hernia. Surg Endosc. 2013;27(1):11–8.
28. Light D, Ratnasingham K, Banerjee A, Cadwallader R, Uzzaman MM, Gopinath B. The role of ultrasound scan in the diagnosis of occult inguinal hernias. Int J Surg. 2011;9(2):169–72.
29. MacMahon P, Hodnett P, Koulouris G, Eustace S, Kavanagh E. Hip and groin pain: radiological assessment. Open Sports Med J. 2010;4:108–20.
30. Weis JM, Levy PC. Copy, paste, and cloned notes in electronic health records: prevalence, benefits, risks, and best practice recommendations. Chest. 2014;145(3):632–8.

5. Groin Pain: An Overview of the Broad Differential Diagnosis

Charles Ma and Archana Ramaswamy

Introduction

Groin pain or inguinodynia has a broad differential diagnosis. Different processes, including but not limited to anatomic pathology, neuromuscular, urinary conditions, trauma, and postsurgery, can cause activation of pain fibers in the inguinal region and the subsequent sensation of pain. It is important to separate primary inguinodynia from secondary inguinodynia. This chapter first describes common causes of primary inguinodynia and then briefly discusses groin pain after surgery. Evaluation and management are addressed in detail in subsequent chapters.

Inguinal Hernias and Femoral Hernias

The etiology of inguinal pain can be straightforward if a groin bulge is the chief complaint and a hernia is palpated on exam. Common exacerbating factors to note in the history include major lifting or coughing. Pain occurring later in the day, after prolonged standing or straining, is also consistent with a hernia.

Inguinal hernias are the most common, accounting for 70–75 % of all hernias [1]. They are divided into indirect and direct forms, with the indirect form being the most common. The pathophysiology behind an indirect hernia is a patent processus vaginalis that failed to degenerate after descent of testes during fetal development. This potential space allows intra-abdominal contents to pass from the deep inguinal ring to the superficial inguinal ring. Direct hernias are protrusions within Hasselbach's triangle, directly through a weakened posterior wall of the

B.P. Jacob et al. (eds.), *The SAGES Manual of Groin Pain*,
DOI 10.1007/978-3-319-21587-7_5

inguinal canal. Finally, femoral hernias account for a small percentage of hernias. Classically quoted in textbooks as the third most common type of primary hernia, femoral hernias account for 20 % of hernias in females and 5 % in males [2]. They occur distal to the inguinal ligament through a defect in the femoral ring, which is bound anteriorly by the inguinal ligament, posteriorly by the iliopectineal ligament, medially by the lacunar ligament, and laterally by the femoral vein. Though altogether representing less than 10 % of all hernias, femoral hernias tend to present more emergently with strangulation or incarceration of bowel [3]. Definitive treatment of all hernias is surgical. Options include open repair with or without mesh and laparoscopic repair with mesh.

Hip and Groin Pain in the Athlete

Athletes are a population of special consideration when it comes to hip and groin pain, a symptom not uncommonly experienced by those engaged in activities such as soccer, rugby, football, and ice hockey. The aforementioned sports involve extensive use of the adductors and hip flexors. Osteitis pubis, fractures, stress fractures, joint disorders producing referred hip pain, bursitis, hernias, muscular pain, tendonitis, tendon or ligament injury, and nerve impingement are just some of the afflictions that may result in groin pain [4]. Separating etiologies into extra- and intra-articular disease processes can help narrow the differential. Extra-articular causes include iliac apophysis injury, iliopsoas tendinosis, bursitis, snapping iliopsoas tendon, and athletic pubalgia. Intra-articular causes include acetabular labral tears and femoroacetabular impingement [5]. Fortunately, most groin pain is the result of muscle strain and will resolve with rest. When conservative management (including physical therapy) fails and other etiologies have been excluded, the diagnosis of a sports hernia is made as one of exclusion.

First described by Gilmore when three professional soccer players presented to him with unclear groin pain refractory to medical management, sports hernias, also known as "Gilmore's groin," "athletic pubalgia," or "groin disruption," represent a small subset of groin pain experienced by high-performance athletes [6]. It is a condition of chronic inguinal pain caused by weakness in the abdominal wall without a palpable hernia. Its true prevalence is difficult to pinpoint, as the diagnosis remains a clinical entity that is poorly understood, with estimates ranging from 5 to 28 % [7]. The etiology, pathophysiology, and surgical treatments have all been variously described in the literature. Pain is located at the confluence of the origin of the rectus abdominis muscle, the adductor longus tendon on the pubic bone, and the insertion of the inguinal ligament on the pubic bone [8]. Pain onset is typically insidious, exacerbated by activity and improved with rest. Various surgical techniques have been reported, ranging from standard inguinal hernia repair with or without mesh, to incorporation of rectus reattachment in combination with adductor release in select cases. A commonly found area of pathology reported in the literature is the posterior inguinal wall along the transversalis fascia [9].

Referred Groin Pain from Lumbar Disc Herniation

Referred groin pain in the absence of low back or radicular pain is found in a small subset of patients with singular lumbar disc herniation. A retrospective study of 512 subjects diagnosed with singular lower lumbar disc herniation (L4-L5 and L5-S1) at Kakegawa City General Hospital between July 1990 and December 1993 reported a 4.1 % incidence of groin pain, especially in the subset of patients with L4-5 involvement [10]. A subsequent prospective study in 2010 found evidence supporting degenerated intervertebral disc as an etiology for referred groin pain: ten subjects with groin pain and single disc degeneration found on MR underwent evaluation of changes in pain scale after local hip joint block, pain provocation on discography, and anesthetic discoblock. All ten subjects had a negative hip joint block, while five showed pain on discography and improvement in pain with discoblock, and definitive improvement after surgical fusion [11]. The proposed mechanism based on physiology studies in rats is the existence of overlapping segments of dorsal root innervation for the sensory nerve endings in the lower lumbar discs, with some of the sensory nerves from the L5 intervertebral disc coming from upper dorsal root ganglions of L2, which supply the genitofemoral and ilioinguinal nerves [12]. Thus, it is possible for patients to feel referred groin pain corresponding to the L2 dermatome.

Spermatic Cord and Testicular Causes

The urologic etiologies for groin pain are quite extensive, including but not limited to epididymitis, hematocele, hydrocele, varicocele, malignancy, orchitis, Fournier's disease, and testicular torsion.

Categorizing etiologies into urgent conditions and emergent conditions requiring surgical intervention helps to elucidate workup and management. Of the aforementioned etiologies, appendage torsion, epididymitis, and testicular torsion make up most of the presentations of an acute scrotum [13]. Appendage torsion occurs primarily in prepubescent males, while the latter two occur more commonly in adolescents [14]. The true frequency of these three conditions is difficult to describe due to variations in age distribution and study settings in the medical literature. A recent retrospective review of 523 pediatric emergency department visits presenting with an acute scrotum found only a 3.25 % incidence of testicular torsion, while epididymitis, appendage torsion, and scrotal pain of unknown etiology accounted for 32.3 %, 7.7 %, and 34 %, respectively [15]. In contrast, a prior retrospective review of 238 cases of acute scrotal pain encountered in a similar pediatric emergency department setting published in 1995 reported incidences of testicular torsion, torsion of a testicular appendage, and epididymitis to be 16 %, 46 %, and 35 %, respectively [16]. Despite the reported disparity, the final diagnosis of scrotal pain of unknown etiology is not uncommon as previously mentioned. Physical exam findings such as Prehn's sign (relief of pain with scrotal elevation) and assessment of cremasteric reflex are used in combination with ultrasound Doppler imaging to make the appropriate diagnosis, though surgical exploration remains the only definitive modality for assessing testicular torsion [17]. When vasculature of the testicle is compromised, prompt surgical intervention within 6 h of pain onset has demonstrated greater than 90 % rate of salvage [18].

Gynecologic Causes

Analogous to urologic etiologies in males, various gynecologic conditions can also have groin pain as a presenting symptom. A 2014 retrospective study of 290 females of reproductive age presenting with right lower quadrant abdominal pain found gynecologic pathology as the etiology in 12.8 % [19]. The differential diagnosis includes but is not limited to ectopic pregnancy rupture, ovarian cyst rupture, corpus hemorrhagicum cyst rupture, and adnexal torsion. Similar to management of male urologic conditions, preserving fertility remains the goal in management.

Ovarian cyst rupture can produce pelvic or groin pain secondary to blood from the ruptured follicle irritating the peritoneum. A pregnancy test and beta hCG can help quickly diagnose an ectopic pregnancy [20]. The spectrum of conditions causing ovarian cyst formation and subsequent rupture can range from benign physiologic conditions, such as ovulation in the case of corpus hemorrhagicum, to malignant processes. Large cystic lesions such as benign mature cystic teratomas, hemorrhagic cysts, and cystadenomas increase the risk for ovarian torsion by predisposing the ovary to swing around its vascular pedicle [21]. Ultrasound imaging is commonly used to elucidate the diagnosis. Hemodynamic instability can occur in all settings, and surgical options include cyst excision and oophorectomy [22].

Secondary Inguinodynia

Ironically, chronic postoperative groin pain is one of the major complications of inguinal hernia repairs with significant long-term pain seen in a small proportion of patients after surgery. Stimulation, entrapment, or injury to the nerves during hernia dissection can produce long-term sequelae of neuralgia, paresthesia, hypoesthesia, or hyperesthesia. The genital branch of the genitofemoral nerve, ilioinguinal nerve, and iliohypogastric nerve are at risk with an open approach, while the lateral femoral cutaneous nerve, anterior femoral cutaneous nerve, and genital or femoral branch of the genitofemoral nerve are at risk with a laparoscopic approach. Chronic groin pain can be potentially disabling, with significant impact on quality of life.

The true incidence of chronic groin pain after inguinal hernia repair is hard to determine, with varied incidence reported in the current literature. A prospective series of 419 subjects after open hernia repair found 19 % had reported residual pain at 1-year follow-up, with 6 % reporting moderate to severe pain. Recurrent hernia and high pain score at 1- and 4-week post-op were identified as predictors of developing moderate to severe pain [23]. Mesh use, nerve division, use of lightweight meshes, and laparoscopic repair have all been studied for potentially reducing post-herniorrhaphy pain, with only the latter two having shown potential benefit [24].

A recurrence must be ruled out when confronted with this complication. Remnant cord lipomas from original surgery must also be distinguished from a recurrence. Neuropathic pain can be managed like other chronic pain conditions. Anti-inflammatory medications, tricyclic antidepressants, nerve blocks, and acupuncture are all viable modalities, with no strong evidence for one predominant modality in terms of efficacy [25]. Operative intervention is a viable option if spontaneous resolution has not occurred by 1 year. Neurectomy and neuroma excision, adhesiolysis, muscle or tendon repair, and foreign body removal are all possibilities, with inconclusive evidence despite favorable reported outcomes [26].

Secondary groin pain has also been reported in the literature after surgery in orthopedic procedures involving the lumbar spine. Injury to the lumbar plexus is a well-known complication of lateral lumbar interbody fusion, a new and increasingly popular alternative for interbody arthrodesis for degenerative spine disease that involves a lateral transpsoas approach to the lumbar spine. Reported incidence in literature ranges from 0.7 to 25 % [27, 28]. Inconsistencies in defining postoperative neurogenic injury and small sample size in the current literature have contributed to this large variation. Operative time, inclusion of the L4–L5 level, and use of recombinant human bone morphogenetic protein-2 (rhBMP-2) have been identified as possible independent risk factors for iatrogenic nerve injury [29].

Conclusion

A comprehensive history and physical exam is the first step toward differentiating primary from secondary groin pain. Though an inguinal hernia is the most common cause of groin pain, other causes include sports hernia, referred hip pain, spermatic cord, and testicular causes, and various gynecologic etiologies. Secondary inguinodynia after surgery is also an increasingly recognized complication of inguinal and spine surgery.

References

1. Dabbas N, Adams K, Pearson K, Royle G. Frequency of abdominal wall hernias: is classical teaching out of date? JRSM Short Rep. 2011;2(1):5.
2. Russell RC, Williams NS, Bulstrode CJ. Bailey & Love's short practice of surgery. 23rd ed. London: Hodder Arnold; 2000.
3. Bay-Nielsen M, Kehlet H, Strand L, Malmstrøm J, Andersen FH, Wara P, Danish Hernia Database Collaboration, et al. Quality assessment of 26,304 herniorrhaphies in Denmark: a prospective nationwide study. Lancet. 2001;358(9288):1124–8.
4. Diesen DL, Pappas TN. Sports hernias. Adv Surg. 2007;41:177–87.

5. Laor T. Hip and groin pain in adolescents. Pediatr Radiol. 2010;40(4):461–7.
6. Gilmore J. Groin pain in the soccer athlete: fact, fiction, and treatment. Clin Sports Med. 1998;17(4):787–93.
7. Preskitt JT. Sports hernia: the experience of Baylor University Medical Center at Dallas. Proc (Bayl Univ Med Cent). 2011;24(2):89–91.
8. Cavalli M, Bombini G, Campanelli G. Pubic inguinal pain syndrome: the so-called sports hernia. Surg Technol Int. 2014;24:189–94.
9. Swan Jr KG, Wolcott M. The athletic hernia: a systematic review. Clin Orthop Relat Res. 2007;455:78–87.
10. Yukawa Y, Kato F, Kajino G, Nakamura S, Nitta H. Groin pain associated with lower lumbar disc herniation. Spine (Phila Pa 1976). 1997;22(15):1736–9. Discussion 1740.
11. Oikawa Y, Ohtori S, Koshi T, Takaso M, Inoue G, Orita S, et al. Lumbar disc degeneration induces persistent groin pain. Spine (Phila Pa 1976). 2012;37(2):114–8.
12. Takahashi Y, Morinaga T, Nakamura S, Suseki K, Takahashi K, Nakajima Y. Neural connection between the ventral portion of the lumbar intervertebral disc and the groin skin. J Neurosurg. 1996;85(2):323–8.
13. Ben-Chaim J, Leibovitch I, Ramon J, Winberg D, Goldwasser B. Etiology of acute scrotum at surgical exploration in children, adolescents and adults. Eur Urol. 1992;21(1):45–7.
14. Davis JE, Silverman M. Scrotal emergencies. Emerg Med Clin North Am. 2011;29(3):469–84.
15. Beni-Israel T, Goldman M, Bar Chaim S, Kozer E. Clinical predictors of testicular torsion as seen in the pediatric ED. Am J Emerg Med. 2010;28(7):786–9.
16. Lewis AG, Bukowski TP, Jarvis PD, Wacksman J, Sheldon CA. Evaluation of the acute scrotum in the emergency department. J Pediatr Surg. 1995;30(2):277–81. Discussion 281–2.
17. Sharp VJ, Kieran K, Arlen AM. Testicular torsion: diagnosis, evaluation, and management. Am Fam Physician. 2013;88(12):835–40.
18. Visser AJ, Heyns CF. Testicular function after torsion of the spermatic cord. BJU Int. 2003;92(33):200–3.
19. Hatipoglu S, Hatipoglu F, Abdullayev R. Acute right lower abdominal pain in women of reproductive age: clinical clues. World J Gastroenterol. 2014;20(14):4043–9.
20. Barash JH, Buchanan EM, Hillson C. Diagnosis and management of ectopic pregnancy. Am Fam Physician. 2014;90(1):34–40.
21. Chang HC, Bhatt S, Dogra VS. Pearls and pitfalls in diagnosis of ovarian torsion. Radiographics. 2008;28(5):1355–68.
22. Evsen MS, Soydinc HE. Emergent gynecological operations: a report of 105 cases. J Clin Exp Invest. 2010;1(1):12–5.
23. Callesen T, Bech K, Kehlet H. Prospective study of chronic pain after groin hernia repair. Br J Surg. 1999;86(12):1528–31.
24. O'Dwyer PJ, Alani A, McConnachie A. Groin hernia repair: postherniorrhaphy pain. World J Surg. 2005;29(8):1062–5.
25. Ferzli GS, Edwards ED, Khoury GE. Chronic pain after inguinal herniorrhaphy. J Am Coll Surg. 2007;205(2):333–41.

26. Aasvang E, Kehlet H. Surgical management of chronic pain after inguinal hernia repair. Br J Surg. 2005;92(7):795–801.
27. Rodgers WB, Gerber EJ, Patterson J. Intraoperative and early postoperative complications in extreme lateral interbody fusion: an analysis of 600 cases. Spine (Phila Pa 1976). 2011;36(1):26–32.
28. Tohmeh AG, Rodgers WB, Peterson MD. Dynamically evoked, discrete-threshold electromyography in the extreme lateral interbody fusion approach. J Neurosurg Spine. 2011;14(1):31–7.
29. Lykissas MG, Aichmair A, Sama AA, Hughes AP, Lebl DR, Cammisa FP, Girardi FP. Nerve injury and recovery after lateral lumbar interbody fusion with and without bone morphogenetic protein-2 augmentation: a cohort-controlled study. Spine J. 2014;14(2):217–24.

6. Groin Pain Etiology: The Inguinal Hernia, the Occult Inguinal Hernia, and the Lipoma

Ibrahim M. Daoud and Katherine Dunn

Inguinal Hernia

Epidemiology/Etiology The inguinal hernia is one of the most common reasons that a general practitioner would refer a patient to a general surgeon. Inguinal hernias can present with a wide array of symptoms, including groin pain, burning, aching, or worsening pressure in the groin throughout the day. Those with hernias may also complain of a lump or a bulge on the affected side. On the other hand, many patients do not complain of a bulge, but instead present with a chief complaint of groin pain—unaware of the vast differential diagnosis list involved. Though the differential diagnosis for groin pain is quite long and can include such diagnoses as chronic appendicitis, diverticulitis, urologic diseases, and gynecological processes, an inguinal hernia is a common cause not to be overlooked [1]. This chapter focuses on inguinal hernias, as well as on the occult inguinal hernia, and the lipoma of the spermatic cord or round ligament.

Abdominal wall hernias account for 4.7 million ambulatory care visits each year, more than 600,000 of which are inguinal hernias that undergo repair [2]. Inguinal hernias present with a 9:1 male predominance, are more common on the right, and are most commonly in the 40–59 year age group. Indirect hernias are twice as likely to be present when compared with direct hernias [3]. In women, as in men, indirect hernias are the most common inguinal hernia. Femoral hernias, however, are relatively more common in women when compared to men, comprising 20 % of all groin hernias in women [1].

B.P. Jacob et al. (eds.), *The SAGES Manual of Groin Pain*,
DOI 10.1007/978-3-319-21587-7_6

Other pertinent hernias that could contribute to groin pain include Spigelian hernias, obturator hernias, and Pantaloon hernias. Pantaloon hernias occur when there is both a direct and indirect hernia component. A Spigelian hernia is usually small and presents with bowel incarceration or strangulation in approximately 25 % of cases [1]. This defect is seen as a protrusion through the transversals fascia lateral to the edge of the rectus muscle, medial to the Spigelian line, and midway between the umbilicus and pubis at the level of the semicircular line of Douglas [1]. The obturator hernia is rare and often presents as a surgical emergency as a peritoneal pouch with accompanying small bowel, which may be incarcerated or strangulated as it follows the course of the obturator vessels through the obturator fossa. Such hernias are five times more common in women than men and most often present between the ages of 50 and 90 [1].

Diagnosis Though study data are limited, inguinal hernia can most often be diagnosed with a detailed history and physical examination. One such study by van den Berg et al. showed that history and physical alone can detect an inguinal hernia with a sensitivity of 75 % and specificity of 96 % [4]. A history consistent with symptoms of gurgling and burning pain in the groin area would raise suspicion of inguinal hernia. Worsening symptoms or groin bulge with performance of Valsalva maneuvers or any activities such as heavy lifting, straining, or coughing that serve to increase intra-abdominal pressure help support this diagnosis. These activities could also cause a groin bulge to increase in size. If a patient reports that this bulge disappears in the supine position, there should be a high clinical suspicion of a groin hernia [5].

Physical exam should be performed with the patient in the upright position. It should include a close inspection of the inguinal and femoral regions for visible bulges. Palpation of the region should include a Valsalva maneuver from the patient in an attempt to elicit a palpable herniation [2]. If physical exam is inconclusive for inguinal hernia, there are several radiographic modalities such as CT, MRI, and ultrasound that can be utilized; these are discussed in further depth in the occult inguinal hernia section.

Treatment The main debate in elective treatment of unilateral inguinal hernia repair revolves around open versus laparoscopic herniorrhaphy. The open, tension-free, Lichtenstein repair is one of the most common general surgery procedures in the world and is the most accepted form of unilateral herniorrhaphy [6]. Laparoscopic repair, including both TEP and TAPP repair, is recognized as superior in bilateral hernia repair and in cases of recurrent hernia. Argument can also be made to perform a unilateral hernia repair laparoscopically, leading to less postoperative pain, quicker return to physical activity, lower incidence of chronic groin pain, and similar recurrence rates [7].

The main disadvantage of unilateral laparoscopic repair involves the long learning curve in mastering the delicate laparoscopic dissection techniques and groin anatomy. It has been said that one becomes confident with the procedure with 80 cases and mastery comes with 250 cases [7]. Other disadvantages include the need for general anesthesia, and the complications with laparoscopic repair, though rare, can also be more serious, as they include vascular or visceral injury [8].

With unilateral repair it is still, therefore, up for debate whether to perform laparoscopic or open repair. This should depend on surgeon preference and comfort level. The type of repair also depends on individual patient needs, including if the patient has any contraindications for laparoscopic surgery or general anesthesia, in which case open herniorrhaphy would be preferred.

Occult Inguinal Hernia

Epidemiology/Etiology Inguinal hernias, as discussed above, are often diagnosed with history and physical exam alone, and treated accordingly. Occult hernias, which include direct, indirect, femoral, and obturator hernias, can present with a story consistent with that of a groin hernia but without the physical exam findings to support the diagnosis [9]. This is when radiographic studies may be of assistance. Additionally, occult hernias can often be discovered at the time of laparoscopic hernia repair.

In those who present with groin pain, aching, discomfort, or intermittent groin swelling with equivocal or negative physical exam findings, it is important to consider occult inguinal hernia as a possible diagnosis [10]. The definition of occult inguinal hernia is not well defined in the literature, and it is often left open to a wide range of interpretations. In a 2013 study by van den Heuvel et al., there is a distinction made between true occult inguinal hernia, which is repairable at the time of surgery, and incipient hernia, which defines a small defect with a shallow hernia sac in which there is no herniation of intra-abdominal contents [11]. In this study, they found that the incidence of a contralateral "occult" inguinal hernia was 13 % when TAPP repair of a clinically palpable inguinal hernia was performed. Of these, 8 % were true occult hernias and 5 % were incipient. True occult hernias were repaired at the time of exploration, and the incipient hernias were followed closely, 21 % of which became symptomatic, requiring additional surgery.

In a 2012 study by Garvey, it was found that of those with symptoms suggestive of hernia, in the absence of clear physical exam findings, 33 % of patients who underwent CT examination were found to have an occult inguinal hernia. This was then confirmed in the operating room with 94 % accuracy. As discussed below, CT may not be the best imaging modality, but this figure of 33 % serves to show the approximate incidence of those with occult inguinal hernia who present with groin pain [10].

In discussing occult hernia, women are an important population to consider. Groin pain can be a common symptom in women with a differential diagnosis similar to men, including urologic, gastrointestinal, or musculoskeletal causes with the addition of gynecologic disorders [12]. The population of women with chronic pelvic pain is also important to consider, as pelvic pain often includes the inguinal region [13]. Hernias are often smaller in females, leading to an undetectable clinical impulse on exam due to the absence of a processes vaginalis [14]. Of the approximately 20 million hernia repairs performed to date, only 6–8 % of these have been performed in women. It has been suggested, however, that occult hernias may be relatively common in women suffering from groin pain, especially those who experience worsening of symptoms with activity. Given the normal physical exam findings, these women can often have a prolonged symptomatic period before a correct diagnosis of groin hernia is achieved. As in men, it is important to consider and diagnose a hernia before it presents as a surgical emergency [15].

Diagnosis The diagnosis of occult inguinal hernia can be tricky, as there is often groin pain and suspicion of a hernia but no discernible physical exam findings by general practitioner or surgeon. A meta-analysis by Robinson et al. served to evaluate herniography, CT, MRI, and ultrasound in finding occult inguinal hernias in those presenting with groin pain. Herniography proved to be the most accurate modality, with an overall sensitivity of 91 % and specificity of 83 %. Conversely, CT showed a sensitivity of 80 % and specificity of 65 %. Ultrasound, being largely operator dependent and with limited available data for this meta-analysis, has a sensitivity of 86 % and specificity of 77 % [9]. Towfigh and colleagues recently published a review of their series and found that when an occult hernia is suspected, an MRI was the best image modality to order [16].

All of the above diagnostic modalities have unique drawbacks. Herniography, though seemingly the most accurate, is the most invasive, second only to surgery, and utilizes contrast medium, which can elicit an allergic reaction. It is also of no use in determining alternative causes for the complaint of groin pain, and is thus rarely used in clinical practice. CT carries with it the risk of radiation exposure, and the patient is unable to stand for the study [10]. Ultrasound is largely operator dependent, though inexpensive and noninvasive. With ultrasound, there is also added benefit of the capability to image the patient in various positions with certain maneuvers to better elicit the hernia impulse.

Not included in the meta-analysis discussed above is the study by Garvey, which looks exclusively at the use of CT in the diagnosis of occult inguinal hernia. This was chosen because it helps to evaluate hernia in obese patients, where ultrasound may be limited. CT is more affordable than MRI, for which there is still limited data on its use in occult hernia diagnosis. CT was found to have an accurate diagnosis 94 % of the time in a carefully selected group of patients. The author of this one study, however, continues to use ultrasound as his preferred imaging method, and reserves CT for obese patients [10].

Although most of the studies mentioned above focus on a mixed population, they favor diagnosis in men. Imaging to diagnose women with occult hernia is similar, with herniography as a popular method in Europe [12]. Given the invasive nature of herniography, Grant et al. looked at ultrasound specifically for the diagnosis of groin hernia in women with normal or inconclusive physical exam findings. The ultrasound was used to look for occult direct, indirect, and femoral hernias. The main benefit of ultrasound is that fact that the Valsalva maneuver can be performed, often making the hernia apparent [12]. Though the literature reports that indirect inguinal hernias consist of 70 % of the groin hernia diagnoses in women, the study by Grant et al. found that direct hernias were more common in those with normal exam findings. Of the hernias found on ultrasound and confirmed in the operating room, 48 % were direct hernias. This discordance is likely due to the fact that direct hernias are more difficult to detect on physical examination. This study showed that ultrasound in women with groin pain has a 95 % sensitivity and 75 % specificity [12].

In the population of women with chronic pelvic pain, and a story suggestive of occult hernia, laparoscopy may be the most effect diagnostic tool. In a single center study done of 365 women with chronic pelvic pain ranging from 6 months to 20 years, only 2 % had normal findings on laparoscopy. These patients had suspicion of occult hernia based on signs and symptoms of inguinal pain radiating to the labia or thigh and reproduction of pain of the internal ring on external palpation or by bimanual exam. Of those with abnormal laparoscopic findings, 77 % indirect hernias were identified, 65 % of which had a large internal ring with incarcerated fat. Additional findings included direct hernias in 20 % of the patients, femoral hernias in 40 %, obturator hernia in 2 %, and bilateral hernias in 40 %. Overall, after repair, 74.69 % of the patients reported complete relief of their pain, 17.83 % noted significant improvement, and the remainder showed no change [13].

It is clear that more studies are needed to prove the best means of diagnosing occult hernia, as everything from ultrasound to diagnostic laparoscopy has been utilized. It seems, however, that modality of choice should depend on physician preference, patient body habitus, suspicion of occult hernia based on symptoms reported, and lack of any other clear diagnosis.

Treatment Laparoscopic repair is advantageous, as those with a known inguinal hernia may often be found to have an existing occult hernia, notably as a femoral or obturator hernia. There is often a low preoperative detection of femoral and obturator hernias. The dissection during a laparoscopic hernia repair may more easily identify these defects and prevent them from causing continued groin pain or becoming a surgical emergency [17]. Occult hernia is found between 9 and 36 % of the time on the contralateral side during laparoscopic repair [8].

Despite longer operative times, Pawanindra et al. proposed bilateral exploration and repair in all cases of TEP repair for unilateral hernia, as they found contralateral occult hernia in 25 % of the cases [8]. They noted that this should be done only in high volume centers in the hands of advanced laparoscopic surgeons. With limited data on this matter, it seems that contralateral exploration and repair are warranted in one where there are risk factors for hernia development or high clinical suspicion of an occult contralateral hernia. It is reasonable to offer bilateral exploration and repair as an option to patients who may wish to avoid further surgery [8].

Lipoma

Epidemiology/Etiology Lipoma of the spermatic cord and round ligament is understood as an extension of the preperitoneal fat, and not as a true benign neoplasm, in the majority of the anatomical and surgical literature. The pathogenesis is largely unknown, but it is thought that this projection of fat through the deep inguinal ring may cause it to dilate and predispose one to indirect hernia. These cord lipomas most often do not have a peritoneal sac, but can nonetheless cause symptoms identical to that of a groin hernia [18]. By this definition of cord lipoma, they are, in effect, all "indirect" in nature. Spermatic cord lipomas as direct extensions of preperitoneal fat were found in the absence of a hernia with an incidence of 36–75 % on male autopsies, and lipomas of the cord and round ligament are found with an incidence of 21–73 % during herniorrhaphy [19, 20].

These lipomas have often been considered as an incidental finding at the time of hernia repair. One such study by Carilli et al. showed that there was a 72.5 % incidence of incidental cord lipoma found with an indirect hernia at the time of open repair [21]. The incidence of cord lipoma was greater with larger hernias, and it has also been suggested that excessive body weight may predispose one to such a lipoma [21].

Cord or round ligament lipoma occurring in conjunction with an inguinal hernia is more likely to be missed when performing laparoscopic herniorrhaphy, especially TAPP repair [22]. There are often times when TAPP repair is to be performed for a clinically palpable mass, and upon visualization, the peritoneum appears normal. In these several instances, incision of the peritoneum and exploration have revealed an inguinal cord lipoma [19]. This raises the question of the significance of potentially overlooked lipoma in relation to groin pain with the increasing popularity of laparoscopic herniorrhaphy [23]. These lipomas do occur with significant incidence, and they can cause hernia-type symptoms even without the presence of a true inguinal hernia [18]. It is often important to remember the potential presence of a lipoma if a patient is still experiencing pain after inguinal herniorrhaphy, especially when done laparoscopically.

Diagnosis Spermatic cord lipomas are diagnosed in much the same way as an inguinal hernia. On imaging studies, they may be misdiagnosed as inguinal hernia. As they consist of preperitoneal fat, lipomas of the round ligament or spermatic cord are not reliably diagnosed with a herniogram [24, 25]. The one difference in imaging is on CT examination, where an inguinal cord lipoma is mesenteric fat passing through the inguinal ring, and an indirect inguinal hernia often has a radiographically visible sac [20]. There is limited literature on the diagnosis of inguinal cord lipoma alone, as it is mostly found on surgical repair of an inguinal hernia. This is an important diagnosis to consider during the time of herniorrhaphy because an undiagnosed and untreated cord lipoma may cause groin pain to persist or predispose the patient to a recurrent hernia after repair. Exploration of the cord, therefore, should take place if initial laparoscopic inspection is negative for hernia and the patient has a convincing history consistent with hernia-type symptoms [19].

Treatment Though the lipoma, in most instances, has no pathological changes suggesting it is a true lipoma, but rather an extension of extraperitoneal fat protruding through the inguinal canal, it may still cause hernia-type symptoms warranting treatment [21]. The distinction between indirect inguinal hernias and inguinal cord lipomas is not necessarily important; if they are symptomatic, they both should be treated with open or laparoscopic surgical repair [20].

Conclusion

Inguinal hernias, whether occult or obvious, and lipomas of the spermatic cord or round ligament are important etiologies to consider in the diagnosis of groin pain. A supportive clinical history and a well-performed physical exam can diagnose inguinal hernias the majority of the time. Imaging may be useful when there is a history indicative of hernia but an equivocal physical exam. Given the multiple imaging modalities available with different benefit and risk profiles, the choice of MRI, CT, or ultrasound is often provider specific. MRI, however, has recently been shown as potentially the best modality for diagnosis of occult hernia. Additionally, diagnostic laparoscopy serves an important purpose in diagnosing the occult hernia. This is especially the case in women with chronic pelvic pain, in whom it is beneficial because diagnosis and repair can be performed at the same time. Lipomas of the cord and round ligament cause similar pain to that of a hernia and should be diagnosed and treated in the same fashion. In all cases, when a patient is symptomatic from a hernia or lipoma of the cord, it should be repaired via a laparoscopic or open approach. Laparoscopic repair clearly has its benefits with bilateral hernias and when contralateral occult hernias may be suspected. Often the fashion of repair depends on surgeon preference and proficiency as well as patient body habitus, previous abdominal surgeries, risk factors, and preference.

References

1. Daoud IM. General surgical aspects. In: Steege JF, Metzger DA, Levy BS, editors. Chronic pelvic pain: an integrated approach. 1st ed. Philadelphia: WB Saunders; 1998. p. 329–36.
2. LeBlanc KE, LeBlanc KA. Inguinal hernias: diagnosis and management. Am Fam Physician. 2013;87(12):844–8.
3. Hamadani FT, Bergman S. Inguinal hernia. In: Ashley SW, editor. Scientific American surgery. Decker intellectual properties; 2014. http://www.sciamsurgery.com.
4. van den Berg JC, de Valois JC, Go PM, Rosenbusch G. Detection of groin hernia with physical examination, ultrasound, and MRI compared with laparoscopic findings. Invest Radiol. 1999;34(12):739–43.
5. Evers BM. Small intestine. In: Townsend Jr CM, Beauchamp RD, Evers BM, Mattox KL, editors. Sabiston textbook of surgery: the biological basis of modern surgical practice. 18th ed. Philadelphia: Saunders/Elsevier; 2008. p. 873–916.
6. Vidović D, Kirac I, Glavan E, Filipović-Čugura J, Ledinsky M, Bekavac-Bešlin M. Laparoscopic totally extraperitoneal hernia repair versus open Lichtenstein hernia repair: results and complications. J Laparoendosc Adv Surg Tech A. 2007;17(5):585–90.
7. Putnis S, Berney CR. Totally extraperitoneal repair of inguinal hernia: techniques and pitfalls of a challenging procedure. Langenbecks Arch Surg. 2012;397(8):1343–51.
8. Pawanindra L, Philips P, Chander J, Ramteke VK. Is unilateral laparoscopic TEP inguinal hernia repair a job half done? The case for bilateral repair. Surg Endosc. 2010;24(7):1737–45.
9. Robinson A, Light D, Kasim A, Nice C. A systematic review and meta-analysis of the role of radiology in the diagnosis of occult inguinal hernia. Surg Endosc. 2013;27(1):11–8.
10. Garvey JFW. Computed tomography scan diagnosis of occult groin hernia. Hernia. 2012;16(3):307–14.
11. van den Heuvel B, Beudeker N, van den Broek J, Bogte A, Dwars BJ. The incidence and natural course of occult inguinal hernias during TAPP repair: repair is beneficial. Surg Endosc. 2013;27(11):4142–6.
12. Grant T, Neuschler E, Hartz W. Groin pain in women: use of sonography to detect occult hernias. J Ultrasound Med. 2011;30(12):1701–7.

13. O'Connell B, Bernstein D, Daoud IM. Laparoscopic inguinal hernia repair for occult (non-palpable) groin hernias in women with chronic pelvic pain. Paper presented at SAGES, 22–25 Apr 2009, Phoenix, AZ.
14. Herrington JK. Occult inguinal hernia in the female. Ann Surg. 1975;181(4):481–3.
15. Spangen L, Smedberg SG. Nonpalpable inguinal hernia in women. In: Bendavid R, Abrahamson J, Arregui ME, Flament JB, Phillips EH, editors. Abdominal wall hernias: principles and management. New York: Springer; 2001. p. 625–9.
16. Miller J, Cho J, Michael MJ, Saouaf R, Towfigh S. Role of imaging in the diagnosis of occult hernias. JAMA Surg. 2014;149(10):1077–80.
17. Lowen MS, Daoud IM, Frenzel CA. Treatment of femoral hernias by a single surgeon: a systematic approach. Paper presented at SAGES, 9–12 Apr 2008, Philadelphia, PA.
18. Lilly MC, Arregui ME. Lipomas of the cord and round ligament. Ann Surg. 2002;235(4):586–90.
19. Heller CA, Marucci DD, Dunn T, Barr EM, Houang M, Dos Remedios C. Inguinal canal "lipoma". Clin Anat. 2002;15(4):280–5.
20. Fataar S. CT of inguinal canal lipomas and fat-containing inguinal hernias. J Med Imaging Radiat Oncol. 2011;55(5):485–92.
21. Carilli S, Alper A, Emre A. Inguinal cord lipomas. Hernia. 2004;8(3):252–4.
22. Gersin KS, Heniford BT, Garcia-Ruiz A, Ponsky JL. Missed lipoma of the spermatic cord. a pitfall of transabdominal preperitoneal laparoscopic hernia repair. Surg Endosc. 1999;13(6):585–7.
23. Nasr AO, Tormey S, Walsh TN. Lipoma of the cord and round ligament: an overlooked diagnosis? Hernia. 2005;9(3):245–7.
24. Spangen L, Anderson R, Ohlsson L. Non-palpable inguinal hernia in the female. Am J Surg. 1988;54(9):574–7.
25. Hall C, Hall PN, Wingate JP, Neoptolemos JP. Evaluation of herniography in the diagnosis of occult abdominal wall hernia in symptomatic adults. Br J Surg. 1990;77(8):902–6.

7. Groin Pain Etiology: Athletic Pubalgia Evaluation and Management

Gregory J. Mancini

Introduction

Athletic pubalgia (sports hernia) is a cluster of distinct injuries that are grouped together because of the common location of pain, overlapping activity triggers, and lack of physical exam findings. The chronic painful symptoms that occur in otherwise healthy, athletic, and young individuals add a psychosocial layer to an already complicated medical condition. Most injuries to athletes result from a single action or collision. There are obvious physical findings of this injury such as swelling or a contusion. These injuries are routine and fully heal with time and basic care. But athletic pubalgia is much more insidious. It develops slowly over time without pain until a relatively minor event halts the activity. There is rarely any outward sign of the injury. The pain may not be present with walking or light physical activity, but manifests at full athletic speeds. In athletic pubalgia, the routine 2–4 weeks of rest is often not sufficient for full resolution of symptoms. The resulting confusion and misinformation have clouded the understanding of athletic pubalgia for athletes, trainers, coaches, parents, and even most medical professionals. This chapter aims to show that athletic pubalgia can be subdivided into three distinct entities, each with its individual treatment recommendations. Most cases of athletic pubalgia are found to be an occult inguinal hernia, osteitis pubis, or a regional nerve entrapment syndrome. The description, diagnostic methodology, and treatment options for each are hence detailed.

B.P. Jacob et al. (eds.), *The SAGES Manual of Groin Pain*,
DOI 10.1007/978-3-319-21587-7_7

Occult Hernia

Background An occult inguinal hernia is a true hernia of the myopectineal orifice that is indicated by symptoms of groin pain, worsened by activity, but not clinically apparent on physical exam or basic imaging. This entity is a common clinical conundrum posed to surgeons on nearly a daily basis. As an example, a middle-aged male is sent by his primary care physician with left inguinal pain limiting his daily work activities, but on exam no hernia can be found. This has been traditionally labeled a groin sprain, and six weeks of lifting restrictions and scheduled oral NSAIDs are recommended and prescribed. An occult hernia is often termed athletic pubalgia, not because of its symptoms without physical exam findings, but rather its presentation in the age and demographics of the patient. If the prior example is changed to a healthy 18-year-old male soccer player who had left inguinal pain only while playing, but not at rest, the label of sports hernia will be given. It is estimated that occult hernia comprise 10–15 % of inguinal hernia disease, and therefore all patients presenting with symptomatic inguinal region pain should be considered to potentially have an occult hernia [1]. Occult hernia can be a cord lipoma or indirect hernia sac that tracks along the spermatic cord within the inguinal canal creating compression on the ilioinguinal or genitofemoral nerves. Similarly, an occult hernia can be a weak transversalis fascia allowing the floor of the inguinal canal to bulge, compressing the nerves. An intact superficial inguinal ring will limit the physician's examination and thereby mask the true hernia, making it difficult to diagnose.

Diagnosis Determining the presence of an occult hernia is difficult based on physical exam alone. Clinical suspicion begins with a thorough review of the patient's duration, location, and triggers of the pain symptoms. Up to one-third of patients with groin pain will have occult hernia as the pathologic cause of their symptoms. Patients with occult hernia often have physical triggers of pain that can be provoked by a position change such as bending over or increase of intra-abdominal pressure such as Valsalva maneuver.

Imaging is an important adjunct to assist the identification of an occult hernia. There are several different imaging modalities with each having their relative strengths and weaknesses. Ultrasound is a low-cost and low-risk diagnostic imaging test. For occult hernia, the sensitivity and predictive values are greatly dependent on the ultrasound technologist's expertise and the patient position during the exam. Performing the ultrasound exam while the patient is standing and performing a Valsalva can enhance detection of the occult hernia [2]. Multiple published studies from 1981 to the present show a sensitivity of 70–97 % for occult hernia. The positive predictive value ranges from 90 to 95 % [3–5].

Computed tomography (CT) scan of the abdomen, to include the pelvis, is another diagnostic imaging option used to detect occult hernia. This modality is less operator dependent but has added cost and radiation exposure to the patient. Though the patient is supine for this test, a Valsalva maneuver during the scanning process can enhance hernia detection. Figure 7.1 shows the cross-sectional image of an occult bilateral inguinal hernia, the left being more obvious than the right. Garvey et al. showed that in 158 patients with groin pain, no hernia on exam, and then a subsequent CT scan, 54 patients (33 %) had evidence of an occult inguinal hernia. At surgery, 49 were confirmed to have a hernia, 3 had cord lipoma, and 2 had no inguinal pathology. This study modality yields a positive predictive value of 92 %, a negative predictive value of 96 %, and an overall accuracy of 94 % [6].

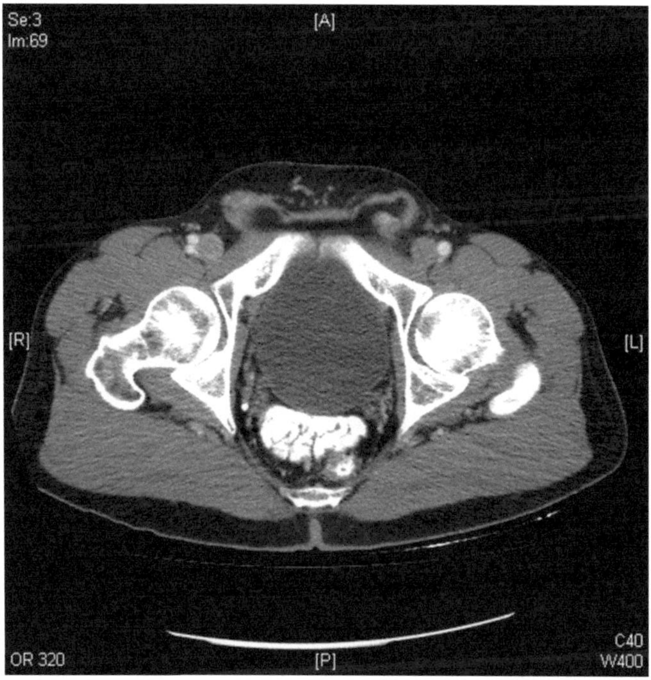

Fig. 7.1. Cross-sectional image of an occult bilateral inguinal hernia, *left* more obvious than *right*.

Magnetic resonance imaging (MRI) of the abdomen has a significant role in the diagnostic evaluation of athletic pubalgia. It has a cost higher than both ultrasound and CT, but no ionizing radiation of CT. Its sensitivity to demonstrate soft tissue edema differences in T2-weighted images is critical to identify non-hernia causes of groin pain. As for occult hernia detection, MRI has been shown to have a sensitivity and specificity figures of 94.5 and 96.3 % [7].

Treatment Treatment of occult hernia is fairly straightforward. This can be done as an open or laparoscopic technique. Laparoscopy is often suggested as a bridge between a diagnostic and therapeutic modality for occult hernia. This is a false logic, as a diagnostic laparoscopy will miss fat-containing hernias that give a normal contour to the pelvic floor. The peritoneum must be taken down in either a transabdominal preperitoneal (TAPP) or totally extraperitoneal (TEP) technique to ensure all hernia sites and pathologies are evaluated. By combining thorough patient history, physical exam, and the optimal imaging modality, the risk of missing an occult hernia can be less than 5 %.

Osteitis Pubis

Background Osteitis pubis is an important clinical entity that deserves significant consideration in any patient who presents with groin pain without obvious hernia on exam. Several clinical features separate osteitis pubis from other groin pain diagnoses. The pain most commonly localizes within the lower abdominal wall and tends to be more medial (between the external ring and the pubic symphysis). As radiographic technology has improved, osteitis pubis is now recognized as a cluster of different injuries to the muscles, tendons, and osseous structures of the lower abdominal wall and pelvis. These include rectus tendinitis, conjoined tendonitis, pubic ramus avulsion fractures, and pubis symphysitis, adductor tendonitis, and gracilis tendonitis. The mechanism of injury in athletic pubalgia combines two physical phenomena: repetitive motion injury and muscle development asymmetry. Individuals at highest risk for the development of osteitis pubis are young athletes in sports that require high-intensity training in which quick changes in speed and direction are required. Another component of this injury mechanism is long-term training in which asymmetric muscle development is promoted. This muscle development imbalance can be either between legs and torso or between right and left sides of the body.

For example, when the foot is planted to accelerate speed or change of direction, the power in the legs must be balanced by the torso to move the entire body in the same direction. As the adductor and gracilis muscles contract, they exert pulling force on the inferior edge of the pubic ramus and pubic symphysis. The pubic symphysis acts to stabilize both halves of the pelvis to the opposing force vector. The rectus muscle then contracts to bring the torso in line with the new vector force, exerting a pulling force on the superior edge of the pubic ramus and symphysis. If the athletic training activity promotes leg muscle development over abdominal wall muscle development (typically the rectus muscle), or promotes right-sided muscle development over the left, a relative pelvic instability can develop. This allows chronic and recurring muscle, tendon, or symphyseal trauma that is collectively known as osteitis pubis. This injury mechanism helps to explain why certain sports and athletic positions have a higher incidence of osteitis pubis: the football player who stops and starts by planting the same pivot foot, the soccer player or punter who plants the left foot and creates the burst kick with the right foot, and the sprinter who explodes from the starting block using the same staggered foot position.

Diagnosis The diagnosis of osteitis pubis begins with a history and physical exam. High-intensity athletes doing year-round training in sports like soccer, football, and track have the highest incidence of injury for the reasons explained above [8]. Commonly, the patient will admit to a chronic and recurring set of symptoms for which they have self-medicated or self-limited their training to allow healing. But upon restarting competitive training, the symptoms recur, and they seek the surgeon to help get back to full speed.

The goal of the physical exam is to best localize the focal area of pain. Osteitis pubis can be divided into three zones for focal pain: suprapubic, intrapubic, and infrapubic. Suprapubic sources of pain include injuries to the rectus muscle, rectus tendon, conjoint tendon, and the periosteum of the pubic rami. Intrapubic sources of pain stem mainly from injury to the pubic symphysis and its fibrocartilaginous interpubic disk. Infrapubic sources of pain include injury to the gracilis muscle, the adductor longus muscle, the tendinous origins of these muscles, and periosteum of the pubic rami.

On examination, the pain can often be elicited by manual palpation. A pubic symphyseal injury can be assessed by performing the spring test. With the patient in supine position, the examiner places direct downward pressure with a hand on each side the pubis. Pain with a rocking motion can indicate instability and inflammation of the fibrocartilaginous interpubic disk. Rectus abdominis tendonitis can be assessed with downward pressure medially and above the pubis and may elicit pain in rectus tendon origin on the pubic crest. It can be difficult to assess for laterality, and this injury can be bilateral in nature. Manual pressure applied slightly more laterally, but medial to the external ring, may indicate conjoint tendonitis. Less severe symptoms not provoked by manual exam may be elicited by a series of exercise tests. A simple bent-knee sit-up, a sitting resistance to thigh adduction maneuver, or a cross-legged resistance to knee lift may produce the typical symptoms [9]. The pain generated by injury to the adductor longus or gracilis typically presents below the inguinal canal, but may radiate to the medial thigh and scrotum. Both the adductor longus and gracilis muscles share their origin on the anterior surface of the inferior pubic ramus. The insertion on the surface of the femur defines the adduction movement that this muscle group has. The adductor or gracilis damage most often occurs at the origin in the pubic rami. This can be muscle and tendon tearing or periosteal microfractures of the pubic bones. On physical exam, pain can typically be elicited by deep palpation of the inferior pubic ramus. A provocative maneuver on examination is a bent-knee raise or a sitting resistance to thigh adduction.

Radiographic evaluation of osteitis pubis is a valuable method to validate the clinical exam in the diagnosis of osteitis pubis. Though modalities such as plain pelvic x-ray, ultrasound, CT, and nuclear medicine study have been used to help make the diagnosis, MRI has become the main imaging modality to both diagnose and confirm resolution of the inflammatory process [10]. Figure 7.2 demonstrates a tendon tear in the adductor longus at the pubic bone. Note the increased tissue edema indicated by the smudged appearance of the tissue. In fact, because of MRI's sensitivity for musculoskeletal edema changes, it may lack usefulness as a screening modality for asymptomatic at-risk individuals. A 2006 study of scholarship male soccer players showed that MRI scans showed moderate to severe bone marrow edema at the pubic symphysis in 11 of the 18 asymptomatic players. Substantial amounts of bone marrow edema at the pubic symphysis can occur in asymptomatic soccer players, and it is only weakly related to the development of osteitis pubis [11]. Therefore, MRI should be used to confirm the clinical suspicions provided by the history and physical exam.

Treatment Treatment of osteitis pubis is a simple prescription that is hard to follow for the patient and sometimes the trainer, coach, or parent. After the diagnosis is made, immediate cessation of strenuous and

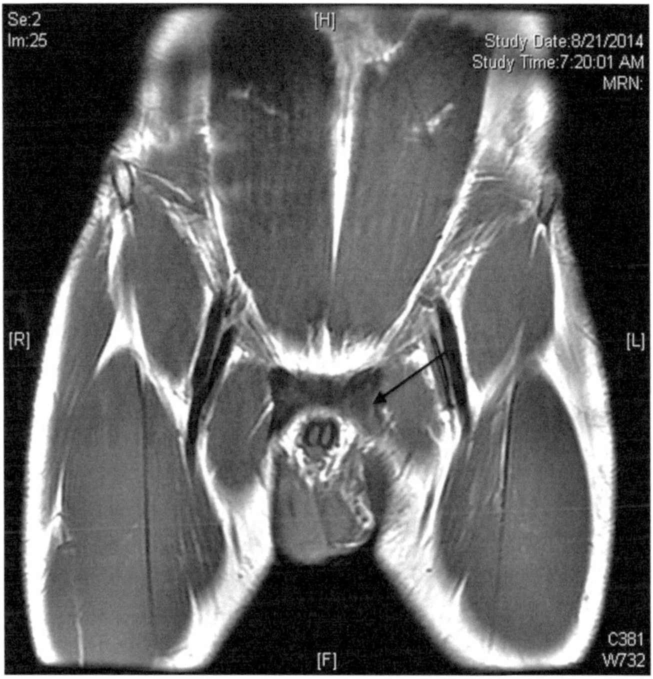

Fig. 7.2. Tendon tear in the adductor longus at the pubic bone. Note the increased tissue edema indicated by the smudged appearance of the tissue.

aggravating activities is mandated. The mainstay treatment is typically nonoperative, most commonly beginning with 6 weeks of rest, though low-impact and cardiac workouts can often be tolerated. Daily scheduled nonsteroidal anti-inflammatory medications are prescribed as tolerated. After the 6 weeks of rest, a rehabilitation program focused on cross-training types of stretching and lifting exercise can be started with a physical therapist [12]. The goal with rehabilitation is to establish muscular balance between the adductor and abdominal regions, thereby reducing the risk of early reinjury. Fricker et al. reported an average time to full recovery after conservative treatment of 9 months for men and 7 months for women [13]. If symptoms do not resolve after rest and rehabilitation programs, a corticosteroid injection treatment may be considered. A 3 mL mixture of 1 % lidocaine, 0.25 % bupivacaine, and 4 mg of dexamethasone injected into the interpubic disk of the pubic symphysis has reported good results in immediate pain relief and progression to full activities in a case series [14].

Unique to the adductor longus tendonitis is the possibility of surgical tendon release. Due to the redundancy in adductor musculature of the thigh, release of the adductor longus at its pubic bone origin is well tolerated with little loss of adduction strength and function. Gill et al. describe the surgical technique [15]. For those whose rest, rehabilitation, and corticosteroid injection fail to permit full recovery, the diagnosis of osteitis pubis must be questioned. Alternative chronic joint inflammatory conditions such as rheumatoid arthritis should be considered. Also, though rare, infectious osteoarthritis should be investigated. These diagnoses may lead to surgical debridement of the pubic bone and the damaged disk.

Nerve Entrapment Inguinal nerve entrapment is a painful condition that is most associated with post-hernia surgery complaints. In the context of athletic pubalgia, nerve entrapment is a primary anatomic problem that should be in the differential diagnosis of athletes with groin pain that limits competitive or training activities. There are three distinct nerves in the inguinal region that have well-documented pathology and treatment strategies: ilioinguinal, genitofemoral, and obturator nerve entrapment syndromes. We will consider each separately.

Ilioinguinal Nerve Ilioinguinal nerve entrapment was described by Kopell in the *New England Journal of Medicine* in 1962 [16]. The ilioinguinal nerve originates from the L1–2 nerve roots and has both motor and sensory functions. The motor innervation of the transversus abdominis and the internal oblique muscles generates muscle tone in the lower lateral abdominal wall. The sensory function gives touch and temperature sensation to the skin over the inguinal ligament, labia majora or scrotum, and the medial thigh. The pain syndrome can be caused by irritation, injury, or trauma to the nerve as it exits the retroperitoneum and pierces both the transversus abdominis and internal oblique muscles to travel within the inguinal canal. It is not a problem limited to athletes, but rather the entrapment may be an anatomic variant whose injury is worsened by intense physical training. A second mechanism is injury to the nerve and can be related to tears in the overlying external oblique aponeurosis that entrap the nerve. This injury has been coined "hockey players' hernia." Regional dermatome mapping can be done in the office to demonstrate if a specific nerve distribution correlates with the patient's symptoms. Figure 7.3 is an office dermatome mapping that shows ilioinguinal nerve distribution as a possible source of pain.

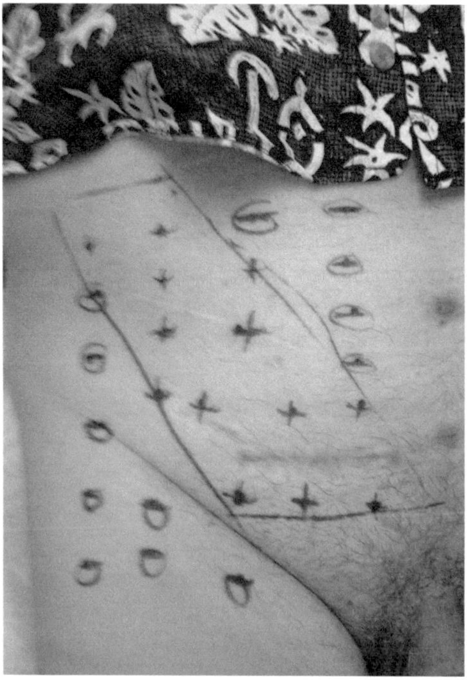

Fig. 7.3. Office dermatome mapping that shows ilioinguinal nerve distribution as a possible source of pain.

Knockaert et al. describe three hallmark clinical findings that suggest ilioinguinal nerve entrapment in patients. First is unilateral groin pain that radiates from the anterior spine of the anterior iliac crest to the scrotum or labia majora and medial thigh. The second feature is a cutaneous hyper-, hypo-, or dysesthesia in the same nerve distribution as the pain. The third finding is a reproducible trigger point located 2–3 cm below and medial to the anterosuperior iliac spine. This trigger point should be relieved by injection of a local anesthetic [17].

Genitofemoral Nerve The genitofemoral nerve is a less likely source of pain in athletic pubalgia than any other nerve entrapment syndrome. Most reports of genitofemoral pain symptoms can be attributed to previous surgery in the groin region, such as appendectomy, a Pfannenstiel incision, or an inguinal hernia incision. Its sensory dermatome overlap with the ilioinguinal nerve makes differentiating the two nerve injuries quite difficult. Like the ilioinguinal nerve, the genitofemoral nerve arises from the ventral rami of L1–2 and follows the psoas muscle into the pelvis. The nerve bifurcates, and the genital branch accompanies the spermatic vessels through the inguinal canal. Its branches pierce the internal spermatic fascia to supply muscular fibers to the cremaster muscle, and its sensory fibers terminate in the skin of the scrotum or labia majora. The location of the nerve's bifurcation is variable, but typically occurs in the retroperitoneum, such that injury to the nerve trunk is rare. Most of the symptoms and triggers of genitofemoral nerve entrapment are therefore correlated to the genital branch of the nerve.

Like the ilioinguinal nerve, genitofemoral nerve pain may be constant and radiate to the groin region, and a hyperesthesia to the skin of the region may be present. Pain may also be aggravated by activities such as walking, bending over, or hyperextension of the thigh and ameliorated by lying flat and flexion of the thigh. Likewise, the trigger point pain should be relieved by injection of a local anesthetic [18].

Obturator Nerve Obturator nerve entrapment can be a difficult diagnosis to make. Its anatomic course through the pelvis protects it from injury in common surgical procedures, unlike the ilioinguinal and genitofemoral nerves. The obturator nerve arises from the anterior divisions of L2–4 nerves.

It descends through the fibers of the psoas major muscle and emerges from its medial border, running behind the common iliac arteries toward the obturator foramen. It then enters the thigh through the obturator canal and splits into anterior and posterior divisions. The anterior division descends between the adductor longus and adductor brevis muscles, giving off motor branches to the adductor longus, adductor brevis, and gracilis muscles. It then pierces the fascia lata terminating in the cutaneous branches, giving sensation to the medial thigh. The posterior division passes anteriorly to innervate the adductor magnus. As the primary motor nerve to this muscle group, the obturator nerve is critical for leg adduction.

The clinical presentation of obturator nerve entrapment is pain, paresthesia, or hyperesthesia of the medial thigh, below the inguinal ligament. Due to the distinct dermatome involvement, obturator neuralgia is rarely confused with ilioinguinal or genitofemoral neuralgias. But its pain localization below the inguinal ligament can make clinical differentiation from adductor tendonitis quite difficult. Bradshaw et al. described obturator neuropathy in athletes as a result of fascial entrapment as the nerve enters the thigh, specifically in the adductor compartment [19].

Induced by exercise, the pain has a characteristic clinical pattern of medial thigh pain commencing in the region of the adductor muscle origin and radiating distally along the medial thigh, with strenuous exercise. An anatomic study on cadaver limbs by Harvey and Bell reinforced the concept that obturator neuropathy is caused by an entrapment syndrome due to the angle that the nerve pierces the adductor muscles and travels between the adductor fascial compartments [20]. In athletes, congenital anatomic nerve variants combined with physical training that augments adductor muscle development may be the main mechanism for obturator nerve entrapment syndrome in athletic pubalgia.

In addition to the sensory abnormalities, motor deficits can be found in advanced cases of obturator nerve entrapment. Physical exam may reveal asymmetry between affected and non-affected sides, with weakness and atrophy of adductor muscles on the affected. MRI will rarely show nerve-related injury, but is helpful to rule out adductor or gracilis tendonitis involvement. In cases where the MRI is normal, but clinical suspicion is high, the best test to confirm obturator neuropathy is needle electromyography (EMG). Kimura et al. noted that fibrillation potentials or high-amplitude, long-duration complex motor unit potentials were consistent with chronic denervation of the hip adductor muscle group, but not in other lower extremity muscles [21].

Treatment of Nerve Entrapment Syndromes

Nerve pain is generally as hard to treat as it can be to diagnose. The longer the pain symptoms have persisted, the more difficult it is to achieve adequate cessation of pain. In fact, treatment strategies can be divided into acute and chronic/recurring pain categories. For all three neuralgia syndromes, acute pain is best treated with activity cessation and NSAIDs for 3–6 weeks, followed by a strength and flexibility rehabilitation program that leads to competitive activity resumption. Regional nerve blocks that are diagnostic of nerve entrapment can be modified with long-acting local anesthetics (bupivacaine) and a corticosteroid (prednisolone tebutate) for sustained pain relief. For chronic and recurring pain that prevents full return to training and competition, nerve surgery will be required. For the ilioinguinal and genitofemoral nerves, neurectomy, often together, achieves good results. Since these nerves are mostly sensory in function, resection at the level of the transversus abdominis has a reported success rate of 70–85 % [18, 22]. This success rate mirrors the surgical cure rates published for chronic inguinodynia in post-hernia surgery nerve injury. For the obturator nerve, since its main function is motor innervation, neurectomy would not be tolerated. In this case, surgical neurolysis, or nerve decompression, is the best option. This technique requires careful dissection of the nerve as it courses through the different fascial compartments of the adductor muscle groups. Release of the tendon and fascial fibrotic bands around the nerve allows release of the nerve from its entrapment. In a case series of 29 elite athletes, all with clinical obturator nerve entrapment symptoms and validated with abnormal EMGs, all 29 had significant recoveries in 2–6 weeks of neurolysis and returned to competition [19].

Summary: Putting It All Together

Athletic pubalgia has been described here as three distinct clinical entities: as an occult hernia, osteitis pubis, or a regional nerve entrapment syndrome. As clinicians, we are tasked in evaluating the patient, accounting for the signs and symptoms, and making the right diagnosis to help the patient make a full recovery. My goal is to take the presumptive diagnosis of a sports hernia and more clearly define it as one of the three diagnoses. My evaluation process begins by interviewing to the patient's complaints to listen for clues that point to a hernia, a musculoskeletal injury, or nerve-related pain. I then begin the conversation by saying that chronic groin pain can be very difficult to treat and a quick fix is not likely to occur. If the patient, parent, coach, or trainer does not walk out the door at this point, I will walk them through the diagnostic testing process as well as the likely timeline for treatment, rehabilitation, and return to competitive training. If the patient has had no prior inguinal or pelvic surgery in the past, my first test of choice is an MRI of the abdomen and pelvis (to the mid-thigh). In my practice, two-thirds of patients with no prior inguinal or pelvic surgery are most likely to have osteitis pubis as the diagnosis. MRI is the best modality for this, and MRI will show inguinal hernias as well. In contrast, athletes with presumptive athletic pubalgia who have had prior inguinal or pelvic surgery are more likely to have a hernia recurrence, meshoma, or regional nerve entrapment syndrome. I therefore obtain a CT of the abdomen and pelvis with Valsalva completed. Positive findings on CT or MRI will guide the medical, surgical, and rehabilitation plans, as previously described. Normal CT and MRI will prompt a more thorough neurologic exam to

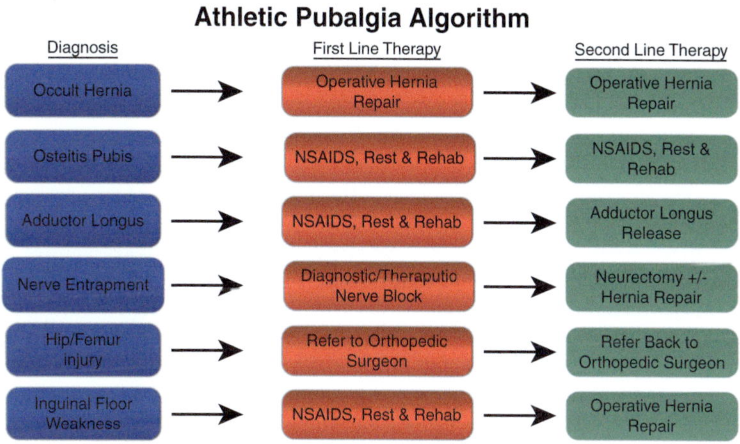

Fig. 7.4. Treatment algorithm for athletic pubalgia.

include dermatome mapping of the sensory of pain symptoms. If the pain correlates with the sensory dermatome distribution of a specific nerve, trigger point injections can be given for both diagnostic and therapeutic effect. The toughest dilemma comes when the patient's exam has no major physical findings, a normal CT and MRI, little relief from the trigger point injections, and continued patient pain that limits competitive performance. This is the point at which the scope of the work-up should be widened and alternative clinical opinions sought. Figure 7.4 demonstrates a treatment algorithm for athletic pubalgia.

References

1. Taylor DC, Meyers WC, Moylan JA, Lohnes J, Bassett FH, Garrett Jr WE. Abdominal musculature abnormalities as a cause of groin pain in athletes. Inguinal hernias and pubalgia. Am J Sports Med. 1991;19(3):239–42.
2. Depasquale R, Landes C, Doyle G. Audit of ultrasound and decision to operate in groin pain of unknown aetiology with ultrasound technique explained. Clin Radiol. 2009;64(6):608–14.
3. Deitch EA, Soncrant MC. Ultrasonic diagnosis of surgical disease of the inguinal-femoral region. Surg Gynecol Obstet. 1981;152(3):319–22.
4. Robinson P, Hensor E, Lansdown MJ, Abrose NS, Chapman AH. Inguinofemoral hernia: accuracy of sonography in patients with indeterminate clinical features. Am J Roentgenol. 2006;187(5):1168–78.

5. Light D, Ratnasingham K, Banerjee A, Cadwallader R, Uzzaman MM, Gopinath B. The role of ultrasound scan in the diagnosis of occult inguinal hernias. Int J Surg. 2011;9(2):169–72.
6. Garvey JF. Computed tomography scan diagnosis of occult groin hernia. Hernia. 2012;16(3):307–14.
7. van den Berg JC, de Valois JC, Go PM, Rosenbusch G. Detection of groin hernia with physical examination, ultrasound, and MRI compared with laparoscopic findings. Invest Radiol. 1999;34(12):739–43.
8. Beatty T. Osteitis pubis in athletes. Curr Sports Med Rep. 2012;11(2):96–8.
9. Rodriguez C, Miguel A, Lima H, Heinrichs K. Osteitis pubis syndrome in the professional soccer athlete: a case report. J Athl Train. 2001;36(4):437–40.
10. Ekberg O, Sjöberg S, Westlin N. Sports-related groin pain: evaluation with MR imaging. Eur Radiol. 1996;6(1):52–5.
11. Lovell G, Galloway H, Hopkins W, Harvey A. Osteitis pubis and assessment of bone marrow edema at the pubic symphysis with MRI in an elite junior male soccer squad. Clin J Sport Med. 2006;16(2):117–22.
12. Mora SA, Mandelbaum BR, Szalai LJ, Potter NB, Naik A, Ryan J, et al. Extraarticular sources of hip pain. In: Byrd JWT, editor. Operative hip arthroscopy. 2nd ed. Heidelberg: Springer; 2005. p. 70–99.
13. Fricker P, Taunton J, Ammann W. Osteitis pubis in athletes: infection, inflammation or injury? Sports Med. 1991;12(4):266–79.
14. Holt MA, Keene JS, Graf BK, Helwig DC. Treatment of osteitis pubis in athletes, results of corticosteroid injections. Am J Sports Med. 1995;23(5):601–6.
15. Gill TJ, Carroll KM, Makani A, Wall AJ, Dumont GD, Cohn RM. Surgical technique for treatment of recalcitrant adductor longus tendinopathy. Arthrosc Tech. 2014; 3(2):e293–7.
16. Kopell HP, Thompson WA, Postel AH. Entrapment neuropathy of the ilioinguinal nerve. New Engl J Med. 1962;266:16–9.
17. Knockaert DC, D'Heygere FG, Bobbaers HJ. Ilioinguinal nerve entrapment: a little-known cause of iliac fossa pain. Postgrad Med J. 1989;65(767):632–5.
18. Starling JR, Harms BA. Diagnosis and treatment of genitofemoral and ilioinguinal neuralgia. World J Surg. 1989;13(5):586–91.
19. Bradshaw C, McCrory P, Bell S, Brukner P. Obturator nerve entrapment: a cause of groin pain in athletes. Am J Sports Med. 1997;25(3):402–8.
20. Harvey G, Bell S. Obturator neuropathy. An anatomic perspective. Clin Orthop Relat Res. 1999;363:203–11.
21. Kimura J. Electrodiagnosis in diseases of nerve and muscle: principles and practice. 2nd ed. Philadelphia: F.A. Davis; 1989. p. 506–9.
22. Hahn L. Clinical findings and results of operative treatment in ilioinguinal nerve entrapment syndrome. Br J Obstet Gynaecol. 1989;96(9):1080–3.

8. Groin Pain Etiology: Hip-Referred Groin Pain

Joshua C. Campbell and Guy D. Paiement

Introduction

The differential diagnosis for groin pain originating from the hip is extensive and includes many disease processes, ranging from degenerative to autoimmune and from traumatic to genetic. It is important for any physician to have a working knowledge of these clinical entities since the presenting symptom, namely, groin pain, overlaps with so many specialties.

More than two-thirds of patients with an intra-articular hip pathology will present with groin pain, and many other extra-articular processes around the hip will present with similar distribution of pain with very subtle or no differences [1]. History taking is hard detective work and the physical exam is critical. An absolute prerequisite is a good knowledge of not only hip anatomy but also the structures around this articulation.

Anatomy Around the Hip

The bony architecture of the hip is important to both its function and its durability. The normal acetabulum is anteverted 15° and abducted approximately 45° in the coronal plane. The femoral neck is anteverted approximately 15° relative to the condylar axis of the distal femur, and the angle between the femoral neck and shaft is around 125° in adults [2]. This relationship is important to preservation of motion between bony structures and the function of the soft tissue structures surrounding the joint. Surrounding the acetabulum is the labrum, which increases the stability of the hip joint through a suction seal effect. Recent biomechanical

B.P. Jacob et al. (eds.), *The SAGES Manual of Groin Pain*,
DOI 10.1007/978-3-319-21587-7_8

studies show that the labrum additionally contributes to both cartilage nourishment and synovial fluid lubrication [3].

The hip capsule is innervated by femoral, sciatic, and obturator nerves [1]. Hip pain may present as pain reported in any of these nerve distributions. This often makes symptoms of hip pathology somewhat vague and nonspecific.

Several muscles are in proximity of the hip joint and may be the source of pain about the hip. These include the hip abductors (glutei maximus, medius, and minimus) and the tensor fascia lata with the iliotibial band that runs from the anterior and lateral iliac crest along the side of the thigh to insert onto the anterolateral tibia. Flexors crossing the hip include the iliopsoas, which originates within the pelvis, exiting below the inguinal ligament to attach on the lesser trochanter, and the rectus femoris, which lies directly anterior to the hip, with its direct head attaching just above the anterior hip capsule. Lastly, the adductor muscles (adductors magnus, longus, and brevis) lie medial to the hip, originating from the pubic rami and inserting on the medial femur.

Basics of Evaluation

A detailed history is crucial in the differential diagnosis. It is important to ask the patient about any types and changes of physical activities at work and any history of trauma, however minor or remote it may seem. Ideally, a patient should be able to provide a detailed history of the pain, including when and how it started as well as what makes the pain better or worse. It is important to establish objective measures of the symptoms: "I used to run 3 miles, but now I can barely walk 3 blocks," or "I cannot sleep on my back anymore." A complete medical, occupational, and family history is important, as many conditions have familial (Gaucher's disease), developmental (hip dysplasia), environmental (caisson disease), or exposure-related risk factors (avascular necrosis).

A complete physical evaluation should include inspection and palpation of all bony prominences: anterior superior iliac spine (ASIS), anterior inferior iliac spine (AIIS), pubic symphysis, ischial tuberosity, sacroiliac joints, and greater trochanters, with special emphasis on tenderness at these areas. Close attention to the exact location of the pain can narrow the differential diagnosis dramatically. Abnormal active and passive hip range of motion may also point the clinician in the right direction (Table 8.1; Figs. 8.1, 8.2, and 8.3). Comparison with the

Table 8.1. Range of motion of the normal hip (from Thompson [4]).

Extension	20°
Flexion	>120°
Adduction	20°
Abduction	40°
Internal rotation hip in extension	30°
Internal rotation hip in flexion at 90°	20°
External rotation hip in extension	50°
External rotation hip in flexion at 90°	30°

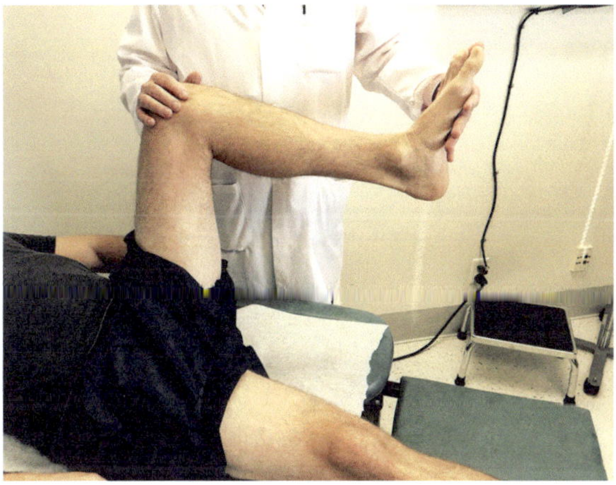

Fig. 8.1. The physical examination maneuver to determine range of motion about the hip. The extent of flexion of the hip is assessed in neutral rotation.

contralateral side (if asymptomatic) will help to detect otherwise subtle clinical signs. Lastly, an evaluation of gait is critical to detect any limp or abnormal posture.

When thinking about groin pain referred from the hip, it is helpful to think of the differential as problems related to either the soft tissue or skeleton. Further separating these into architectural versus physiological causes helps to clarify thinking. Although considerable overlap occurs with many processes, it remains a useful analytical framework. For the sake of brevity, trauma will be excluded from the discussion; however, suspicion of fracture following even minor trauma should remain high. Any fracture of or about the hip or pelvis should be treated with protected weight bearing and immediate referral to an orthopedic surgeon.

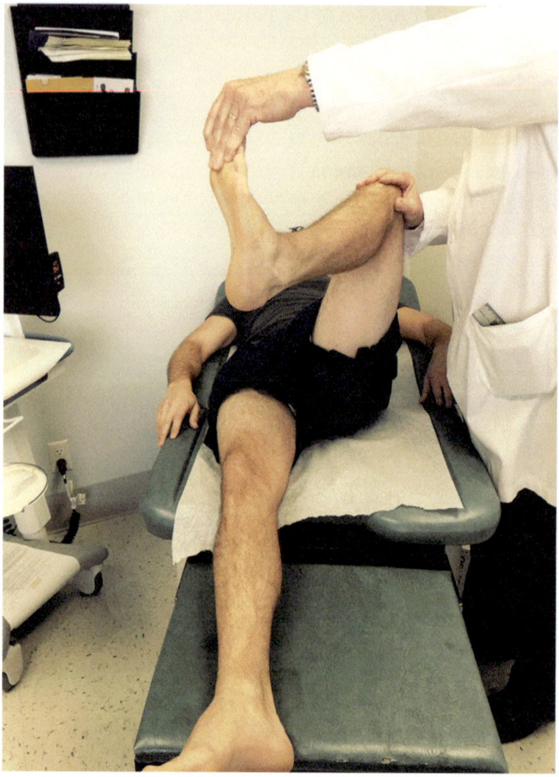

Fig. 8.2. The extent of external rotation of the hip is assessed with the hip in 90 degrees of flexion.

Groin Pain from the Bone

Architectural Problem

Osteoarthritis

Presentation Osteoarthritis (OA) is extremely common in an aging patient population. It is estimated to affect 60 million Americans by the year 2020 [5]. In any patient over 50 years, it should be high on one's differential as the cause of groin pain. Most patients with OA present

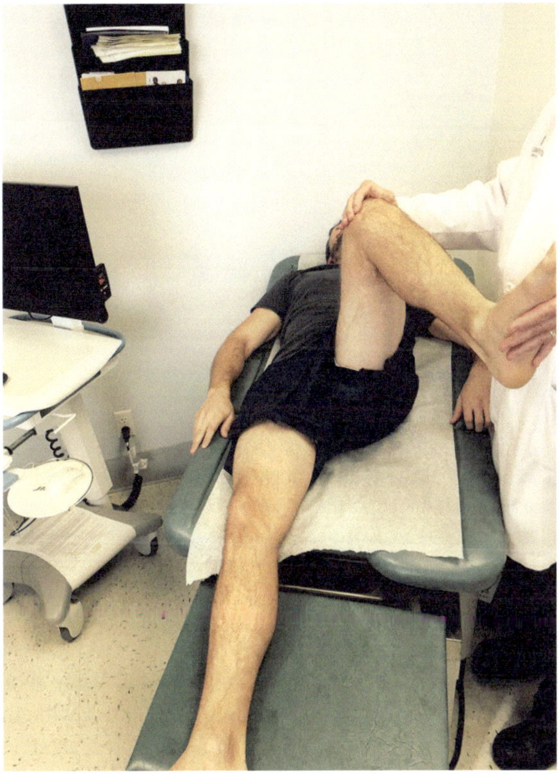

Fig. 8.3. The extent of internal rotation of the hip is assessed with the hip in 90 degrees of flexion.

with a slowly progressive pain in the groin, hip, or thigh, typically worse in the morning and at night. Patients describe an aching pain that improves with light activity and is worse with strenuous activity. Patients commonly report difficulties with initial motion after prolonged periods of rest, with improvement after a few steps. More advanced hip OA eventually results in stiffness and difficulty with activities of daily living.

Physical Exam The classic finding of crepitus with range of motion is rare. Patients may present with a very stiff joint, often with back pain that is more severe than the hip pain. Alternatively, some patients present with severe pain on weight bearing, with an almost normal range of motion. An important and early clinical sign of hip OA is decrease in internal rotation. Passive external rotation during flexion of the hip is known as Drehmann's sign and is indicative of this loss of internal rotation. Eliciting a positive

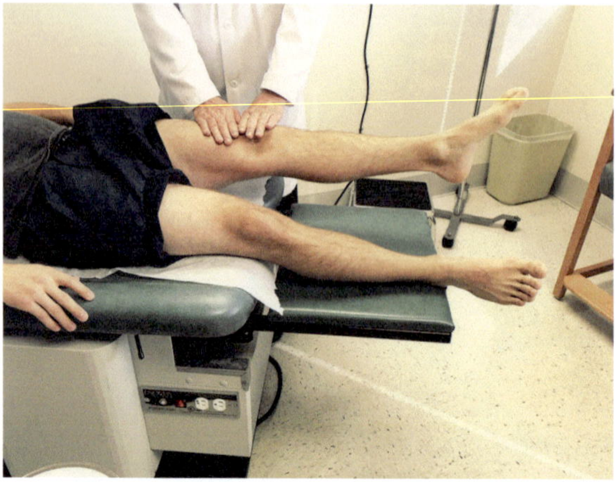

Fig. 8.4. The Stinchfield test. The patient performs a forced straight-leg raise against downward resistance at the thigh placed by the examiner. Pain in the groin with this maneuver is considered positive.

Stinchfield test, which results in pain at the hip with resisted straight-leg raise, is sensative however has low specificity (Fig. 8.4).

Diagnostic Exams The most useful radiological study is a standing low anteroposterior (AP) pelvis (including both hips) with the patient bearing weight equally on both sides (Fig. 8.5). A supine lateral radiograph of the affected side ("frog leg lateral") will complete the examination. These two simple views will help to elucidate more than 90 % of hip-referred groin pain originating from the bone. OA will have an obvious appearance on plain x-rays, and no further imaging is needed to arrive at this diagnosis (Fig. 8.6).

Differential Patients, especially young ones, with no obvious OA but bony abnormalities should receive a consultation with an orthopedic surgeon. Some of these pathologies can be treated early (e.g., impingement or hip dysplasia), leading to decreased rates of degeneration about the hip.

Appropriate Treatment/Referral Osteoarthritis is very common, and its first line of treatment is simple: nonsteroidal anti-inflammatory drugs (NSAIDs), stretching, physical therapy (including pool therapy), and weight loss. Once these treatments have been exhausted, the patient should probably be referred to a specialist. Steroid injection under ultrasound or other imaging modalities should be considered if the pain

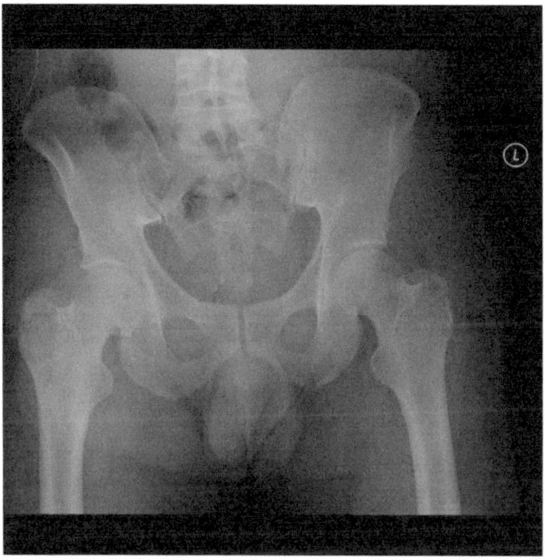

Fig. 8.5. Radiograph of a normal pelvis.

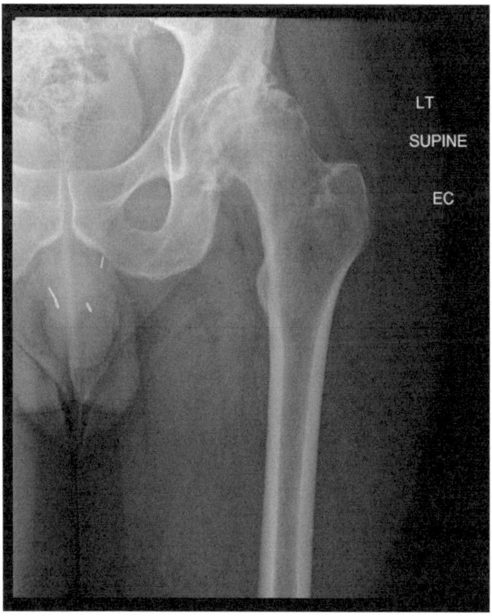

Fig. 8.6. Radiograph of the left hip showing changes typical of advanced osteoarthritis. Note joint space narrowing, subchondral sclerosis, osteophyte and cyst formation in the femoral head and acetabulum.

is very acute. Hip arthroplasty is an effective and reproducible procedure but should be carefully weighed against the risks of the procedure. Finnish Registry data suggest a 15-year revision-free survival rate of 71–86 % for total hip arthroplasty [6]. Not unlike other complex procedures, outcomes are better when performed by high-volume surgeons in high-volume centers.

Femoroacetabular Impingement

Presentation Femoroacetabular impingement (FAI) is a developmental abnormality of either the femoral head-neck junction and/or the acetabulum, either of which leads to abnormal hip function. These patients are generally young and/or active. They fall into two broad categories: cam-type impingement (loss of femoral head-neck offset) and pincer-type impingement (acetabular over-coverage). These biomechanical abnormalities lead to tears of the acetabular labrum (discussed in the next section) and delamination of the cartilage. This is theorized to be the precursor of the so-called idiopathic OA; however, it is not yet clear if surgical intervention has any influence on development of OA later in life [7].

Physical Exam Groin pain with anterior impingement is exacerbated with high flexion, adduction, and internal rotation at the hip (Fig. 8.7). Alternatively, posterior impingement is made worse with extension and external rotation [1]. Either of these may be combined with or exclusively present with labral-type symptoms, often with a popping and catching sensation with motion, which causes pain.

Diagnostic Exams Low AP weight-bearing pelvis (including both hips) with a supine lateral radiograph of the affected side ("frog leg lateral") is recommended. Radiographic measurements are taken to assess for these abnormalities, as they are often subtle (Fig. 8.8).

Differential Cam-type impingement is classically described among young athletic males. Given this population, it is important to rule out muscular strain or even sports hernia. Femoral hernia should be considered among women, even if pincer-type impingement is noted. The strict definition of cam versus pincer type impringement is somewhat of an oversimplification, however, with as much as 80 % of cases being considered a combined mechanism [9].

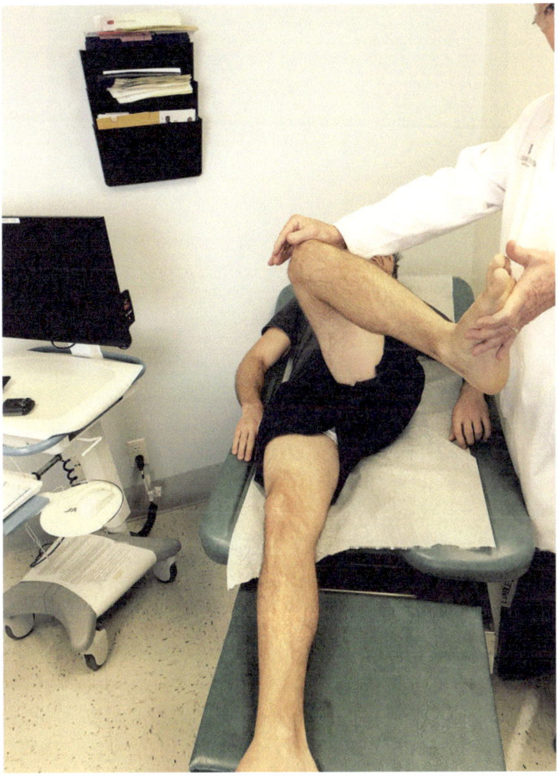

Fig. 8.7. The impingement maneuver consisting of flexion, adduction, and internal rotation. Pain or a pinching sensation in the groin during this maneuver is considered to be indicative of impingement.

Appropriate Treatment/Referral Both open and laparoscopic treatments have been effective in the control of symptoms. If FAI is suspected on clinical or radiographic examination, referral to an orthopedic surgeon who has experience with FAI is important, as early intervention may delay or prevent progression of the degeneration of the joint.

Labral Tear

Presentation The acetabular labrum has been shown to have a role in maintaining appropriate synovial fluid pressure for adequate lubrication of the hip joint [3]. The best analogy is a rubber gasket in a hydraulic

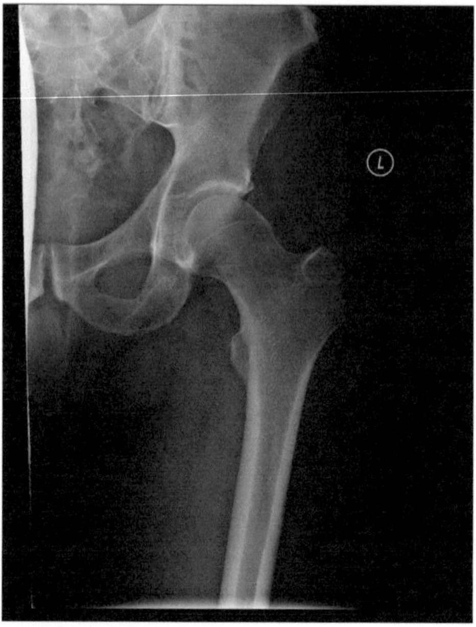

Fig. 8.8. Anteroposterior (AP) radiograph of the left hip showing the typical cam deformity of the proximal femur with os acetabulum (an accessory bone unrelated to the pathology).

joint. As such, the labrum has received new attention regarding its potential role in preserving the hip cartilage. Patients with tears of the labrum often present with deep-seated hip or groin pain or report a popping or clicking sensation with motion.

Physical Exam Painful range of motion is present, most pronounced with flexion or extension of the hip in abduction, combined with a rotational movement. Rolling the hip through this range of motion often produces pain and a popping sensation for a patient with labral pathology.

Diagnostic Exams X-rays may occasionally show a small calcification at the acetabular rim, indicating a calcified labrum from recurrent trauma and degeneration. However, most labral pathologies are not diagnosed with plain radiographs. Magnetic resonance (MR) arthrogram is the imaging study of choice for diagnosis and when a hip joint preservation procedure is a consideration (Fig. 8.9).

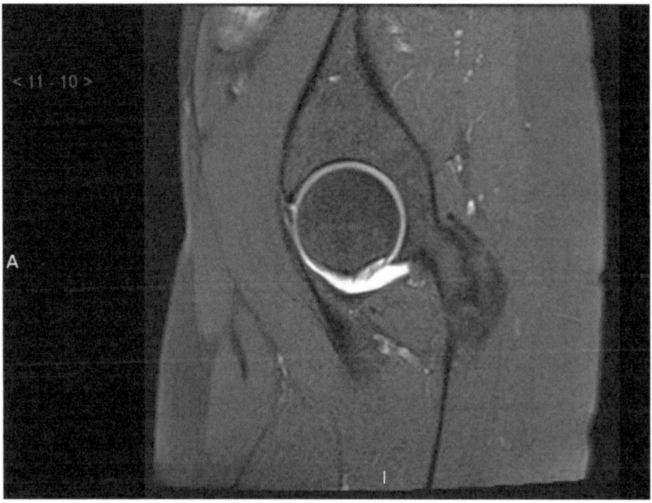

Fig. 8.9. T2-weighted MR arthrogram showing contrast tracking between the acetabular rim and the labrum indicative of a labral tear.

Differential Up to 90 % of patients with labral pathology have some degree of FAI according to some authors [7]. As such, an evaluation should include radiographic measurements, including the alpha angle and center edge angle to evaluate for this entity.

Appropriate Treatment/Referral Referral to an orthopedic surgeon with experience treating FAI is indicated due to the common coexistence of these pathologies. Although not an emergency, early intervention may decrease long-term degeneration of the joint.

Hip Dysplasia

Presentation Whereas symptoms of acetabular impingement occur due to actual or effective over-coverage of the femoral head, hip dysplasia represents the other side of this spectrum: under-coverage of the femoral head leads to increased stresses on the chondral surfaces. Congenital hip dysplasia in its extreme form will lead to dislocation of the hip among infants; however, the disease process lies on a spectrum, and many patients with dysplastic hips may be asymptomatic for many years prior to diagnosis. Additional conditions such as spondyloepiphyseal dysplasia and achondroplasia commonly lead to malformations of the hip that lead to increased rates of osteoarthritis. As such, any patient affected with dwarfism should be evaluated for orthopedic causes with any presentation of groin pain.

Physical Exam Painful range of motion, positive Stinchfield test (see the section on Osteoarthritis above), and classic signs of early osteoarthritis are present among these patients (see Fig. 8.4).

Diagnostic Exams Plain x-rays are often adequate to diagnose both the architectural problems and any early degenerative process occurring of the joints. However, patients without radiographically obvious degenerative changes but clinically evident hip dysplasia should continue to receive further orthopedic evaluation.

Appropriate Treatment/Referral As hip dysplasia is a congenital problem, patients with this malformation often present at relatively young ages with the beginning of degenerative changes. If caught early enough, operative options exist to increase the coverage of the femoral head with periacetabular osteotomies that may lead to decreased rate of degeneration of the hip. Once a patient has progressed to advanced degeneration of the joint, the only effective option is joint arthroplasty, which, although successful in pain relief and restoration of more normal biomechanics, has a limited life span. Revision surgery has much less reliably positive results and a higher complication rate. Referral to an orthopedic surgeon with experience with these procedures is important to delay progression as long as possible.

Occult Fracture

Presentation Although it is uncommon for a patient to initially present to the doctor's office with a hip fracture, it is possible that a patient may have had a prior workup that was falsely negative and is now presenting with groin pain that is in fact due to a missed hip fracture. This scenario may be seen among patients who suffered a trauma or fall with no clear x-ray evidence of a fracture. Although most emergency departments or urgent care centers will adequately work up a nondisplaced fracture seen on x-ray with computed tomography (CT) scan or MRI, an occult hip fracture is an important diagnosis to consider among patients who have a history of trauma or fall and pain, but with no obvious x-ray evidence of fracture. Nondisplaced fractures of the femoral neck, pubic rami, and sacrum are common following falls in elderly patients.

Physical Exam Patients present with groin and hip pain, typically exacerbated with any movement about the hip. Among patients with pubic ramus fractures, palpation of the pubic symphysis is often particularly painful. It is also important to palpate the sacrum, as tenderness to palpation may represent a fracture of the sacral ala.

Diagnostic Exams Among those with nondiagnostic x-rays, CT scan will help demonstrate nondisplaced and minimally displaced fractures about the hip. However, MRI is preferred, specifically for femoral neck fractures, as it is 100 % sensitive in the detection of radiographically occult femoral neck fractures [8]. A black line within the bone on T1-weighted images indicates a nondisplaced fracture.

Appropriate Treatment/Referral Nondisplaced pubic ramus and sacral ala fractures may be treated with simple pain control and radiographic follow-up to ensure that no unrecognized instability is present. If x-ray examination remains stable following mobilization, the patient does not require protected weight bearing; however, a walking aid should be recommended to ensure stability. On the other hand, nondisplaced femoral neck and intertrochanteric fractures require strict non-weight bearing and immediate referral to an orthopedic surgeon, as displacement may lead to a more difficult surgical treatment or displacement of the fragment.

Physiological

Septic Hip

Presentation A septic joint typically presents with acute onset hip and/or groin pain that is exacerbated by movement. Patients may or may not demonstrate erythema and swelling, due to the extent of the soft tissue surrounding the hip. A history of recent sexual contacts should be obtained among those who are sexually active, as gonococcal infections are known to present with monoarticular septic joints. Consideration of this diagnosis should also be considered among immunocompromised patients.

Physical Exam Patients report a painful joint, with dramatic increase in pain with any motion. It is this sign of irritable range of motion, with even small movements, that is the most reliable of the clinical signs.

Fevers, chills, and leukocytosis may be seen; however, their absence does not exclude the diagnosis of a septic hip.

Diagnostic Exams X-rays most often do not show any abnormalities, with the exception of long-standing cases of osteomyelitis in which a sequestrum and involucrum have had time to evolve. MRI can be useful if considering osteomyelitis. However, in the case of a simple septic joint, MRI will provide no more information other than the presence and size of the effusion. Aspiration and targeted surgical and antibiotic treatment are used as the standard of treatment. Synovial fluid is routinely sent for crystal examination, cell counts, and cultures. Cell counts above 50,000 white blood cells (WBC) and with greater than 75 % polymononuclear (PMN) cells are considered indicative of infection and are typically taken to the operating room for joint irrigation and debridement [9]. It is important to note that immunosuppressed patients may have an infected joint space in the presence of lower cell counts. Culture examination, although a critical part of the examination, does not trump the need for operative decompression and irrigation of the septic hip.

With any prosthetic joint in which infection is a concern, erythrocyte sedimentation rate (ESR) and C-reactive protein (CRP) need to be checked in addition to a complete blood count (CBC). ESR above 30 mm/h and of CRP greater than 1 mg/dL should raise suspicion for prosthetic joint infection. As in native hips, aspiration of the joint with cell counts is the gold standard for diagnosis, however cell counts as low as 1760 cells/μL are suggestive of a periprosthetic infection [2].

Differential It is important to consider crystalline arthropathy in the differential diagnosis of a septic joint, as its clinical presentation is nearly indistinguishable from that of a septic joint. Additionally, even if crystals are seen on joint aspiration, in the scenario of high number of WBCs and a high percentage of PMNs, the physician should consider the possibility of a superimposed infectious process. Cultures should be followed for a minimum of 3 days to rule out this possibility. However, if the suspicion is high for an infection, one should expeditiously proceed with operative treatment, i.e., decompression, irrigation, and debridement of the infected joint.

Appropriate Treatment/Referral In any patient with an examination concerning for septic hip, it is important to initiate an orthopedic consultation immediately. A septic joint is considered an operative emergency due to the unchecked inflammatory reaction that occurs as a result of the infection within the joint space. Metalloproteases and other destructive enzymes are released by immune cells in response to the infection, thereby causing irreversible damage to articular cartilage. Failure to intervene early in the process leads to cartilage destruction and may lead to chronic osteomyelitis if the infection spreads into the adjacent bone. As such, should cell counts be diagnostic or cultures be positive, urgent irrigation and debridement of the joint, either arthroscopically or open, should be performed.

Stress Fracture and Pathological Fracture

Presentation Stress fractures are most common among running athletes. Typically, patients present with pain of insidious onset, often associated with increases or changes in training. The pain is worse with weight bearing and with increased activity. Stress fractures are more common among female athletes, especially those with low body weight and amenorrhea.A stress fracture forms when the body is unable to remodel at the rate necessary to deal with the increased repetitive stresses imposed on it. These can occur in the sacrum, pubic rami, and femoral neck. Although only 10 % of stress fractures occur at the femoral neck, they require immediate attention. There is a risk of fracture displacement which may lead to complications such as a vascular necrosis (AVN) [10].Another subset of patients who may present with groin or thigh pain are elderly patients with known osteoporosis who have been on bisphosphonate therapy for several years. These patients, ironically, may also have stress fractures. Bisphosphonates have become extremely common for the treatment of osteoporosis, as they have been shown to decrease the rate of hip fracture by up to 51 % in some series [11]. Bisphosphonates function by incorporating into the mineral content of bone, consequently decreasing osteoclastic activity. However, due to the coupling of osteoclastic (bone resorbing) and osteoblastic (bone building) cell activities, this leads to the decreased ability of the bone to remodel. Whereas this is helpful in decreasing the rate of bone resorption in patients with osteoporosis, the reduced rate of bone turnover does not allow for adequate remodeling of bones with high stresses. The subtrochanteric femur is particularly prone to this process. These patients present with thigh or groin pain of insidious onset that is worse with activity and on rising from a seated position. Patients with this process are prone to fractures of the subtrochanteric femur with very minimal trauma, often as little as a twisting moment while walking.

Physical Exam Range of motion and examination of the hip are essentially normal. Weight bearing may demonstrate some pain. Clinically, stress fractures of the femoral neck may be difficult to differentiate from early AVN. Consideration of the risk factors and advanced imaging may be required to differentiate between the fractures and AVN.

Diagnostic Exams For stress fractures, x-rays may show a periosteal reaction in the area of the stress fracture. However, x-rays are not sensitive or specific for this process.

In the case of bisphosphonate-related stress fractures, the x-ray may show lateral breaking of the subtrochanteric femur (Fig. 8.10). Although bone scans are 100 % sensitive for stress fractures, they lack specificity [10]. MRI is highly sensitive and specific for stress fracture. On T2-weighted imaging, edema will be seen in association with the area of the stress reaction; a dark line will represent the fracture line on T1-weighted imaging (Fig. 8.11).

Differential Consideration of a pathological fracture due to either primary or metastatic infiltration of the bone should be considered among patients who present with pain of insidious onset associated with weight bearing. Any lytic or blastic lesion on x-ray should be worked up, which may include an open or fluoroscopically guided biopsy, especially in patients with risk factors for malignancy.

Appropriate Treatment/Referral Any patient with confirmed or suspected stress fractures should be referred to an orthopedic surgeon for evaluation. In addition to activity modification, patients require protected weight bearing until their pain resolves. With regard to those patients with stress fractures involving the superior aspect of the femoral neck (tension-side femoral neck stress fractures), or those with complete stress fractures that are still nondisplaced, urgent operative intervention may be indicated to prevent displacement.

Inflammatory Arthritis

Presentation Although less common than OA, rheumatologic disease must be considered in any differential of joint pain. Rheumatoid arthritis (RA), systemic lupus erythematosus (SLE), and ankylosing spondylitis may all present with polyarticular complaints that may involve the hip. Generally, patients with inflammatory arthritides present with symmetrical joint complaints. It is important to note that these will rarely cause isolated hip pain without other systemic complaints. Evaluation for low back pain and other articular complaints is important for accurate diagnosis. Classically, patients with RA will complain of at least 1 h of morning stiffness.

A similar clinical picture is present among those with SLE, with the addition of further systemic complaints such as skin rashes with sun exposure and possible renal complications. Consideration to the diagnosis of SLE should be given with any workup for rheumatological disease.

In patients with ankylosing spondylitis, involvement of the lumbar spine needs to be evaluated. Patients will report chronic low back or hip pain that waxes and wanes, with limited range of motion, sometimes with groin pain and pain down the inner thigh.

Physical Exam The disease predominantly affects the cervical spine, wrist, elbows, knees, hands, and feet. It is not classically described as affecting the hip and typically spares the lumbosacral spine and distal interphalangeal joints [9]. The presence of rheumatoid nodules on the extensor surface of the forearm is considered pathognomonic and is seen in up to 30 % of patients with the disease. Examination is otherwise often nondiagnostic and needs to be combined with other modalities for accurate diagnosis.

Diagnostic Exams X-rays may show periarticular osteopenia, with loss of joint space and minimal to no osteophytes. Laboratory examination shows elevated inflammatory markers (ESR and CRP). If the clinical presentation fits with an inflammatory condition, it is not unreasonable to consider sending specific blood tests checking for rheumatoid factor (RF), antinuclear antibody (ANA), and HLA-B27 antigen. It may be reasonable to defer the workup to a rheumatologist.

Appropriate Treatment/Referral In situations where an inflammatory arthritis is suspected, referral to a rheumatologist for a complete yet targeted workup and treatment is warranted. In severe cases with advanced joint disease, total joint arthroplasty is an option for treatment, and referral to an orthopedic surgeon should be considered in patients with advanced disease.

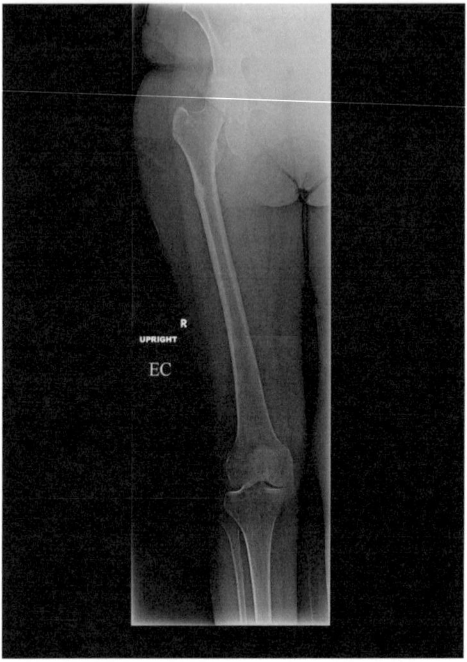

Fig. 8.10. Radiograph of the right femur showing breaking of the proximal lateral cortex. There is an associated stress fracture characteristic of a bisphosphonate-related fracture of the proximal femur.

Avascular Necrosis

Presentation Patients with AVN present with groin pain of insidious onset, most often with no other symptoms. As AVN progresses to subchondral collapse, pain can be severe. Risk factors may include prior fracture or dislocation, alcoholism, steroid use, and caisson disease (the bends or decompression sickness). AVN has also been associated with use of protease inhibitors for treatment of human immunodeficiency virus [12]. Less common causes of AVN are sickle cell disease or lysosomal storage diseases. As such, groin pain of insidious onset in these patient populations should prompt high clinical suspicion. Assessment of risk factors is an important part of the history taking in patients with suspected AVN.

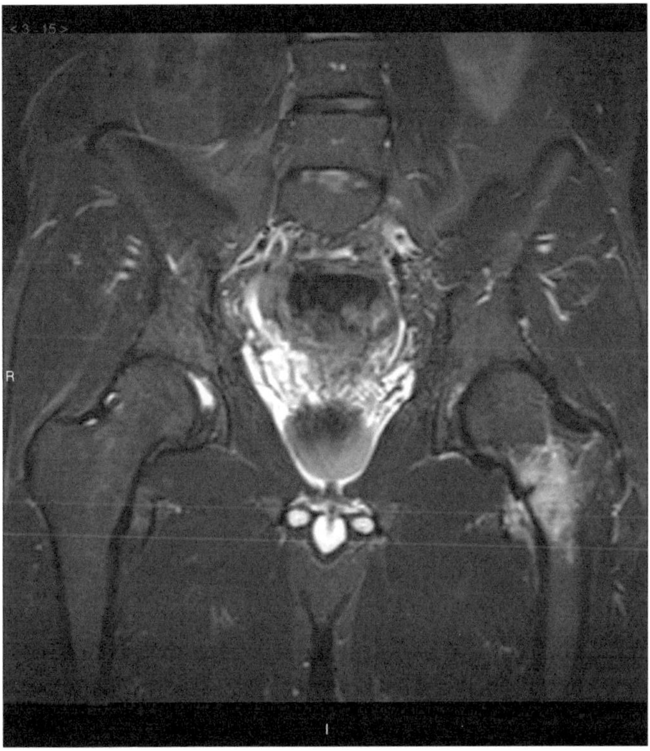

Fig. 8.11. Coronal STIR sequence MRI showing marked left femoral neck bone edema (*white*) with inferior femoral neck cortical disruption (*black line*) typical of a compression-side stress fracture.

Physical Exam Early examination of patients with AVN is essentially normal. As the physiological changes progress toward squaring of the femoral head and subchondral collapse, the range of motion of the hip becomes increasingly limited.

Diagnostic Exams X-rays are appropriate in the initial diagnosis, classically graded on the Ficat scoring system, ranging from a normal examination to advanced collapse of the femoral head with resulting joint degeneration [13]. However, positive x-ray findings are generally noted late in the disease progress. MRI has the highest specificity and sensitivity for AVN. Early signs include diffuse edema seen on T2 imaging that evolves to a focal crescent beneath the chondral surface of the femoral head with advanced disease.

Differential Early AVN may present in a similar fashion to transient osteoporosis (see below). Subtle differences do exist between their MRI findings.

Appropriate Treatment/Referral Once a hip has progressed to AVN, there is no proven effective treatment for reversal. Some centers advocate for bisphosphonate therapy and surgical core decompression. Although both of these modalities have shown to be somewhat effective in the management of the pain related to the diagnosis, neither has consistently shown to have an effect on the progression of AVN to subchondral collapse. Once a hip has progressed to subchondral collapse, the only option for the patient is total hip arthroplasty.

Transient Osteoporosis

Presentation Transient osteoporosis presents as unilateral hip pain of insidious onset. Diffuse groin and hip pain without inciting event is common. The pain progresses over the course of weeks, generally to the point of non-weight bearing. This entity was originally described in 1959, among women in their third trimester of pregnancy [14]. Today, it is recognized more commonly among middle-aged men [15]. The natural history of the disease is for pain to resolve over the course of 2–3 months, with bone remineralization expected within approximately 6 months without intervention.

Physical Exam Patients may present with sufficient pain to not permit weight bearing. Range of motion and other physical exam findings are typically unremarkable.

Diagnostic Exams X-rays are often normal at the onset of pain and progress to show diffuse loss of trabecular bone mineralization of the femoral head without detectable cortical disruption after several weeks of symptoms. Importantly, there should not be involvement in the acetabulum; if there is, then inflammatory or infectious causes should be considered.

MRI shows diffuse low signal on T1-weighted imaging and high signal on T2-weighted imaging. Of note, these findings are nearly identical to early AVN of the femoral head; therefore, the two processes are difficult to differentiate. However, the changes are typically more diffuse for transient osteoporosis as compared to the more classic "band-like pattern" of femoral head involvement seen in AVN [16].

Differential Early AVN and inflammatory or infectious causes should be considered.

Appropriate Treatment/Referral In patients with radiographically recognizable lesions, patients should practice protected weight bearing for 4–6 weeks and be supplemented with vitamin D and calcium. Although relatively rare, fracture of the demineralized femoral head and neck is reported, and the patient should be aware of such a complication. If pain is uncontrollable with mild analgesics, consideration may be given to bisphosphonate therapy and/or core decompression via an operation. That said, there is poor evidence to strongly recommend either option. Small case series report early resolution of clinical symptoms with both bisphosphonate therapy and core decompression therapy [17, 18].

Groin Pain from the Soft Tissue

Architectural Problems

Snapping Hip Syndrome

Presentation Patients with snapping hip syndrome present with painful popping sensations in the groin with movement. The problem is common and is estimated to occur in 5–10 % of the general population [10]. The differential for a snapping hip includes intra-articular and extra-articular causes.

Intra-articular causes include labral tears, osteochondral lesions, or loose bodies. Rarely, injuries to the ligamentum teres following dislocation may also present with intra-articular snapping.

An external snapping hip pops or catches while in hip flexion. It is typically related to either the iliotibial band or tensor fascia lata catching over the greater trochanter. Alternatively, the gluteus tendon may also catch over the trochanter. This catching leads to inflammation and pain over the lateral greater trochanter and may present similar to bursitis.

Internal snapping hip by contrast presents with a popping or catching sensation upon external or internal rotation. Here, it is typically the iliopsoas tendon or the iliofemoral ligament that is catching over the lesser trochanter, anterior aspect of the hip joint, or the superior pubic ramus. Internal snapping hip may also occur as a result of a poorly positioned acetabular cup after total hip arthroplasty.

Physical Exam An audible or palpable popping with associated pain is noted in flexion or in rotation of the hip. Often, the involved tendons may develop tendonitis and are tender to palpation.

Diagnostic Exams Careful history needs to exclude intra-articular causes of snapping hip, such as labral tear or a loose body. Snapping hip relieved by intra-articular steroid injection are often caused by labral tears [19]. MRI is commonly employed to look for evidence of tendon impingement or snapping; edema around or within the tendons may indicate this. Some authors advocate for the use of dynamic ultrasound in the diagnosis of snapping hip. The sensitivity and specificity of this is operator dependent and may not be available at all institutions.

Differential Intra-articular anesthetic and steroid injections are an excellent diagnostic tool to differentiate intra-articular causes of snapping hip from extra-articular causes. If pain resolves with intra-articular injection, then the pain is likely of intra-articular origin.

Appropriate Treatment/Referral The mainstay of treatment for both internal and external snapping hip is physical therapy and NSAIDs combined with activity modification. Among patients resistant to this treatment, tendon lengthening or tendon release has been shown to improve symptoms. For snapping hip of intra-articular cause, referral to an orthopedic surgeon is appropriate.

Lateral Femoral Cutaneous Neuralgia

Presentation Lateral femoral cutaneous nerve pain, also known as "meralgia paresthetica," presents with an uncomfortable, numb, and/or painful sensation in the anterolateral upper thigh. It is reported to be more common among the obese and diabetic patient groups. It also may present during pregnancy and with hypothyroidism due to the associated peripheral edema [20]. The cause of symptoms is theorized to be the pannus direct compression of the nerve, either due to a pannus or entrapment by the inguinal ligament as the nerve crosses under or through it. Additionally, this condition may occur as a result of hip replacement performed through an anterior approach or anterior iliac crest bone grafting, where the lateral femoral cutaneous nerve may be injured. Direct trauma to the nerve, such as a shear injury from a seat belt, is another possible cause of meralgia paresthetica.

Physical Exam Examination of the hip for any prior scars or operations is advisable. Tapping of the inguinal ligament laterally, 1 cm medial to the ASIS where the nerve crosses, elicits a Tinel's sign, with stinging or burning into the anterolateral thigh [21]. Hypesthesia and/or allodynia of a patch of skin along the upper lateral thigh is consistent with the dermatomal findings for this neuralgia. Extension of the thigh may also aggravate symptoms as it places the nerve on stretch.

Diagnostic Exams The diagnosis of meralgia paresthetica is clinical and does not require imaging. X-rays will be unrevealing. MRI may reveal edema and swelling of the nerve in extreme cases, best seen on T2 images proximal and laterally along the inguinal ligament; this is subtle and not universally present. Ultrasound may show swelling of the nerve between the inguinal ligament and deep circumflex iliac artery, with flattening of the nerve as it courses under the inguinal ligament. Sensory nerve conduction velocities may be ordered to confirm the diagnosis if questions exist [21].

Differential It is important to rule out lumbar disk herniation as a cause of symptoms. Any focal weakness or other symptoms in the L2 distribution should prompt evaluation for this. Additionally, intrapelvic masses have been known to compress this nerve along its course and should be considered in the differential.

Appropriate Treatment/Referral Referral to a pain specialist for corticosteroid injections should be considered in patients in whom meralgia paresthetica is suspected. Typically, this is undertaken with the use of ultrasound guidance. These can be both confirmatory of the diagnosis and therapeutic. Small series have shown good results and high rates of resolution over the course of 1–2 months [22]. If there is no long-term improvement, neurectomy may be considered.

Physiological Problems

Gluteus Medius Tendonitis

Presentation Tears and tendonitis of the gluteus medius tendon have only recently become recognized as causes of hip pain. They can also present with primary complaint of groin pain. Likened to the "rotator cuff tear of the hip," some authors of small series advocate arthroscopic debridement and repair for large tears [22], although limited evidence currently exists to advocate for or against these procedures. Additionally, its role in greater trochanteric bursitis is becoming clearer, as some believe the presence of tendonitis of either the gluteus medius or minimus is the primary pathology in greater trochanteric bursitis [23].

Physical Exam Patients with gluteus medius tendonitis present with pain in the hip or groin, exacerbated with activity. Patients have focal tenderness to palpation and pain with resisted abduction. In severe tendonitis or in cases in which a gluteus medius tear is present, patients may demonstrate weakness of the abductors and a positive Trendelenburg gait. Testing will reveal they are unable to keep their hips level during single-leg stance of the affected side.

Diagnostic Exams X-ray is unrevealing, and the diagnosis is clinical. MRI has a role if there is concern for a gluteus medius tear. MRI should be considered in cases of tendonitis nonresponsive to multimodality treatment, such as physical therapy and steroid injections.

Differential Greater trochanteric bursitis is almost identical in presentation, and as understanding of the process grows, it is increasingly becoming inseparable from gluteus medius tendonitis.

Appropriate Treatment/Referral The first line of treatment is a dedicated course of NSAID therapy combined with physical therapy. Referral to an orthopedic surgeon for steroid injections is warranted if the patient does not respond to 4–6 weeks of physical therapy. If patients fail to improve, it is reasonable to consider an MRI to evaluate for a tear or other causes for the patient's pain.

Greater Trochanteric and Iliopsoas Bursitis

Presentation With greater trochanteric bursitis, patients will typically present with lateral hip and leg pain radiating down the side of the leg and into the knee, along the iliotibial (IT) band. Patients have difficulty sleeping on the affected side. They have increased pain with flexion of the hip, as this tightens the IT band against the greater trochanter.

Iliopsoas bursitis may occur in internal snapping hip syndrome with or without the presence of an audible snap (see Snapping Hip Syndrome above). Due to either acute or repetitive trauma from the recurrent dragging of the tendon over the lesser trochanter, AIIS, or iliopectineal eminence, the bursa between the tendon and the pelvic brim becomes inflamed. Patients report a history of anterior hip or groin pain exacerbated by flexion and extension of the hip. Although most commonly associated with an audible snap, up to 31 % of patients have no history of snapping or popping.

Physical Exam Patients have focal tenderness to palpation over the lateral trochanter. The Ober test is used to evaluate tightness of the IT band, which is thought to be the cause of the problem. The patient lies on his/her contralateral side, the affected leg is extended and allowed to fall behind the patient; inability to adduct beyond the midline is considered a positive test [23].

Patients with iliopsoas bursitis present with anterior hip or groin pain, exacerbated by flexion or extension of the hip. Tenderness to palpation over the course of the iliopsoas tendon deep in the femoral triangle is common, but not universal, and when seen is considered pathognomonic for the process.

Diagnostic Exams X-ray and ultrasound are not sensitive or specific for any of the bursitis conditions around the hip. Occasionally, calcification may be present in the bursa, which indicates chronic inflammation and calcium deposition. MRI does have some role, as it is the most sensitive and specific for these conditions; however, diagnosis of bursitis is mostly clinical, and confirmatory imaging is not usually required.

Appropriate Treatment/Referral With all bursitis, the mainstay of treatment is NSAID therapy and physical therapy for stretching and strengthening. Corticosteroid injection may be performed with or without the use of ultrasound guidance in greater trochanteric bursitis; for iliopsoas bursitis, ultrasound guidance is required due to the proximity to neurovascular structures. In the event that conservative treatment is unsuccessful, iliopsoas tenotomy or surgical lengthening of the IT band may be considered.

Muscle Sprain or Strain

Presentation Muscle strain injuries are common in the young athletic population. Patients typically will have an injury during eccentric contraction of the muscle and suffer immediate pain. This is reported as a chronic and nagging pain that lasts an extended period of time with delayed recovery. Adductor strains, in particular, present with medial groin pain that worsens with resisted adduction. They are particularly common among hockey players and represent as many as 10 % of all injuries to hockey players at the professional level [24]. By contrast, rectus femoris strains will present with anterior pain and swelling, approximately 8–10 cm below the AIIS, and may be seen among sprinting athletes and soccer players [25].

Physical Exam After adductor strain, pain is localized medially and in the groin. Pain is exacerbated by forced adduction and direct palpation of the muscle belly. The attachment of the adductor muscles at the inferior pubic ramus is focally tender to palpation.

After rectus femoris strains, pain and swelling localize anteriorly, directly over the hip. Pain is exacerbated with extension of the hip and flexion of the knee and worsens with resisted flexion of the hip.

Diagnostic Exams AP x-ray should be obtained to rule out avulsion type fractures associated with such strains. Occasionally, calcification of the tendon or tendinous insertion may be seen in some patients. MRI is more helpful in clinical situations in which the diagnosis is unclear, however, it is not required for the diagnosis. If obtained, focal edema involving the muscle is seen on T2-weighted signals. In more severe cases, tears in the muscle may also be seen.

Differential It is important to consider the possibility of avulsion fracture of the tendinous insertions when evaluating these injuries, especially in patients who have not yet reached skeletal maturity or patients with massive muscle volume, such as certain professional athletes. These are often findings seen in the spectrum of sports hernias.

Appropriate Treatment/Referral Rest, ice, compression, and elevation are the mainstays of treatment. Activity modification is recommended until pain subsides. Return to activity should start with range of motion exercises and gradually increase in intensity over several weeks. Physical therapy should focus on stretching and strengthening exercises. The use of NSAIDs should be avoided in the management of the pain associated with muscular strain due to concern over long-term healing [25]. In patients without improvement despite a period of immobilization followed by physical therapy, adductor tenotomy may be considered. However, although small series have shown improvement in symptoms in 100 % of patients, as few as 63 % of patients are able to return their prior level of competitive sport [24].

Sacroiliac Joint Pain

Presentation Although typically presenting with low back pain radiating down the back of the leg, sacroiliac (SI) joint pain may also present as pain radiating into the groin. Due to its proximity to the lumbar spine and hip, it is very commonly mistaken for pain coming from either of these sources. The diagnosis remains somewhat controversial, but some small series have shown it to be a notable cause of residual pain in patients with prior lumbar fusion [26]. Patients typically report pain while lying on their side, in a prolonged seated position, upon initial rising, and with sitting down. Patients with altered gait mechanics (such as favored weight bearing of an injured or painful extremity) are prone to SI joint pain and dysfunction.

Physical Exam Palpation over the SI joint may reproduce the patient's pain. Provocative tests include compression of either the posterior superior iliac spine or the ASIS, medially directed force over the iliac crests, pain with resisted external rotation, and FABER test (forced flexion, abduction, external rotation or the "figure of 4" position) [27]

Diagnostic Exams Occasionally, x-rays show spurring inferiorly at the SI joint; however, most will not have radiographic findings. A fluoroscopically guided SI joint injection with anesthetic and steroids is both diagnostic and therapeutic.

Differential SI joint pain is difficult to differentiate from back or hip pain and is often overlooked as a source of pain. Any differential including SI joint pain should rule out lumbar causes and any intra-articular hip pathology that may be causing the symptoms. Sacroiliitis, either infectious or arthritic, is also part of the differential diagnosis (see Inflammatory Arthritis above).

Appropriate Treatment/Referral NSAIDs are the mainstay of treatment. This is typically combined with physical therapy for core strengthening and pelvic stability exercises. Fluoroscopically guided injections of the SI joint are also helpful in more severe cases. If all else fails, fusion of the SI joint is an option, but this is reserved for severe cases in which repeated injections and courses of physical therapy are unsuccessful in relieving pain.

References

1. Fernandez E, Gastaldi P. Hip pain from the orthopedic point of view. Eur J Radiol. 2012;81(12):373–9.
2. Miller MD, Thompson SR, Hart JA. Review of orthopaedics. 6th ed. Philadelphia, PA: WB Saunders Elsevier; 2012.
3. Ferguson SJ, Bryant JT, Ganz R, Ito K. An in vitro investigation of the acetabular labral seal in hip joint mechanics. J Biomech. 2003;36(2):171–8.
4. Thompson JC. Netter's concise orthopaedic anatomy. 2nd ed. Philadelphia, PA: WB Saunders Elsevier; 2010.
5. Garstang SV, Stitik TP. Osteoarthritis: epidemiology, risk factors, and pathophysiology. Am J Phys Med Rehabil. 2006;85(11 Suppl):S2–11.
6. Mäkelä KT, Eskelinen A, Paavolainen P, Pulkkinen P, Remes V. Cementless total hip arthroplasty for primary osteoarthritis in patients aged 55 years and older: results of the 8 most common cementless designs compared to cemented reference implants in the Finnish arthroplasty register. Acta Orthop. 2010;81(1):42–52.
7. Bedi A, Kelly BT. Femoroacetabular impingement. J Bone Joint Surg Am. 2013;5(1):82–92.
8. Evans P, Wilson C, Lyons K. Comparison of MRI with bone scanning for suspected hip fracture in elderly patients. J Bone Joint Surg (Br). 1994;76(1):158–9.
9. Flynn JM. OKU 10: orthopaedic knowledge update. Rosemont, IL: American Academy of Orthopaedic Surgeons; 2011.
10. Henning PT. The running athlete: stress fractures, osteitis pubis, and snapping hips. Sports Health. 2014;6(2):122–7.
11. Ainhorn TA, Bogdan Y, Tornetta 3rd P. Bisphosphonate-associated fractures of the femur: pathophysiology and treatment. J Orthop Trauma. 2014;28(7):433–8.
12. Ries MD, Barcohana B, Davidson A, Jergesen HE, Paiement GD. Association between human immunodeficiency virus and osteonecrosis of the femoral head. J Arthroplasty. 2002;17(2):135–9.
13. Ficat RP. Idiopathic bone necrosis of the femoral head. Early diagnosis and treatment. J Bone Joint Surg (Br). 1985;67(1):3–9.
14. Curtiss Jr PH, Kincaid WE. Transitory demineralization of the hip in pregnancy: a report of three cases. J Bone Joint Surg Am. 1959;41-A:1327–33.
15. Korompilias AV, Karantanas AH, Lykissas MG, Beris AE. Transient osteoporosis. J Am Acad Orthop Surg. 2008;16(8):480–9.
16. Guerra JJ, Steinberg ME. Distinguishing transient osteoporosis from avascular necrosis of the hip. J Bone Joint Surg Am. 1995;77(4):616–24.
17. Kibbi L, Touma Z, Khoury N, Arayssi T. Oral bisphosphonates in treatment of transient osteoporosis. Clin Rheumatol. 2008;27(4):529–32.

18. Hofmann S, Engel A, Neuhold A, Leder K, Kramer J, Plenk Jr H. Bone-marrow oedema syndrome and transient osteoporosis of the hip. An MRI-controlled study of treatment by core decompression. J Bone Joint Surg (Br). 1993;75(2):210–6.
19. Yamamoto Y, Hamada Y, Ide T, Usui I. Arthroscopic surgery to treat intra-articular type snapping hip. Arthroscopy. 2005;21(9):1120–5.
20. Harney D, Patjin J. Meralgia paresthetica: diagnosis and management strategies. Pain Med. 2007;8(8):669–77.
21. Martinoli C, Miguel-Perez M, Padua L, Gandolfo N, Zicca A, Tagliafico A. Imaging of neuropathies about the hip. Eur J Radiol. 2013;82(1):17–26.
22. Tagliafico A, Serafini G, Lacelli F, Perrone N, Valsania V, Martinoli C. Ultrasound-guided treatment of meralgia paresthetica (lateral femoral cutaneous neuropathy): technical description and results of treatment in 20 consecutive patients. J Ultrasound Med. 2011;30(10):1341–6.
23. Aaron DL, Patel A, Kayiaros S, Calfee R. Four common types of bursitis: diagnosis and management. J Am Acad Orthop Surg. 2011;19(6):359–67.
24. Nicholas SJ, Tyler TF. Adductor muscle strains in sport. Sports Med. 2002;32(5):339–44.
25. Anderson K, Strickland SM, Warren R. Hip and groin injuries in athletes. Am J Sports Med. 2001;29(4):521–33.
26. Cohen SP. Sacroiliac joint pain: a comprehensive review of anatomy, diagnosis, and treatment. Anesth Analg. 2005;101(5):1440–53.
27. Maigne JY, Aivaliklis A, Pfefer F. Results of sacroiliac joint double block and value of sacroiliac pain provocation tests in 54 patients with low back pain. Spine. 1996;21(16):1889–92.

9. Groin Pain Etiology: Spine and Back Causes

Charles H. Li, Victor W. Chang, Irene Wu, and Daniel C. Lu

Introduction

Groin pain is a common cause of complaints in the primary care clinic setting [1]. Groin injuries are responsible for approximately 5 % of all athletic injuries and account for 10 % of visits to sports medicine clinics [2]. In the clinical evaluation of groin pain, it is important to elicit the history of onset [3]. Although the differential is wide, altered sensation or weakness can raise suspicion for neurological causes such as peripheral nerve entrapment, herniated disc, or lumbar disc degeneration [3]. Fractures or malignancy are always within the differential and should be suspected when there is a history of pain at rest or at night. Sacroiliac joint dysfunction can be a chronic cause of groin and lower back pain that is commonly underdiagnosed [4].

Sacroiliac (SI) Joint Dysfunction

The SI joint is a synovial joint with hyaline cartilage on the sacral side of the joint [4]. It is believed that the joint is entirely innervated by the sacral dorsal rami [5]. The joint is mostly a bony structure supported by a number of ligaments and the surrounding muscles. The joint shares these muscles with the hip joint and is subject to all the same shear forces that the hip experiences [4]. SI joint pain is commonly due to trauma or strain. Repetitive motions associated with athletic activities can also cause repetitive shear.

B.P. Jacob et al. (eds.), *The SAGES Manual of Groin Pain*,
DOI 10.1007/978-3-319-21587-7_9

Physical Exam Most of SI joint pain is referred to the buttocks (94 %), lower lumbar region (72 %), and lower extremities (50 %) [6]. Pain that is localized to the groin is an unusual presentation, seen in around 14 % cases [6]. There are extensive innervations in the hip and groin area, making physical examination difficult and nonspecific. Three provocative SI joint movements can detect SI joint dysfunction with a sensitivity of 77–87 % [7]. Common tests of SI joint function include Laguere, Gillette, Patrick, and the Gaenslen tests [7–9]. Further, radiographic exam may be helpful in corroborating physical exam findings, but radiology alone is not sufficient for diagnosis [4]. Elgafy et al. [10] showed that CT scans for SI joint dysfunction had a sensitivity of 57.5 % and a specificity of 69 %. SI joint blocks can be used as a diagnostic tool and have been associated with a positive predictive value of 60 % when used with three physical exam tests [6, 11].

Treatment Patients with SI joint dysfunction should be treated with a multimodal approach. Results of treatment consisting of physical therapy, orthotics, joint blocks, surgery, and neuroaugmentation have been highly variable [4]. Physical therapy exercises focus on movements that can strengthen the hip muscles and stabilize the pelvis [12]. Placement of an orthotics belt has also been useful in some treatments as a way of limiting further motion that can cause increased injury to the joint [13]. Several studies have also noted relief of symptoms and improvement in function with intraarticular injections of the SI joint [14–17]. This can further be followed with radiofrequency rhizotomies of the innervation to the SI joint for more lasting analgesia. There is no class I evidence to support this procedure.

Historically, surgical treatment was used only when the SI joint was proven to be unstable [18]. Traditional techniques for SI joint fusion involved large open procedures, which introduced a great amount of morbidity and were of limited clinical benefit. With the recent advent of minimally invasive techniques to fuse the SI joint, there is growing interest among spine surgeons to pursue this technique. However, to date there is not a great deal of high-level evidence to support this procedure. Neuroaugmentation is a new modality for treatment of SI joint pain, and case reports [19, 20] have suggested that it may be a standard treatment option in the future.

Lumbar Disc Degeneration

Lumbar disc degeneration is a known cause of persistent groin pain [21]. The groin is innervated by the genitofemoral and ilioinguinal nerves; degeneration of the spinal canal can cause referred pain to the groin. In particular, patients with herniated discs (most commonly in the L4–L5 or L5–S1 levels) have been known to report groin pain [21]. These discs can cause compression on transversing sacral nerve roots (S1–S3). Additionally, herniation at the L1/L2 levels is rarer but can cause characteristic symptoms of groin pain, manifesting as buttock pain and anterolateral thigh pain [22].

Diagnostic Workup Studies have shown that around 21 % of patients with lower lumbar disc degeneration or herniation have had associated symptoms of groin pain [23]. Making a diagnosis of discogenic groin pain is difficult due to the nature of the disease presentation. There has been controversy in the use of discography to diagnose discogenic pain [24]. MRI is commonly used as a noninvasive approach to diagnosing lower back pain from degenerative disc disease [25], and ultrasound imaging has been shown to have a 90 % sensitivity and 75 % specificity for finding disc degeneration when combined with discography [26].

Lumbar Stenosis

Stenosis is defined as the narrowing of the spinal canal, usually to an absolute diameter of less than 75 mm^2 as characterized by imaging [27]. Similar to degeneration, nerve root stenosis at the L1/L2 levels will affect the L2/L3 nerve roots, manifesting in a positive femoral stretch test and anterolateral thigh pain [28]. Imaging is not definitive; in one study, more than 30 % of patients had images consistent with lumber stenosis but did not feel any of the associated symptoms [29]. Diagnosis is made through physical examination using similar tests reported for SI joint dysfunction, as well as through exclusion of other possible diagnoses. Once diagnosed, the most effective treatment for spinal stenosis involves patient education, therapy, exercise, and training [27]. For symptoms of pain, exercises that focus on strengthening the muscles involved in thoracic extension and lumbar rotation were found to be most effective in relieving pain [30, 31], presumably because these types of exercises were the most important for increasing flexibility in the groin region. Intervention for severe spinal stenosis includes epidural steroid injections and surgery. Injections have increased in popularity in recent years, but their efficacy has been controversial [32]. Surgical versus nonsurgical approaches have recently been evaluated by the Spine Patient Outcome Study (SPORT) [33]; the authors concluded that patients with degenerative spondylolisthesis treated with surgery had significantly better outcomes at four years compared to those who were managed noninvasively. The outcomes for patients with stenosis without spondylolisthesis were not as clearly defined.

Herniated Disc

Herniated discs are one of the most common discogenic causes of groin pain. The most common sites of herniation are at the L4–5 and L5–S1 levels. Additionally, other sites of herniation that will manifest as groin pain include the L1/L2 and S3/S4 levels. S3/S4 involvement likely is not due to direct S3/4 disc herniation, rather by S3/4 nerve root compression by more rostral disc herniations (i.e., L4–5 or L5–S1). L1/L2 disc herniation will localize to the inner thigh, while a herniation that affects the S3/S4 level will localize to the scrotal region. It is believed that decreased hydration of the annular disc leads to decreased ability of the disc to cushion load. This dehydration can be due to age, genetics, and environmental factors. A sharp stabbing pain that radiates down to the extremities below the knees is highly suggestive of herniation [34]. On physical exam, increased pressure on the annular fibers of the disc will help distinguish herniation from low back pain, which is typically made worse by twisting motions of the lower back muscles. The straight leg raise is usually indicative of a pinched nerve or nerve root.

Nonsurgical approaches to the management of a herniated disc are similar to those for other forms of disc degeneration. These approaches include physical therapy, focused exercises, and epidural injections. The natural history of lumbar herniated disc is that a majority of patients will resolve their symptoms without intervention given enough time. Surgery may be indicated in severe cases that cause significant pain or disability and also in cauda equina syndrome [34].

Spondylolisthesis

Spondylolisthesis refers to anterior subluxation of the vertebral body that is caused by a defect in the pars interarticularis [35]. Spondylolisthesis falls into three categories: spondylolysis, isthmic, and degenerative [36].

Isthmic is the most common form, occurring in 4–8 % of the general population and is found twice as often in males compared to females [37]. Although isthmic spondylolisthesis can usually be detected in childhood, patients usually do not present with symptoms until later in life. Presenting symptoms include pain in the lumbar area that can manifest as groin pain that radiates into the buttocks and thighs [37]. The pain can be exacerbated with weight lifting maneuvers or Valsalva. Higher grade spondylolisthesis can also manifest in hamstring tightness [37]. Degenerative spondylolisthesis usually presents much later in life and is caused by long-standing instability in the lumbar segments. The instability is most frequently due to arthritis, malfunction of the ligaments stabilizing the lumbar joints, or ineffective muscle stabilization [38]. The treatment for spondylolisthesis is dependent on the extent of listhesis. Most patients who present are asymptomatic and can be managed nonoperatively using modalities such as steroid injections, brace therapy, and restriction of heavy lifting and intense athletic activities [37]. When nonoperative therapies fail, surgical intervention may be needed. Generally, the indications for surgery include (1) persistence of debilitating pain and function, (2) progression of listhesis greater than 30 %, and (3) cosmetic deformities that result in functional disability [37]. Typical surgical procedures involve decompression and fusion of the segments undergoing listhesis.

Neoplasm

Tumors that arise from the nerve root can also cause radicular symptoms due to mass effect that can radiate into the groin, resulting in compression of the nerve roots. A lesion arising from any of the lower thoracic or upper lumbar nerve roots can cause symptoms along their corresponding dermatomal distribution, which can manifest as groin pain. The most common types of neoplasms encountered are benign nerve sheath tumors: schwannomas or neurofibromas [39, 40]. Such lesions are typically slow growing, and the onset of symptoms can be fairly insidious. MRI with gadolinium-enhanced sequences is the diagnostic imaging modality of choice to identify these lesions. Such tumors can arise anywhere along the course of the nerve, and can be intradural, extradural, or both. Surgical resection can be curative in such benign lesions. In addition, stereotactic radiosurgery is another option in treating these lesions, depending on their location [40].

Metastasis to the spinal canal or column can also manifest as groin pain [41]. Similar to nerve sheath tumors, those lesions that involve the lower thoracic or upper lumbar areas can cause nerve root compression or irritation, which would be referred to the corresponding dermatome. Such lesions typically seed into the vertebral body and can cause bony destruction leading to pathological fractures and/or compression of neurological structures. MRI with gadolinium contrast as well as CT of the area can be helpful in making a diagnosis and identifying the lesion. Treatment options in such patients must be weighed on a case-by-case basis. Overall disease burden, prognosis, histology of the tumor, and the general medical condition of the patient must be weighed given the potential morbidity from surgery. In patients who are good surgical candidates, surgery can be fairly effective in improving pain and neurological function. Other treatments such as stereotactic radiosurgery, palliative radiation, or palliative chemotherapy are additional options to consider in this difficult patient population [41].

Summary

The differential for groin pain from spinal causes can be fairly extensive. Presenting signs and symptoms can be helpful for identifying these conditions, while MRI of the lumbar spine is a very effective diagnostic tool for identifying any potential causes. Treatment for patients who have identifiable pathology on MRI that correlates with their symptoms can be fairly efficacious. Differentiating spinal and back pathologies from inguinal etiologies is challenging, but the characteristics, distribution, symptoms, signs, and imaging help to appropriately guide the evaluation and subsequent therapy.

References

1. Bradshaw CJ, Bundy M, Falvey E. The diagnosis of longstanding groin pain: a prospective clinical cohort study. Br J Sports Med. 2008;42(10):851–4.
2. Suarez JC, Ely EE, Mutnal AB, Figueroa NM, Klika AK, Patel PD, et al. Comprehensive approach to the evaluation of groin pain. J Am Acad Orthop Surg. 2013;21(9):558–70.
3. Hackney RG. (iv) Groin pain in athletes. Orthop Trauma. 2012;26(1):25–32.
4. Hansen HC, Helm 2nd S. Sacroiliac joint pain and dysfunction. Pain Physician. 2003;6(2):179–89.
5. Fortin JD, Kissling RO, O'Connor BL, Vilensky JA. Sacroiliac joint innervation and pain. Am J Orthop (Belle Mead NJ). 1999;28(12):687–90. Review.

6. Slipman CW, Jackson HB, Lipetz JS, Chan KT, Lenrow D, Vresilovic EJ. Sacroiliac joint pain referral zones. Arch Phys Med Rehabil. 2000;81(3):334–8.
7. Broadhurst NA, Bond MJ. Pain provocation tests for the assessment of sacroiliac joint dysfunction. J Spinal Disord. 1998;11(4):341–5.
8. Werner CM, Hoch A, Gautier L, König MA, Simmen H-P, Osterhoff G. Distraction test of the posterior superior iliac spine (PSIS) in the diagnosis of sacroiliac joint arthropathy. BMC Surg. 2013;13:52.
9. Gorton GE, Stout JL, Bagley AM, Bevans K, Novacheck TF, Tucker CA. Gillette functional assessment questionnaire 22-item skill set: factor and rasch analyses. Dev Med Child Neurol. 2011;53(3):250–5.
10. Elgafy H, Semaan HB, Ebraheim NA, Coombs RJ. Computed tomography findings in patients with sacroiliac pain. Clin Orthop Relat Res. 2001;(382):112–8.
11. Manchikanti L, Staats PS, Singh V, Schultz DM, Vilims BD, Jasper JF, et al. Evidence-based practice guidelines for interventional techniques in the management of chronic spinal pain. Pain Physician. 2003;6(1):3–81.
12. DonTigny RL. Function and pathomechanics of the sacroiliac joint. A review. Phys Ther. 1985;65(1):35–44.
13. Vleeming A, Buyruk HM, Stoeckart R, Karamursel S, Snijders CJ. An integrated therapy for peripartum pelvic instability: a study of the biomechanical effects of pelvic belts. Am J Obstet Gynecol. 1992;166(4):1243–7.
14. Kennedy DJ, Shokat M, Visco CJ. Sacroiliac joint and lumbar zygapophysial joint corticosteroid injections. Phys Med Rehabil Clin N Am. 2010;21(4):835–42.
15. Maugars Y, Mathis C, Berthelot JM, Charlier C, Prost A. Assessment of the efficacy of sacroiliac corticosteroid injections in spondylarthropathies: a double-blind study. Br J Rheumatol. 1996;35(8):767–70.
16. Hanly JG, Mitchell M, MacMillan L, Mosher D, Sutton E. Efficacy of sacroiliac corticosteroid injections in patients with inflammatory spondyloarthropathy: results of a 6 month controlled study. J Rheumatol. 2000;27(3):719–22.
17. Pulisetti D, Ebraheim NA. CT-guided sacroiliac joint injections. J Spinal Disord. 1999;12(4):310–2.
18. Dreyfuss P, Dreyer SJ, Cole A, Mayo K. Sacroiliac joint pain. J Am Acad Orthop Surg. 2004;12(4):255–65.
19. Calvillo O, Esses SI, Ponder C, D'Agostino C, Tanhui E. Neuroaugmentation in the management of sacroiliac joint pain. Report of two cases. Spine (Phila Pa 1976). 1998;23(9):1069–72.
20. Calvillo O, Guevara U, Chahadeh H. Neuroaugmentation in the management of pelvic pain syndromes. Tech Reg Anesth Pain Manag. 2006;10(1):7–11.
21. Oikawa Y, Ohtori S, Koshi T, Takaso M, Inoue G, Orita S, et al. Lumbar disc degeneration induces persistent groin pain. Spine (Phila Pa 1976). 2012;37(2):114–8.
22. Lee S. Choi pp. L1–2 disc herniations: clinical characteristics and surgical results. J Korean Neurosurg Soc. 2005;38:196–201.
23. Yukawa Y, Kato F, Kajino G, Nakamura S, Nitta H. Groin pain associated with lower lumbar disc herniation. Spine (Phila Pa 1976). 1997;22(15):1736–9. discussion 1740.
24. Carragee EJ, Lincoln T, Parmar VS, Alamin T. A gold standard evaluation of the "discogenic pain" diagnosis as determined by provocative discography. Spine (Phila Pa 1976). 2006;31(18):2115–23.

25. Zhang Y, Guo T, Guo X, Wu S. Clinical diagnosis for discogenic low back pain. Int J Biol Sci. 2009;5(7):647–58. Erratum in Int J Biol Sci. 2010;6(6):613.
26. Yrjämä M, Tervonen O, Kurunlahti M, Vanharanta H. Bony vibration stimulation test combined with magnetic resonance imaging. Can discography be replaced? Spine (Phila Pa 1976). 1997;22(7):808–13.
27. Backstrom KM, Whitman JM, Flynn TW. Lumbar spinal stenosis-diagnosis and management of the aging spine. Man Ther. 2011;16(4):308–17.
28. Hidalgo-Ovejero AM, García-Mata S, Martínez-Grande M, Maravi-Petri E, Izco-Cabezón T. L5 root compression caused by degenerative spinal stenosis of the L1–L2 and L2–L3 spaces. Spine (Phila Pa 1976). 1998;23(17):1891–4.
29. Boden SD, Davis DO, Dina TS, Patronas NJ, Wiesel SW. Abnormal magnetic-resonance scans of the lumbar spine in asymptomatic subjects. A prospective investigation. J Bone Joint Surg Am. 1990;72(3):403–8.
30. Murphy DR, Hurwitz EL, Gregory AA, Clary R. A non-surgical approach to the management of lumbar spinal stenosis: a prospective observational cohort study. BMC Musculoskelet Disord. 2006;7:16.
31. Creighton DS, Krauss J, Marcoux B. Management of lumbar spinal stenosis through the use of translatoric manipulation and lumbar flexion exercises: A case series. J Man Manip Ther. 2006;14(1):1E–0.
32. Buenaventura RM, Datta S, Abdi S, Smith HS. Systematic review of therapeutic lumbar transforaminal epidural steroid injections. Pain Physician. 2009;12(1):233–51.
33. Weinstein JN, Lurie JD, Tosteson TD, Zhao W, Blood EA, Tosteson ANA, et al. Surgical compared with nonoperative treatment for lumbar degenerative spondylolisthesis. Four-year results in the Spine Patient Outcomes Research Trial (SPORT) randomized and observational cohorts. J Bone Joint Surg Am. 2009;91(6):1295–304.
34. Humphreys SC, Eck JC. Clinical evaluation and treatment options for herniated lumbar disc. Am Fam Physician. 1999;59(3):575–82, 587–8.
35. Majid K, Fischgrund JS. Degenerative lumbar spondylolisthesis: trends in management. J Am Acad Orthop Surg. 2008;16(4):208–15.
36. Kalichman L, Kim DH, Li L, Guermazi A, Berkin V, Hunter DJ. Spondylolysis and spondylolisthesis: prevalence and association with low back pain in the adult community-based population. Spine (Phila Pa 1976). 2009;34(2):199–205.
37. Ganju A. Isthmic spondylolisthesis. Neurosurg Focus. 2002;13(1):E1.
38. Kalichman L, Hunter DJ. Diagnosis and conservative management of degenerative lumbar spondylolisthesis. Eur Spine J. 2008;17(3):327–35.
39. Rustagi T, Badve S, Parekh AN. Sciatica from a foraminal lumbar root schwannoma: case report and review of literature. Case Rep Orthop. 2012;2012:3. Article ID 142143.
40. Grimm S, Chamberlain MC. Adult primary spinal cord tumors. Expert Rev Neurother. 2009;9(10):1487–95.
41. Ramchandren S, Dalmau J. Metastases to the peripheral nervous system. J Neurooncol. 2005;75(1):101–10.

10. Groin Pain Etiology: Spermatic Cord and Testicular Causes

Juzar Jamnagerwalla and Howard H. Kim

Introduction

Chronic groin pain can persist for months and even years. Symptoms can be vague and often linger as the patient seeks care from multiple providers. Although historically urologists were the primary specialists in managing men with chronic groin pain, now various specialties are involved, using a multidisciplinary approach. The etiologies of chronic groin pain are not limited to testicular and spermatic cord causes and can be referred from other sources within the pelvis, abdomen, and lower extremities. As urological, gynecological, orthopedic, and general surgeons often collaborate for complex cases of chronic groin pain, each specialist should be familiar with the different etiologies and treatment. Furthermore, as more patients are diagnosed with chronic pelvic pain syndrome (CPPS), understanding the anatomy and pathophysiology of the genitourinary system can help physicians accurately diagnose and treat patients with chronic groin pain.

"Groin pain" is often used interchangeably with pain of testicular, epididymal, spermatic cord, scrotal, inguinal, and pelvic origin; in this chapter, although the discussion broadly encompasses the concept of urological groin pain, further distinction of anatomic origin is made when appropriate to keep consistent with the referenced studies.

Definition

Groin pain can be acute or chronic. The acute scrotum is characterized by pain, erythema, or swelling, with onset measured in hours to days. Chronic testicular pain is defined as intermittent or constant, unilateral

B.P. Jacob et al. (eds.), *The SAGES Manual of Groin Pain*,
DOI 10.1007/978-3-319-21587-7_10

Table 10.1. Common urological causes of acute and chronic groin pain.

Acute
Testicular torsion
Infectious (epididymo-orchitis)
Torsion of the appendix testis
Fournier's gangrene
Nephrolithiasis
Chronic
Testicular masses
Chronic pelvic pain syndrome/chronic scrotal pain syndrome
Infectious (epididymitis/prostatitis)
Varicocele
Hydrocele
Epididymal cysts/spermatocele
Post-vasectomy pain syndrome

or bilateral pain for at least 3 months' duration that compromises the patient's daily activities to such a degree that he seeks medical attention [1]. Common urological causes of acute and chronic groin pain are listed in Table 10.1. Referred groin pain is covered in other chapters.

While there are many identifiable causes, groin pain of unclear etiology is often given the nonspecific diagnosis of chronic epididymitis. But with better understanding of pelvic floor anatomy, there has been a concerted effort toward a more nuanced approach to chronic groin pain, or more expansively, CPPS. In 2010, the European Association of Urology (EAU) released guidelines that broadly defines chronic pelvic pain as nonmalignant pain perceived in pelvic structures that is present in a continuous or recurrent fashion for at least 6 months, with the caveat that if non-acute and central sensitization pain mechanisms are well documented the pain can be considered chronic irrelevant of time period [2]. Chronic scrotal pain syndrome is a subset of CPPS characterized by persistent or episodic scrotal pain with symptoms suggestive of urinary tract or sexual dysfunction [2]. The EAU classification system of chronic urogenital pain syndromes is shown in Table 10.2.

Epidemiology

The true incidence of chronic groin pain in the general population is difficult to assess. Bartoletti et al. found the prevalence and incidence of CPPS in men aged 25–50 years to be 13.8 % and 4.5 %, respectively, with 18 % of those patients already diagnosed with chronic scrotal pain

Table 10.2. European Association of Urology classification of chronic pelvic pain syndromes.

Axis I Region		Axis II System	Axis III End-organ as pain syndrome as identified from Hx, Ex, and Ix	Axis IV Referral characteristics	Axis V Temporal characteristics	Axis VI Character	Axis VII Associated symptoms	Axis VIII Psychological symptoms
Chronic	Specific	Urological	Prostate	Suprapubic	*Onset*	Aching	*Urological*	*Anxiety*
pelvic	disease-		Bladder	Inguinal	Acute	Burning	Frequency	About pain or
pain	associated		Scrotal	Urethral	Chronic	Stabbing	Nocturia	putative cause
	pelvic		Testicular	Penile/clitoral	*Ongoing*	Electric	Hesitance	of pain
	pain		Epididymal	Perineal	Sporadic		Dysfunctional	Catastrophic
	OR		Penile	Rectal	Cyclical		flow	thinking about
	Pelvic		Urethral	Back	Continuous		Urge	pain
	pain		Post-vasectomy	Buttocks	*Time*		Incontinence	*Depression*
	syndrome	Gynecological	Vulvar	Thighs	Filling		*Gynecological*	Attributed to
			Vestibular		Emptying		Menstrual	pain or impact
			Clitoral		Immediate post		Menopause	of pain
			Endometriosis-associated		Late post		*Gastrointestinal*	Attributed to
			Chronic pelvic pain		*Trigger*		Constipation	other causes
			syndrome with cyclical		Provoked		Diarrhea	Unattributed
			exacerbations		Spontaneous		Bloatedness	*PTSD*
			Dysmenorrhea				Urge	*symptoms*
		Gastrointestinal	Irritable bowel				Incontinence	Reexperiencing
			Chronic anal				*Neurological*	Avoidance
			Intermittent chronic anal				Dysaesthesia	
							Hyperaesthesia	

(continued)

Table 10.2. (continued)

Axis I Region	Axis II System	Axis III End-organ as pain syndrome as identified from Hx, Ex, and Ix	Axis IV Referral characteristics	Axis V Temporal characteristics	Axis VI Character	Axis VII Associated symptoms	Axis VIII Psychological symptoms
	Peripheral nerves	Pudendal pain syndrome				Allodynia	
	Sexological	Dyspareunia				Hyperalgesia	
		Pelvic pain with sexual dysfunction				*Sexological*	
						Satisfaction	
	Psychological	Any pelvic organ				Female dyspareunia	
	Musculoskeletal	Pelvic floor muscle				Sexual avoidance	
		Abdominal muscle				Erectile dysfunction	
		Spinal				Medication	
		Coccyx				*Muscle*	
						Function impairment	
						Fasciculation	
						Cutaneous	
						Trophic changes	
						Sensory changes	

From Engeler et al. [2], with the kind permission of the EAU
Hx History, *Ex* Examination, *Ix* Investigation, *PTSD* Post-traumatic stress disorder

[3]. The prevalence of chronic epididymitis in men visiting urology clinics in Canada was estimated to be about 0.9 % [4]. In 2005, Strebel et al. found that urologists had an average of 6.5 new encounters for chronic scrotal pain per month, and 2.5 % of all urological visits led to a diagnosis of chronic scrotal pain syndrome [5]. The incidence of chronic scrotal pain was estimated to be about 350–450 per 100,000 men aged 25–85 years [5]. Multiple publications have reported the peak age of chronic pelvic pain in their cohort to be at ages 40–49, with range from 20 to 83 [6–10]; however, these numbers were based on trials of men seeking interventions for chronic pain, and not direct epidemiological measurements.

Anatomy

Testicular pain is mediated by scrotal and spermatic branches of the genitofemoral and ilioinguinal nerves, as well as by sympathetic fibers along the testicular artery [8]. The genital branch of the genitofemoral nerve supplies the cremaster muscle and scrotal skin, and a branch of the ilioinguinal nerve supplies the skin of the upper scrotum and base of the penis. There is significant sensory overlap among the ilioinguinal, iliohypogastric, and genitofemoral nerves [11]. The course of the nerves through the inguinal canal can be seen in Fig. 10.1 [12]. Spermatic cord traction during scrotal surgery may trigger peritoneal stimulation [13]. The superior and inferior spermatic nerves provide autonomic innervation [14]. The superior spermatic nerve originates from the celiac and aortic plexuses and descends along the testicular vessels and forms the major nerve supply of the testis [14]. Sympathetic fibers arise from the thoracic segments 10 and 11, whereas the parasympathetic fibers arise from the vagus nerve [14]. The inferior spermatic nerve travels with the ductus deferens and the epididymis to the lower pole of the testis [14]. Sympathetic fibers arise from the inferior mesenteric and hypogastric plexuses and parasympathetic fibers branch from the pelvic nerve [14]. Animal models indicate that the nerve supply to the testis helps to regulate its endocrine function, but the precise function of testicular innervation in humans remains unclear. Epididymal innervation consists of a high density of sympathetic nerve endings in the corpus and cauda of the epididymis, with progressive concentration approaching the ductus deferens, consistent with their contractile role during ejaculation [15].

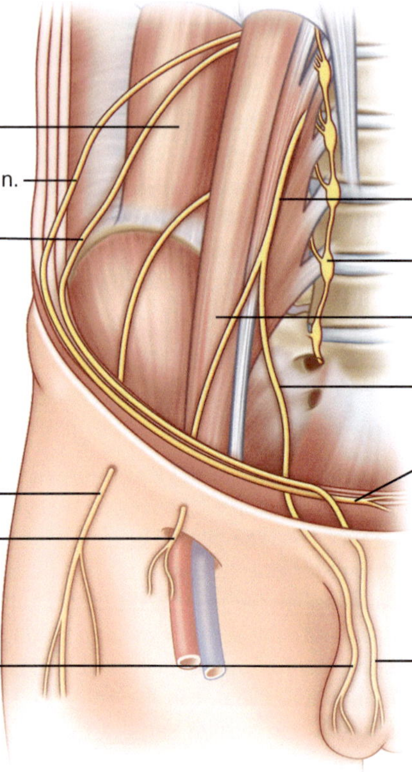

Fig. 10.1. Pelvic nerves in relationship to the inguinal canal. Note the location of the genital branch of the genitofemoral nerve, ilioinguinal nerve, and iliohypogastric nerve as they travel through the inguinal ring (redrawn from Kapoor et al. [12] with kind permission of Medscape Reference from WebMD).

Acute Groin Pain

Although the focus of this chapter is spermatic cord and testicular causes of chronic groin pain, a review of acute groin pain highlights the range and complexity of pain in this region. Men with acute scrotal pain must be quickly triaged to identify those who need immediate surgical intervention, as a delayed diagnosis may result in significant morbidity and even mortality.

Testicular Torsion Testicular torsion must be suspected in a man presenting with sudden onset unilateral pain without a history of trauma, often accompanied by nausea and vomiting. This may be seen after an open inguinal hernia repair, especially if the hernia had scrotal extension of its contents. During the operative manipulation, the testicle may be raised into the operative field, i.e., the groin incision, and returning the testicle back down into the scrotum may initiate the torsion. The window of time to save the testis is 4–8 h [16]. Physical exam findings include scrotal edema, erythema, and exquisite diffuse tenderness over the testis. Pain localized over the epididymis may be due to epididymitis or torsion of the appendix testis. Although the cremasteric reflex is often absent in testicular torsion, the presence of the reflex does not rule out torsion [17]. Ultrasound is helpful in the diagnosis of testicular torsion, with specificity approaching 100 % [18, 19]. If suspicion for torsion is high, surgical exploration should proceed without need for ultrasonic verification [20]. Surgery involves either orchiopexy or orchiectomy of the affected testis and orchiopexy of the contralateral testis if it was a spontaneous torsion.

Fournier's Gangrene Fournier's gangrene is an infected, necrotizing fasciitis of the perineal, genital, or perianal regions [21]. Patients present with local discomfort associated with erythema, swelling, and crepitus. Abnormal vital signs and metabolic derangements predict worse prognosis and higher mortality risk [22]. Fournier's gangrene is a clinical diagnosis, requiring emergent surgical intervention if suspicion is high [23]. CT scan may demonstrate subcutaneous emphysema along fascial planes in the scrotum, perineum, and inguinal regions [24]. Treatment includes aggressive fluid resuscitation, broad-spectrum antibiotics, and early extensive debridement of the involved fascial planes [25]. The mortality of Fournier's gangrene even with appropriate treatment is high, approaching 15–40 % [23, 26, 27].

Torsion of the Appendix Testis One of the most common causes of acute scrotal pain in the pediatric population is torsion of the appendix testis. Torsion of the appendix testis is a far more common cause of acute scrotal pain in boys than testicular torsion; one series showed that only 16 % of children presenting with testicular pain had torsion of the testicle as opposed to 46 % diagnosed with torsion of the appendix testis [20]. The appendix testis is a remnant of the Müllerian duct located on the upper pole of the testis. Differentiating torsion of the appendix testis from true testicular torsion can be challenging. Patients with torsion of the appendix testis may have a more insidious onset of pain over several days, with waxing and waning of pain levels. The "blue dot sign" of a palpable, infarcted appendix testis can be seen on exam in up to 21 % of patients [28]. Ultrasound can reliably identify torsion of the appendix testis and differentiate it from testicular torsion [29]. Treatment consists of rest and nonsteroidal anti-inflammatory drugs (NSAIDs).

Acute Epididymitis Acute epididymitis is an inflammation of the epididymis presenting acutely with pain and swelling. Objective findings of acute epididymitis include fever, scrotal erythema, leukocytosis on urinalysis, and positive urine culture. The pathophysiology is unclear but is thought to be secondary to retrograde flow of infection into the ejaculatory ducts [30]. In men under age 35 years, the most common etiology of acute epididymitis is sexually acquired *Chlamydia trachomatis* and *Neisseria gonorrhoeae*, while in men aged 35 years and over, the organisms that cause urinary tract infections (e.g., Gram-negative rods) are the predominant isolates [31, 32]. Men presenting with possible acute epididymitis should have a midstream urine collection along with Gram stain of a urethral smear, although empiric treatment should begin at the time of initial evaluation. Treatment involves bed rest, scrotal support, NSAIDs, and antibiotics.

Orchitis Isolated acute orchitis is relatively rare, as it usually occurs by local spread of infection from the epididymis. Isolated orchitis often has a viral cause, with mumps being the most common etiology. Mumps orchitis is characterized by painful testicular swelling 4–8 days after the appearance of parotitis [33]. Orchitis develops in 15–30 % of men with mumps. Mumps orchitis is not common before puberty [34]. Mumps orchitis is associated with reduced testicular size in up to half of patients and with semen analysis abnormalities in about 25 % [35]. Treatment is largely supportive.

Nephrolithiasis Nephrolithiasis is a common urological problem, with lifetime prevalence of approximately 10 % in men [36]. Although the classic presentation includes flank pain and hematuria, a stone impacted in the distal third of the ureter can cause referred pain to the groin. A stone should be considered in a patient who has groin pain associated hematuria, flank pain, or a history of nephrolithiasis. A non-contrast helical computed tomography (CT) scan is the preferred imaging study, with a specificity of 98 % and sensitivity of 95 % [37]. Spontaneous passage rates of stones are directly related to size: as high as 60 % for 5–7 mm stones and less than 25 % for stones larger than 9 mm [38]. If surgical intervention is indicated, ureteroscopy with laser lithotripsy results in a stone-free rate up to 96 % [39].

Chronic Groin Pain

Chronic pain, whether epididymal or testicular in origin, has been defined as symptoms of at least 3 months' duration [1, 40]. Chronic pain may be of neuropathic origin. When a nerve is sensitized by repeated stimulation, pain can persist even after the initial insult has resolved. This "hard-wiring" is mediated by peripheral and central modulation that reduces the threshold for activation of the action potential and decreases response latency [11]. Reversible causes of chronic groin pain must be ruled out before diagnosing CPPS.

Testicular Mass Testicular cancer is the most common cancer among men between ages 15 and 35 years—an age group that overlaps with that of chronic groin pain [41]. Although the majority present with a painless palpable testicular mass, some report a dull ache or heaviness in the scrotum or lower abdomen. Approximately 10 % of men with testicular cancer present with groin pain [42]. Ultrasound confirms the diagnosis, after which tumor markers are sent prior to prompt inguinal orchiectomy.

Conversely, an incidental impalpable testicular mass may be diagnosed during an evaluation for chronic groin pain. With increasingly finer resolution of ultrasound, masses as small as 1 mm can be detected long before they would be palpable [43]. Among men undergoing scrotal ultrasound for reasons other than for the evaluation of a retroperitoneal mass, Powell and Tarter found the incidence of testicular mass to be 0.38 % [44]. In men undergoing testicular ultrasound for infertility, the incidence of testicular tumors was 0.5 % [45]. Generally, the treatment of any sized testicular mass is radical orchiectomy, but some centers now perform excisional biopsies of small masses under 1 cm in diameter with the aid of an operating microscope, intraoperative ultrasound, and frozen section pathologic analysis. In one review, 19 of 49 cases of incidental testicular masses were found to be malignant [46].

Varicocele Varicocele is a dilation of the pampiniform plexus of spermatic veins, which can result in a dull, aching scrotal pain [47]. Varicoceles affect 15 % of adolescent and adult men [48]. In men seeking treatment for infertility, the prevalence of varicoceles has been found to be as high as 40 % [49]. Varicoceles almost always occur on the left side or are bilateral. The pathophysiology of varicoceles is poorly understood. Anatomy may play a role, as the left gonadal vein drains into the higher pressure left renal vein at almost a 90° angle, as opposed to the right gonadal vein, which drains into the lower pressure inferior vena cava. The left gonadal vein is also longer, with fewer valves compared to the right gonadal vein [48].

Unilateral right-sided varicoceles raise concern for extrinsic compression of the gonadal vein by a retroperitoneal mass. Also, among patients who have undergone inguinal hernia repair, note that the scar tissue and/or the mesh implant may be causing the extrinsic compression resulting in secondary varicocele. In such cases, further imaging is indicated, and workup should be directed to the offending cause of the varicocele.

Varicocelectomy can be performed by the open or laparoscopic approach, but microsurgery remains the gold standard (Fig. 10.2) [50]. Because of the proximity of scrotal lymphatic channels to the ligated veins during varicocelectomy, hydrocele formation is a possible complication. The rates of postoperative hydrocele range from 0 to 1.2 % in the microsurgical group, 5–20 % in the laparoscopic group, and 6–15 % in the open group [51–54]. A recent meta-analysis of randomized trials comparing open, laparoscopic, and microsurgical varicocelectomy showed both the hydrocele formation and varicocele recurrence rates to be significantly lower in the microsurgical group [55]. The recurrence rate after microsurgical varicocelectomy is low, ranging from 0 to 2.5 %, compared to 7–16 % and 17–20 % in the open and laparoscopic groups, respectively [52, 53, 56].

Hydrocele A hydrocele is a fluid collection around the testis contained within the tunica vaginalis and is a common finding in both adults and children. Although usually asymptomatic, a large hydrocele can cause a pulling sensation on the scrotum and significant discomfort. Hydroceles are either communicating or noncommunicating, with the former more often seen in boys and the latter in men. Communicating hydroceles represent a patent processus vaginalis, while noncommunicating hydroceles result from an imbalance of fluid production and reabsorption. The

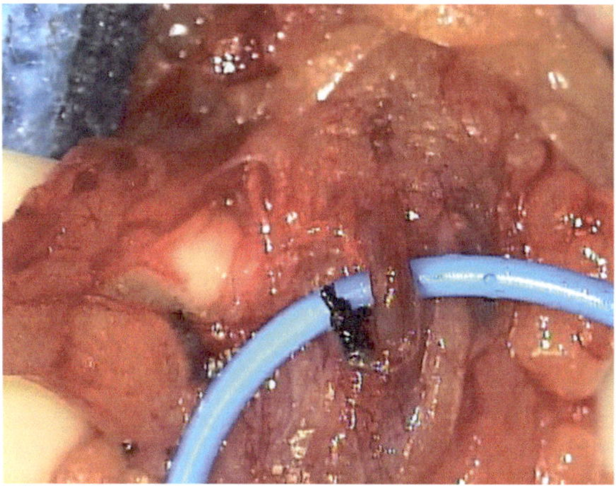

Fig. 10.2. Microsurgical varicocelectomy. The testicular artery is identified and isolated with a vessel loop, lymphatics channels are seen to the right of the artery and are preserved, and all venous structures are ligated with 4–0 silk or clips.

incidence of acquired or noncommunicating hydrocele is estimated at about 1 % in men [57].

Pediatric hydrocelectomy is similar to a pediatric herniorrhaphy. A small inguinal incision is made, the spermatic cord is visualized, and the hernia sac is dissected off the cord, taking care not to injure the spermatic cord structures. High ligation of the hernia sac is recommended to prevent future recurrence [58].

The treatment of adult hydroceles is almost always through a scrotal incision, and a variety of techniques have been described [59]. At our institution, we commonly use the Jaboulay technique, which involves delivering the testis through a scrotal incision, excising the excess portion of the tunica vaginalis, and everting the remnant [60]. Communicating hydroceles in men, although rare, may be a sign of an underlying inguinal hernia. Hydroceles associated with hernias can be a cause of groin pain, and an inguinal exploration with herniorrhaphy is indicated.

Epididymal Causes of Chronic Scrotal Pain Chronic epididymitis is a common cause of chronic scrotal pain. Nickel developed a classification system for chronic epididymitis (Table 10.3) [40]. Workup and treatment

Table 10.3. Classification of chronic epididymitis.

1. Inflammatory chronic epididymitis: pain and discomfort associated with abnormal swelling and induration
   a. Infective
   b. Post-infective (following acute bacterial epididymitis)
   c. Granulomatous (tuberculosis)
   d. Drug induced (amiodarone)
   e. Associated with syndromes (Behçet's disease)
   f. Idiopathic
2. Obstructive chronic epididymitis (i.e., congenital obstruction vs. post-vasectomy scarring)
3. Chronic epididymalgia: pain or discomfort in a normal feeling epididymis with no identifiable etiology

From Nickel et al. [40], reprinted with permission of MedReviews®, LLC. *Reviews in Urology* is a copyrighted publication of MedReviews®, LLC.

of chronic epididymitis is difficult because there is no identifiable cause in the majority of cases.

Granulomatous Epididymitis Tuberculosis should be suspected in men presenting with chronic granulomatous epididymitis, especially if they have a known history or recent exposure. Those with tuberculosis epididymitis should have further evaluation for systemic disease along with treatment with 6 months of triple drug therapy of isoniazid, rifampin, and pyrazinamide [30]. Sarcoidosis is a less common cause of granulomatous epididymitis, with an estimated 0.2–5 % of cases having genitourinary involvement [30, 61].

Drug-Induced Epididymitis The most common drug implicated in the development of epididymitis is the anti-arrhythmic amiodarone. While the exact mechanism is unknown, tissue levels of amiodarone and its metabolites have been shown to be 25–400 times higher in the epididymis than serum, which may lead to fibrosis and lymphocyte infiltration [62]. The incidence of amiodarone-induced epididymitis in men taking high-dose amiodarone is 3–11 %, with discontinuation of the drug generally leading to resolution of symptoms [63].

Idiopathic Chronic Epididymitis In the absence of an identifiable cause of epididymal pain, the diagnosis of idiopathic chronic epididymitis is often made. Idiopathic inflammatory epididymitis involves focal epididymal tenderness with swelling and induration, whereas chronic epididymalgia involves referred epididymal pain as part of

CPPS. In a cohort of 488 men evaluated for CPPS, 47 % had subjective symptoms of testicular pain, of whom only 7.5 % had reproducible tenderness [64], highlighting the importance of properly classifying patients with CPPS. The treatment of CPPS is discussed below, but patients with idiopathic inflammatory epididymitis are often treated with long-acting NSAIDs and rest [30]. If pain persists despite conservative management, surgical options such as microsurgical spermatic cord neurolysis or epididymectomy are considered.

Lesions of the Epididymis Mass lesions can cause noninflammatory epididymal pain, although the majority of these are painless. In a series of 1000 men undergoing ultrasound for testicular pain or swelling, 24 % were found to have epididymal cysts [65]. In a series of men undergoing ultrasound for infertility, the incidence was 7.6 % [45]. Spermatocele is another common cystic epididymal lesion that is usually asymptomatic. Solid masses of the epididymis are usually benign, with adenomatoid histology being the most common [30].

Post-vasectomy Pain Syndrome A small subset of men develop chronic testicular pain after vasectomy that can be debilitating and difficult to treat. Post-vasectomy pain syndrome (PVPS) is defined as a scrotal pain syndrome that follows vasectomy and falls under the second type of chronic epididymitis as described by Nickel (see Table 10.3) [2]. Prospective studies have found that almost 15 % of men who had no scrotal pain before vasectomy have some scrotal discomfort 7 months postoperatively, with 0.9 % having "severe" pain affecting quality of life [66]. Overall the incidence of PVPS ranges from 1 to 52 % [67–70]. The pathogenesis of PVPS is unclear; theories include the extravasation of sperm with resultant sperm granuloma, infection, nerve entrapment, and testicular engorgement from sperm due to long-standing obstruction [67, 71]. Mechanical obstruction may be a significant contributor, as Moss et al. found closed-ended vasectomies have a threefold higher rate of PVPS than open-ended vasectomies [72]. Conservative management with NSAIDs, scrotal support, and limitations in activity is first choice. While many respond to conservative measures, often further intervention is needed. Spermatic cord blocks can provide relief of the pain, and definitive interventions may include microsurgical spermatic cord denervation, vasectomy reversal, epididymectomy, or orchiectomy [73].

Post-inguinal Herniorrhaphy Testicular Pain Similar to the post-vasectomy pain syndrome, patients can develop post-inguinal herniorrhaphy groin pain, also known as inguinodynia. The incidence ranges from 0 to 62.9 %, with up to 10 % of patients falling into the moderate to severe pain group [74]. In a large population study of over 2400 patients who underwent either inguinal or femoral hernia repair, the incidence of groin pain significant enough to interfere with daily activity was as high as 6 % [75]. Inguinodynia differs from hernia-related groin pain, as the pain is new onset after the hernia repair and lasts longer than 3 months. The pain may be secondary to a variety of factors, including nerve trauma from retraction and dissection, neuroma formation after partial or complete transection, or nerve entrapment either by suture material or mesh associated fibrosis [74]. The pain can be classified as neuropathic or non-neuropathic, with approximately 50 % of patients falling into each category [76]. Neuropathic pain tends to be exercise induced with radiation down into the scrotum, and can be relieved by stretching or positioning techniques. Spermatic cord blocks can be diagnostic and therapeutic with up to 80 % of men with neuropathic inguinodynia reporting relief of their pain [76]. Non-neuropathic pain is secondary to a variety of etiologies, including recurrent hernias, periostitis, and spermatic cord congestion [76].

The treatment of inguinodynia begins with conservative treatment including rest and NSAIDS, similar to the treatment of post-vasectomy pain syndrome. In patients who have chronic, debilitating pain despite conservative management, surgical intervention is indicated. Surgical management is dependent on the underlying pathology of the pain. 80–95 % of patients with neuropathic pain have relief from triple neurectomy (ligation of the ilioinguinal, iliohypogastric, and genital branch of the genitofemoral nerve) [77]. Removal of mesh is effective for non-neuropathic pain secondary to spermatic cord compression from local mesh fibrosis. A microsurgical spermatic cord neurolysis can be performed concurrently if there is a significant component of associated spermatic cord/testicular pain. Intraoperative guidelines to prevent inguinodynia during routine hernia and further management are discussed in Chap. 28, "Prevention of Pain: Optimizing the Open Primary Inguinal Hernia Repair Technique."

Chronic Pelvic Pain Syndrome Although a majority of patients evaluated for chronic groin pain have a diagnosable, identifiable source of pain, no cause is identified in up to 25 % of cases [1]. Often these men are given the nonspecific diagnosis of chronic epididymitis and/or chronic prostatitis

(CP). However, with new understanding of the pelvic floor, symptoms previously described as idiopathic chronic scrotal pain are now being classified within CPPS as chronic scrotal pain syndrome (CSPS). The EAU definitions of chronic pelvic pain are listed in Table 10.4 [78].

Isolated pelvic organ complaints may link to a more generalized pelvic floor dysfunction. For example, Planken and colleagues found that of 41 patients with chronic testicular pain, 93 % had at least one symptom

Table 10.4. European Association of Urology definitions of chronic pelvic pain.

Terminology	Description
Chronic pelvic pain	Nonmalignant pain perceived in structures related to the pelvis of both males and females. In the case of documented nociceptive pain that becomes chronic, pain must have been continuous or recurrent for at least 6 months. If nonacute and central sensitization pain mechanisms are well documented, the pain may be regarded as chronic, irrespective of the time period. In all cases, there are often associated negative cognitive, behavioral, sexual, and emotional consequences. Chronic pelvic pain is subdivided into pelvic pain syndromes and non-pelvic pain syndromes
Pelvic pain syndrome	Persistent or recurrent episodic pelvic pain associated with symptoms suggesting lower urinary tract, sexual, bowel, or gynecological dysfunctions. No proven infection or other obvious pathology
Bladder pain syndrome	Suprapubic pain is related to bladder filling accompanied by other symptoms such as increased daytime and nighttime frequency. No proven urinary infection or other obvious pathology
Prostate pain syndrome	Persistent or recurrent episodic prostate pain, associated with symptoms suggestive of urinary tract and/or sexual dysfunction. No proven infection or other obvious pathology. Definition adapted from the NIH consensus definition and classification of prostatitis and includes conditions described as "chronic pelvic pain syndrome"
Scrotal pain syndrome	Persistent or recurrent episodic scrotal pain associated with symptoms suggestive of urinary tract or sexual dysfunction. No proven epididymoorchitis or other obvious pathology
Pelvic floor muscle pain syndrome	Persistent or recurrent episodic pelvic floor pain with associated trigger points either related to the micturition cycle or associated with symptoms suggestive of urinary tract, bowel, or sexual dysfunction. No proven infection other obvious pathology

From Fall et al. [78] with kind permission from the EAU
NIH U.S. National Institutes of Health

suspicious for pelvic floor dysfunction and 22 % had either micturition, defecation, or sexual complaints [79]. Patients who have symptoms of interstitial cystitis often have other disorders of the pelvis, including irritable bowel syndrome and defecatory dysfunction [80–82]. This link may be explained by cross-sensitization between the bladder and colon via primary afferent fibers [83, 84]. Although interstitial cystitis is classically associated with bladder pain and urinary symptoms, some men may present with complaints of isolated groin or genital pain [85].

Yoshioka et al. reported that noxious stimuli applied to the testes in rats resulted in significantly decreased bladder capacity compared to controls [86]. Rats pretreated with capsaicin had normal bladder function even after noxious stimuli, suggesting that testicular primary afferent C-fibers are responsible for the bladder overactivity [86]. The neural cross talk between the bladder and the testes may explain why almost 50 % of patients with CP/CPPS endorse symptoms of chronic testicular discomfort [64]. Urodynamic studies in men with chronic pelvic pain (with almost 40 % complaining of primary testicular discomfort) demonstrated a high percentage of urethral sensitivity, increased sphincter length and tone, and decreased peak urine flow [87].

Despite mounting evidence of CSPS as part of a systemic pelvic floor dysfunction as opposed to a disorder of infectious etiology, men often receive a course of antibiotics as initial therapy. A survey of Swiss urologists in 2005 showed that 98 % believed CSPS was secondary to infectious etiologies [5]. Despite the belief that CSPS is of bacterial origin, studies report only 21 % of those who present with CSPS were found to have a significant bacterial colony count or positive PCR with bacteria [88]. Antibiotics should not be considered first-line treatment for CSPS in the absence of culture proven infection.

NSAIDs may have a limited efficacy for CPPS, but it may be reasonable to try them, given their minimal potential morbidity. An EAU update for the treatment of CPPS/CP indicated level 1b evidence that NSAIDS may be beneficial in symptom reduction [2]. Alpha-blocker therapy was also hypothesized to improve CPPS symptoms, but randomized controlled trials have demonstrated no significant benefit [89]. Pharmacological treatment may be more beneficial in alleviating associated symptoms of CPPS such as urinary complaints.

In men with CPPS refractory to analgesia and muscle relaxant therapy, biofeedback and pelvic floor retraining and relaxation may provide modest symptom improvement [7, 90]. Given that 88 % of CPPS patients have an increased pelvic floor resting tone [79], these conservative therapies may be beneficial.

Surgical Management: Chronic Pain

Conservative measures such as rest, scrotal support, and sitz baths should always be attempted first for chronic groin pain, with surgery reserved for those who have persistent refractory pain that significantly diminishes their quality of life. Medical therapy includes antibiotics, anti-inflammatory agents, phytotherapy, anxiolytics, narcotics, acupuncture, and injection therapy with steroids and anesthetics [40]. A consultation with the pain service can be helpful before surgical intervention is considered.

Microsurgical Spermatic Cord Neurolysis Microsurgical denervation was first described in 1978 by Devine and Schellhammer [91]. Branches from the ilioinguinal nerve, the genital branch of the genitofemoral nerve, and autonomic fibers all merge at the spermatic cord, making it amenable to neurolysis.

Microsurgical denervation is a primary surgical option for men with chronic testicular pain in the absence of identifiable pathology. Potential candidates should undergo spermatic cord blocks preoperatively with demonstrated improvement in pain. If there is no pain relief with the nerve blocks, neurolysis may not be an appropriate therapy and other options should be considered.

Heidenreich et al. showed that 97 % of men who had complete relief of pain with preoperative nerve blocks were pain free at a mean of 34 months after surgery [8]. In another series, 71 % had complete pain relief and 17 % had partial relief [92]. Overall, a review of the literature including more than 600 testicular units demonstrated 82 % pain-free rate after microsurgical spermatic cord neurolysis [93].

The operation involves identification of the external inguinal ring where a 3 cm transverse incision is made. The spermatic cord is isolated and delivered into the field. The external and internal spermatic fasciae are incised. The operating microscope is used to identify the vas bundle. This is preserved. The testicular artery and several lymphatic channels are preserved. All other veins and nerves are ligated or clipped. Some authors advocate dividing the periadventitial tissue of the testicular artery to achieve further denervation, but the success rate is comparable and this step may increase the risk of arterial injury [8].

Vasovasostomy Microsurgical vasovasostomy (or vasoepididymostomy) is used for two primary indications: reversal of sterility or relief of intractable testicular pain after vasectomy (Fig. 10.3) [94]. Some cases of PVPS are

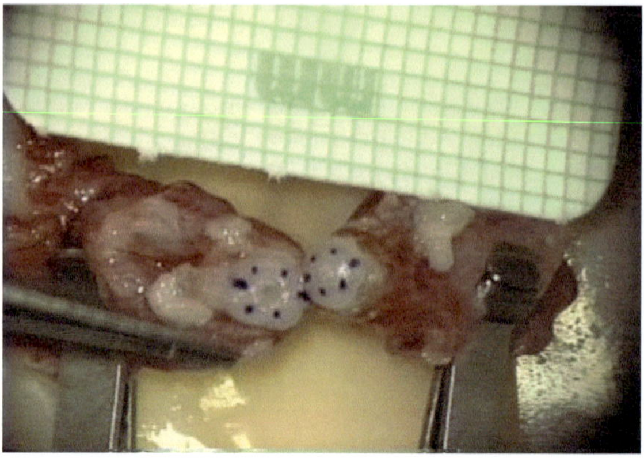

Fig. 10.3. Microsurgical vasovasostomy. The ends of the vas deferens to be reapproximated using 10–0 and 9–0 monofilament sutures (from Kim et al. [94], with kind permission from Elsevier).

thought to be secondary to mechanical obstruction, and vasectomy reversal relieves the pressure. In men undergoing vasectomy reversal for PVPS, 75–100 % had relief of symptoms, with 50–69 % rendered pain free [71, 95–97]. For those who continue to have pain after vasectomy reversal, redoing the reversal can be beneficial. In a series published by Nangia et al., two out of three men with persistent pain after vasovasostomy were pain free after a second vasovasostomy or vasoepididymostomy [96]. It is important to counsel patients after vasectomy reversal that they are no longer sterile and must consider other forms of contraception.

Epididymectomy Epididymectomy has been proposed as an option for chronic epididymal pain refractory to conservative therapy and PVPS. It is significantly less technically demanding than microsurgical denervation or vasovasostomy and less morbid than an orchiectomy. The surgical technique involves a trans-scrotal incision, unless there is concern for a malignant lesion, which would necessitate an inguinal approach. The tunica vaginalis is opened, and the epididymis is dissected free from the testis before the epididymal branch of the testicular artery is ligated. The distal vas deferens is ligated, and the epididymis is removed [30].

Long-term outcomes of epididymectomy depend on the indication for surgery. About 52 % of men undergoing epididymectomy for chronic epididymitis report being pain free after surgery, 22 % have persistent but improved pain, and 25 % have no improvement. Efficacy rates for epididymectomy performed for PVPS are similar, with 53 % being pain free postoperatively [98]. Those with a primary complaint of epididymal pain or discomfort related to a spermatocele or epididymal cyst had better outcomes. One series reported 75 % were pain free after epididymectomy and only 4 % reported no improvement [99]. Epididymectomy is also effective for PVPS with epididymal obstruction or sperm granuloma formation. West et al. found 90 % had long-term relief of scrotal pain [100]. Given the association of chronic epididymitis and groin pain with CPPS, it is not surprising that many fail to respond to epididymectomy. Careful patient selection and counseling are critical; with the exception of findings of epididymal masses or PVPS, other interventions may be more appropriate.

Orchiectomy Orchiectomy should only be considered as a last option among men with intractable pain. There is little data on outcomes as orchiectomy appropriately is not often performed for pain. Orchiectomy for referred pain is unlikely to be successful, and careful patient selection and counseling are critical.

An inguinal surgical approach is preferred. Davis et al. reported superior results compared to the scrotal approach (73 % vs. 55 % complete relief of pain) [1]. Unfortunately, up to 80 % have persistent pain following orchiectomy [101]. The benefit of inguinal orchiectomy likely involves high ligation of the spermatic cord, with resultant denervation and release of nerve entrapment.

Conclusion

There are many urological causes of groin pain, and it is useful to triage groin pain into acute and chronic causes. Urological chronic groin pain has many reversible, treatable causes such as infection, hydrocele, varicocele, or PVPS; however, a significant number of men do not have an obvious etiology and fall under the CPPS domain. A multidisciplinary approach should be used for those without a clear etiology of groin pain, and operative management should be considered only after failure of more conservative multimodal therapies.

References

1. Davis BE, Noble MJ, Weigel JW, Foret JD, Mebust WK. Analysis and management of chronic testicular pain. J Urol. 1990;143(5):936–9.
2. Engeler D, Baranowski AP, Borovicka J, Cottrell A, Dinis-Oliveira P, Elneil S, et al., European Association of Urology. Guidelines on chronic pelvic pain. European Association of Urology. 2015. http://uroweb.org/individual-guidelines/non-oncology-guidelines/. Accessed 9 Apr 2015.
3. Bartoletti R, Cai T, Mondaini N, Dinelli N, Pinzi N, Pavone C, et al. Prevalence, incidence estimation, risk factors and characterization of chronic prostatitis/chronic pelvic pain syndrome in urological hospital outpatients in Italy: results of a multicenter case-control observational study. J Urol. 2007;178(6):2411–5. Discussion 2415.
4. Nickel JC, Teichman JM, Gregoire M, Clark J, Downey J. Prevalence, diagnosis, characterization, and treatment of prostatitis, interstitial cystitis, and epididymitis in outpatient urological practice: the Canadian PIE Study. Urology. 2005;66(5): 935–40.
5. Strebel RT, Leippold T, Luginbuehl T, Muentener M, Praz V, Hauri D. Chronic scrotal pain syndrome: management among urologists in Switzerland. Eur Urol. 2005; 47(6):812–6.
6. Sweeney CA, Oades GM, Fraser M, Palmer M. Does surgery have a role in management of chronic intrascrotal pain? Urology. 2008;71(6):1099–102.
7. Anderson RU, Sawyer T, Wise D, Morey A, Nathanson BH. Painful myofascial trigger points and pain sites in men with chronic prostatitis/chronic pelvic pain syndrome. J Urol. 2009;182(6):2753–8.
8. Heidenreich A, Olbert P, Engelmann UH. Management of chronic testalgia by microsurgical testicular denervation. Eur Urol. 2002;41(4):392–7.
9. Misra S, Ward S, Coker C. Pulsed radiofrequency for chronic testicular pain-a preliminary report. Pain Med. 2009;10(4):673–8.
10. Nickel JC, Siemens DR, Nickel KR, Downey J. The patient with chronic epididymitis: characterization of an enigmatic syndrome. J Urol. 2002;167(4):1701–4.
11. Levine L. Chronic orchialgia: evaluation and discussion of treatment options. Ther Adv Urol. 2010;2(5–06):209–14.
12. Kapoor, VK. Open inguinal hernia repair perioprocedural care. Medscape Reference from WebMD website. [Updated 2014 Aug 6]. From http://emedicine.medscape.com/article/1534281-overview. Accessed 14 Oct 2014.
13. Verghese ST, Hannallah RS, Rice LJ, Belman AB, Patel KM. Caudal anesthesia in children: effect of volume versus concentration of bupivacaine on blocking spermatic cord traction response during orchidopexy. Anesth Analg. 2002;95(5):1219–23.
14. Gerendai I, Banczerowski P, Halasz B. Functional significance of the innervation of the gonads. Endocrine. 2005;28(3):309–18.
15. Vendrely E. Histology of the epididymis in the human adult. In: Bollack CG, Clavert A, editors. Epididymis and fertility: biology and pathology. Progress in reproductive biology, vol. 8. Basel: S. Karger; 1981. p. 21–33.

16. Bartsch G, Frank S, Marberger H, Mikuz G. Testicular torsion: late results with special regard to fertility and endocrine function. J Urol. 1980;124(3):375–8.
17. Ciftci AO, Senocak ME, Tanyel FC, Buyukpamukcu N. Clinical predictors for differential diagnosis of acute scrotum. Eur J Pediatr Surg. 2004;14(5):333–8.
18. Karmazyn B, Steinberg R, Livne P, Kornreich L, Grozovski S, Schwarz M, et al. Duplex sonographic findings in children with torsion of the testicular appendages: overlap with epididymitis and epididymoorchitis. J Pediatr Surg. 2006;41(3):500–4.
19. Wilbert DM, Schaerfe CW, Stern WD, Strohmaier WL, Bichler KH. Evaluation of the acute scrotum by color-coded Doppler ultrasonography. J Urol. 1993;149(6):1475–7.
20. Lewis AG, Bukowski TP, Jarvis PD, Wacksman J, Sheldon CA. Evaluation of acute scrotum in the emergency department. J Pediatr Surg. 1995;30(2):277–81. Discussion 281–2.
21. Smith GL, Bunker CB, Dinneen MD. Fournier's gangrene. Br J Urol. 1998;81(3):347–55.
22. Yeniyol CO, Suelozgen T, Arslan M, Ayder AR. Fournier's gangrene: experience with 25 patients and use of Fournier's gangrene severity index score. Urology. 2004;64(2):218–22.
23. Eke N. Fournier's gangrene: a review of 1726 cases. Br J Surg. 2000;87(6):718–28.
24. Levenson RB, Singh AK, Novelline RA. Fournier gangrene: role of imaging. Radiographics. 2008;28(2):519–28.
25. Shyam DC, Rapsang AG. Fournier's gangrene. J R Coll Surg Edinb. 2013;11(4):222–32.
26. Tahmaz L, Erdemir F, Kibar Y, Cosar A, Yalcyn O. Fournier's gangrene: report of thirty-three cases and a review of the literature. Int J Urol. 2006;13(7):960–7.
27. Jeong HJ, Park SC, Seo IY, Rim JS. Prognostic factors in Fournier gangrene. Int J Urol. 2005;12(12):1041–4.
28. McCombe AW, Scobie WG. Torsion of scrotal contents in children. Br J Urol. 1988;61(2):148–50.
29. Sellars ME, Sidhu PS. Ultrasound appearances of the testicular appendages: pictorial review. Eur Radiol. 2003;13(1):127–35.
30. Tracy C, Steers W. Anatomy, physiology and diseases of the epididymis. AUA Update Ser. 2007;26(12):114–23.
31. Luzzi GA, O'Brien TS. Acute epididymitis. BJU Int. 2001;87(8):747–55.
32. Berger RE, Alexander ER, Harnisch JP, Paulsen CA, Monda GD, Ansell J, et al. Etiology, manifestations and therapy of acute epididymitis: prospective study of 50 cases. J Urol. 1979;121(6):750–4.
33. Casella R, Leibundgut B, Lehmann K, Gasser TC. Mumps orchitis: report of a mini-epidemic. J Urol. 1997;158(6):2158–61.
34. Hviid A, Rubin S, Muhlemann K. Mumps. Lancet. 2008;371(9616):932–44.
35. Bartak V. Sperm count, morphology and motility after unilateral mumps orchitis. J Reprod Fertil. 1973;32(3):491–4.

36. Hiatt RA, Dales LG, Friedman GD, Hunkeler EM. Frequency of urolithiasis in a prepaid medical care program. Am J Epidemiol. 1982;115(2):255–65.
37. Dalrymple NC, Verga M, Anderson KR, Bove P, Covey AM, Rosenfield AT, et al. The value of unenhanced helical computerized tomography in the management of acute flank pain. J Urol. 1998;159(3):735–40.
38. Coll DM, Varanelli MJ, Smith RC. Relationship of spontaneous passage of ureteral calculi to stone size and location as revealed by unenhanced helical CT. AJR. 2002;178(1):101–3.
39. Arrabal-Polo MA, Arrabal-Martin M, Mijan-Ortiz JL, Valle-Diaz F, Lopez-Leon V, Merino-Salas S, et al. Treatment of ureteric lithiasis with retrograde ureteroscopy and holmium: YAG laser lithotripsy vs extracorporeal lithotripsy. BJU Int. 2009; 104(8):1144–7.
40. Nickel JC. Chronic epididymitis: a practical approach to understanding and managing a difficult urologic enigma. Rev Urol. 2003;5(4):209–15.
41. Siegel R, Ward E, Brawley O, Jemal A. Cancer statistics, 2011: the impact of eliminating socioeconomic and racial disparities on premature cancer deaths. CA Cancer J Clin. 2011;61(4):212–36.
42. Richie JP. Detection and treatment of testicular cancer. CA Cancer J Clin. 1993;43(3):151–75.
43. Leopold GR, Woo VL, Scheible FW, Nachtsheim D, Gosink BB. High-resolution ultrasonography of scrotal pathology. Radiology. 1979;131(3):719–22.
44. Powell TM, Tarter TH. Management of non-palpable incidental testicular masses. J Urol. 2006;176(1):96–8. Discussion 99.
45. Pierik FH, Dohle GR, van Muiswinkel JM, Vreeburg JT, Weber RF. Is routine scrotal ultrasound advantageous in infertile men? J Urol. 1999;162(5):1618–20.
46. Sheynkin YR, Sukkarieh T, Lipke M, Cohen HL, Schulsinger DA. Management of nonpalpable testicular tumors. Urology. 2004;63(6):1163–7. Discussion 1167.
47. Chen LK, Chen SS. Risk factors for developing pain in normospermic patients with varicocoele. Int J Androl. 2012;35(2):176–80.
48. Canales BK, Zapzalka DM, Ercole CJ, Carey P, Haus E, Aeppli D, et al. Prevalence and effect of varicoceles in an elderly population. Urology. 2005;66(3):627–31.
49. Kim ED, Lipshultz LI. Role of ultrasound in the assessment of male infertility. J Clin Ultrasound. 1996;24(8):437–53.
50. Jung A. Clinical andrology—EAU/ESAU course guidelines. Andrologia. 2010;42(6):404.
51. Ghanem H, Anis T, El-Nashar A, Shamloul R. Subinguinal microvaricocelectomy versus retroperitoneal varicocelectomy: comparative study of complications and surgical outcome. Urology. 2004;64(5):1005–9.
52. Cayan S, Acar D, Ulger S, Akbay E. Adolescent varicocele repair: long-term results and comparison of surgical techniques according to optical magnification use in 100 cases at a single university hospital. J Urol. 2005;174(5):2003–6. Discussion 2006–7.
53. Al-Kandari AM, Shabaan H, Ibrahim HM, Elshebiny YH, Shokeir AA. Comparison of outcomes of different varicocelectomy techniques: open inguinal, laparoscopic,

and subinguinal microscopic varicocelectomy: a randomized clinical trial. Urology. 2007;69(3):417–20.
54. Hassan JM, Adams MC, Pope JC, Demarco RT, Brock 3rd JW. Hydrocele formation following laparoscopic varicocelectomy. J Urol. 2006;175(3 Pt 1):1076–9.
55. Ding H, Tian J, Du W, Zhang L, Wang H, Wang Z. Open non-microsurgical, laparoscopic or open microsurgical varicocelectomy for male infertility: a meta-analysis of randomized controlled trials. BJU Int. 2012;110(10):1536–42.
56. Al-Said S, Al-Naimi A, Al-Ansari A, Younis N, Shamsodini A, A-sadiq K, et al. Varicocelectomy for male infertility: a comparative study of open, laparoscopic and microsurgical approaches. J Urol. 2008;180(1):266–70.
57. Leung ML, Gooding GA, Williams RD. High-resolution sonography of scrotal contents in asymptomatic subjects. AJR. 1984;143(1):161–4.
58. Grosfeld JL, Minnick K, Shedd F, West KW, Rescorla FJ, Vane DW. Inguinal hernia in children: factors affecting recurrence in 62 cases. J Pediatr Surg. 1991;26(3):283–7.
59. Rioja J, Sanchez-Margallo FM, Uson J, Rioja LA. Adult hydrocele and spermatocele. BJU Int. 2011;107(11):1852–64.
60. Jaboulay M. Chirurgie des centres nerveux des viscères et des membres. Lyon: Storck; 1902 [book in French].
61. Rao PK, Sabanegh ES. Genitourinary sarcoidosis. Rev Urol. 2009;11(2):108–13.
62. Gasparich JP, Mason JT, Greene HL, Berger RE, Krieger JN. Non-infectious epididymitis associated with amiodarone therapy. Lancet. 1984;2(8413):1211–2.
63. Gasparich JP, Mason JT, Greene HL, Berger RE, Krieger JN. Amiodarone-associated epididymitis: drug-related epididymitis in the absence of infection. J Urol. 1985; 133(6):971–2.
64. Schaeffer AJ, Landis JR, Knauss JS, Propert KJ, Alexander RB, Litwin MS, et al. Demographic and clinical characteristics of men with chronic prostatitis: the national institutes of health chronic prostatitis cohort study. J Urol. 2002;168(2):593–8.
65. Jones G, Mclean C, Goldie L, Aitchison M, Deane RF, Palmer MA, et al. Testicular assessment ultrasonography clinic: 1000 cases. Br J Urol. 2000;85(S5):9.
66. Leslie TA, Illing RO, Cranston DW, Guillebaud J. The incidence of chronic scrotal pain after vasectomy: a prospective audit. BJU Int. 2007;100(6):1330–3.
67. Manikandan R, Srirangam SJ, Pearson E, Collins GN. Early and late morbidity after vasectomy: a comparison of chronic scrotal pain at 1 and 10 years. BJU Int. 2004;93(4):571–4.
68. McMahon AJ, Buckley J, Taylor A, Lloyd SN, Deane RF, Kirk D. Chronic testicular pain following vasectomy. Br J Urol. 1992;69(2):188–91.
69. Ahmed I, Rasheed S, White C, Shaikh NA. The incidence of post-vasectomy chronic testicular pain and the role of nerve stripping (denervation) of the spermatic cord in its management. Br J Urol. 1997;79(2):269–70.
70. Schuman LM, Coulson AH, Mandel JS, Massey Jr FJ, O'Fallon WM. Health Status of American Men—a study of post-vasectomy sequelae. J Clin Epidemiol. 1993;46(8):697–958.
71. Myers SA, Mershon CE, Fuchs EF. Vasectomy reversal for treatment of the post-vasectomy pain syndrome. J Urol. 1997;157(2):518–20.

72. Moss WM. A comparison of open-end versus closed-end vasectomies: a report on 6220 cases. Contraception. 1992;46(6):521–5.
73. Tandon S, Sabanegh Jr E. Chronic pain after vasectomy: a diagnostic and treatment dilemma. BJU Int. 2008;102(2):166–9.
74. Hakeem A, Shanmugam V. Inguinodynia following Lichtenstein tension-free hernia repair: a review. World J Gastroenterol. 2011;17(14):1791–6.
75. Franneby U, Sandblom G, Nordin P, Nyren O, Gunnarsson U. Risk factors for long-term pain after hernia surgery. Ann Surg. 2006;244(2):212–9.
76. Loos MJ, Roumen RM, Scheltinga MR. Classifying post-herniorrhaphy pain syndromes following elective inguinal hernia repair. World J Surg. 2007;31(9):1760–5. Discussion 1766–7.
77. Alfieri S, Amid PK, Campanelli G, Izard G, Kehlet H, Wijsmuller AR, et al. International guidelines for prevention and management of post-operative chronic pain following inguinal hernia surgery. Hernia. 2011;15(3):239–49.
78. Fall M, Baranowski AP, Elneil S, Engeler D, Hughes J, Messelink EJ, et al. EAU guidelines on chronic pelvic pain. Eur Urol. 2010;57(1):35–48.
79. Planken E, Voorham-van der Zalm PJ, Lycklama à Nijeholt AA, Elzevier HW. Chronic testicular pain as a symptom of pelvic floor dysfunction. J Urol. 2010; 183(1):177–81.
80. Alagiri M, Chottiner S, Ratner V, Slade D, Hanno PM. Interstitial cystitis: unexplained associations with other chronic disease and pain syndromes. Urology. 1997;49(5A Suppl):52–7.
81. Clauw DJ, Schmidt M, Radulovic D, Singer A, Katz P, Bresette J. The relationship between fibromyalgia and interstitial cystitis. J Psychiatr Res. 1997;31(1):125–31.
82. Koziol JA. Epidemiology of interstitial cystitis. Urol Clin North Am. 1994; 21(1):7–20.
83. Christianson JA, Liang R, Ustinova EE, Davis BM, Fraser MO, Pezzone MA. Convergence of bladder and colon sensory innervation occurs at the primary afferent level. Pain. 2007;128(3):235–43.
84. Pezzone MA, Liang R, Fraser MO. A model of neural cross-talk and irritation in the pelvis: implications for the overlap of chronic pelvic pain disorders. Gastroenterology. 2005;128(7):1953–64.
85. Forrest JB, Schmidt S. Interstitial cystitis, chronic nonbacterial prostatitis and chronic pelvic pain syndrome in men: a common and frequently identical clinical entity. J Urol. 2004;172(6 Pt 2):2561–2.
86. Yoshioka K, Tanahashi M, Uchida W. Behavioral and urological evaluation of a testicular pain model. Urology. 2010;75(4):943–8.
87. Zermann DH, Ishigooka M, Doggweiler R, Schmidt RA. Neurourological insights into the etiology of genitourinary pain in men. J Urol. 1999;161(3):903–8.
88. Strebel RT, Schmidt C, Beatrice J, Sulser T. Chronic scrotal pain syndrome (CSPS): the widespread use of antibiotics is not justified. Andrology. 2013;1(1):155–9.

89. Nickel JC, Krieger JN, McNaughton-Collins M, Anderson RU, Pontari M, Shoskes DA, et al. Alfuzosin and symptoms of chronic prostatitis-chronic pelvic pain syndrome. NEJM. 2008;359(25):2663–73.
90. Clemens JQ, Nadler RB, Schaeffer AJ, Belani J, Albaugh J, Bushman W. Biofeedback, pelvic floor re-education, and bladder training for male chronic pelvic pain syndrome. Urology. 2000;56(6):951–5.
91. Devine Jr CJ, Schellhammer PF. The use of microsurgical denervation of the spermatic cord for orchialgia. Trans Am Assoc Genitourin Surg. 1978;70:149–51.
92. Strom KH, Levine LA. Microsurgical denervation of the spermatic cord for chronic orchialgia: long-term results from a single center. J Urol. 2008;180(3):949–53.
93. Oomen RJ, Witjens AC, Van Wijck AJ, Grobbee DE, Lock MT. Prospective double-blind preoperative pain clinic screening before microsurgical denervation of the spermatic cord in patients with testicular pain syndrome. Pain. 2014;155(9):1720–6.
94. Kim HH, Goldstein M. History of vasectomy reversal. Urol Clin North Am. 2009;36(3):359–73.
95. Horovitz D, Tjong V, Domes T, Lo K, Grober ED, Jarvi K. Vasectomy reversal provides long-term pain relief for men with the post-vasectomy pain syndrome. J Urol. 2012;187(2):613–7.
96. Nangia AK, Myles JL, Thomas AJ. Vasectomy reversal for the post-vasectomy pain syndrome: a clinical and histological evaluation. J Urol. 2000;164(6):1939–42.
97. Shapiro EI, Silber SJ. Open-ended vasectomy, sperm granuloma, and postvasectomy orchialgia. Fertil Steril. 1979;32(5):546–50.
98. Hori S, Sengupta A, Shukla CJ, Ingall E, McLoughlin J. Long-term outcome of epididymectomy for the management of chronic epididymal pain. J Urol. 2009; 182(4):1407–12.
99. Padmore DE, Norman RW, Millard OH. Analyses of indications for and outcomes of epididymectomy. J Urol. 1996;156(1):95–6.
100. West AF, Leung HY, Powell PH. Epididymectomy is an effective treatment for scrotal pain after vasectomy. BJU Int. 2000;85(9):1097–9.
101. Costabile RA, Hahn M, McLeod DG. Chronic orchialgia in the pain prone patient: the clinical perspective. J Urol. 1991;146(6):1571–4.

11. Groin Pain Etiology: Pudendal Neuralgia

Michael Hibner and Catherine Coyne

Introduction

The definition of pudendal neuralgia is pain in the area innervated by the pudendal nerve [1]. Due to such an extensive definition, pudendal neuralgia is often confused with other diseases such as vulvodynia, prostatodynia, vaginismus, levator syndrome, pelvic floor tension myalgia, and painful bladder syndrome.

The pudendal nerve arises from S2–4 sacral plexus. It courses ventrally to the piriformis muscle and exits the pelvis through the greater sciatic foramen. It then wraps around the dorsal surface of the sacrospinous ligament and reenters the pelvis through the lesser sciatic foramen. The nerve continues to travel through the fat of ischiorectal fossa until it enters the pudendal nerve canal also called Alcock's canal. The pudendal nerve then branches into the rectal nerve, perineal nerve, and dorsal clitoral/penile nerve branches. The rectal branch innervates the external anal sphincter, distal anal canal, and perianal skin. It may also provide sensation to the lower edge of the vagina in women. The perineal branch supplies sensation to the labial and scrotal skin, and lower edge of the vagina. Motor branches provide innervation to the pelvic floor, perianal, and urethral muscles. The dorsal clitoral/penile innervates the clitoris and penile skin (Fig. 11.1).

Throughout this chapter, the use of pudendal neuralgia will be referred to as a symptom, rather than as a diagnosis. Pudendal nerve entrapment is defined as compression of the pudendal nerve, typically by scar tissue or surgical material such as sutures or mesh products.

The true prevalence of pudendal neuralgia is unknown. The Portal for Rare Diseases and Orphan Drugs (orpha.net) estimates that 1–5 out of

B.P. Jacob et al. (eds.), *The SAGES Manual of Groin Pain*,
DOI 10.1007/978-3-319-21587-7_11

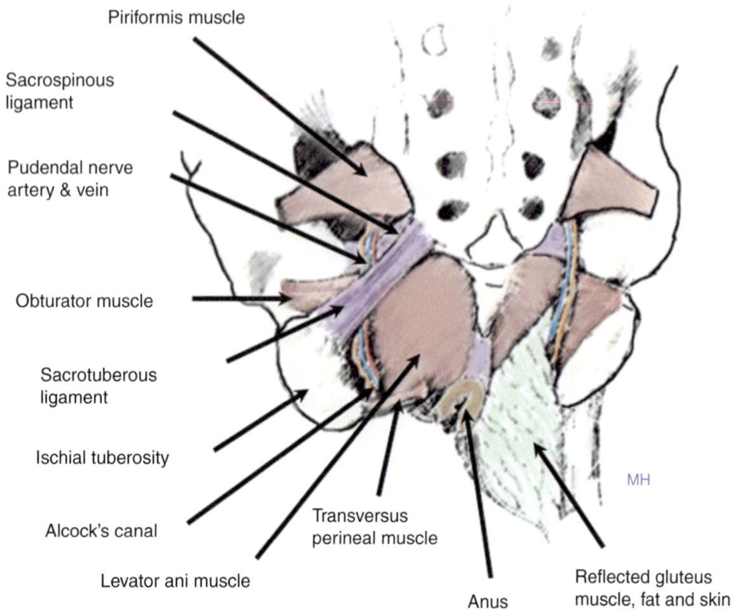

Fig. 11.1. Course of the pudendal nerve. Posterior, close-up view of the sacral region. Important anatomical landmarks such as ligaments and muscles are represented, illustrating the course of the pudendal nerve.

10,000 patients have pudendal neuralgia, and of their population suffering from chronic pain, around 4 % have this condition.

Etiology: Types of Pudendal Neuralgia

Pudendal neuralgia is broadly defined. We define this as pain in the distribution of the pudendal nerve that can arise from a series of separate mechanisms (Fig. 11.2).

Pain that mimics pudendal nerve compression may be caused by pelvic floor muscle spasm. Since the pudendal nerves course through the pelvic floor muscles toward the surface of the skin, any muscle spasm that occurs in the pelvic floor can result in further compression of the nerve and severe pain. Causes of pelvic floor spasm include direct pelvic floor injury, other pain conditions in the pelvis, psychological influences, or idiopathic causes (Fig. 11.3) [2].

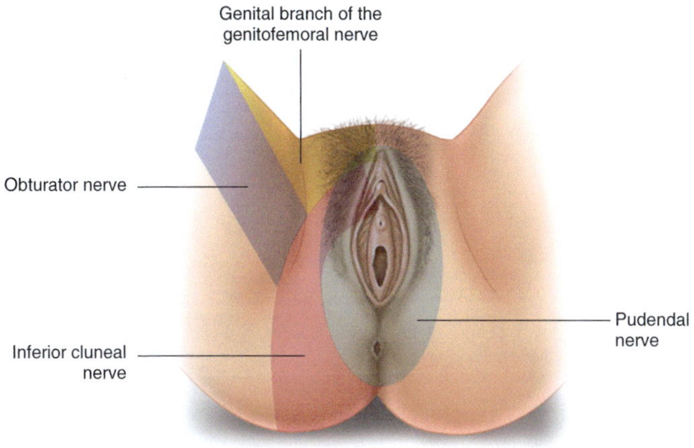

Fig. 11.2. Dermatomal distribution of the perineum. (**a**) Pudendal nerve, (**b**) inferior cluneal nerve (or gluteal branch of posterior femoral cutaneous nerve), (**c**) obturator nerve, and (**d**) genital branch of the genitofemoral nerve and ilio-inguinal nerve.

Pudendal nerve entrapment (PNE) is the direct result of pelvic trauma [3]. Causes of trauma can include surgery, childbirth, bicycle seats, sex, or athletic activities. Pelvic surgery is among the more common causes of PNE today.

Symptoms: What the Patient Experiences

Patients with pudendal neuralgia commonly describe their pain as a burning sensation in the region innervated by the pudendal nerve. Patients often state that the pain is exacerbated by sitting and improved with lying down [4]. The pain can be bilateral or unilateral. However, those with direct pudendal nerve injury tend to have unilateral pain.

It is important to appreciate that this pain is neuropathic in nature and often accompanied by significant allodynia and hyperesthesia. Common complaints or descriptions that arise during consultations include feelings of a foreign body sensation in the rectum or vagina, without an

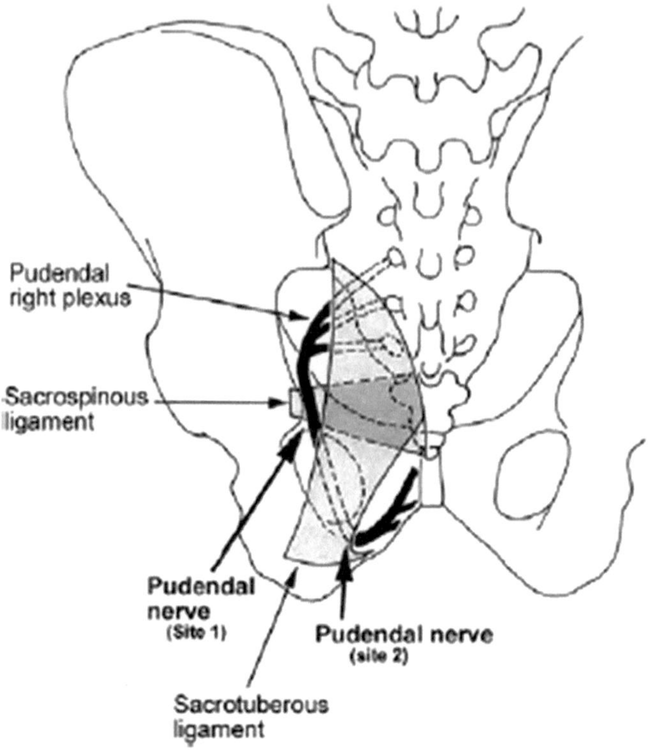

Fig. 11.3. Posterior view of the pelvis shows the anatomical course of the pudendal nerve from a posterior view of the patient (from Kastler et al. [2], with kind permission of Springer Science+Business Media).

actual foreign body present (allotriesthesia). Depending on which branches of the nerve are involved, patients can experience pain with micturition, full/distended bladder, bowel movements, intercourse, or during an orgasm.

How to Diagnose Pudendal Neuralgia

Patient History It is important to determine the events that preceded the symptoms for pudendal neuralgia as well as its progression. Many patients will present after traumatic injury to the pelvis with either gradual onset or immediate pain. Their pain is located in the area of innervation supplied by the pudendal nerve. It is neuropathic in nature and is exacerbated by sitting [4]. The gradual progression in pain for patients usually is minimal in the morning and progressively worsens throughout the activities of daily living, becoming most severe in the evening. Classically, most patients will note they have less pain sensation when sitting on the toilet seat in comparison to a chair.

Almost all patients with pudendal neuralgia have pain with intercourse and postcoital dyspareunia. Patients often state that it is so severe they will refrain from having intercourse altogether, which negatively affects their partners as well. Along with dyspareunia, common complaints also include pain with bowel movements, urination, or sexual arousal, as discussed above.

Physical Examination The physical examination is extremely important in order to rule out other possible causes of pain. Common findings among patients with pudendal neuralgia include pelvic floor muscle spasm and pain on pelvic examination; therefore, it can be difficult to determine any underlying nerve injury until the spasms are treated [5]. Among patients with pudendal neuralgia, the physical examination should always confirm pain within the dermatome supplied by the pudendal nerve. In patients with pudendal nerve entrapment, there will be tenderness over the sacrospinous ligament just medial to the ischial spine. Palpation of this area precipitates a tingling sensation commonly known as Tinel's sign.

Ancillary Testing As noted in Table 11.1, pudendal nerve motor terminal latency (PNMTL) testing may help determine nerve conduction and the integrity of the nerve [6]. Electrical impulses are applied transvaginally or transrectally using a pudendal electrode (St. Mark's) on the tip of the examiner's finger. Increased conduction time signifies damage to the nerve. It is a nonspecific test, as it does not determine the cause or the level of the injury. Also, the results of the study have been shown to have wide range, as the length of this nerve can be variable [7]. The value of this testing therefore remains controversial.

Quantitative sensory threshold testing and mapping have been used for the diagnosis of other peripheral nerve disorders, but have not been validated for the pudendal nerve [8]. These tests gauge the ability to discern between hot and cold temperatures and also between two pressure points. Compressed nerves lose the ability of fine discrimination.

Magnetic resonance imaging (MRI) of the pelvis is superior to other imaging modalities of the pelvis, as it can determine abnormalities of the muscles and soft tissue in great detail [9]. Higher resolution 3 T

Table 11.1. Ancillary testing [6–9].

Test	How to perform	Comments
Pudendal nerve motor terminal latency (PNMTL)	Measures speed of impulse conduction between ischial spine and pelvic floor. Performed with St. Mark's electrode	High inter- and intra-observer variability. Test unreliable due to inability to measure the length of pudendal nerve
Quantitative sensory testing	Warm detection test—based on patient's perception of change in temperature Two point discrimination test—based on patient's ability to distinguish between one and two points	Used in diagnosis of other neuropathic conditions but not validated for pudendal nerve
Magnetic resonance imaging (anatomic)	1.5 and 3 T MR with specific pelvic floor protocols	Pudendal nerve smaller than resolution of MRI. Studies on normal subjects not done
Magnetic resonance imaging (functional)	Magnetic resonance neurography—novel technique to enhance image of neurovascular bundle using water properties of the nerve	Studies on normal subjects not done

MRI can provide more detailed information along the pathway of the pudendal nerve, including scarring, inflammation, and entrapment. Magnetic resonance neurography (MRN) may also help identify the pudendal nerve along its pathway. Often the nerve is too small to be viewed accurately, unless it is grossly damaged and enlarged. In addition, MRN is typically performed in highly specialized centers, as it requires radiologists to be familiar with specific neuroanatomy to accurately interpret the images.

The Nantes Criteria The Nantes Criteria for diagnosing pudendal nerve entrapment was established by Professor Roger Robert in Nantes, France. He is one of the pioneers of surgical treatment of pudendal neuralgia. Table 11.2 lists these criteria, which have been validated [10]. Patients who meet these criteria are more likely to respond positively to surgical options than those who partially meet these criteria.

Table 11.2. Nantes criteria for the diagnosis of pudendal neuralgia (from Labat et al. [10], with kind permission of John Wiley & Sons).

Inclusion criteria

- Pain in the area innervated by the pudendal nerve
- Pain more severe with sitting
- Pain does not awaken patient from sleep
- Pain with no objective sensory impairment
- Pain relieved by diagnostic pudendal block

Exclusion criteria

- Pain located exclusively in the coccygeal, gluteal, pubic, or hypogastric area (without pain in the area of distribution of pudendal nerve)
- Pruritus
- Pain exclusively paroxysmal
- Abnormality on imaging (MRI, CT) which can account for pain

Complementary criteria

- Pain characteristics: burning, shooting, numbing
- Allodynia or hyperesthesia
- Allotriesthesia
- Pain progressively throughout the day
- Pain predominantly unilateral
- Pain triggered by defecation
- Significant tenderness around ischial spine
- Abnormal neurophysiology testing (PNMTL)

Associated signs

- Buttock pain (around ischial tuberosity)
- Referred sciatic pain
- Pain referred to the medial side of the thigh
- Suprapubic pain
- Urinary frequency with full bladder
- Pain after orgasm/ejaculation
- Dyspareunia or pain after intercourse
- Erectile dysfunction
- Normal PNMTL

Differential Diagnosis

Pelvic Floor Tension Myalgia (Also Called Levator Syndrome) This is a diffuse spasm of pelvic floor muscles. Symptoms are associated with pain during and after intercourse [11]. Patients also report hesitancy and sensation of incomplete voiding. On pelvic examination, there are pelvic floor muscle spasms. This may be confirmed by urodynamic testing. Of note, pelvic floor muscle spasms are nonspecific findings that often present among patients with other painful conditions, including pudendal nerve entrapment, endometriosis, and even inguinal hernia.

Painful Bladder Syndrome/Interstitial Cystitis Symptoms of interstitial cystitis are primarily suprapubic discomfort associated with filling of the bladder, including frequency, urgency, and/or nocturia. These symptoms overlap with that of pudendal neuralgia [12]. There is pain relief with bladder emptying. Patients can also experience pain in the labia/scrotum, vagina, and clitoris/penis. Inflammation of the bladder may be incited by exogenous factors, including certain foods. On examination, the patient may demonstrate bladder tenderness. Cystoscopy with hydrodistention should be included as part of the workup. The potassium chloride sensitivity test involves instillation of KCl into the bladder during cystoscopy. Eliciting pain is diagnostic of, though not specific to, interstitial cystitis and predictive of positive response to medical therapy for this disease. After this test, local anesthetic can be instilled in the bladder. Relief of pain with this anesthetic challenge test is also diagnostic.

Vulvodynia The diagnosis of vulvodynia can be enigmatic. It is often chronic, and diagnosis and treatment are difficult, as no single variant exists. As the term denotes, patients suffer from pain of the vulva. This may involve the labia, vagina, and clitoris. It is often due to a dermatologic condition, but similar symptoms can be mimicked by pudendal neuralgia [13].

Provoked Vestibulodynia This disease results in pain at the vulvar vestibule, also referred to as vulvar vestibulitis syndrome. Patients have a burning pain only upon entry to the vagina, such as with sexual intercourse. On examination, palpation with a Q-tip results in tenderness of vestibule.

Vaginismus This is a spasm of the muscles surrounding the vagina. It is typically elicited upon any insertion or penetration, similar to pelvic floor tension myalgia or levator syndrome. On pelvic examination, pelvic floor muscle spasms are obvious in response to digital examination. As explained prior, treatment of pelvic floor muscle spasms must be completed prior to initiating treatment for pudendal neuralgia.

Endometriosis Endometriosis should be included in all diagnoses of the pelvis among women. The pelvic pain is cyclical and typically in the lower abdomen. More severe cases may result in continuous pain. Hormonal therapy may help address the symptoms. Laparoscopy will confirm the diagnosis.

Treatment

Noninvasive Treatment and Modalities Avoiding activities that are causing the pudendal pain is the single most important step to treatment [14]. Scarring around the pudendal nerve can increase with repeat trauma from certain activities, and some of these can also cause pelvic floor muscle spasms, leading to severe pain. High-risk patients include those who perform specific activities such as cycling, gymnastics, ballet, and competitive athletics. These patients can halt progression of their symptoms and reduce the risk of developing chronic pudendal neuralgia if they decrease or stop their activity altogether.

Physical therapy provides excellent benefits in this patient population. Pelvic floor therapists address muscle spasms and improve upon muscle imbalances. Their therapies can help release restrictive connective tissue and improve other symptoms that patient may be experiencing [5]. Pelvic floor therapy applies "hands-on" techniques, improves posture and range of motion, strengthens surrounding muscles, and provides patient education to prevent further injury and trauma to the area. A majority of patients with pudendal neuralgia suffer from significant pelvic floor muscle spasms, with subsequent muscle shortening throughout the pelvic girdle. Manual techniques help release the spasms and result in lengthening of the muscles. Techniques focus on myofascial release, soft and connective tissue mobilization, and trigger point release. Other modalities therapists might include in their treatment approach include biofeedback, ultrasonography, and electrical stimulation.

Pharmacotherapy Medical therapy can play a role in multimodal therapy for the effective treatment of pudendal neuralgia. Table 11.3 lists options for medical therapy. However, little research has been performed to validate the efficacy of any of these medications as a preferred treatment for pudendal neuralgia. Since the symptoms are a result of neuropathy with or without muscle spasm, medical therapy is directed to these entities [15, 16]. No single medical treatment regimen is currently recommended; the patient may need to attempt multiple different combination treatments for symptomatic control.

Botulinum Toxin Injections If patients with significant pelvic floor muscle spasms do not have improvement with physical therapy, the next line of treatment is botulinum toxin injections [2, 17]. These have been found to be very effective in decreasing the muscle spasms of the pelvic floor. Doses between 50 and 400 units of botulinum toxin have been

Table 11.3. Medications for treatment of pudendal neuralgia.

Medication	Dose	Comments
Oral muscle relaxants (e.g., diazepam, cyclobenzaprine, carisoprodol, tizanidine)	Variable	Marginally effective
Diazepam and Baclofen vaginal suppository	Diazepam 5 mg Baclofen 4 mg Start qHS, increase to BID	Appears effective No studies available
Belladonna and opium rectal suppository	Belladonna 16.2 mg Opium 30 mg	Appears effective No studies available
Gabapentin	Titrate to 2400–3600 mg/day in three divided doses	Significant side effects No studies available for pudendal neuralgia
Pregabalin	Start at 75 mg BID and titrate up or down	Side effects No studies available for pudendal neuralgia

reported to block these spasms. The timeline for effect of the toxin is 5 days postoperatively with a maximum effect occurring 2 weeks postoperatively. Response to botulinum toxin will vary per patient. Some will improve after a single dose and will not require further treatment. However, the majority of patients will need repetitive injections every 3 months as the effects of the toxin wear off. Approximately 70–80 % of patients have significant improvement of pain after the toxin injection.

Pudendal Nerve Block Pudendal nerve blocks can be completed either unguided or guided. Unguided, they are performed through the vagina, perineum, or buttock. Guidance can be completed with the assistance of ultrasound, fluoroscopy, or computed tomography (CT) imaging [18]. Guided blocks are used to both diagnose and treat pudendal neuralgia. If a patient experiences temporary relief of pain after the block, this establishes that the pain is directly related to the pudendal nerve or the area innervated by the pudendal nerve. A positive block with temporary relief of the pain does not confirm pudendal nerve compression. If the block is negative, meaning the patient does not receive any pain improvement, it rules out the pudendal nerve as the cause of pain.

Along with local anesthetic injection, patients also experience long-term relief with injected steroids. One study found that 92 % of patients experienced some relief after undergoing a steroid injection [19]. In our clinical practice, this number seems inflated; approximately 30–40 % of our patients experience long-term relief with the addition of steroids to the nerve injection.

Surgical Treatment Approach

Transgluteal Pudendal Neurolysis Transgluteal pudendal neurolysis is currently the most common procedure performed for surgical decompression of the pudendal nerve. It allows surgeons to access the region between the sacrospinous and sacrotuberous ligaments in order to visualize the longest segment of the pudendal nerve. This is also a common area for pudendal nerve compression [20].

To perform this operation, the patient is placed in the prone, jackknife position. An incision is made on the buttock of the affected side, immediately over the sacrotuberous ligament. The area is dissected until the ligament is identified and the space between the sacrotuberous and sacrospinous ligaments is found, revealing the pudendal neurovascular bundle (Fig. 11.4). The initial dissection is usually difficult, particularly in patients who have significant scarring in this area, who have undergone previous operation in the area, or who experience thinning of the nerve.

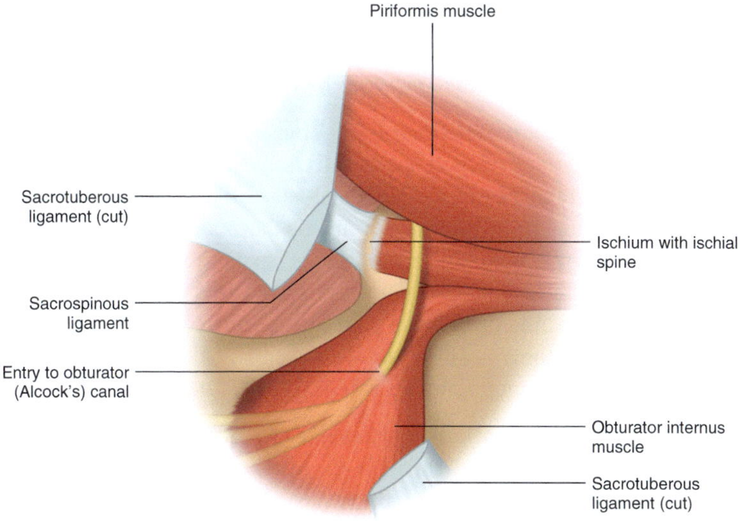

Fig. 11.4. View of the *left* pudendal nerve during transgluteal pudendal neurolysis. Patient is in prone jackknife position.

Once the nerve is identified, scar tissue is carefully removed from around the nerve.

When the pudendal nerve is cleared of scar tissue, the sacrospinous ligament is transected and the nerve is transposed anteriorly, thus decreasing the pressure off the nerve and allowing it to run a more direct, straight course anteriorly. Once the nerve is transposed, it can be wrapped in a nerve wrap product of the surgeon's choice, further decreasing the reoccurrence of scar tissue build up. The sacrotuberous ligament is subsequently repaired and the multilayered surgical wound is closed.

In our practice, approximately 75 % of patients have at least 20 % improvement in pain after this operation (ranging from 100 % improvement to 20 % improvement). This is similar to reports by Robert et al., among others [20]. Patients usually report improvement within 4 months of the operation, and maximum improvement can be expected in 18–24 months postoperatively.

Transischiorectal Pudendal Neurolysis Transischiorectal pudendal neurolysis is preformed transvaginally in women and through a perineal incision in men. One advantage of this procedure is that it does not require transection of the sacrotuberous ligament, and therefore may decrease the risk of instability of the sacroiliac joint postoperatively. The biggest drawback seems to be the limited visualization and poor access to the nerve itself. Particularly in male patients, a perineal incision is painful and is difficult to heal. In a French study, 83 % of patients had resolution of their pain after this procedure. Controversially, this high rate of success has not been confirmed by other studies.

Transperineal Pudendal Neurolysis Transperineal pudendal neurolysis is a procedure developed specifically for patients with entrapment of the terminal branches of the pudendal nerve. Thus, first the patients experiencing pain from the main trunk of the pudendal nerve, found between the sacrotuberous and sacrospinous ligaments, must be excluded [21]. This procedure is most widely applied to patients who experience isolated clitoral/penile or perineal pain. It is performed with the patient placed in the lithotomy position and the incision made just lateral to the labia majora in women or lateral to the scrotum in men.

Endoscopic Transperitoneal Pudendal Neurolysis The pudendal nerve can be approached through the abdominal cavity during a laparotomy, laparoscopy, or robotic-assisted laparoscopy [22]. With this approach, the trunk of the nerve found between the sacrospinous and sacrotuberous ligaments can be accessed. Visualizing the nerve distal to this area would require extensive transection of the levator ani muscle, increasing possible complications and postsurgical pain and recovery. There are several surgeons who perform transperitoneal pudendal neurolysis, but due to their small number and the mixed patient outcomes, it is difficult to validate its complete effectiveness.

Specific Case Discussion: Pudendal Neuralgia as a Complication of Surgical Implantation of Vaginal Mesh

In 2011, the Food and Drug Administration released a warning of potential complications of transvaginal mesh used in the treatment of incontinence and prolapse [23].

When examining the type of pain that patients may experience after vaginal mesh implantation, clinicians find that the majority will experience pudendal neuralgia. Logically, this could be directly related to the technique with which the mesh biomaterial is required to be implanted. For example, several mesh products are designed to anchor directly to the sacrospinous ligament. In cases where patients experience pudendal neuralgia after a vaginal mesh procedure, we initially advocate a conservative treatment approach. This includes first exploring the options of pelvic floor physical therapy, botulinum toxin injections to the pelvic floor, and nerve blocks. If conservative treatment fails, we recommend removing the mesh completely. Several studies have found that removal of mesh is beneficial in patients who experience post-vaginal mesh procedure pain [24]. Although most of the mesh can be removed, there is a small portion of the mesh located posterior to the sacrospinous ligament that cannot be removed through either a vaginal or abdominal approach. If the patient's pain persists, they might benefit from a transgluteal neurolysis procedure to remove the part of the mesh posterior to the sacrospinous ligament.

References

1. Hibner M, Castellanos M, Desai N, Balducci J. Pudendal neuralgia. In: Arulkumaran S, editor. Global library of women's medicine [Internet]. David Bloomer; 2011. Available from: http://www.glowm.com/section_view/heading/Pudendal Neuralgia/item/691
2. Kastler B, Clair C, Boulahdour Z, Puget J, De Billy M, Fergane B. Pudendal nerve infiltration under CT guidance. In: Kastler B, editor. Interventional radiology in pain treatment. Berlin, Heidelberg: Springer; 2007. p. 113–7.

3. Hibner M, Desai N, Robertson LJ, Nour M. Pudendal neuralgia. J Minim Invasive Gynecol. 2010;17(2):148–53.
4. Robert R, Prat-Pradal D, Labat JJ, Bensignor M, Raoul S, Rebai R, et al. Anatomic basis of chronic perineal pain: role of the pudendal nerve. Surg Radiol Anat. 1998; 20(2):93–8.
5. Prendergast SA. Causes of pelvic pain. Clin Obstet Gynecol. 2003;46(4):773–82.
6. Le Tallec de Certaines H, Veillard D, Dugast J, Estèbe J-P, Kerdraon J, Toulouse P, et al. Comparison between the terminal motor pudendal nerve terminal motor latency, the localization of the perineal neuralgia and the result of infiltrations. Analysis of 53 patients. Ann Readapt Med Phys. 2007;50(2):65–9 [Article in French].
7. Tetzschner T, Sørensen M, Rasmussen OO, Lose G, Christiansen J. Reliability of pudendal nerve terminal motor latency. Int J Colorectal Dis. 1997;12(5):280–4.
8. Walk D, Sehgal N, Moeller-Bertram T, Edwards RR, Wasan A, Wallace M, et al. Quantitative sensory testing and mapping: a review of nonautomated quantitative methods for examination of the patient with neuropathic pain. Clin J Pain. 2009; 25(7):632–40.
9. Chen A, Kalinkin O, Castellanos ME, Hibner M. Common MRI findings in patients with symptoms of pudendal neuralgia. Paper presented at: The International Pelvic Pain Society on Pelvic Pain 2012 Annual Fall Meeting on Pelvic Pain; 2012 Oct 29–20; Chicago, Illinois.
10. Labat JJ, Riant T, Robert R, Amarenco G, Lefaucheur JP, Rigaud J. Diagnostic criteria for pudendal neuralgia by pudendal nerve entrapment (Nantes criteria). Neurourol Urodyn Wiley Online Library. 2008;27(4):306–10.
11. Butrick CW. Pelvic floor hypertonic disorders: identification and management. Obstet Gynecol Clin North Am. 2009;36(3):707–22.
12. Possover M, Forman A. Voiding dysfunction associated with pudendal nerve entrapment. Curr Bladder Dysfunct Rep. 2012;7(4):281–5.
13. Shafik A. Pudendal canal syndrome as a cause of vulvodynia and its treatment by pudendal nerve decompression. Eur J Obstet Gynecol Reprod Biol. 1998;80(2): 215–20.
14. Antolak SJ. Genitourinary pain and inflammation. In: Potts JM, editor. Current clinical urology. Totowa, NJ: Humana Press; 2008. p. 39–56.
15. Benson JT, Griffis K. Pudendal neuralgia, a severe pain syndrome. Am J Obstet Gynecol. 2005;192(5):1663–8.
16. Backonja M, Glanzman RL. Gabapentin dosing for neuropathic pain: evidence from randomized, placebo-controlled clinical trials. Clin Ther. 2003;25(1):81–104.
17. Rao A, Abbott J. Using botulinum toxin for pelvic indications in women. Aust N Z J Obstet Gynaecol. 2009;49(4):352–7.
18. Filippiadis DK, Velonakis G, Mazioti A, Alexopoulou E, Malagari A, Brountzos E, et al. CT-guided percutaneous infiltration for the treatment of Alcock's neuralgia. Pain Physician. 2011;14(2):211–5.
19. Fanucci E, Manenti G, Ursone A, Fusco N, Mylonakou I, D'Urso S, et al. Role of interventional radiology in pudendal neuralgia: a description of techniques and review of the literature. Radiol Med. 2009;114(3):425–36.

20. Robert R, Labat J-J, Bensignor M, Glemain P, Deschamps C, Raoul S, et al. Decompression and transposition of the pudendal nerve in pudendal neuralgia: a randomized controlled trial and long-term evaluation. Eur Urol. 2005;47(3):403–8.
21. Hruby S, Dellon L, Ebmer J, Höltl W, Aszmann OC. Sensory recovery after decompression of the distal pudendal nerve: anatomical review and quantitative neurosensory data of a prospective clinical study. Microsurgery. 2009;29(4):270–4.
22. Possover M. Laparoscopic management of endopelvic etiologies of pudendal pain in 134 consecutive patients. J Urol. 2009;181(4):1732–6.
23. Haylen BT, Sand PK, Swift SE, Maher C, Moran PA, Freeman RM. Transvaginal placement of surgical mesh for pelvic organ prolapse: more FDA concerns—positive reactions are possible. Int Urogynecol J. 2012;23(1):11–3.
24. Ridgeway B, Walters MD, Paraiso MFR, Barber MD, McAchran SE, Goldman HB, et al. Early experience with mesh excision for adverse outcomes after transvaginal mesh placement using prolapse kits. Am J Obstet Gynecol. 2008;199(6):703.e1–7.

12. Chronic Pelvic Pain in Women

M. Jonathon Solnik and Matthew Thomas Siedhoff

Introduction

Our ability to provide optimal care for women who suffer from chronic pelvic pain (CPP) has traditionally been limited, in part, by the complexity of the presentation and the relative lack of understanding of the mechanisms involved and data to support consistent therapeutic options that relieve pain. Although some women can prove to be challenging with regard to applying suitable diagnostic and treatment paradigms, identifying a clinician who is comfortable accepting chronic patients who can be perceived as "difficult to manage" can likewise prove to be problematic.

Inconsistencies in nomenclature, along with the lack of a consistently utilized definition of CPP, affect our ability to determine the prevalence of this disorder in women, and they also contribute to the global clinical problem. The American College of Obstetricians and Gynecologists proposed defining CPP as noncyclic pain of 6 or more months' duration that localizes to the anatomic pelvis, anterior abdominal wall at or below the umbilicus, the lumbosacral back, or the buttocks, and is of sufficient severity to cause functional disability or lead to medical care [1]. One general estimation of prevalence was 4 % across different populations of women, which included pain from nongynecologic origins [2]. Using a common set of definitions, a systematic review of high-quality prevalence studies published by the World Health Organization cited a range of 16.8–81 % for dysmenorrhea (pain that occurs with menstruation), 2.1–24 % for women with noncyclic pain (pain that occurs outside of the menstrual cycle), and 8–21.1 % for dyspareunia (pain with intercourse) [3].

Notwithstanding, a substantial number of women are exposed to potentially nontherapeutic surgical procedures; approximately 40 % of diagnostic laparoscopies and 20 % of all hysterectomies are performed

B.P. Jacob et al. (eds.), *The SAGES Manual of Groin Pain*,
DOI 10.1007/978-3-319-21587-7_12

for the indication of pelvic pain [4, 5]. The observation that the majority of women who have a "negative" laparoscopy will continue to experience chronic pain suggests that more careful and systematic evaluation before or instead of a laparoscopy might be more productive and beneficial to the patient.

The purpose of this chapter is to provide clinicians who see female patients with a précis and guide to allow for more effective triage of CPP and to implement appropriate, contemporary diagnostics and therapeutic interventions. Critical to this process is a real understanding of the pathogenesis behind chronic pain and how this may be associated with the spectrum of related disorders.

History and Background

The last half-century has seen significant evolution in the understanding of chronic pelvic pain. Conceptualization was largely Cartesian in the 1950s, where the degree of tissue damage should correlate with the degree of pain experienced, and anything more was the result of psychological distress. Beginning in the 1960s, Wall and Melzack began developing the *gate control theory*, a concept represented by the nervous system having both the ability to carry nociceptive input from the periphery and the ability of central systems to temper that input with descending modulation. The degree to which these "gates" are open or closed relates to the amount of discomfort experienced by the patient. Relative gate closure could thus explain how the basketball player can play her championship game through a knee injury without thinking, but experiences significant pain once the competition is complete. The gate control theory is still a useful concept, but many now have grown to accept a concept of *central sensitization* in understanding chronic pain. In this paradigm, repeated noxious stimuli "ramp up" the patient's pain signaling over time, and coupled with a genetic predisposition or traumatic life experience, results in a general pain hypersensitivity, regardless of the stimulus. This helps explain the frequent finding of multiple chronic pain syndromes comorbid in a single patient, such as dysmenorrhea, vulvodynia, interstitial cystitis (IC)/painful bladder syndrome (PBS), irritable bowel syndrome (IBS), temporomandibular joint disorder (TMJ), migraines, and fibromyalgia [6]. These conditions are known as "functional" pain syndromes, meaning that there is no readily identifiable anatomic or physiologic abnormality, but rather the disease is defined by the symptoms the patient suffers.

While theories of pain perception evolved, gynecologists began incorporating laparoscopy into their diagnostic armamentarium. Endometriosis, for example, had long been understood as a cause of pelvic pain, but the presence of milder disease could now be investigated without the morbidity of a laparotomy. This led to great attention paid to endometriosis, adhesions, simple ovarian cysts, hernias, and other variations as pain etiologies. However, it became apparent over time that these findings could be completely incidental in some cases. On the other hand, patients with pristine pelvic anatomy could have the same constellation of symptoms that were thought to be caused by laparoscopic findings in others. Over time, enthusiasm about endometriosis and adhesions as concrete "causes" of pelvic pain has waned, but the remnants of these impressions left us with three significant problems [7]:

1. The significance of laparoscopic findings is exaggerated, either by well-intentioned physicians or in the desperate patient's interpretation. The result can be that every twinge of pain can be translated as the development or rupture of a follicular cyst or the spread of endometriosis throughout the abdomen and pelvis like metastatic cancer.
2. These diagnoses are flogged with unhelpful repeated laparoscopic lysis of adhesions, excision or ablation of endometriosis, or ovarian cystectomies. In the worst case, belief that the root of pain is housed in gynecologic organs ("Doctor, I feel like I want to just rip it all out of me") results in serial removal of the uterus and ovaries at a young age and leaves the patient with an unaddressed chronic pain syndrome.
3. When minimal or no abnormalities are detected on laparoscopy, the patient is made to feel "crazy" or that her pain is "all in her head."

This chapter outlines some of the more common contributors to gynecologic pain syndromes (Table 12.1). Put in the context of central sensitization, where the patient's pain response can be globally abnormal no matter the stimulus, these contributors can be understood as *peripheral pain generators*. The aim of treatment is thus to (1) turn down the overall "master volume dial" through an individualized regimen of medication (e.g., antidepressants, neuroleptics, etc.), psychotherapy, and/or alternative strategies (e.g., mindfulness-based

Table 12.1. Peripheral pain generators.

Gynecologic
Endometriosis
Pelvic inflammatory disease, chronic endometritis, chronic salpingitis
Pelvic congestion syndrome
Ovarian remnant syndrome
Residual ovary syndrome
Vulvodynia
Vulvar vestibulitis
Vaginismus
Urologic
Interstitial cystitis/painful bladder syndrome
Chronic urethritis
Gastrointestinal
Irritable bowel syndrome: constipation-dominant, diarrhea-dominant, mixed
Chronic constipation
Dyspepsia
Musculoskeletal
Pelvic floor tension myalgia, including levator spasm, piriformis syndrome
Abdominal wall trigger points
Low back pain
Temporomandibular joint disorder
Fibromyalgia
Neurologic
Peripheral neuropathies
Headaches, migraines

meditation, yoga, hypnosis, etc.); and (2) treatment of the peripheral generators—endometriosis, dysmenorrhea, bladder pain, bowel dysfunction, and pelvic floor muscular tension.

Endometriosis

The presence of endometrial glands and stroma documented outside of the uterine corpus occurs in approximately 15 % of reproductive-aged women [8]. Notwithstanding the relatively high prevalence of this disorder, may women remain asymptomatic. This feature is somewhat critical to the surgeon who incidentally notes endometriotic lesions at the time of surgery for non-pain indications. It is absolutely acceptable to leave these lesions undisturbed; however, if preoperative symptoms are suggestive of endometriosis, which is subsequently encountered, then surgical management is recommended. Typical symptoms include progressively worsening menstrual cramps that often begin quite painfully at menarche (dysmenorrhea), deep pain with intercourse (dyspareunia), dyschezia (pain with bowel movements), or chronic, noncyclic pain that occurs regardless of timing during the menstrual cycle. Endometriosis can also contribute to infertility. Although we are focusing on reproductive-aged women, endometriosis has been documented in many stages of life, from premenarcheal to postmenopausal years.

Endometriosis is characteristically described as a disease that affects peritoneal surfaces and the ovary (seen as endometrioma, or "chocolate cysts"), but it also represents an inflammatory process, whereby chronic fibrosis and infiltrating disease can occur. Sampson originally described his theory of retrograde menstruation in the early twentieth century, but most menstruating women will demonstrate this phenomenon [9]. It has been suggested that women with endometriosis then have a deficiency in their cell-mediated immune response, and foreign bodies, such as menstrual effluent passed retrograde through the oviducts, remain untargeted after being exposed to peritoneal surfaces [10]. Repeated exposure, facilitated stromal infiltration, and an activated inflammatory response result in the development of persistent lesions that can manifest in a myriad of ways from small peritoneal blebs to peritoneal fibrosis.

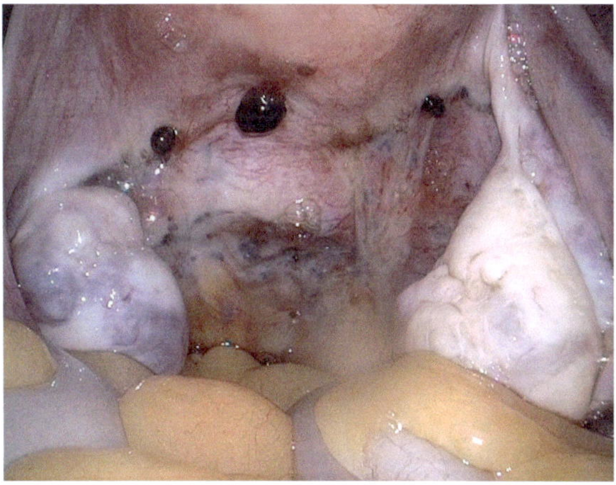

Fig. 12.1. Endometriosis as peritoneal blebs in the midline posterior cul-de-sac.

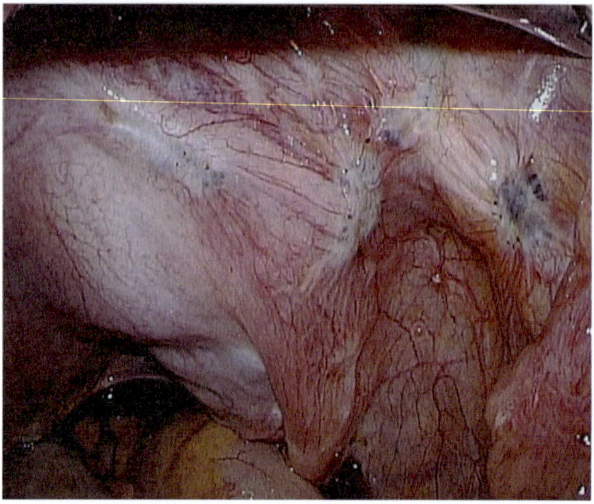

Fig. 12.2. Endometriosis as powder burn lesions in right ovarian fossa.

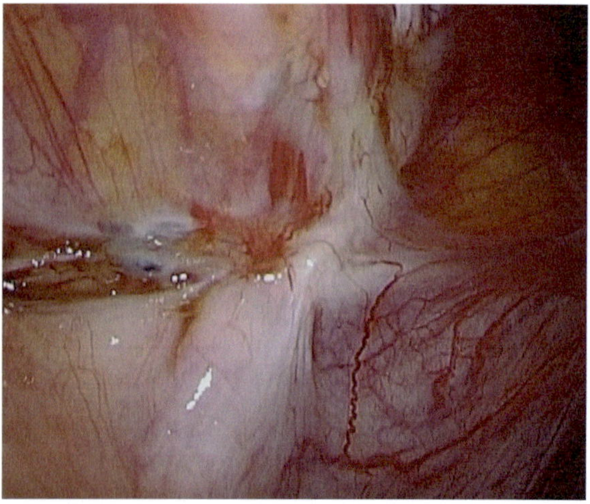

Fig. 12.3 Endometriosis as fibrotic lesions on pelvic brim (*white ridge*).

See Figs. 12.1, 12.2, 12.3, and 12.4 for varying types of disease seen at laparoscopy.

Endometriosis is a disease state most readily diagnosed with a good history and physical examination. A tender, retroverted, and fixed uterus can be suggestive of deeply infiltrating endometriosis (DIE) that distorts normal and mobile pelvic anatomy. However, physical findings may

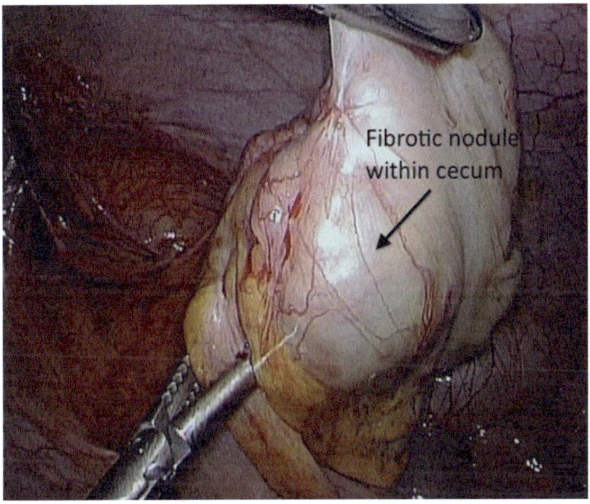

Fig. 12.4. Fibrotic cecal nodule with proven endometriosis (obliterated appendix).

also be unremarkable. Patients with progressively worsening pain complaints who are often refractory to first-line agents raise a diagnostic flare. This process is often aided by simple measures such as transvaginal ultrasound (TVUS). TVUS is used to evaluate for ovarian or the occasional bladder infiltrative disease, which is clinically useful because treatment options, to be discussed later, would then differ. Pelvic exam is a relatively poor predictor for DIE, but TVUS is consistently more effective than more sophisticated and expensive tools such as MRI [11]. Surgery with histopathologic confirmation has been the historical gold standard for diagnosing endometriosis; however, treatment can be initiated based on clinical suspicion alone without visual confirmation.

Traditional first-line therapies used to manage pain from endometriosis include pharmacological agents that suppress endogenous steroid hormone production [12–15]. Table 12.2 provides a summary of medical options.

Most operative strategies used to treat women with symptomatic endometriosis focus on conservative measures designed to preserve reproductive options. Hysterectomy with bilateral adnexectomy and excision of infiltrating disease, however, may have the potential to eliminate pain in women with refractory symptoms [16]. Any therapy short of hysterectomy predisposes women to recurrent symptoms. Laparoscopic ablation or excision of endometriotic lesions has been evaluated in a number of clinical trials, with up to 80 % of women

Table 12.2. Medical options for treating endometriosis-associated pain.

Medication	Route of administration	Notes
Combined oral contraceptives (COCs)	Cyclic or continuous oral daily use	Lack high-quality clinical trials No agent seems to be more effective than another
Progestins	Continuous oral daily use (Norethindrone acetate 5–20 mg daily; Medroxyprogesterone acetate (5–100 mg daily)) Intramuscular depot form (150 mg IM q 3 months) Intrauterine (levonorgestrel-intrauterine system; LNGIUS)	Reduction in pain compared to placebo in clinical trials Depot form no more effective with more side effects Intrauterine administration effective even for DIE with fewer side effects
Androgens	Daily oral, vaginal or intrauterine	Not all forms available in the US Untoward androgenic side effects especially with oral forms
Gonadotropin-releasing hormone agonists (GnRH-a)	Intramuscular depot (leuprolide acetate 3.75 mg IM [1 month], 11.25 mg [3 month])	Reduction in pain compared to placebo in clinical trials Comparable to COCs and progestins Positive response in empiric trial Hormonal add-back therapy required after 6 months of use to minimize side effects (vasomotor symptoms, reduction in bone mineral density); can be initiated immediately
Aromatase inhibitors	Daily oral use	Limited data, but may be effective in reducing pain for DIE Use with progestins or GnRH-a to prevent ovarian follicle development

DIE Deeply invasive endometriosis

reporting significant improvement in pain symptoms [17]. Notwithstanding adequate trial design, these studies documented a 30 % placebo effect; the difference, however, was that the therapeutic benefit was much longer-lasting for those in the treatment arms [18].

DIE is not consistently recognizable by the unfamiliar surgeon, and since depth of infiltration correlates with pain symptoms, excising these lesions seems more logical [19]. Ovarian endometrioma, often associated with DIE, are better managed by enucleating the cyst rather than ablation [20, 21]. Data to support postoperative suppression are limited, although the benefits of surgery may be extended with combined oral contraceptives (COCs), progestins, or gonadotropin-releasing hormone agonists (GnRH-a).

Adenomyosis

Adenomyosis is an enigmatic disorder characterized anatomically as endometrial glands and stroma existing within the myometrium diffusely. Focal lesions are referred to as an adenomyoma. Although traditionally gynecologists have considered adenomyosis a cause of heavy, prolonged, and/or painful menses, it is apparent from studies of hysterectomy specimens for a spectrum of benign disorders that adenomyosis is an extremely common entity, found in approximately 25–65 % of hysterectomy specimens.

Until relatively recently, adenomyosis was something diagnosed based on clinical suspicion and confirmed only at the time of hysterectomy. Advances in uterine imaging have provided the clinician with the opportunity to diagnose this entity with reasonable accuracy. In the relatively small uterus, TVUS is an effective means of identifying adenomyosis, assuming adequate sonographer skill and real-time evaluation of the study, as opposed to review of still images [21]. Magnetic resonance imaging (MRI) may be an effective secondary tool used to confirm adenomyosis, especially when the uterus is large or associated with concomitant uterine myoma.

The relationship between adenomyosis and abnormal uterine bleeding (AUB) or CPP remains unclear, in particular because many trials were performed to evaluate symptom reduction without performing a hysterectomy. The levonorgestrel intrauterine system (LNG-IUS) has proved to be more effective than other modalities in quality of life measures when compared to hysterectomy in a randomized trial [22].

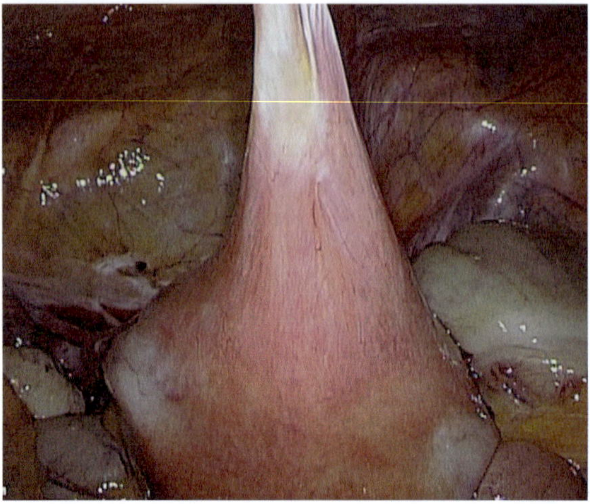

Fig. 12.5. Dense uterine adhesion to anterior abdominal wall from prior cesarean section.

Adhesions

There is a tendency to associate pelvic adhesions with CPP, but there is very little evidence that a clear relationship exists. The exception may be for dense connections involving viscera, but the lack of control groups for surgical studies weakens the cause-and-effect association, and the therapeutic benefits are more difficult to quantify [23]. As an example, Fig. 12.5 demonstrates a thick adhesion between the uterus and anterior abdominal wall after Cesarean section. Clinical trials to determine the incidence of such adhesions, its correlation with pain, and recurrence after adhesiolysis would require second-look laparoscopic evaluation; such trials have not been approved to date. Current recommendations focus on the implementation of microsurgical techniques to minimize the risk of *de novo* adhesions. When adhesions are encountered during surgical exploration in a woman with CPP, seek to identify an underlying cause of adhesions such as endometriosis, and divide only adhesions necessary to accomplish the surgical objectives.

Myofascial and Musculoskeletal Pain

Patients with myofascial pain complain of aching pain that is worse with activity or at the end of the day, and improved with rest or relaxation such as a warm bath. Just as lumbar muscles can be the source of pain in low back pain, pelvic floor muscles can be the source in patients with pelvic pain [24]. Dyspareunia is a common presenting symptom. Evaluation begins with a careful examination of the lower back and sacroiliac joints, and then progresses to a complete abdominal exam. The patient is asked to locate any discrete locus of pain and examine herself with palpation until pain is elicited. At this point, a Carnett maneuver (having the patient contract her rectus muscles by raising her head into a half situp) can be helpful to help distinguish a visceral from a somatic source. Finally, the examiner moves to the pelvic exam, using a single digit to palpate the levator muscles (iliococcygeus, pubococcygeus, puborectalis) just inside the introitus. Moving further back, the obturator is palpated as a square-shaped muscle just deep to levator ani. Finally, the piriformis can be palpated as a rubber-band-like structure that emerges during isometric contraction when the knee is externally rotated at the hip against the examiner's hand. In each case, the examiner asks the patient if the muscle is tender, if pain is felt at the site or migrates elsewhere, and if the pain experienced is similar to symptoms offered in the history, such as dyspareunia. Trigger point injections can be helpful for abdominal wall pain, but the treatment of myofascial pain is generally physical therapy, where a variety of strength, balance, relaxation, and control modalities are employed. In pelvic pain, referral should be made to a therapist specifically trained to address women's health and pelvic floor issues.

Pelvic Congestion Syndrome

Similar to the case for pelvic adhesions, high-quality data linking pelvic pain with radiological evidence for pelvic congestion of the gonadal veins are lacking. It has been suggested that faulty valves within these vessels create flow disruption and resultant distention, similar to what has been described for varicoceles in men. Pelvic congestion syndrome (PCS) is not an uncommon finding in women; however, it is not clear if the finding is associated with the complaint of pain. Controlled studies are needed to provide better guidance regarding treatment outcomes [25]. To date, multiple procedures have been described in observational trials and include internal iliac/ovarian vein embolization, sclerotherapy, and ovarian vein ligation.

Ovarian Remnant Syndrome

Women at risk for ovarian remnant syndrome (ORS), whereby a functional portion of the ovary is inadvertently left in place after oophorectomy, include those in whom the fibrovascular attachments are obscured from dense pelvic adhesions, endometriosis, or reproductive malignancy [26]. Resulting pain symptoms are typically cyclic and are accompanied by a solid pelvic mass with ovarian follicular development. Follicle stimulating hormone (FSH) and serum estradiol levels are in the premenopausal range if bilateral salpingo-oophorectomy was undertaken; however, hormone levels are not conclusively diagnostic. Nonsurgical attempts to manage pain with steroid hormone suppression should first be considered, given the predisposing surgical risks. Properly identifying and resecting the remaining disease can also prove to be extremely difficult, and has not been evaluated sufficiently in clinical trials.

Pain of Neurosensory Origin

Chronic pain can arise from virtually any organ system, implicating a framework for a common origin within the neurosensory system. This concept is corroborated by surgical research, whereby the presumed underlying pathology was adequately treated, but symptoms persisted or recurred soon after the targeted operation [27].

Two major areas of interest, as mentioned previously, include IBS and IC/PBS. Notwithstanding similarities such as pelvic pain, some form of visceral dysfunction, and the label of diagnosis of exclusion, there is no consensus upon which a diagnosis for either of these can be confirmed. Treatment has thus focused on lifestyle modifications or reducing inflammatory exposure and sensory input. Pharmacological agents such as tricyclic antidepressants (TCAs) represent the most widely studied neuroleptic agents. Although they are routinely used in CPP management paradigms, there is no high-level evidence to support their use [28]. One small trial evaluating nortriptyline in women with CPP resulted in a 50 % dropout rate because of intolerable side effects.

Neuroablative techniques can be performed surgically by transecting specific nerve or nerve bundles or percutaneously by injecting sclerosing agents. These procedures are typically reserved for women with refractory pain where the distribution is aligned within an identifiable nerve or nerve plexus. Nevertheless, the role for surgical management of pain that is likely neurosensory in origin remains limited. Laparoscopic uterosacral nerve ablation (LUNA) has been described as a technique that interrupts pain conduction to the uterus, but is of limited to no use for managing pain of gynecologic origin. Presacral neurectomy (PSN) is a procedure that excises a segment of sympathetic nerve bundles at the level of the superior hypogastric plexus. When used as a surgical adjuvant in the setting of endometriosis, PSN may add a component of relief to those with midline pain [29].

Hysterectomy clearly represents one of the more aggressive surgical maneuvers used to treat CPP, but as a stand-alone procedure it may result in failure for many women not properly evaluated for the likely cause of pain. Endometriosis-associated pain, refractory to conservative surgical measures, pain associated with menstrual bleeding, and pain suspected to be of uterine origin may provide women relief from their symptoms. If performed hastily, hysterectomy has the potential to worsen or induce new pain symptoms as a result of surgical trauma. Prior to considering hysterectomy, alternatives as discussed above should first be considered.

Vaginal Cuff Pain

Just as neuropathic pain can develop following an abdominal incision [30], the vaginal cuff can be a source of pain following hysterectomy [31]. Vaginal cuff pain usually presents as new dyspareunia following hysterectomy, whereas dyspareunia that was present before and remains after surgery is more likely related to other factors such as pelvic floor tension myalgia. The character of discomfort from vaginal cuff pain is often burning, stinging, or simply sharp, radiating pain when contact is made. Cuff pain is diagnosed by tenderness elicited with careful cotton-swab palpation along the length of the cuff. In evaluating post-hysterectomy dyspareunia, moving directly to a bimanual exam can provide confusing information, as it can be difficult to distinguish if pain is arising from the cuff itself or pathology beyond the cuff, such as an adherent ovary. Treatment follows in-line with other neuropathic modalities with systemic medications such as tricyclic antidepressants or

GABAergic agents, in addition to local anesthesia, either patient-administered or as an in-office injection. Vaginal apex revision can be considered in refractory cases, but patients should be advised that recurrence of pain is common several years after such a reoperation [32].

Comorbid Pain Conditions

An understanding of chronic pain as a problem of central sensitization, whereby hypersensitive responses to diverse stimuli are the reflection of abnormal processing of pain signals by the nervous system, may explain why functional pain syndromes tend to cluster in patients. Peripheral pain generators encountered in evaluating patients with chronic pelvic pain are listed in Table 12.1, with gastrointestinal and urologic symptoms being some of the most common adjacent-organ contributors. Indeed, IBS has been documented in over one-third of patients with CPP [33, 34], and nearly two-thirds of these women demonstrate features of IC/PBS [35]. These conditions are defined by the symptoms experienced by the patient (e.g., alternating constipation/loose stool, urinary frequency, dysuria, nocturia). After more traditional diagnoses are excluded (e.g., inflammatory bowel disease, celiac disease, urinary tract infection), tools such as the Rome Criteria [36] or the Pelvic Pain and Urinary/Frequency (PUF) symptom scale can be helpful [37], although they are more sensitive than specific by design. For IC, the bladder or urethra may be tender on unidigit exam, and a cystoscopy with hydrodistention can be useful. The bladder is filled to capacity and then reexamined to look for subepithelial petechiae or Hunner's ulcers. If a low bladder capacity is observed, a therapeutic distention for approximately 10 min can be helpful. Potassium sensitivity testing is rarely needed or helpful in patients with bladder pain. Treatment of IC generally involves avoiding bladder irritants such as tobacco or caffeine, administering tricyclic antidepressants (especially amitriptyline, exploiting its anticholinergic activity), GABAergics (e.g., gabapentin), pentosan polysulfate, bladder analgesics (e.g., phenazopyridine), and occasionally repeated bladder instillations with a cocktail of lidocaine, dimethyl sulfoxide, and/or heparin. IBS is generally managed with dietary modification and medications to either slow (e.g., loperamide) or speed (e.g., lubiprostone) transit time, together with agents to change the character of stool to a more favorable consistency (e.g., polyethylene glycol 3350). It is common for bladder symptoms, nausea, or diarrhea to accompany severe dysmenorrhea, and treating the latter hormonally may also improve urologic or gastrointestinal symptoms.

Psychological Factors

Mood disorders, a history of sexual abuse, and sexual dysfunction are all commonly encountered in patients with pelvic pain [38–41]. The careful clinician cannot ignore these important influences, but must also tread lightly. Patients with pelvic pain are frequently accustomed to being made to feel their symptoms lie on the first side of an artificial psychiatric/organic divide. Asking too soon about depression, anxiety, or whether a patient has seen a therapist can create a barrier difficult to overcome. By its nature, sexual abuse or current sexual dysfunction may be difficult to talk about freely in a traditional medical setting. After establishing patient–clinician trust—not necessarily on the first visit—the clinician can preempt apprehension with an explanation of psychological factors having a symbiotic, rather than causal, relationship with pelvic pain. For example, although a history of sexual abuse is more common in patients with CPP than without, clearly not all abuse victims develop chronic pain, and there are many women with pain and no history of abuse. With depression, pain thresholds are lowered even in people without chronic pain. It makes intuitive sense that struggling with daily pain could easily lead to a depressed disposition.

Determining the cause is less important than simply treating pain and mood symptoms to the degree that they are present. Sensitively suggesting consideration of enlisting a therapist's help can be presented as augmenting treatment of pain symptoms instead of conveying that a woman's discomforts are simply supratentorial.

Recognizing catastrophization, the belief that things are as bad as they can be and are unlikely to improve, is likewise important. This trait is often seen in patients with CPP [42] and presents one of the more refractory obstacles in treating these women. Catastrophizing is often supported by well-meaning family members and spouses who reinforce the sick role with kind attention and devoted attempts to help. These situations especially are best served with a multidisciplinary approach to treatment.

Discussion

Nonacute pelvic pain represents a spectrum of disorders not unlike many others, whereby the diagnosis remains elusive and the treatment is fraught with episodes of trial and error. In recent years, our fundamental understanding of pain mechanisms has helped to provide not only a better awareness among providers who care for women, but an improved capacity to have a positive impact on these patients who are indeed suffering. Salient features of the historical intake along with a focused exam, without the need for expansive diagnostic studies, often direct us to treatment options that are typically nonsurgical and can be applied readily. When addressing elective surgery, not all clinicians will be able to offer each option, and so knowing when to operate and when to refer will only enable her care. Ultimately, an honest discussion with a patient in pain—listening to her concerns and allowing her to be active in her care—becomes our obligation and may be therapeutic in and of itself.

Key Points

- A complete medical and psychosocial history, as well as a pain-oriented physical examination, should be completed before diagnostic laparoscopy is performed.
- Neuropathic and musculoskeletal components of chronic pelvic pain often require treatment both before and after appropriate pelvic surgery.
- A minimally invasive surgical approach is particularly appropriate for chronic pain patients.
- Laparoscopic treatment of endometriosis is more effective than diagnostic exploration alone.
- Resection of deeply infiltrating endometriosis is effective treatment of organ-specific symptoms.
- Improvement in pain symptoms following GnRHa treatment does not prove the existence of endometriosis. Many painful conditions (e.g., irritable bowel, inguinal hernia) vary with the menstrual cycle, and elimination of hormonal variation can change symptom profile, regardless of the presence of endometriosis.
- Complete skeletonization of the infundibulopelvic vessels, especially in difficult oophorectomy, reduces the risk of adja-

cent-organ injury and ORS. Ovarian remnants should be removed with careful opening of avascular spaces and identification of retroperitoneal structures.

- Though not a cure for all components of a woman's CPP, and in some cases a causative agent of pain, hysterectomy can be an effective treatment. A comprehensive evaluation must first be conducted prior to considering hysterectomy.

References

1. ACOG Committee on Practice Bulletins—Gynecology. ACOG practice bulletin no. 51. Chronic pelvic pain. Obstet Gynecol. 2004;103(3):589–605.
2. Howard FM. Chronic pelvic pain. Obstet Gynecol. 2003;101(3):594–611.
3. Latthe P, Latthe M, Say L, Gülmezoglu M, Khan KS. WHO systematic review of prevalence of chronic pelvic pain: a neglected reproductive health morbidity. BMC Public Health. 2006;6:177.
4. Howard FM. The role of laparoscopy in chronic pelvic pain: promise and pitfalls. Obstet Gynecol Surv. 1993;48(6):357–87.
5. Farquhar CM, Steiner CA. Hysterectomy rates in the United States 1990–1997. Obstet Gynecol. 2002;99(2):229–34.
6. Fenton BW, Brobeck L, Witten E, Von Gruenigen V. Chronic pelvic pain syndrome-related diagnoses in an outpatient office setting. Gynecol Obstet Invest. 2012;74(1):64–7.
7. Warren JW, Morozov V, Howard FM. Could chronic pelvic pain be a functional somatic syndrome? Am J Obstet Gynecol. 2011;205(3):199.e1–5.
8. Giudice LC. Clinical practice. Endometriosis. N Engl J Med. 2010;362(25):2389–98.
9. Sampson JA. Peritoneal endometriosis due to the menstrual dissemination of endometrial tissue into the peritoneal cavity. Am J Obstet Gynecol. 1927;14:411–69.
10. Young VJ, Brown JK, Saunders PT, Home AW. The role of the peritoneum in the pathogenesis of endometriosis. Hum Reprod Update. 2013;19(5):558–69.
11. Abrao MS, Gonçalves MO, Dias Jr JA, Podgaec S, Chamie LP, Blasbalg R. Comparison between clinical examination, transvaginal sonography and magnetic resonance imaging for the diagnosis of deep endometriosis. Hum Reprod. 2007;22(12):3092–7.
12. Davis L, Kennedy SS, Moore J, Prentice A. Modern combined oral contraceptives for pain associated with endometriosis. Cochrane Database Syst Rev. 2007;18(3):CD001019.
13. Guzick DS, Huang LS, Broadman BA, Nealon M, Hornstein MD. Randomized trial of leuprolide versus continuous oral contraceptives in the treatment of endometriosis-associated pelvic pain. Fertil Steril. 2011;95(5):1568–73.
14. Fedele L, Bianchi S, Zanconato G, Portuese A, Raffaelli R. Use of a levonorgestrel-releasing intrauterine device in the treatment of rectovaginal endometriosis. Fertil Steril. 2001;75(3):485–8.

15. Ling FW. Randomized controlled trial of depot leuprolide in patients with chronic pelvic pain and clinically suspected endometriosis. Pelvic Pain Study Group. Obstet Gynecol. 1999;93(1):51–8.
16. Fedele L, Bianchi S, Zanconato G, Berlanda N, Borruto F, Frontino G. Tailoring radicality in demolitive surgery for deeply infiltrating endometriosis. Am J Obstet Gynecol. 2005;193(1):114–7.
17. Sutton CJ, Ewen SP, Whitelaw N, Haines P. Prospective, randomized, double-blind, controlled trial of laser laparoscopy in the treatment of pelvic pain associated with minimal, mild, and moderate endometriosis. Fertil Steril. 1994;62(4):696–700.
18. Jones KD, Haines P, Sutton CJ. Long-term follow-up of a controlled trial of laser laparoscopy for pelvic pain. JSLS. 2001;5(2):111–5.
19. Fauconnier A, Chapron C, Dubuisson JB, Vieira M, Dousset B, Bréart G. Relation between pain symptoms and the anatomic location of deep infiltrating endometriosis. Fertil Steril. 2002;78(4):719–26.
20. Hart RJ, Hickey M, Maouris P, Buckett W. Excisional surgery versus ablative surgery for ovarian endometriomata. Cochrane Database Syst Rev. 2008;2:CD004992.
21. Dueholm M. Transvaginal ultrasound for diagnosis of adenomyosis: a review. Best Pract Res Clin Obstet Gynaecol. 2006;20(4):569–82.
22. Ozdegirmenci O, Kayikcioglu F, Akgul MA, Kaplan M, Karcaaltincaba M, Haberal A, Akyol M. Comparison of levonorgestrel intrauterine system versus hysterectomy on efficacy and quality of life in patients with adenomyosis. Fertil Steril. 2011;95(2):497–502.
23. Swank DJ, Swank-Bordewijk SC, Hop WC, van Erp WF, Janssen IM, Bonjer HJ, Jeekel J. Laparoscopic adhesiolysis in patients with chronic abdominal pain: a blinded randomised controlled multi-centre trial. Lancet. 2003;361(9365):1247–51.
24. Gyang A, Hartman M, Lamvu G. Musculoskeletal causes of chronic pelvic pain: what a gynecologist should know. Obstet Gynecol. 2013;121(3):645–50.
25. Tu FF, Hahn D, Steege JF. Pelvic congestion syndrome-associated pelvic pain: a systematic review of diagnosis and management. Obstet Gynecol Surv. 2010;65(5):332–40.
26. Magtibay PM, Nyholm JL, Hernandez JL, Podratz KC. Ovarian remnant syndrome. Am J Obstet Gynecol. 2005;193(6):2062–6.
27. Stratton P, Sinali N, Segars J, Koziol D, Wesley R, Zimmer C, et al. Return of chronic pelvic pain from endometriosis after raloxifene treatment: a randomized controlled trial. Obstet Gynecol. 2008;111(1):88–96.
28. Onghena P, Van Houdenhove B. Antidepressant-induced analgesia in chronic nonmalignant pain: a meta-analysis of 39 placebo-controlled studies. Pain. 1992;49(2):205–19.
29. Proctor ML, Latthe PM, Farquhar CM, Khan KS, Johnson NP. Surgical interruption of pelvic nerve pathways for primary and secondary dysmenorrhoea. Cochrane Database Syst Rev. 2005;4:CD001896.
30. Loos MJ, Scheltinga MR, Roumen RM. Surgical management of inguinal neuralgia after a low transverse Pfannenstiel incision. Ann Surg. 2008;248(5):880–5.

31. Rhodes JC, Kjerulff KH, Langenberg PW, Guzinski GM. Hysterectomy and sexual functioning. JAMA. 1999;282(20):1934–41.
32. Lamvu G, Robinson B, Zolnoun D, Steege JF. Vaginal apex resection: a treatment option for vaginal apex pain. Obstet Gynecol. 2004;104(6):1340–6.
33. Williams RE, Hartmann KE, Sandler RS, Miller WC, Steege JF. Prevalence and characteristics of irritable bowel syndrome among women with chronic pelvic pain. Obstet Gynecol. 2004;104(3):452–8.
34. Choung RS, Herrick LM, Locke 3rd GR, Zinsmeister AR, Talley NJ. Irritable bowel syndrome and chronic pelvic pain: a population-based study. J Clin Gastroenterol. 2010;44(10):696–701.
35. Tirlapur SA, Kuhrt K, Chaliha C, Ball E, Meads C, Khans KS. The 'evil twin syndrome' in chronic pelvic pain: a systematic review of prevalence studies of bladder pain syndrome and endometriosis. Int J Surg. 2013;11(3):233–7.
36. Drossman DA, Dumitrascu DL. Rome III. New standard for functional gastrointestinal disorders. J Gastrointestin Liver Dis. 2006;15(3):237–41.
37. Parsons CL, Dell J, Stanford EJ, Bullen M, Kahn BS, Waxell T, Koziol JA. Increased prevalence of interstitial cystitis: previously unrecognized urologic and gynecologic cases identified using a new symptom questionnaire and intravesical potassium sensitivity. Urology. 2002;60(4).573–8.
38. Walker E, Katon W, Harrop-Griffiths J, Holm L, Russo J, Hickok LR. Relationship of chronic pelvic pain to psychiatric diagnoses and childhood sexual abuse. Am J Psychiatry. 1988;145(1):75–80.
39. Steege JF, Stout AL, Somkuti SG. Chronic pelvic pain in women: toward an integrative model. Obstet Gynecol Surv. 1993;48(2):95–110.
40. Slocumb JC, Kellner R, Rosenfeld RC, Pathak D. Anxiety and depression in patients with the abdominal pelvic pain syndrome. Gen Hosp Psychiatry. 1989;11(1):48–53.
41. Reiter RC. Occult somatic pathology in women with chronic pelvic pain. Clin Obstet Gynecol. 1990;33(1):154–60.
42. Martin CE, Johnson E, Wechter ME, Leserman J, Zolnoun DA. Catastrophizing: a predictor of persistent pain among women with endometriosis at 1 year. Hum Reprod. 2011;26(11):3078–84.

13. Imaging for Evaluation of Groin Pain

Joseph M. Miller, Shane D. Smith, David N. Ishimitsu, and Rola Saouaf

While inguinal hernias can typically be diagnosed with a history and physical examination alone, patients who present with atypical symptoms or subtle examination findings may require one or more imaging studies to determine the true cause of their pain. The high prevalence of inguinal hernia among the general populace implies the possibility that coincident pathology—whether musculoskeletal, neurologic, urologic, or gynecologic—must be entertained as the true cause of groin pain. Moreover, it is important to understand that patients with classic symptoms of an inguinal hernia may have such a small defect as to be practically undetectable on physical exam. Imaging diagnosis of the so-called occult or hidden hernia is difficult, and is therefore given focused attention at the end of this chapter. The radiologic evaluation of patients with groin pain after herniorrhaphy presents unique challenges and is detailed in a separate chapter.

Imaging Modalities

The four modern branches of imaging consist of *computed tomography (CT)/radiography (X-ray), magnetic resonance (MR), ultrasound (US),* and *nuclear imaging*. Each has their own strengths and weaknesses with respect to the kinds of pathologies they can discriminate. Knowing the limitations of a given study in advance can improve diagnostic accuracy and help set patient expectations accordingly, thus preventing the frustration elicited by unanticipated follow-up exams. Regardless of the modality chosen, the more information provided to the radiologist, the more specific the interpretation can be: while "pain" is a

B.P. Jacob et al. (eds.), *The SAGES Manual of Groin Pain*,
DOI 10.1007/978-3-319-21587-7_13

valid indication, lateralizing and characterizing the pain (e.g., "sharp left groin pain with defecation") can assist with selection of proper image acquisition protocols and winnow down the list of potential pathologies that may present with similar appearances.

Computed Tomography/Radiography (X-ray)

An X-ray tube generates radiation by smashing high-energy electrons into a durable piece of metal that can resist melting from the consequent heat exchange. The process is known as the photoelectric effect, and underlies a number of modern technologies like solar panels and digital cameras. X-rays passing through the body have a probability of being scattered or absorbed that is related to tissue density; that probability of being attenuated by tissue allows for the discrimination of fluid, fat, soft tissue, and bone. The amount of attenuation can be quantified on a relative scale quantified by the Hounsfield Unit, where air is arbitrarily defined as −1000 HU and water as 0 HU, with the implication of an upper bound of +1000 HU for bone. While these values are only explicitly measured for CT, they are still an important factor in conventional radiography.

On conventional radiographs, the differentiation of two structures relies on a sufficient discrepancy between their attenuations: air-filled bowel can be easily seen against the background of the peritoneum, while fluid-filled bowel is too similar to distinguish. Musculoskeletal abnormalities that may cause groin pain, such as femoral acetabular impingement (FAI), spondylosis/spondylolisthesis of the lumbosacral spine, sacroiliitis, slipped-cap femoral epiphysis, hip osteoarthritis, calcific tendonitis, and fracture, can all be reasonably eliminated from a differential diagnosis on the basis of radiography, provided that the proper views are acquired and that technical factors (such as patient's body habitus) allow (Figs. 13.1, 13.2, and 13.3). The gold standard in the evaluation of inguinal hernia, i.e., herniography—the injection of contrast material into the peritoneal cavity followed by radiography of the groins—is rarely performed these days outside of a few specialized centers, though it is frequently referenced in the radiologic literature. In addition to the drawbacks of its invasiveness, it has been obviated by dramatic technical advances in dynamic CT and MR acquisition.

CT relies on the same principles as plain film radiography. Instead of acquiring a single image with a set amount of radiation, a CT scanner

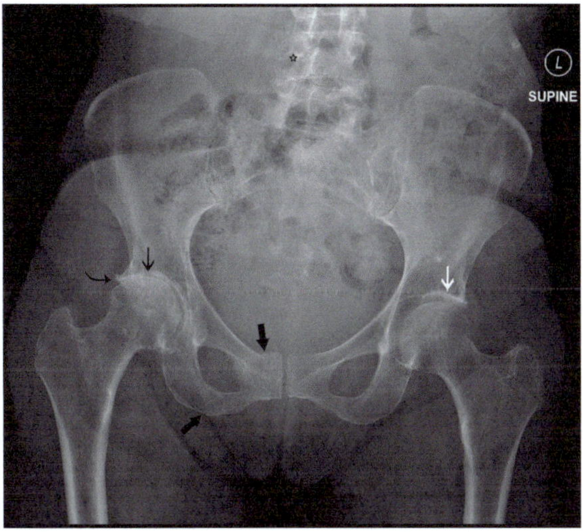

Fig. 13.1. Anteroposterior (AP) view of the pelvis is acquired with internal rotation of the legs, laying out the femoral necks to evaluate for possible fracture. There is severe osteoarthritic change about the right hip with preferential loss of the superior, weight-bearing portion of the joint space (*black arrow*) and osteophytosis (*curved black arrow*) and relative preservation of the medial space. In contrast, the left hip is normal in appearance (*white arrow*). Chronic superior and inferior pubic rami fractures are seen on the right (*thick black arrows*) with interruption of the smooth cortical line and callus formation. Significant degenerative changes of the lumbar spine are partially visualized (*asterisk*).

acquires many images from many angles, each of which requires much less radiation than a single conventional radiograph because they are not meant to stand alone. A sophisticated processing algorithm then integrates these individual "projections" into a three-dimensional attenuation map that can be viewed in the axial coronal or sagittal plane, or some oblique combination thereof. The advance of technology is such that current generation CT scanners can produce diagnostic images in any conceivable plane without appreciable loss of resolution. Pathologies such as appendicitis and degenerative spine disease in particular benefit from modern high-resolution multiplanar imaging (Fig. 13.4). Other less common causes of groin pain such as diverticulitis, abdominal aortic aneurysm, myositis ossificans, adductor tendonitis, prostatitis, and pelvic inflammatory disease are well visualized by computed tomography,

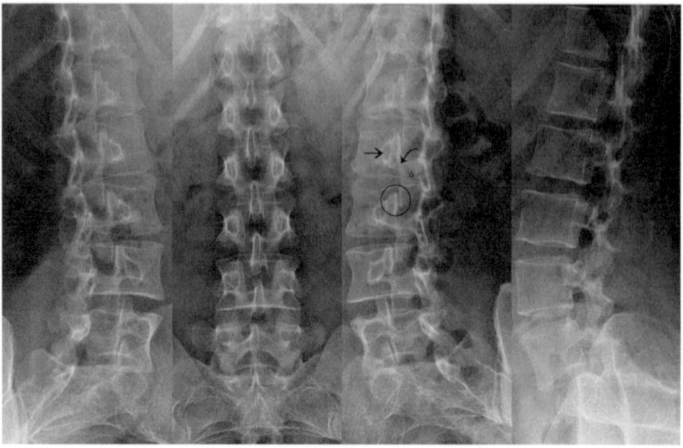

Fig. 13.2. A four view series of the lumbar spine combines the traditional two view series (anteroposterior and lateral) with left and right oblique views. While the AP and lateral views are sufficient for evaluating for vertebral body height and alignment, oblique views reveal the "scotty dog" appearance of the posterior structures, allowing for evaluation of the pedicle (*arrow*), lamina (*asterisk*), pars articularis (*curved arrow*), and facets (*circle*). Spondylolysis, or interruption of the pars articularis, may predispose to neuroforaminal stenosis and subsequent neurogenic groin pain.

although full characterization may require follow-up evaluation with contrast or another imaging modality.

The overall radiation dose of a CT study is a function of the dose associated with each acquisition, and the total number of acquisitions needed to generate the three-dimensional attenuation map. Patient factors such as body width, lean muscle mass, and the presence of intervening hardware (e.g., spinal fusion, hip arthrodesis, etc.) influence total dose. Modern technologies such as exposure control and iterative reconstruction can automatically reduce dose to minimally necessary levels. For comparison's sake, the American College of Radiology (ACR) maintains a set of Appropriateness Criteria. Their most recent Radiation Dose Assessment lists a typical abdominal CT as the rough equivalent of ten single view radiographs [1]. This is in comparison to natural background levels of radiation, which averages an equivalent of about three radiographs per year in the United States [2]. The significant drop in radiation associated with CT has allowed for the application of dynamic acquisition in some diagnostic circumstances.

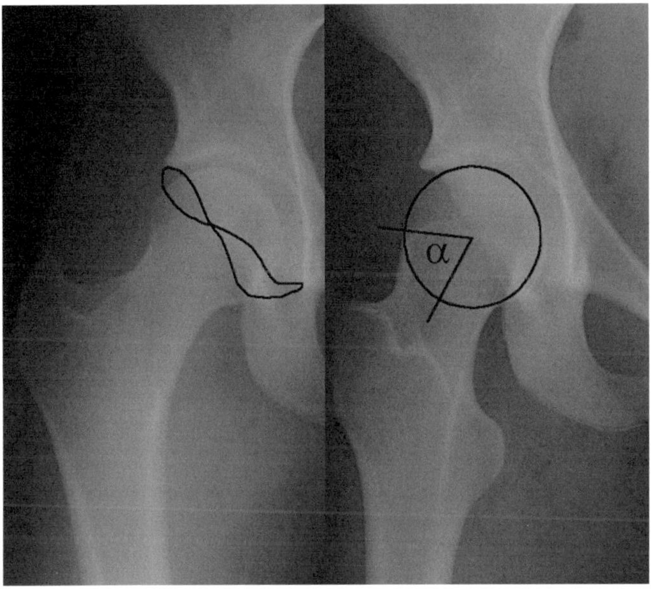

Fig. 13.3. Two views of the right hip demonstrating the findings of mixed-type femoroacetabular impingement. Pincer-type deformity can be diagnosed on the anteroposterior view if there is evidence of acetabular overcoverage: in this case the anterior acetabular wall is somewhat more lateral than the posterior wall, forming a figure-of-eight. The modified Dunn view (patient lying supine with feet flat on the table) allows for the evaluation of cam-type deformity, which is due to asphericity of the femoral head. A circle is drawn within the confines of the femoral head and an alpha angle measured between the axis of the femoral neck and the point where the cortex of the neck first meets the head. This alpha angle measured 65°, while normal is considered less than or equal to 50°.

Intravenous iodinated contrast material has also advanced substantially since its early uses. Recent research suggests that the nonionic, low osmolarity contrast formulations currently employed in CT are not a causative factor in the development of nephropathy, particularly among patients with normal renal function at baseline [3, 4]. Rather, baseline glomerular filtration rate appears to be the primary determinant of acute kidney injury. Current iodinated contrast materials have been found to not represent an independent risk factor for AKI even among patients with impaired GFR (below 30 mL/min/1.73 m^2) [5]. It bears mentioning that this research is relatively recent and requires independent confirma-

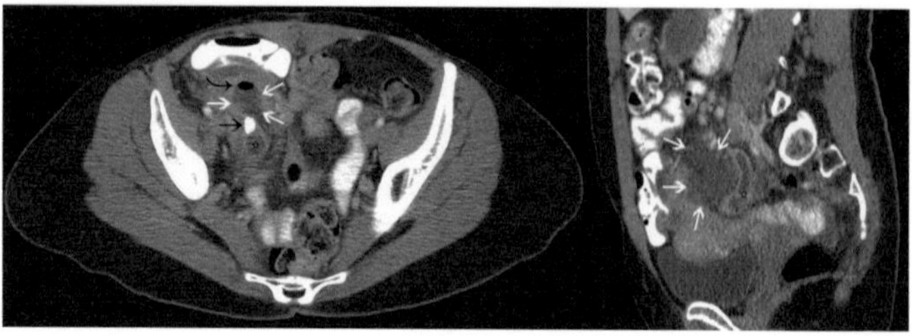

Fig. 13.4. Axial and coronal CT images displaying a case of perforated appendicitis. The tubular shape of the appendix (*asterisks*) is clearly seen in the coronal view with a high-density fecalith (*black arrow*) best seen on the axial view. Gas (*curved black arrow*) within the surrounding fluid collection (*surrounded by white arrows*) is consistent with peri-appendiceal abscess formation.

tion before current practice guidelines will change [6]. The suggestion, however, is that the administration of iodinated contrast material should not be avoided in otherwise healthy individuals. While intravenous contrast material is not required for most protocols, it increases the ability of CT to evaluate for all manner of infectious, inflammatory, and neoplastic processes (Figs. 13.5, 13.6, and 13.7). Likewise, abdominopelvic evaluation by CT benefits substantially from routine oral contrast administration (Fig. 13.8). While rectal contrast agents also have utility, improvements in image resolution and multiplanar reformatting are leading to decreased reliance on such invasive administration.

CT contrast reactions are infrequent occurrences that are incompletely understood. In the majority of patients, reactions to iodinated contrast are not allergies in the traditional sense of IgE mediation, yet may present with anaphylactoid airway edema or other severe physiologic consequences all the same [7]. As per the above discussion of contrast-induced nephropathy, the incidence of severe adverse reactions has decreased substantially since the switch was made to low osmolarity contrast material [8]. A history of asthma or prior contrast reaction increases the risk of acute reaction. Pretreatment with oral corticosteroids is recommended for at-risk patients [9]. Common pretreatment regimens for high-risk individuals may involve oral administration of 50 mg prednisone at 13, 7, and 1 h before contrast administration, with or without oral administration of 50 mg diphenhydramine and 25 mg ephedrine at 1 h before contrast administration [10], though facility-specific variations abound. Recent studies suggest that multiple exposures to contrast material may be necessary for severe reaction to occur [11].

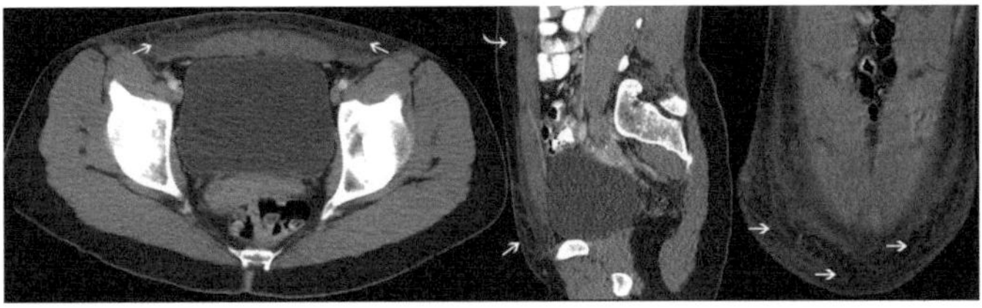

Fig. 13.5. Axial, sagittal, and coronal CT images with features of delayed onset muscle soreness (DOMS), a form of exercise-induced muscular pain that, when involving the abdominal wall or pelvic musculature, may result in groin pain. With severe exertion, rhabdomyolysis and subsequent acute renal failure may occur. The soft tissues and fascial planes of the abdominal wall are diffusely edematous (*white arrows*) as compared to the unaffected tissues (*curved white arrow*) found more superiorly. Findings must be differentiated from fasciitis on the basis of clinical presentation and laboratory results.

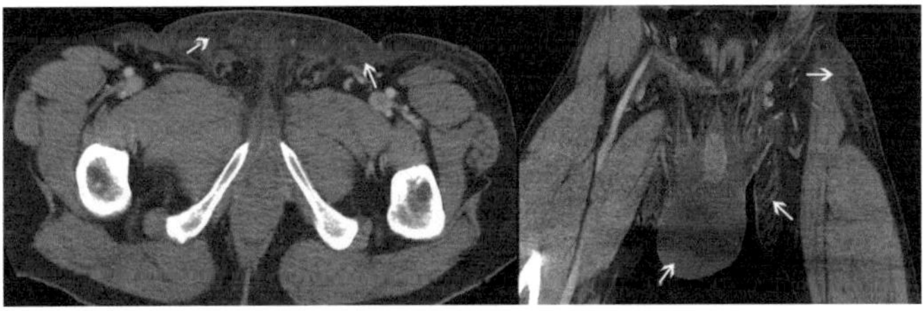

Fig. 13.6. Axial and coronal CT images of Fournier's necrotizing fasciitis. Diffuse abdominal wall edema is again seen (*white arrows*), primarily involving the scrotum. Although subcutaneous air is sensitive for fasciitis, as in this case it is not always present. In contrast to prior patient, the clinical presentation here was that of sepsis secondary to pelvic infection.

Magnetic Resonance

Magnetic resonance scanners utilize low-energy light to interact with the hydrogen atoms found throughout most organic tissues. The electromagnet inside an MR scanner is always "on," and typically operating at 1.5 T of field strength (roughly 10,000 times the strength of the Earth's natural field) although 3 T scanners are becoming more widely available in routine clinical imaging. This magnetic field provides energy to hydrogen atoms, forcing them to line up along the direction of the magnet much in the way that a compass needle will line up with the Earth.

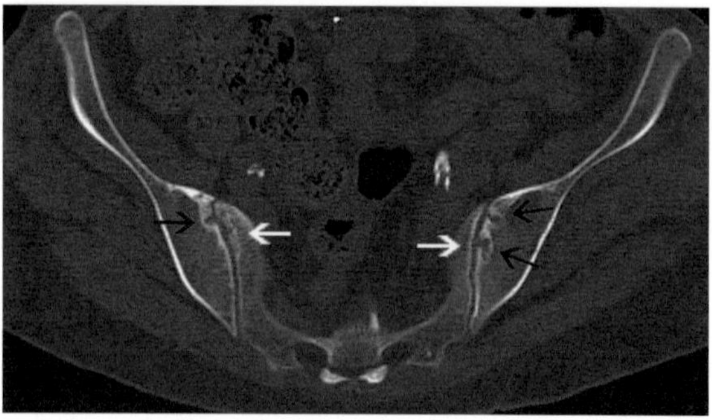

Fig. 13.7. Axial CT of the pelvis demonstrates bony erosions of the bilateral iliac wings (*black arrows*) with sparing of the sacrum (*white arrows*), consistent with sacroiliitis. Ferguson view of the pelvis ("pelvic outlet radiograph" not shown) will accentuate the sacroiliac joints and may reveal sacroiliitis without the need for CT.

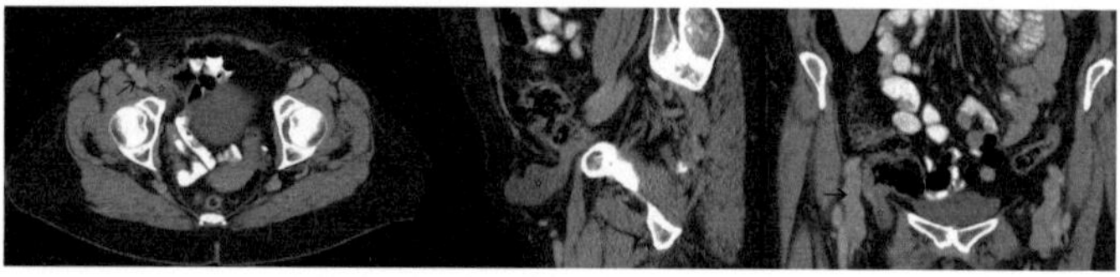

Fig. 13.8. Axial, sagittal, and coronal CT reveal loop of bowel (*asterisks*) exiting the peritoneal cavity below the inguinal ligament and medial to the femoral vessels (*black arrows*), diagnostic of femoral hernia.

Once the hydrogen atoms line up, the machine can communicate with them by sending out radio-frequency pulses that only specific atoms are able to respond to, forcing them to change direction and oppose the magnetic field. This opposition is relatively unstable and the flipped hydrogen atoms will eventually switch back to their natively aligned state, emitting their own radio-frequency pulse and communicating back with the scanner as they do so. Since every tissue is different, the rate of this process varies dramatically throughout the body and allows for tissue discrimination. Foreknowledge of how fat, water, and other body substances will behave under these conditions allows for the targeting by specific "sequences" such as T1-weighted sequences for fat, T2-weighted sequences for water, and *short-tau inversion recovery (STIR)* for edema, among many others. Compare this to computed tomography, where there is only one parameter (i.e., density) that significantly impacts tissue discrimination.

MR excels at differentiating between many subtle soft tissue and musculoskeletal pathologies responsible for groin pain, such as those seen in iliopsoas tendinosis, bursitis, osteitis pubis, and athletic pubalgia (Figs. 13.9, 13.10, and 13.11). Avascular necrosis of the femoral head in particular is apparent on MR long before it is demonstrable by CT (Fig. 13.12). Yet MR is not always a definitive examination, particularly with respect to certain musculoskeletal pathologies in which the relative lack of hydrogen in bone can necessitate correlation with conventional radiography or CT. While radiographically occult stress fractures are clearly identified on MR, many other benign and malignant osseous lesions may be indistinguishable from each other on the basis of MR alone. Definitive evaluation of acetabular labral tears is via MR arthrography, wherein contrast material is directly injected into the hip joint (Fig. 13.13).

Dynamic MR has become the primary evaluation of pelvic floor dysfunction now that most centers have stopped performing colpography and defecography, which involve the respective administrations of vaginal or rectal contrast agents followed by fluoroscopic visualization during Valsalva maneuver (Fig. 13.14). Dynamic MR can demonstrate ligamentous laxity and organ prolapse (rectocele, cystocele, enterocele), as well as less conspicuous pathologies such as vesicovaginal and rectovaginal fistulae. Occult inguinal hernias also benefit from evaluation with dynamic MR, as will be discussed later in this chapter. Other gynecologic sources of groin pain such as endometriosis, uterine fibroids, and ovarian masses/cysts are well evaluated on MR but can often be more readily and economically demonstrated with ultrasound.

Intravenous MR contrast material is fundamentally different than the iodinated material used for CT. The risk of contrast reaction is significantly lower with gadolinium-based MR contrast agents, and there is no known cross-reactivity between MR and CT contrast agents, although patients with atopic tendencies in general are at increased risk for either [6]. The unique risk of the gadolinium contrast agents used for MR is a condition known as nephrogenic systemic fibrosis, of which only a few hundred cases have been identified. Patients with severe or end-stage renal failure (GFR below 30 ml/min/1.73 m^2) seem to be most at risk, and risk appears to be dose dependent [12]. Inability to remove the gadolinium chelates at a normal rate is theorized to lead to free gadolinium accumulation in the blood, followed by precipitation within the skin, retroperitoneum, heart, etc., with secondary sclerotic change [13]. While

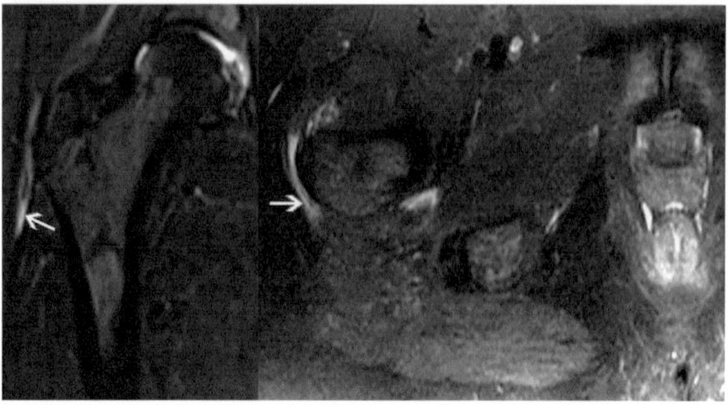

Fig. 13.9. Coronal and axial T2-weighted MR through the pelvis show increased fluid signal (*white arrows*) lateral to the greater trochanter, consistent with greater trochanteric bursitis.

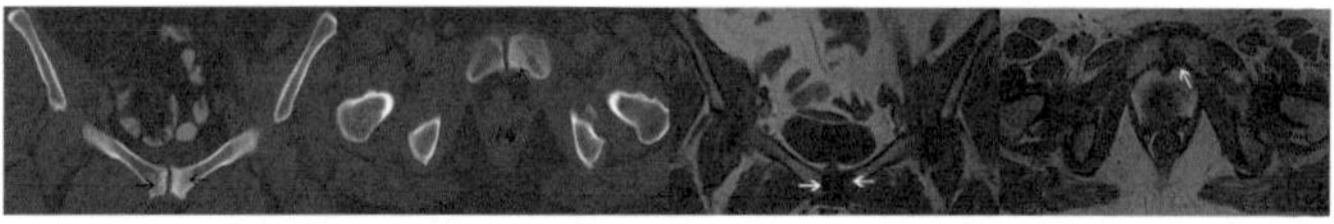

Fig. 13.10. Coronal and axial CT of the pelvis demonstrate diffuse subchondral sclerosis (*black arrows*) of the pubic symphysis. Corresponding coronal and axial T1-weighted MR demonstrate focal hypointensity (*white arrows*), consistent with osteitis pubis.

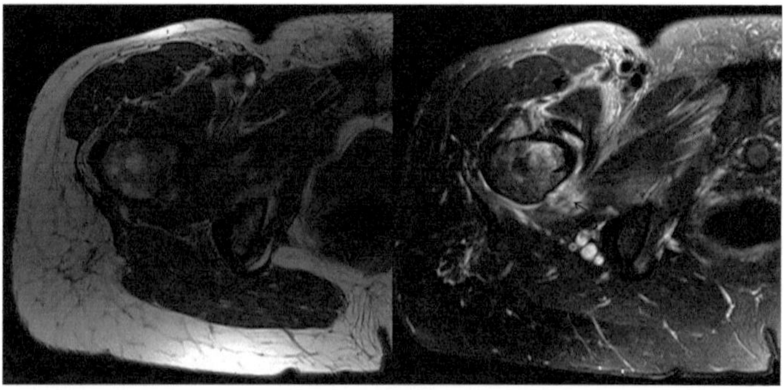

Fig. 13.11. Axial T1- and T2-weighted MR through the pelvis demonstrating significant edema (*black arrow*) and trace fluid within the adductor compartment, consistent with low-grade adductor strain.

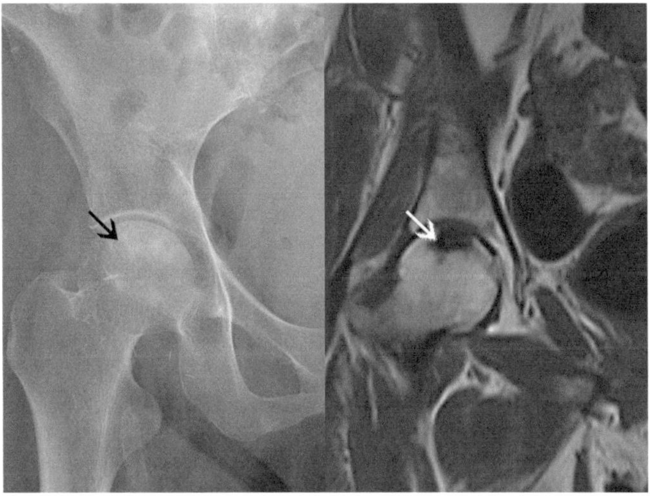

Fig. 13.12. Anteroposterior radiograph of the right hip reveals subtle sclerosis (*black arrow*) representative of avascular necrosis. Coronal T1-weighted MR demonstrates serpiginous hypointensity (*white arrow*), confirming the diagnosis.

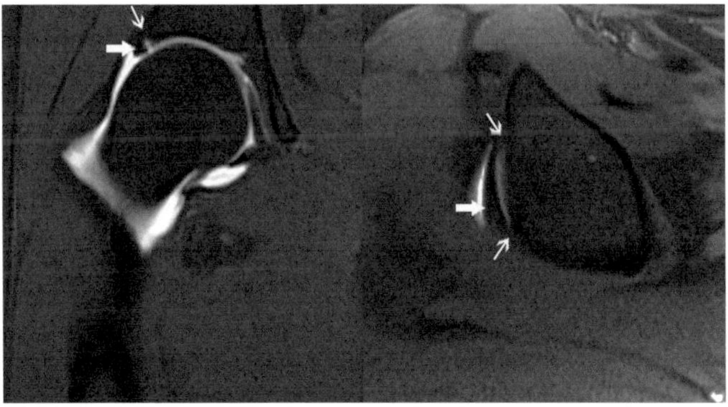

Fig. 13.13. Coronal and axial T1-weighted MR arthrogram of the hip after intracapsular injection of gadolinium-containing contrast agent. A hyperintense fluid cleft (*white arrows*) is seen separating the labrum (*large white arrows*) from the chondral surface of the acetabulum. Findings represent superior labral tear in this patient with cam-type deformity and femoroacetabular impingement.

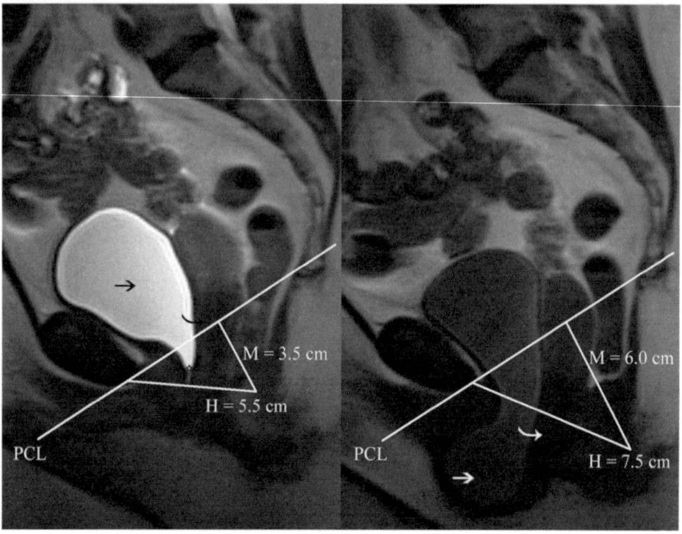

Fig. 13.14. Sagittal T2-weighted MR images. The pubococcygeal line (PCL) extends from the pubic symphysis to the final coccygeal joint and roughly demarcates the pelvic floor. The H-line is measured from the pubic symphysis to the anorectal junction's posterior wall and should measure less than 5 cm; the M-line is the perpendicular drawn to connect the PCL and the H-line, and should measure less than 2 cm. All organs should remain above the PCL. As seen on the leftmost image, even at rest there is pelvic floor relaxation with pathologic H-line and M-line measurements. The bladder (*black arrow*) and uterus (*curved black arrow*) are mostly above the PCL, although there is mild prolapse of the urethra in particular (*asterisk*). With Valsalva, there is significant elongation of both H-line and M-line beyond their already pathologic baselines, as well as severe prolapse of the bladder (*white arrow*) and the uterus (*curved white arrow*), both dropping below the H-line.

protocols are facility specific, lower doses of contrast are often administered in patients with abnormal renal function (GFR below 60 ml/min/1.73 m^2) and contrast is withheld altogether in patients with severe or end-stage renal disease. While intravenous contrast is useful for the evaluation of neoplasms in particular, it is not required for most differential diagnoses related to groin pain, and the availability of diffusion-weighted MR sequences may provide a useful alternative in patients with contraindications to gadolinium-based agents.

Unlike CT, there are contraindications to undergoing MR itself with respect to implanted medical devices. Even MR safe devices can create the sensation of tugging, particularly when entering or exiting the machine [14]. Devices with functional circuitry (e.g., pacemakers, infusion pumps, etc.) may be disrupted by the oscillating magnetic fields [15, 16]. MR safe devices without circuitry (e.g., orthopedic implants) may heat up during the application of certain sequences [17, 18]. Patients with anxiety or claustrophobia may be unable to undergo MR imaging without sedation, and may even require general anesthesia.

Ultrasound

Ultrasound technology is based on a property known as piezoelectricity, wherein mechanical deformation of a material results in the generation of an electrical current and vice versa. In the typical US transducer, multiple piezoelectric crystals are placed in a shaped array and a voltage applied, causing the crystals to vibrate. The specific characteristics of the voltage, the crystal structure, and the configuration of the array determine such technical factors as the frequency of vibration, as well as its coherence, depth of penetration, and field of view. Sound waves travel at different speeds in different materials, and so the interface between two different body tissues often results in reflection of at least some aspect of the sound wave. The US transducer probe rapidly switches back and forth between generating sound waves and then listening for the eventual echoes. In regard to evaluation of groin pain, US is most readily used for the evaluation of intra-pelvic organs (Fig. 13.15). The use of Doppler US allows for the additional evaluation of vascularity within visualized tissues (Fig. 13.16). There have been recent papers documenting frictional heat deposition by Doppler US, with the suggestion of a theoretical risk to the developing fetus with the use of this modality [19]. The ACR currently lists no contraindications to US in pregnancy, but does suggest limited fetal exposure [20, 21]. As a result, many radiology practices will avoid imaging the fetus when examining the mother for a complaint such as groin pain.

The primary limitations of US are its operator dependence and inability to image the pelvis completely: a negative US examination does not necessarily confirm the absence of disease. Additionally, technical factors such as bowel gas and body habitus can make an US examination completely nondiagnostic, whereas CT and MR evaluations are rarely so. Yet, US is a widely available and cost-effective way to evaluate for common causes of groin pain, which often elevates it as a first-line mode of

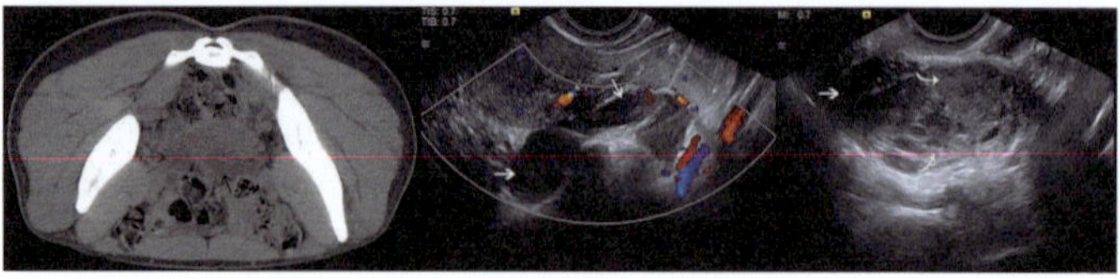

Fig. 13.15. Axial CT of the pelvis with corresponding sagittal Doppler and coronal US images of the left adnexa. While CT is able to show a multiloculated low-density fluid collection (*black arrows)* within the pelvis, further characterization is difficult. Follow-up US images demonstrate that the fluid is contained within tubular structures (*white arrows*) in continuity with a dilated, heterogenous appearing ovary (*curved white arrows*), allowing for diagnosis of tubo-ovarian abscess.

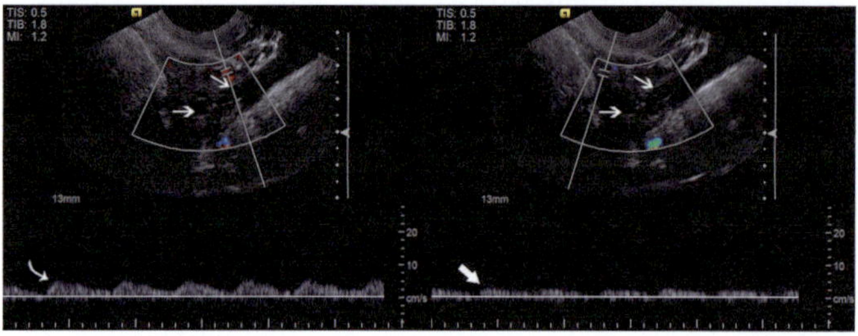

Fig. 13.16. Sagittal Doppler US of the ovary demonstrating normal arterial waveform on the left; normal venous waveform on the right. Morphologically, typical ovaries are of homogenous echotexture with hypoechoic follicles (*white arrows*) scattered throughout. Normal ovarian size varies significantly from patient to patient, and asymmetry between the left and right ovaries is often more diagnostically useful. Arterial waveform (*curved white arrow)* should have rapid systolic rise and gradual diastolic fall without inversion below the baseline. Likewise, venous flow (*large white arrow*) should remain uniform without inversion below the baseline.

evaluation, particularly when genitourinary pathologies are suspected. US is often a definitive examination for groin pain due to ovarian and testicular masses or torsion, ectopic pregnancy, uterine fibroids, epididymo-orchitis, pelvic inflammatory disease, hydrocele, varicocele, and pelvic congestion (Figs. 13.17, 13.18, and 13.19).

US also has numerous applications in the initial evaluation of musculoskeletal pathologies, though it may not be definitive. Pathologies such as tendinosis and tendinopathy, bursitis, and intramuscular contusions of the abdominal wall are easily and rapidly evaluated with US.

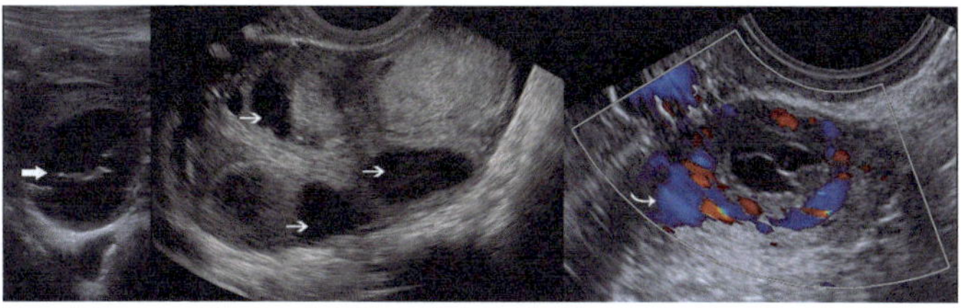

Fig. 13.17. Transverse US and Doppler US of a complex ovarian cyst with large central septation (*large white arrow*), an ovary with numerous large thin-walled theca lutein cysts (*white arrows*) in follicular hyperstimulation, and a normal corpus luteum cyst surrounded by its pathognomic "ring of fire" (*curved white arrow*) of peripheral increased vascularity.

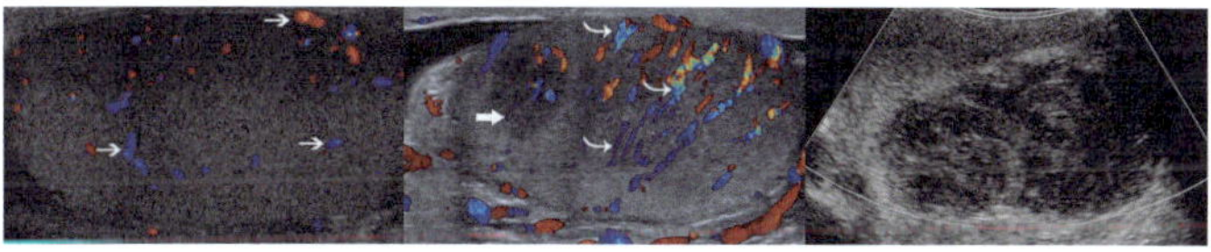

Fig. 13.18. Sagittal Doppler US images. Normal testicle has homogenous echotexture and scattered vascularity (*white arrows*). By contrast, testicular lymphoma presents as an ill-defined, hypoechoic focus (*large white arrow*) with significant hypervascularity (*curved white arrows*). Scrotal hematoma can also appear irregular and heterogenous, but should not demonstrate internal vascularity.

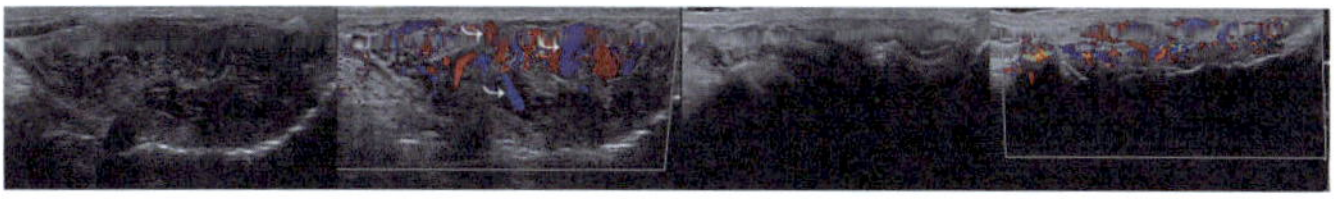

Fig. 13.19. Bilateral sagittal US and Doppler US images through the epididymis comparing the enlarged, hypoechoic, and hypervascular (*curved white arrows*) appearance of epididymitis with the normal contralateral anatomy.

Nuclear

Nuclear imaginginvolves the intravenous administration of radionuclides bound to target-specific chemicals that are known to accumulate in a given organ or at the site of a presumed pathology such as infection. The studies most applicable to groin pain include the *indium-111 white blood cell* and *gallium-67* scans typically utilized for the evaluation of pelvic abscess/inflammation and osteomyelitis, respectively. Technetium-99m bone scans are also utilized to evaluate for osseous pathology (Fig. 13.20).

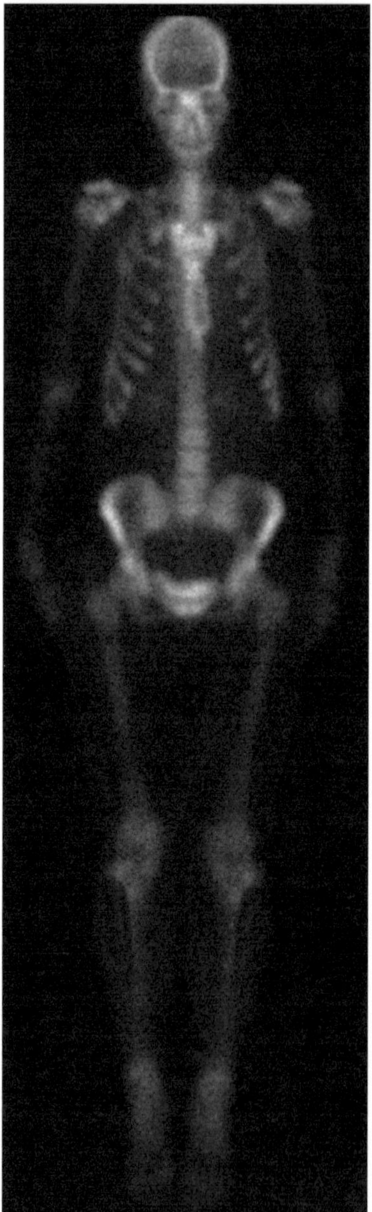

Fig. 13.20. Normal bone scan.

Imaging Evaluation of Occult Hernias

Occult hernia is defined as a clinically symptomatic defect in the abdominal wall (whether direct, indirect, or femoral) that presents without clear physical examination finding [22]. Surgical exploration of all patients with groin pain would result in an unacceptably high negative rate; as such, evaluation for an occult hernia is primarily reliant on imaging for diagnosis. There has been controversy however about the most appropriate way to engage in radiologic workup, and no standards have been established.

Studies have highlighted the positive predictive value of US in the diagnosis of inguinal hernia [23, 24]; however, few have offered surgical exploration in the setting of negative US evaluation, preventing the estimation of true negative predictive values. Dynamic examination of the groin with Valsalva maneuver further strengthens the ability of US to find small defects, but there is reason to believe that physically undetectable hernias would be equally difficult to demonstrate with the limited resolution of US. While it is certainly reasonable to begin radiologic assessment of suspected occult inguinal hernia with US, the test should not be considered definitive on account of its operator dependence and inability to ensure visualization of the entire pelvis [25].

CT of the pelvis solves this visualization problem and can often elucidate unconsidered pathologies as the source of groin pain. Yet while CT is often considered a definitive test for occult inguinal hernia (Fig. 13.21), [26], the common focus of study design on the hernias that "require" surgery—indicative of hernias more prone to complication—likely biases results toward detection of larger hernias. In a recent study directly comparing all patients operated on for uncomplicated groin

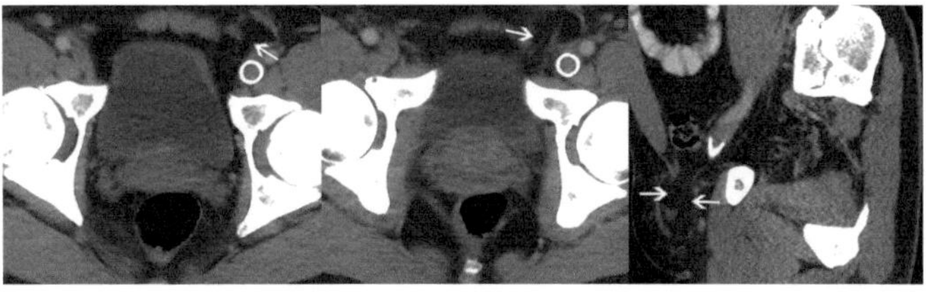

Fig. 13.21. Axial and sagittal CT images acquired through the pelvis in a patient with a pair of unilateral direct hernias (*white arrows*). While these hernias were clearly palpable on physical exam, smaller defects may require MR for detection.

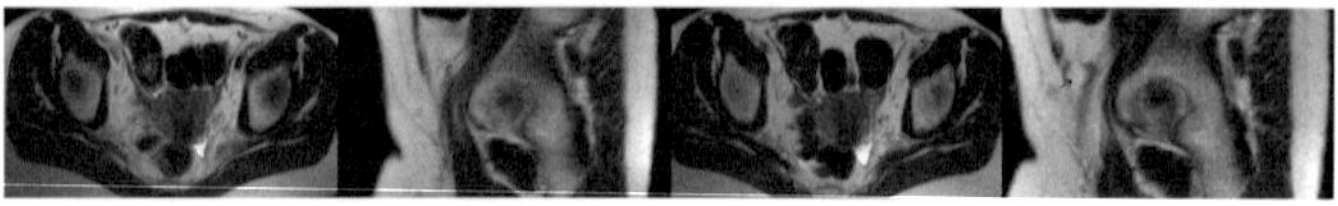

Fig. 13.22. Static axial and sagittal T2-weighted MR, and dynamic axial and sagittal T2-weighted MR of occult femoral hernia not detected on CT or US. Valsalva maneuver forces a small amount of fat (*black arrows*) through the femoral triangle on the left.

pain, CT was found to be substantially weaker than MR in the evaluation of occult hernia, with MR correctly identifying 91 % of the hernias overlooked by CT (Fig. 13.22) [27].

The recommendation is that MR be considered the definitive nonoperative test for inguinal hernia in patients with history strongly suggestive of such pathology but lacking clear physical findings. US and CT still have utility in the evaluation of groin pain due to occult inguinal hernia; however, in light of negative US or CT result, further diagnostic workup with MR is indicated as symptoms persist [27].

References

1. American College of Radiology. ACR appropriateness criteria radiation dose assessment introduction [Internet]. 2007. Last review date Feb 2015. http://www.acr.org/~/media/ACR/Documents/AppCriteria/RadiationDoseAssessmentIntro.pdf. Accessed 9 Mar 2015.
2. National Council on Radiation Protection and Measurements. Ionizing radiation exposure of the population of the United States [Internet]. NCRP report no. 160; 2009. http://www.ncrponline.org/Publications/Press_Releases/160press.html. Accessed 9 Mar 2015.
3. McDonald RJ, McDonald JS, Bida JP, Carter RE, Fleming CJ, Misra S, et al. Intravenous contrast material-induced nephropathy: causal or coincident phenomenon? Radiology. 2013;267(1):106–18.
4. Davenport MS, Khalatbari S, Dillman JR, Cohan RH, Caoili EM, Ellis JH. Contrast material-induced nephrotoxicity and intravenous low-osmolality iodinated contrast material. Radiology. 2013;267(1):94–105.
5. McDonald JS, McDonald RJ, Carter RE, Katzberg RW, Kallmes DF, Williamson EE. Risk of intravenous contrast material-mediated acute kidney injury: a propensity score-matched study stratified by baseline-estimated glomerular filtration rate. Radiology. 2014;271(1):65–73.
6. American College of Radiology. ACR manual of contrast media v9 [Internet]. 2013. http://www.acr.org/~/media/ACR/Documents/PDF/QualitySafety/Resources/Contrast%20Manual/2013_Contrast_Media.pdf. Accessed 9 Mar 2015.

7. Brockow K. Contrast media hypersensitivity—scope of the problem. Toxicology. 2005;209(2):189–92.
8. Wysowski DK, Nourjah P. Deaths attributed to X-ray contrast media on U.S. death certificates. AJR Am J Roentgenol. 2006;186(3):613–5.
9. Pasternak JJ, Williamson EE. Clinical pharmacology, uses, and adverse reactions of iodinated contrast agents: a primer for the non-radiologist. Mayo Clin Proc. 2012;87(4):390–402.
10. Greenberger PA, Patterson R. The prevention of immediate generalized reactions to radiocontrast media in high-risk patients. J Allergy Clin Immunol. 1991;87(4):867–72.
11. Kim MH, Lee SY, Lee SE, Yang MS, Jung JW, Park CM, et al. Anaphylaxis to iodinated contrast media: clinical characteristics related with development of anaphylactic shock. PLoS One. 2014;9(6):e100154.
12. Collidge TA, Thomson PC, Mark PB, Traynor JP, Jardine AG, Morris S, et al. Gadolinium-enhanced MR imaging and nephrogenic systemic fibrosis: retrospective study of a renal replacement therapy cohort. Radiology. 2007;245(1):168–75.
13. Abraham JL, Thakral C, Skov L, Rossen K, Marckmann P. Dermal inorganic gadolinium concentrations: evidence for in vivo transmetallation and long-term persistence in nephrogenic systemic fibrosis. Br J Dermatol. 2008;158(2):273–80.
14. Expert Panel on MR Safety: Kanal E, Barkovich AJ, Bell C, Borgstede JP, Bradley WG Jr, Froelich JW, et al. ACR guidance document on MR safe practices: 2013. J Magn Reson Imaging. 2013;37(3):501–30.
15. Levine GN, Gomes AS, Arai AE, Bluemke DA, Flamm SD, Kanal E, et al., American Heart Association Committee on Diagnostic and Interventional Cardiac Catheterization, American Heart Association Council on Clinical Cardiology, American Heart Association Council on Cardiovascular Radiology and Intervention. Safety of magnetic resonance imaging in patients with cardiovascular devices: an American Heart Association scientific statement from the Committee on Diagnostic and Interventional Cardiac Catheterization, Council on Clinical Cardiology, and the Council on Cardiovascular Radiology and Intervention: endorsed by the American College of Cardiology Foundation, the North American Society for Cardiac Imaging, and the Society for Cardiovascular Magnetic Resonance. Circulation. 2007;116(24):2878–91.
16. Gimbel JR. Unexpected asystole during 3T magnetic resonance imaging of a pacemaker-dependent patient with a 'modern' pacemaker. Europace. 2009;11(9):1241–2.
17. Konings MK, Bartels LW, Smits HF, Bakker CJ. Heating around intravascular guidewires by resonating RF waves. J Magn Reson Imaging. 2000;12(1):79–85.
18. Hartnell GG, Spence L, Hughes LA, Cohen MC, Saouaf R, Buff B. Safety of MR imaging in patients who have retained metallic materials after cardiac surgery. AJR Am J Roentgenol. 1997;168(5):1157–9.
19. Sheiner E, Abramowicz JS. A symposium on obstetrical ultrasound: is all this safe for the fetus? Clin Obstet Gynecol. 2012;55(1):188–98.
20. American College of Radiology. ACR–SPR–SRU practice parameter for performing and interpreting diagnostic ultrasound examinations [Internet]. Amended 2014. http://www.acr.org/~/media/ACR/Documents/PGTS/guidelines/US_Performing_Interpreting.pdf. Accessed 9 Mar 2015.

21. American College of Radiology. ACR–ACOG–AIUM–SRU practice parameter for the performance of obstetrical ultrasound [Internet]. Amended 2014. http://www.acr.org/~/media/ACR/Documents/PGTS/guidelines/US_Obstetrical.pdf. Accessed 9 Mar 2015.
22. Towfigh S. Inguinal hernia. In: Cameron JL, Cameron AM, editors. Current surgical therapy. 11th ed. Philadelphia, PA: Elsevier Saunders; 2013. p. 531–5.
23. Lilly MC, Arregui ME. Ultrasound of the inguinal floor for evaluation of hernias. Surg Endosc. 2002;16(4):659–62.
24. Bradley M, Morgan J, Pentlow B, Roe A. The positive predictive value of diagnostic ultrasound for occult herniae. Ann R Coll Surg Engl. 2006;88(2):165–7.
25. Robinson A, Light D, Nice C. Meta-analysis of sonography in the diagnosis of inguinal hernias. J Ultrasound Med. 2013;32(2):339–46.
26. Garvey JF. Computed tomography scan diagnosis of occult groin hernia. Hernia. 2012;16(3):307–14.
27. Miller JM, Cho J, Michael MJ, Saouaf R, Towfigh S. Role of imaging in the diagnosis of occult hernias. JAMA Surg. 2014;149(10):1077–80.

14. Perioperative Pain Management: Multi-modalities to Prevent Postoperative Chronic Pain

Brian J. Dunkin

Introduction

Inguinal hernia surgery is the most common operation done by a general surgeon with approximately 770,000 repairs performed in 2003. Among the most feared complications of this common surgery is the chronic pain that occurs in 11 % of patients, one-third of whom report limitations in daily leisure activities [1]. One component of this problem may be inadequate control of acute pain [2]. As a result, it is important that surgeons employ excellent pain management strategies for their hernia patients not only to ensure a good perioperative experience but also to avoid long-term problems.

This chapter describes the concept of multimodal pain therapy and provide examples of medications that can be used in this approach. It will also provide recommendations for pain management in the pre-, intra-, and postoperative periods.

Multimodal Pain Therapy

In order to understand how to use multiple modalities to treat postoperative pain, we must first understand how surgical pain is perceived. The process begins when noxious stimuli activate specialized nerve cells (nociceptors) at the site of surgery. Nociceptive pain is so intense that it elicits an autonomic response resulting in a withdrawal reflex. There are four types of nociceptive pain (heat, cold, intense mechanical force, chemical irritants), but in surgery, it is the intense mechanical force pathway that is activated (Fig. 14.1). The nociceptors transmit a signal along the nociceptive

B.P. Jacob et al. (eds.), *The SAGES Manual of Groin Pain*,
DOI 10.1007/978-3-319-21587-7_14

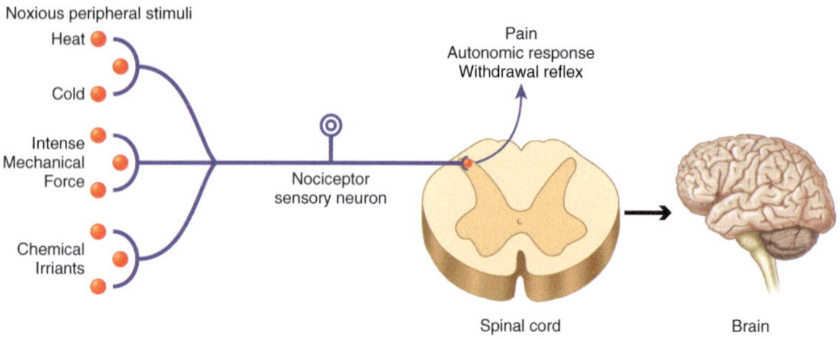

Fig. 14.1. Nociceptive pain pathway.

nerve fibers in the periphery to the dorsal horn of the spinal cord. The signal then continues up the spinal cord and is transmitted to multiple parts of the brain. There are also descending inhibitory or excitatory pathways that travel down the brain and back to the dorsal horn of the spinal cord to decrease or increase the pain signal via interneurons. Intense mechanical force during surgery causes tissue damage as well, which initiates an immune response that liberates inflammatory mediators. These mediators also activate pain receptors that transmit signals via the nociceptive nerve fibers. This inflammatory pathway is responsible for patients feeling pain beyond the duration of the surgical event and results in a hypersensitivity at the surgical site with allodynia (reduced pain threshold) and hyperalgesia (increased response to painful stimuli). Multimodal pain therapy uses medications and local anesthetics to block or modulate pain signals along the entire pain pathway (Fig. 14.2). This results in more effective therapy while minimizing the side effects of any one drug.

Multimodal Analgesics for Acute Pain Management

There are multiple analgesics that can be used to modulate the pain pathway. Beginning in the periphery and working toward the central nervous system, these include (Table 14.1):

Local Anesthetics

Local anesthetics are aminoamide or aminoester compounds that temporarily block the sodium channels in the nociceptive nerves, preventing conduction of the pain signal. They may be infiltrated into

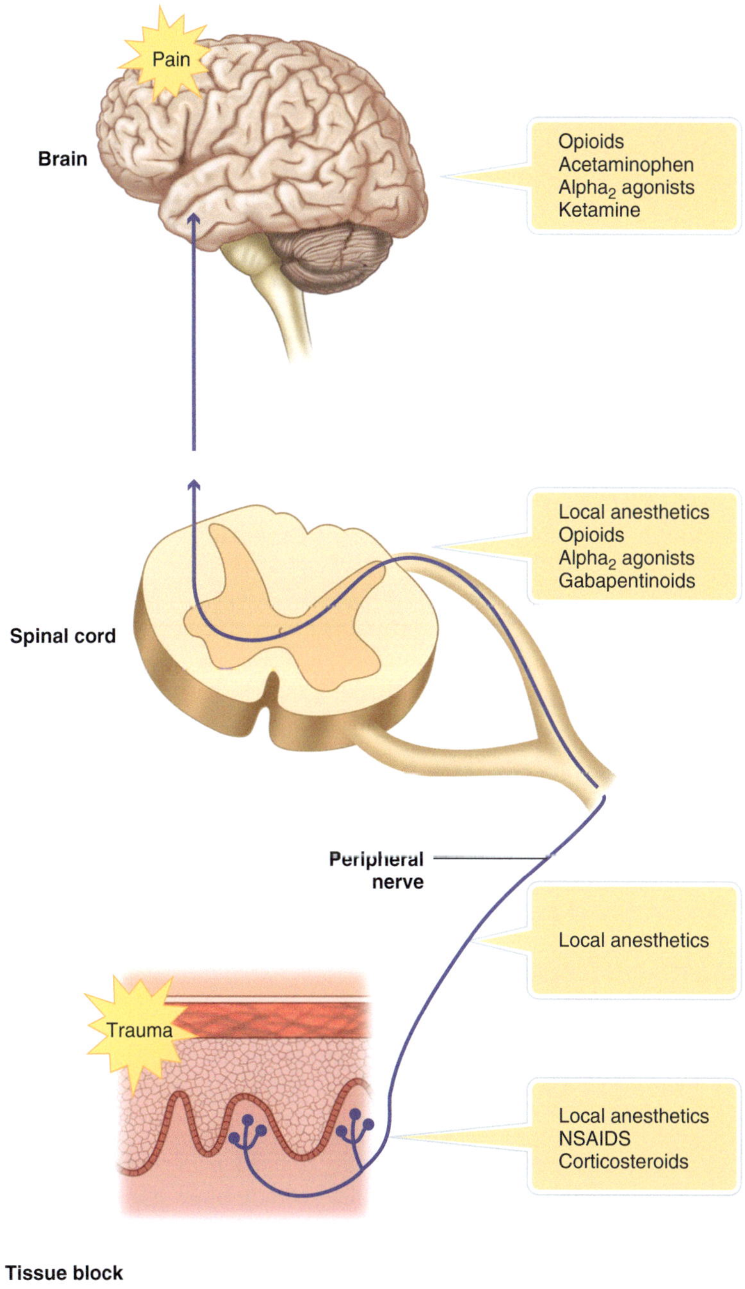

Fig. 14.2. Comprehensive view of pain pathway with multimodal points of modulation.

Table 14.1. Summary of analgesics for multimodal pain therapy.

Agent	Mode of Action	Dose	Safety
Local anesthetics	Nerve conduction blockade	Ropivacaine 0.2 % infiltration Bupivacaine 0.1–0.25 % infiltration Bupivacaine liposome injectable suspension (EXPAREL) 1.3 % infiltration	Neurotoxicity and cardiotoxicity
NSAIDs	COX-2 inhibition	IV: ketorolac 15–30 mg q6 h PO: celecoxib 200 mg BID	Wound site and GI bleeding[a], renal toxicity[b]
Gabapentinoids	α2δ ion channel blockade	Gabapentin 600 mg po TID Pregabalin 100 mg po BID	Sedation[c]
α-agonists (Clonidine)	Enhanced monoamine-mediated analgesia	Patch: 0.1 mg/24 h	Hypotension and bradycardia
Acetaminophen (APAP)	GABA inhibition, serotonergic interaction	PO: 650 mg q6 h IV: 1000 mg q6 h	Hepatotoxicity
Opioids	μ-receptor agonist	Varied depending on formulation	Sedation, hypotension, respiratory depression, nausea
Ketamine	Nonselective NMDA antagonism	1 mg/kg/h infusion	Hallucinations, confusion

[a]Ketorolac should not be administered preoperatively
[b]More selective COX-2 inhibitors like celecoxib are associated with lower incidence of these complications
[c]Adjust carefully in renal failure patients

tissue to block nerves locally, regionally to block pain perception from an area of the body, or around the spinal cord to block transmission of pain signals to the brain. They may also be short acting (lidocaine: 1–2 h duration), intermediate acting (bupivacaine: 3–6 h duration), or long acting (liposomal bupivacaine: 72 h duration). Local anesthetics may also be continuously pumped into the surgical site using an elastomeric or electrical pump in an effort to prolong the duration of their effect. For hernia surgery, local anesthetics may be applied using four different techniques: *inguinal nerve block* (discrete nerve block at the site of the ilioinguinal, iliohypogastric, and/or genitofemoral nerve); *field block* (infiltration into the superficial and deeper structures in the field of surgery, which may result in a block of the ilioinguinal, iliohypogastric, and/or genitofemoral nerve); *infiltration* (injection of local anesthetic into the cutaneous/subcutaneous/deeper structures of the surgical field); and *instillation* (local anesthetic application without needles (e.g., spray) into the surgical site).

Techniques for Administration of Local Anesthetics

Inguinal Nerve Block

The ilioinguinal and iliohypogastric nerves can be blocked by a directed administration of local anesthetic pre- or intraoperatively. Beginning two fingerbreadths superior and medial to the anterior superior iliac spine, a 1.5 inch needle is introduced deep into the abdominal wall, targeting the layers between the internal and external oblique fascia (Fig. 14.3). Depth is estimated by feeling for the "pop" of the needle across the external oblique fascial layer and 10 mL of long-acting local anesthetic (bupivacaine or liposomal bupivacaine) is infiltrated.

Transversus Abdominis Plane Block

This procedure is done under ultrasound or laparoscopic guidance and used to block T11–L1 cutaneous, myofascial, and peritoneal nerves, including the iliohypogastric and ilioinguinal nerves. It provides excellent pain relief for surgeries involving the infra-umbilical abdominal wall.

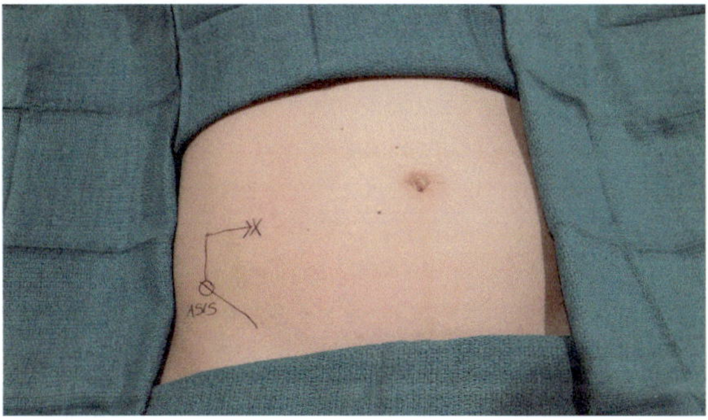

Fig. 14.3. Landmarks for inguinal nerve block.

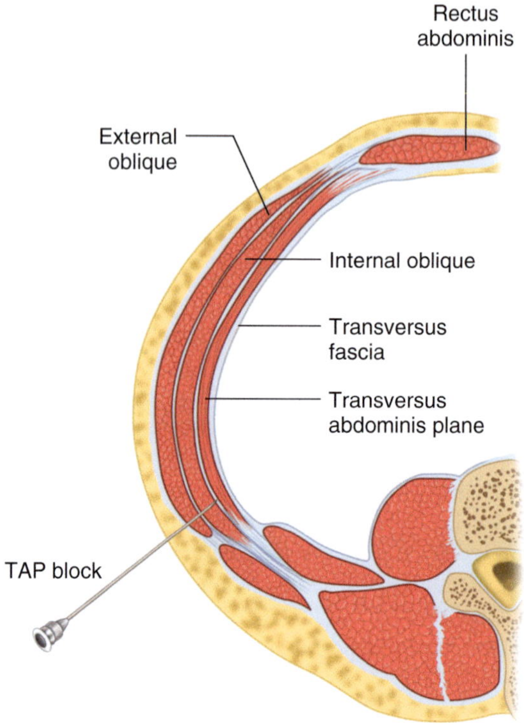

Fig. 14.4. Transversus abdominis plane (TAP) block.

Under ultrasound guidance, a blunt-tipped 22-gauge needle is inserted into the abdominal wall two fingerbreadths above the iliac crest along the mid-axillary line and advanced into the plane between the transversus abdominis and internal oblique muscles (Fig. 14.4). After aspirating to exclude a vascular puncture, 5 mL of local anesthetic is infiltrated while ultrasound imaging confirms separation of the two muscles and infiltration into the correct plane. Once confirmed, an additional 15 mL of anesthetic is infiltrated. The genital branch of the genital–femoral nerve runs on the posterior aspect of the spermatic cord and is most commonly identified where the cord crosses the pubic tubercle. Some surgeons infiltrate local anesthetic into the spermatic cord at this area in an attempt to block this nerve.

During laparoscopic surgery, the transversus abdominis plane (TAP) block can be done without ultrasound, using intra-abdominal visualization. The needle is passed using the same external landmarks and advanced under laparoscopic monitoring into the plane just above the transversus abdominis muscle. A test injection should demonstrate an obvious bulge without elevation of the peritoneum.

Field Block

Local anesthetic is infiltrated into the surgical field. This begins with infiltration of the dermal layer of the skin beyond the planned area of incision. Each layer encountered is then infiltrated with additional anesthetic. The subcutaneous layer is infiltrated using a fanning technique extending both superiorly and inferiorly from the wound. The external oblique aponeurosis, internal oblique fascia, and transversus abdominis fascia are also injected.

Instillation

Local anesthetic may be applied to the surgical area by simple instillation, without the use of a needle. The surgeon essentially bathes the field in local anesthetic. This is done at the time of closure in open inguinal surgery and prior to port removal for TEP or just after peritoneal closure in TAPP (needle puncture of peritoneum required).

Nonsteroidal Anti-Inflammatory Drugs

Nonsteroidal Anti-Inflammatory Drugs (NSAIDS) play a key role in multimodal analgesia as adjuncts or alternatives to opioids. They modulate the inflammatory pain pathway by blocking cyclo-oxygenase (COX) enzymes. When cells die or are damaged, arachidonic acid is released from the cell walls and acted upon by one of two COX enzymes. The COX-1 enzyme is normally expressed in the body and is considered a "homeostatic" isoform that converts arachidonic acid to prostaglandins that are used to promote platelet aggregation, renal blood flow, and gastric mucosal protection. The COX-2 enzyme is minimally expressed normally, but greatly upregulated following trauma or surgery. It promotes conversion of arachidonic acid to prostaglandin E_2 (PGE_2), which activates nociceptors and initiates the inflammatory cascade. NSAIDs are classified based on their effect on COX-1 versus COX-2. For pain control, it is ideal to have selective COX-2 inhibition without interfering with COX-1, which can increase the risk of bleeding, gastric ulceration, or renal damage. Figure 14.5 details the COX-2 versus COX-1 inhibition of common NSAIDS used in the USA.

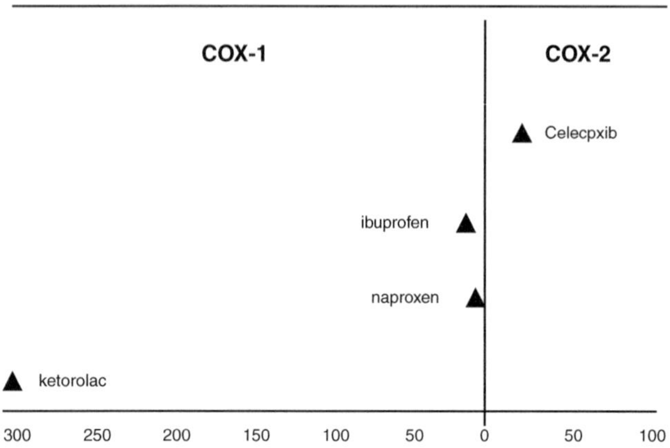

Fig. 14.5. Inhibition of COX-2 relative to COX-1.

Corticosteroids

Corticosteroids are potent anti-inflammatory drugs that can reduce the intensity of postsurgical pain. They stabilize lysosomal membranes in injured cells, decreasing the release of arachidonic acid, which in turn reduces the production of prostaglandins and leukotrienes. The clinical analgesic effect of corticosteroids is delayed in comparison to NSAIDs, taking up to 4 h. However, the duration of analgesia provided by a single dose can last up to 3 days [3]. In low doses, corticosteroids are also potent antiemetics.

Gabapentinoids

Gabapentin and pregabalin are anticonvulsant analgesics approved by the FDA for the treatment of neuropathic pain such as fibromyalgia or post-herpetic neuralgia. However, they are increasingly being used for acute surgical pain management. Both drugs have a high affinity for presynaptic calcium channels, which decreases the influx of calcium and decreases the release of excitatory neurotransmitters in the spinal and supraspinal pain pathways. They also decrease the excitability of peripheral nociceptive nerves.

Clonidine

Alpha-2 receptor agonists (e.g., clonidine) provide sedation, anxiolysis, and analgesia through central actions in the dorsal horn of the spinal cord and the brainstem. Coadministration of clonidine and an opioid produces more effective analgesia and reduction of adverse events than higher doses of either drug alone. The ability of clonidine to potentiate opioid-mediated analgesia is particularly useful in patients with opioid tolerance. In the USA, it is available in oral, transdermal, and epidural formulations. Administration of clonidine can cause alterations in a patient's hemodynamics, including an initial hypertensive phase followed by hypotention and bradycardia.

Acetaminophen

Acetaminophen (APAP) is a synthetic, centrally acting analgesic for mild to moderate pain. Contrary to popular belief, it does not act as an NSAID and has negligible COX inhibition peripherally. The exact mechanism of action is unknown, but it is believed to activate serotonergic descending inhibitory pain pathways. Oral APAP has been available in the USA since the 1950s. An IV formulation (Ofirmev) was approved by the FDA in 2010. Intravenous APAP achieves maximal plasma concentration more rapidly and predictably than oral or rectal, and the magnitude of plasma concentration is much higher. For this reason, it is often used in the perioperative setting. In clinical trials, IV APAP is superior to oral APAP and equivalent to 30 mg of IV ketorolac for moderate postsurgical pain [4]. The onset of analgesia for IV APAP is 5–10 min, with peak effect in 1 h. Duration of effect is 4–6 h. The effectiveness of IV APAP for postsurgical pain seems to be best when given preemptively before incision [5].

Opioids

Opioids are naturally occurring, semisynthetic, or synthetic compounds that produce analgesic effects by binding to opioid receptors in the central nervous system. They are classified as "strong" or "weak," depending on the strength of their clinical effect, which has historically been measured against the effect of morphine. Table 14.2 lists a

Table 14.2. Common opioids used in the USA.

Opioid	Clinical "strength"
Morphine	Strong
Oxycodone	Strong
Hydromorphone	Strong
Meperidine	Strong
Fentanyl	Strong
Methadone	Strong
Hydrocodone	Weak
Codeine	Weak
Tramadol	Weak

few common opioids used in the USA for surgical pain. It should be noted that the most common weak opioid prescribed in the USA is hydrocodone, which is available only in formulations that combine it with APAP. This must be kept in mind when using IV or oral APAP as part of a multimodal pain management strategy in order to avoid APAP overdose.

Ketamine

Ketamine is a nonopioid, centrally acting dissociative anesthetic. At subanesthetic doses, it provides rapid and highly potent analgesia without many of the adverse effects observed with opioids. It binds to and antagonizes the NMDA receptors in the central nervous system. Ketamine is particularly useful in opioid-tolerant patients. However, major complications can be associated with its use, including hyperdynamic cardiovascular responses and psychomimetic reactions.

Perioperative Pain Management Strategies

The choice of anesthetic technique as well as preoperative, intraoperative, and postoperative management of medications, affects postoperative pain. Proper management of all three phases of the perioperative experience will lead to a better surgical experience for the patient. The following is a summary of pain management strategies based on review of the literature by the authors of this chapter and by consensus recommendation by PROSPECT (PROcedure SPECific postoperative pain management consortium; http://www.postoppain.org/) and the Surgical Pain Consortium (http://surgicalpainconsortium.com/) [6, 7].

Preoperative Pain Management

Eighty-five percent of hernia surgery is done as outpatient. As a result, coordination of preoperative preparation of patients for surgery requires clear communication with the patient and all members of both the preoperative care unit and the operative team.

Recommended Preoperative Strategies

- Systemic analgesics
  - *COX-2-selective NSAIDs*. Preoperative administration of a COX-2 inhibitor (celecoxib or meloxicam in the USA) 1 h before incision is recommended for its analgesic efficacy and minimal effect on platelet aggregation. Use must include an assessment of individual patient risks (contraindicated in patients with cardiovascular morbidity, actual or recent gastroduodenal ulcer history, renal or hepatic dysfunction, or aspirin-sensitive asthma).
  - *Acetaminophen (APAP)*. APAP may be given orally or IV. Because of the rapid and more predictable achievement of plasma concentrations, the IV formulation is often preferred in the perioperative setting. IV APAP seems to have a better analgesic effect when given preemptively 30–60 min before incision. Because many hernia surgery patients receive a weak opioid combined with APAP orally (e.g., Lortab—hydrocodone plus APAP) for postoperative pain control, the timing and use of APAP perioperatively must be carefully considered to avoid overdosing.
- Local anesthetics
  - *Inguinal nerve block including TAP block/field block/infiltration*. Studies of local anesthetic administration for inguinal hernia repair do not clearly differentiate between inguinal nerve block versus field block. As a result, either can be used for anesthetic infiltration. Local anesthetics can be injected preoperatively or intraoperatively with the same control of early postoperative pain. Long-acting local anesthetics are recommended over short-acting.

Possible Preoperative Strategies in Select Patients

- *Gaba agonists.* Patients with a history of tolerance to opioids may be considered for the preoperative use of gabapentin or pregabalin.
  - Gabapentin 600–900 mg the night before surgery, followed by 600–900 mg TID for up to 72 h.
  - Pregabalin 75–150 mg the night before surgery, followed by 75–150 mg BID for up to 72 h.

Nonrecommended Preoperative Strategies

- *NSAIDs with significant COX-1 inhibition.* Ketorolac is an extreme outlier in its inhibitory effect on COX-1 relative to COX-2. As a result, the FDA warns against its use as a prophylactic analgesic prior to any major surgery. Because pre- and/or intraoperative infiltration of local anesthetics provides good analgesia in the immediate postoperative period, ketorolac should be administered immediately at the end of the procedure when the risk of postoperative bleeding can be more accurately assessed by the surgeon.
- *Local anesthetic plus epinephrine.* Adding epinephrine to the local anesthetic does not significantly decrease the risk of toxicity in the dosages associated with inguinal hernia repair and does not improve its analgesic effect. It may, however, result in undesirable cardiovascular side effects.

Intraoperative Pain Management

Recommended Intraoperative Strategies

- Anesthetic technique
  - *Local anesthesia* (inguinal nerve block or TAP block/field block/infiltration techniques), with or without intravenous sedation, is recommended because it is associated with less postoperative pain and provides additional recovery benefits (earlier ambulation, less urinary retention) compared with spinal, epidural, or general anesthesia. Intraoperative use of local anesthetic injection techniques post-incision is as effective as preoperative administration.

  - *General anesthesia*. If general anesthesia is used, it should be combined with local anesthetic techniques to reduce postoperative pain.
- Systemic analgesia
  - *Ketorolac*. If a COX-2 inhibitor is not used preoperatively, then IV ketorolac may be administered intraoperatively at the end of the procedure when the risk of postoperative bleeding has been assessed to be low by the surgeon.
  - *Acetaminophen (APAP)*. If APAP was not administered preoperatively, then it may be given intravenously intraoperatively. Because of the same concerns listed for APAP in the preoperative state, the timing and use of APAP perioperatively must be carefully considered to avoid APAP overdosing. In addition, because IV APAP costs more than ketorolac, many centers prefer the latter if there is no contraindication to use of a NSAID for the patient.

Nonrecommended Intraoperative Strategies

- Anesthetic technique
  - *Spinal anesthesia*. While this technique provides good early postoperative analgesia, it is associated with factors that can delay discharge, including delay in early ambulation, hypotention, and urinary retention.
  - *Epidural anesthesia*. Similar issues to spinal anesthesia with more technical challenges to administer.
  - *Paravertebral nerve block*. Has only marginal analgesic benefit over local anesthetic techniques and is more complex to perform.
  - *Local anesthetic plus epinephrine*. Adding epinephrine to the local anesthetic does not significantly decrease the risk of toxicity in the dosages associated with inguinal hernia repair and does not improve its analgesic effect. It may, however, result in undesirable cardiovascular side effects.
  - *Local anesthetic instillation*. Despite some evidence that this technique decreases postoperative pain, the data are currently limited.
  - *Pre-peritoneal instillation of local anesthetic (laparoscopic hernia repair)*. Current literature does not demonstrate efficacy of this technique.

Postoperative Pain Management

Recommended Postoperative Strategies

- *NSAIDS.* Conventional or COX-2-selective NSAIDs should be used in the postoperative setting. If they have not been administered pre- or intraoperatively, then they should be started as early in the postoperative period as possible. Because of the associated risks of bleeding and gastroduodenal ulcer with conventional NSAIDs, many surgeons prefer using a COX-2-selective formulation. NSAIDS are best taken on a scheduled basis, not as needed, and may be continued for 1–2 weeks postoperatively.
- *Acetaminophen (APAP).* APAP is recommended for routine pain therapy in combination with conventional NSAIDs/COX-2-selective inhibitors or weak opioids. It may be taken on a scheduled basis if the patient is not taking any other APAP formulation, or as needed if included in another formulation such as a weak opioid (e.g., hydrocodone plus APAP).
- *Weak opioids.* Weak opioids are recommended when conventional NSAIDs or COX-2-selective inhibitors plus APAP are not sufficient or contraindicated. The most commonly prescribed weak opioid in the USA is hydrocodone, where it is only available in formulations that combine it with APAP. This must be taken into consideration when using this weak opioid in combination with IV or oral APAP.
- *Strong opioids.* Strong opioids are recommended as rescue analgesia for severe pain in addition to the use of nonopioid agents. They are not recommended for first-line analgesia because of side effects that may delay early ambulation.

Nonrecommended Postoperative Strategies

- *Continuous wound infusion with local anesthetic.* Continuous infusion of an intermediate-acting local anesthetic using either an elastomeric or electric pump has shown longer duration of postoperative analgesia compared to infiltration alone. However, in hernia surgery, it is unclear as to whether the infusion should be in the subcutaneous or subfascial space and whether use of this system is superior to TAP block or nerve/field block using long-acting liposomal bupivacaine. The expense and inconvenience of the system are also a barrier to use.

- *Bolus wound infusion of local anesthetic*. Some surgeons have placed a catheter into the would during closure of the hernia surgery site to enable a single bolus injection of local anesthetic postoperatively. This has not been shown to be effective.
- *Transcutaneous electrical nerve stimulation (TENS)*. This technique involves placing electrodes on either side of the hernia incision following the operation and stimulating with a relatively low pulse amplitude in the recovery room. After discharge, the pulse amplitude is adjusted to deliver the maximum electrical current that is comfortably tolerated. Study of TENS units for open inguinal hernia surgery has not demonstrated efficacy.

Not Routinely Recommended Pre-, Intra-, or Postoperatively

- *Corticosteroids*. Not recommended due to limited procedure-specific evidence and because herniorraphy per se is not associated with a high incidence of postoperative nausea and vomiting.
- *Clonidine*. Not recommended because there is no procedure-specific evidence, and there are potential side effects, including hypotension, sedation, dizziness, and bradycardia. It may delay early ambulation.
- *Ketamine*. Not recommended due to associated side effects that may hinder early ambulation, despite some evidence of analgesic efficacy in other procedures. Ketamine is associated with a risk of adverse effects on the central nervous system.
- *Magnesium*. Magnesium is an antagonist of N-methyl-d-aspartate glutamate receptors, which can alter the perception and duration of pain. As a result, its use has been studied for managing postoperative pain. Currently, there is no evidence that magnesium administered preemptively before open hernia incision and in conjunction with use of NSAIDs plus intraoperative nerve block decreased postoperative pain [8].
- *Gabapentin/Pregabalin*. Not recommended due to the lack of procedure-specific evidence, despite analgesic efficacy in other procedures.

Summary

Understanding how to employ multimodal pain management strategies for hernia surgery patients will help to ensure a good perioperative experience and may decrease the potential for chronic pain long term. Use of these strategies is an imperative for responsible surgeons performing these common operations.

References

1. Nienhuijs S, Stall E, Strobbe L, Rosman C, Groenewoud H, Bleichrodt R. Chronic pain after mesh repair of inguinal hernia: a systematic review. Am J Surg. 2007;194(3):394–400.
2. Aasvang EK, Gmaehle E, Hansen JB, Gmaehle B, Forman JL, Schwarz J, et al. Predictive risk factors for persistent postherniotomy pain. Anesthesiology. 2010;112(4):957–69.
3. Raeder J, Dahl V. Clinical application of glucocorticoids, antineuropathics, and other analgesic adjuvants for acute pain management. In: Sinatra R, de Leon-Casasola O, Ginsberg B, Viscusi E, editors. Acute pain management. New York: Cambridge University Press; 2009. p. 377–90.
4. Gorocs TS, Lambert M, Rinne T, Krekler M, Modell S. Efficacy and tolerability of ready-to-use intravenous paracetamol solution as monotherapy or as an adjunct analgesic therapy for postoperative pain in patients undergoing elective ambulatory surgery: open, prospective study. Int J Clin Pract. 2009;63(1):112–20.
5. Arici S, Gurbet A, Türker G, Yavaşcaoğlu B, Sahin S. Pre-emptive analgesic effects of intravenous paracetamol in total abdominal hysterectomy. Agri. 2009;21(2):54–61.
6. Joshi GP, Rawal N, Kehler H, PROSPECT collaboration, Bonnet F, Camu F, Fischer HB, et al. Evidence-based management of postoperative pain in adults undergoing open inguinal hernia surgery. Br J Surg. 2012;99(2):168–85.
7. Sinatra RS, Larach S, Ramamoorthy S. Surgeon's guide to postsurgical pain management: colorectal and abdominal surgery. West Islip, NY: Professional Communications; 2012.
8. Tramer MR, Glynn CJ. An evaluation of a single dose of magnesium to supplement analgesia after ambulatory surgery: randomized controlled trial. Anesth Analg. 2007;104(6):1374–9.

Part II
Secondary Groin Pain

15. Chronic Groin Pain Following Anterior Hernia Surgery

Jennifer S. Schwartz, David S. Strosberg, and David B. Renton

Introduction

Inguinal herniorrhaphy is one of the most common general surgery operations performed in the United States at nearly 600,000 repairs annually. An anterior approach is the most common method for surgical repair, and may be performed as either a tissue repair or tension-free repair [1].

Tissue repairs were the first type of repair for inguinal hernias. Since the creation of the Bassini repair in 1887, at least 70 tissue repairs have been described in the literature. This type of repair uses the patient's native tissues to close the hernia defect. Types of tissue repairs include the Shouldice, Bassini, and McVay repairs. The Shouldice repair is based on a multilayer imbricated repair of the posterior wall of the inguinal canal, and has the lowest recurrence of tissue-based repairs in highly selected patient populations. In a Cochrane review, the rate of recurrence in specialized centers for a Shouldice repair is cited between 0.4 and 1.6 %; however, in nonspecialist centers recurrence is as high as 10 % [2]. The Bassini repair, the most popular type of repair prior to the introduction of tension-free repairs, involves suturing the transversus abdominis and internal oblique musculoaponeurotic arches to the inguinal ligament. The McVay repair, or Cooper's ligament repair, approximates the transversus abdominis aproneurosis to Cooper's ligament. This operation may also be used for femoral hernias, as the femoral space is closed with this repair. Tissue repairs are rarely used due to higher recurrence

B.P. Jacob et al. (eds.), *The SAGES Manual of Groin Pain*,
DOI 10.1007/978-3-319-21587-7_15

rates, cited as high as 4–6 % [3], and prolonged postoperative pain and recovery time. However, a tissue repair is useful when prosthetic mesh is contraindicated, including situations of ischemic bowel where resection is necessary, in the presence of ascites, or following a Cesarean section.

Mesh-Based Repairs

Tension-free, or mesh-based, repairs have been the gold standard for inguinal hernia repairs since the early 1990s due to the lower recurrence rate. Tension is eliminated with the placement of a synthetic mesh to bridge the defect, thereby reducing the rates of recurrence to less than 1 % compared to the 4–6 % recurrence rate with tissue repair [3]. Types of tension-free repairs include the Lichtenstein repair, plug and patch, and sandwich technique. The Lichtenstein repair encompasses the placement of a prosthetic mesh in the inguinal canal and re-creation of a new mesh internal inguinal ring. Of note, the ilioinguinal nerve and genital branch of the genitofemoral nerve pass through this newly created ring, and care must be taken to protect these nerves from entrapment during the repair. The plug and patch technique, an extension of the Lichtenstein repair, provides an additional cone-shaped plug of polypropylene mesh that is placed in the hernia defect, which occludes the hernia with Valsalva. This is currently the most common type of anterior herniorrhaphy performed. The sandwich technique utilizes an underlay patch, a plug type connector, and an onlay patch that covers the posterior inguinal floor (Fig. 15.1) [4].

Complications

Complications of inguinal herniorrhaphy are multifold. Intraoperative complications are noted at less than 2 %. Postoperative complications are as high as nearly 20 %, including urinary retention, urinary tract infection, orchitis, surgical site infection, neuralgia, or (rarely) life-threatening complications. Long-term complications are nearly 18 %, and may include seroma formation, chronic orchitis, chronic infection, chronic pain, or recurrence [5].

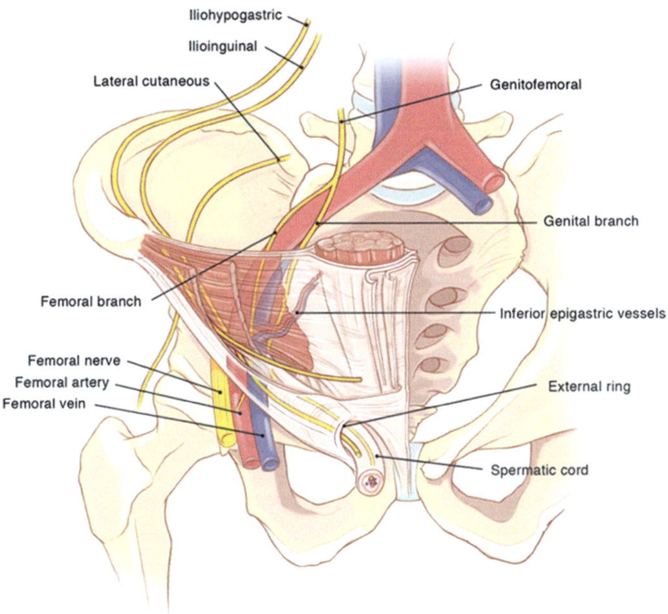

Fig. 15.1. Borders of the inguinal canal (from Wagner et al. [4], with kind permission © McGraw-Hill Education).

Chronic Post-herniorrhaphy Groin Pain: Definition

Chronic post-herniorrhaphy groin pain is defined as pain lasting >3 months following hernia repair. Studies show an incidence of chronic pain of 11 % [6]. Inguinodynia can be neuropathic or non-neuropathic. Neuropathic pain can be caused by nerve entrapment, stretching of nerves, and partial or complete division of nerves with neuroma formation. The three nerves most commonly involved are the iliohypogastric nerve, the ilioinguinal nerve, and the genital branch of the genitofemoral nerve. Non-neuropathic causes include periosteal reaction and mechanical pressure caused by folded mesh [7].

Treatments

Treatments for chronic groin pain include nonoperative interventions such as pain control with or without narcotic pain medications, and injection-based therapies such as nerve blocks and radiofrequency neurolysis. Additionally, operative intervention has been used for refractory chronic groin pain. Surgical treatments include single nerve resection, triple neurectomy, and mesh removal.

Nonsteroidal anti-inflammatory agents are often first-line therapy for chronic groin pain. NSAIDs are nonselective COX inhibitors that decrease inflammation. These are particularly helpful in the postoperative period. Scheduled NSAIDs for 2 weeks improved pain symptoms in 25 % of patients with chronic groin pain after inguinal hernia repair in one study [7]. For moderate to severe chronic groin pain, opioid analgesia may be required. Treatment with this regimen is recommended for a minority of patients and in conjunction with a pain specialist. Antidepressants are another first-line therapy used for chronic neuropathic pain. At low doses, antidepressants work by blocking neurotransmitter uptake at the presynaptic terminal and function as an analgesic. Tricyclic antidepressants followed by SSRIs are the most commonly used antidepressants. Antiepileptic drugs such as pregabalin, gabapentin, and topiramate have been used for neuropathic pain. The mechanism of action is by modulation of calcium and sodium channels that stabilize neurons involved in rapid firing, thus affecting the intensity of neuropathic pain [8]. However, there are few studies that investigate the efficacy of these treatment modalities for inguinodynia following hernia repair.

Injection-based therapies are another treatment alternative for chronic groin pain following inguinal hernia repair. Multiple studies have been performed looking at the effect of nerve blocks with local anesthetics, often under ultrasound guidance. In a single study, 43 subjects were evaluated for chronic moderate to severe inguinal pain status post open hernia repair. Each was given an ilioinguinal and/or iliohypogastric nerve block with a long-acting local anesthetic (bupivacaine) and a corticosteroid (triamcinolonacetonide). There was an average of two injections per subject. Post-procedure, 32–55 % of subjects reported resolution of their moderate to severe neuropathic pain after 20 months [9]. Other studies have demonstrated patients receiving at least temporary relief from nerve blocks with local anesthetic. Ilioinguinal or iliohypogastric nerve blocks can be an effective treatment modality for chronic groin pain, though multiple treatments may be required. Long-term success of injection-based therapies is still unclear. This treatment is often used for diagnosis of affected nerve and prior to surgical intervention.

Radiofrequency neurolysis (RFN) has become a more common procedure in interventional pain management used for chronic inguinodynia. It has been shown to have some longer lasting pain relief in patients with refractory inguinal neuralgia. Although evidence is limited, one small retrospective review evaluated 42 patients and compared radiofrequency ablation to local infiltrative therapy. RFN showed longer lasting pain relief, with the mean duration of pain relief 12.5 months versus 1.6 months compared to an injection-based therapy control group. Patients required from 1 to 3 radiofrequency neurolysis procedures. Local nerve infiltration may be used to aid the identification of inguinal neuralgia and which nerve is affected prior to RFN treatments. The use of this therapy has been limited secondary to the need for extensive knowledge of the inguinal anatomy and expertise in the technology required to perform this procedure, found predominantly in pain management specialists and radiologists (Fig. 15.2) [4, 10, 11].

Surgical interventions have been reserved for patients with severe chronic inguinal neuropathic pain who have been refractory to nonoperative management. Identification of the involved nerve is often performed using local infiltration. There is no gold standard operative intervention for chronic inguinal pain. Mesh and suture removal, resection of a single nerve, and triple neurectomy have all been proposed as potential treat-

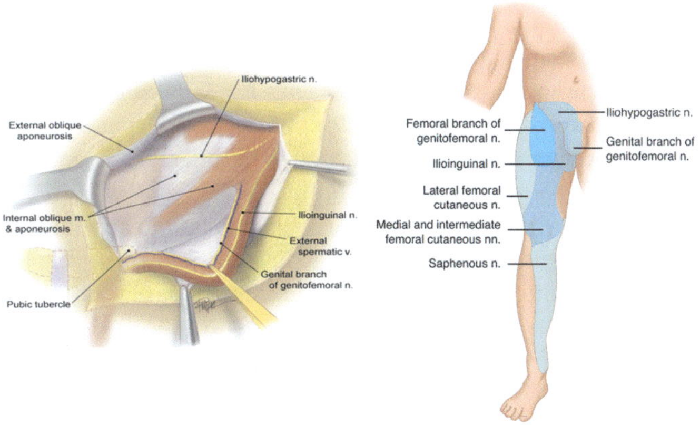

Fig. 15.2. *Left* Retroperitoneal neuroanatomy (from Chen et al. [11] with kind permission Springer Science+Business Media); *Right* corresponding dermatomal sensory distribution of the lumbar plexus (from Wagner et al. [4], with kind permission © McGraw-Hill Education).

ments. Some studies even advocate routine preventive resection of the ilioinguinal nerve at the time of the original hernia repair.

Mesh and suture removal has been proposed as an option, alone or in combination with neurectomy. Reoperative surgery and mesh removal can be very challenging. Identification of the ilioinguinal, iliohypogastric, and genitofemoral nerve and removal of any sutures through the nerve or freeing it from entrapment of mesh can completely or partially relieve chronic neuropathic pain. Additionally, if the nerve can be identified preoperatively, it can be resected and ligated with or without mesh removal. It is essential to ligate the nerve in order to prevent neuroma formation. Although mesh removal can be effective for pain, it causes a high hernia recurrence rate [12].

Triple Neurectomy

Triple neurectomy has become a promising surgical technique for chronic inguinal neuropathic pain after inguinal hernia repair. It involves ligation of the ilioinguinal, iliohypogastric, and genitofemoral nerve. The incision is made through the previous hernia repair, and the external oblique aponeurosis is divided. First, the ilioinguinal nerve is identified between the lateral border of the prosthetic mesh and the anterior superior iliac spine. It may easily be hidden if attached to the inguinal ligament, upper external oblique aponeurosis, within the fat-filled grooves of the internal oblique muscle, or simply under the retractor. The nerve is sharply transected, and the proximal end is buried within the internal oblique muscle to prevent future scarring. Next, the iliohypogastric is identified between the external and internal oblique aponeurosis. The intramuscular segment is followed lateral to the internal ring and divided proximal to the surgical field of the original hernia repair. The iliohypogastric nerve is the most vulnerable to injury due to the inability to visualize it during the hernia repair. The inguinal segment of the genital branch of the genitofemoral nerve can be identified by entering the internal ring through its inferior crus. After transection, the proximal ligated cut end is allowed to retract into the preperitoneal space [13]. In a study of 415 patients, 85 % had complete resolution of pain, with the remaining 15 % having significant improvement of pain after the triple neurectomy performed [14]. Other studies, although smaller, show success rates of this procedure of 80–95 %. Triple neurectomy appears to be a

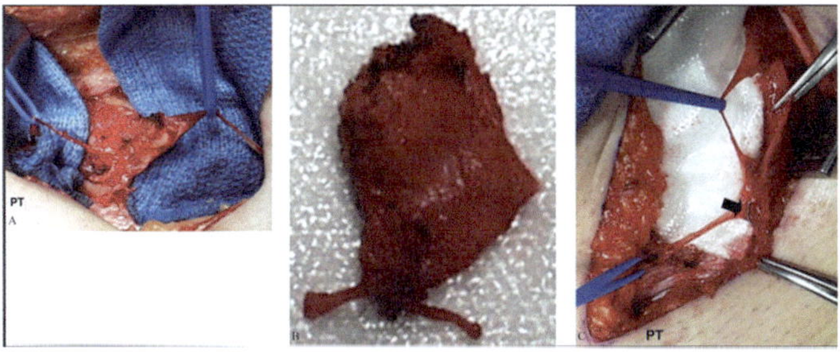

Fig. 15.3. Injuries to the IM segment of the iliohypogastric nerve, left groin, after inguinal hernia repair. (**a**). Nerve entrapped during a tissue repair. (**b**). Nerve sutured to mesh plug. (**c**). Nerve trapped by staple (*arrow*) adherent to upper edge of mesh patch held within forceps (Lichtenstein repair). PT, pubic tubercle (from Amid and Hiatt [14]).

very effective treatment for chronic neuropathic pain after inguinal hernia repair, although this procedure can be quite technically challenging (Fig. 15.3) [14].

Laparoscopic triple neurectomy is performed using a retroperitoneal approach, as described by Santos and Towfigh at Cedars Sinai Medical Center. It is typically performed for patients with inguinodynia following laparoscopic inguinal hernia repair or open posterior inguinal hernia repair. The patient is positioned supine if bilateral neurectomy is performed, or in the lateral decubitus position if unilateral. The ports are placed in the same fashion as a laparoscopic adrenalectomy, with a supraumbilical Hasson and two to three subcostal ports. The retroperitoneum is accessed following detachment of the colon at the white line of Toldt. Once accessed, the 12th rib is identified superiorly, femoral nerve inferiorly, iliac crest laterally, and ureter and medial half of the psoas muscle medially. The iliohypogastric and ilioinguinal nerves arise from the posterolateral border of the psoas muscle caudal to the 12th rib. Care must be taken not to mistake the 12th intercostal nerve for the iliohypogastric nerve, or the lateral femoral cutaneous nerve for the ilioinguinal nerve. The genitofemoral nerve exists from the mid-psoas muscle and branches distally, with the ureter lateral. The nerves are transected at their exit from the psoas, and proximal ends implanted into the muscle, while the distal end is cut 5 cm distally to prevent communication [15].

Conclusion

While treatment modalities vary widely and include medication, injection-based therapy, radiofrequency ablation, and surgical intervention, the most effective treatment for chronic neuropathic pain is prevention. Meticulous identification of all three nerves with careful preservation is essential in preventing the development of chronic pain following inguinal hernia repair.

References

1. Malangoni MA, Rosen MJ. Hernia. In: Townsend Jr CM, Beauchamp RD, Evers BM, Mattox KL, editors. Sabiston textbook of surgery: the biological basis of modern surgical practice. 19th ed. Philadelphia, PA: Elsevier Saunders; 2012. p. 1114–40.
2. Amato B, Moja L, Panico S, Persico G, Rispoli C, Rocco N, Moschetti I. Shouldice technique versus other open techniques for inguinal hernia repair. Cochrane Database Syst Rev. 2012;4:CD001543.
3. Amid P. Groin hernia repair: open techniques. World J Surg. 2005;29(8): 1046–51.
4. Wagner JP, Brunicardi FC, Amid PK, Chen DC. Inguinal hernias. In: Brunicardi FC, Andersen DK, Billiar TR, Dunn DL, Hunter JG, Matthews JB, Pollock RE, editors. Schwartz's principles of surgery. 10th ed. New York, NY: McGraw Hill Medical; 2014. p. 1495–521.
5. Neumayer L, Giobbie-Hurder A, Jonassen O, Fitzgibbons Jr R, Dunlop D, Gibbs J, et al, Veterans Affairs Cooperative Studies Program 456 Investigators. Open mesh versus laparoscopic mesh repair of inguinal hernias. N Engl J Med. 2004;350(18):1819–27.
6. Nienhuijs S, Staal E, Strobbe L, Rosman C, Groenewoud H, Bleichrodt R. Chronic pain after mesh repair of inguinal hernia: a systematic review. Am J Surg. 2007;194(3):394–400.
7. Palumbo P, Minicucci A, Nasti AG, Simonelli I, Vietri F, Angelici AM. Treatment for persistent chronic neuralgia after inguinal hernioplasty. Hernia. 2007;11(6):527–31.
8. Hansen MB, Andersen KG, Crawford ME. Pain following the repair of an abdominal hernia. Surg Today. 2010;40(1):8–21.
9. Thomassen I, van Suijlekom JA, van de Gaag A, Ponten JE, Neinhuijs SW. Ultrasound-guided ilioinguinal/iliohypogastric nerve blocks for chronic pain after inguinal hernia repair. Hernia. 2013;17(3):329–32.
10. Kastler A, Aubry S, Piccand V, Hadjidekov G, Tiberghien F, Kastler B. Radiofrequency neurolysis versus local nerve infiltration in 42 patients with refractory chronic inguinal neuralgia. Pain Physician. 2012;15(3):237–44.
11. Chen DC, Amid PK. Technique: lichtenstein. In: Jacob BP, Ramshaw B, editors. The SAGES manual of hernia repair. New York, NY: Springer; 2013. p. 41–54.

12. Ferzli G, Edwards E, Khoury G. Chronic pain after inguinal herniorrhaphy. J Am Coll Surg. 2007;205(2):333–41.
13. Amid PK, Chen DC. Inguinal neurectomy for nerve entrapment: triple neurectomy. In: Jones DB, editor. Master techniques in surgery: Hernia. Philadelphia, PA: Lippincott Williams and Wilkins/Wolters Kluwer; 2013. p. 141–8.
14. Amid P, Hiatt J. New understanding of the causes and surgical treatment of postherniorrhaphy inguinodynia and orchalgia. J Am Coll Surg. 2007;205(2):381–5.
15. Santos D, Towfigh S. Laparoscopic retroperitoneal triple neurectomy: a new technique for persistent herniorraphy neuralgia (abstract). SAGES 2011, 30 Mar-2 Apr 2011, San Antonio, Texas. P336. abstract archive. http://www.sages.org/meetings/annual-meeting/abstracts-archive/laparoscopic-retroperitoneal-triple-neurectomy-a-new-technique-for-post-herniorraphy-neuralgia/. Accessed 5 Mar 2015.

16. Chronic Groin Pain Following Posterior Hernia Surgery

Edward L. Felix

Introduction

Today most posterior inguinal hernia repairs are performed laparoscopically, but the origin of this approach dates back to Annandale [1] in 1876 with the first published report of an open posterior approach to the hernia sac. The approach was slow to gather attention until Cheatle (1921) [2] and later Henry (1936) [3] suggested it might be used for both femoral and inguinal hernias. It was not until 1959, however, that Nyhus [4] began to popularize an open posterior primary repair of hernia defects. Mesh soon became a staple of the repair as patch reinforcement, but with further refinements by Rignault [5], Stoppa [6], and later Wantz [7], the patch was replaced by a large or sometimes giant mesh covering the entire posterior floor.

Using the principles of the open posterior approach, a laparoscopic approach was born in the early 1990s with the birth of advanced laparoscopic techniques [8–10]. Two distinct approaches were successfully developed, the transabdominal preperitoneal (TAPP) and the totally extraperitoneal (TEP). Both approaches utilize a large mesh covering all three potential defects. The techniques have been modified over time, and the use of fixation and mesh type continue to be debated and will be discussed later in the chapter.

Results of laparoscopic approaches have now been extensively studied in both retrospective and prospective randomized reports [11–14]. In the hands of experienced laparoscopic surgeons, recurrence rates are equal or lower than open anterior approaches, but long-term chronic pain is reduced by the laparoscopic approach when compared to anterior open approaches [15–17]. Whether (TAPP or TEP) approach is used does

B.P. Jacob et al. (eds.), *The SAGES Manual of Groin Pain*,
DOI 10.1007/978-3-319-21587-7_16

not seem to alter the incidence of chronic pain. As will be discussed, fixation and mesh type may influence results.

Although the predominant posterior approach today is laparoscopic, open posterior approaches are being performed [18–20]. The mechanisms by which these approaches cause chronic pain are in general similar to the laparoscopic approaches except in those cases where unique meshes are utilized that may have their own problems.

Anatomy

Understanding the anatomy of the groin is integral in understanding why patients develop chronic pain after posterior inguinal hernia repair. The location of the nerves of the groin that puts them at risk and the mechanisms by which they can be injured or irritated explain why chronic pain develops, how it can be treated, and how it can be prevented. One must first understand the conventional anatomy [21] (Fig. 16.1), and then accept that as many as 25 % of patients have some variation of the location of the named nerves [22] that puts them at jeopardy for injury during the procedure or from irritation after the procedure is completed.

The named nerves that are at risk for injury in the retroperitoneum from dissection during a posterior hernia repair or after the repair are the femoral nerve, genitofemoral nerve and its branches (femoral and genital), and the lateral cutaneous nerve. In most patients, the nerves run below the iliopubic tract. How to avoid these nerves has been well described by multiple authors [23–25], but, unfortunately, as previously mentioned, the location of the nerves is variable. In an excellent cadaver study [22], in as many as 25 % patients, the nerves are not out of harm's way. They run above the iliopubic tract where they can be injured by dissection and anchoring hardware such as staples or tacks. In addition, anterior nerves that should not be at jeopardy for injury, on rare occasion, present within reach of posterior fixation of mesh. Cases have been reported of tacks penetrating the entire wall, injuring a superficial nerve.

Increasing the chance for chronic pain are idiopathic reactions to mesh that can result in irritation of any of the posterior nerves, including the obturator nerve, although it is well inferior but can be exposed during placement of the mesh. If the mesh bunches up or wrinkles, it may become thickened and hardened, acting as a potential pressure point on any of the nerves. The result will be chronic pain aggravated by activity or motion.

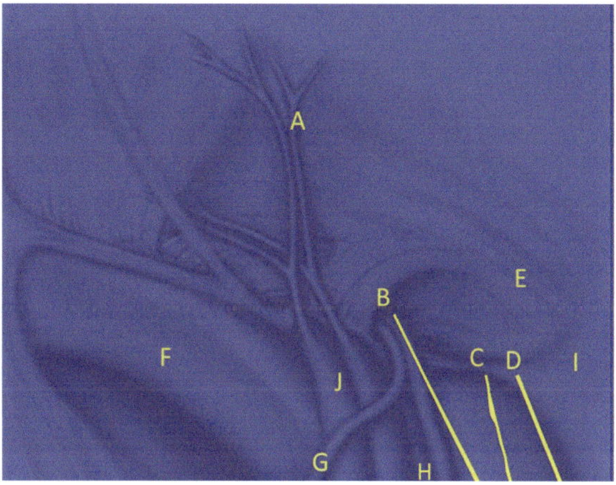

Fig. 16.1. The anatomy of posterior floor as viewed through a laparoscope. (**a**) Inferior epigastric vessels. (**b**) Genital branch of the genital–femoral nerve. (**c**) Genital–femoral nerve. (**d**) Lateral cutaneous nerve. (**e**) Indirect hernia defect. (**f**) Pubis with Cooper ligament. (**g**) Vas deferens. (**h**) Testicular vessels. (**i**) Iliopubic tract. (**j**) External iliac vessels.

To summarize, nerve injury and chronic pain can be reduced by careful dissection and by being aware of where nerve injuries are most likely to occur. When mesh is fixed with anchors or mesh hardens, however, chronic pain can develop even after taking the proper precautions. Despite the potential for the development of chronic pain after posterior hernia repair, the incidence is quite low and in fact less than that after open inguinal hernia repair in randomized studies [15, 16].

Differential Diagnosis of Chronic Pain Following Posterior Hernia Repair

Pain after posterior inguinal hernia repair can be divided into acute and chronic, lasting more than 90 days. The etiology and treatment are usually quite different, and therefore one must be able to differentiate the two scenarios. Acute pain occurs from minutes to days after the repair. Except for severe unrelenting pain in the distribution of the femoral nerve, most pain arising within days after the surgery will slowly decrease and disappear. It is usually due to minor irritation of the posterior nerves from the surgical dissection, hematoma, or seroma. Anti-inflammatory medications and watchful waiting are the treatments of choice. If, however, pain is debilitating and in the distribution of a named nerve, such as the femoral nerve, injury to that nerve from a tack or staple should be suspected. Immediate re-exploration and removal of the offending anchor are required. If pain persists more than 3 months, it can be classified as chronic. Fortunately, this is uncommon, occurring in less than 4 % [26, 27] of patients undergoing laparoscopic posterior hernia repair.

Determining the etiology of a patient's chronic pain after posterior repair is essential to guiding the treatment. The nature, timing, and location of the pain as well as inciting factors will help determine the cause and possible management. Several questions are important when determining the cause.

1. When did the pain begin? Immediately after surgery, persisting, essentially unchanged, or did it begin weeks or months after the repair?
2. Where is the pain located? Is it pinpoint or diffuse? Is it in the distribution of a single named nerve?
3. What is the character of the pain?
4. Is the pain elicited by any activity in particular?

Determining when and how the pain began helps determine the cause. If it develops soon after the operation, the cause was something that happened during the operation such as an injury of a nerve from a staple; if it develops months after the procedure, it is most likely a reaction to the mesh or even a recurrence of the hernia. If the offending agent is a staple, the pain usually begins soon after the repair and persists or worsens with time. Pain from a folded-up mesh begins much later and has a new onset.

The location of the pain guides the observer toward the proper nerve distribution. There is a fair amount of crossover, however, so location can be misleading. If the patient has pinpoint tenderness, it is usually in the location of the offending tack or staple used for fixation. This area should be marked prior to any exploration and will guide the surgeon at reoperation, as will be described later.

The type of pain may also help. Whether it is musculoskeletal or neuropathic. Finally, *questioning the patient about activities* that elicit or increase the pain may lead one to the correct location and precipitating event.

The workup should include a full history as just outlined and a complete exam of the area, including a directed palpation looking for recurrence and trigger points. Again, if a single point is found, it should marked prior to exploration. Ultrasounds, CT scans, and MRI have been reported to be helpful in looking for occult recurrence [28].

Treatment of Chronic Pain following Posterior Inguinal Hernia Repair

Unless postoperative pain is severe and occurring immediately after the procedure in the distribution of a major nerve as previously outlined, the patient should be observed and treated symptomatically. Because most postoperative pain will decrease and disappear over time, it is important that the surgeon explain that waiting 30–90 days is essential before more aggressive action is taken. The surgeon must explain that acting too early or invasively may do more harm to the patient than good. This time taken to talk with the patient to alleviate fears should prevent the patient from losing trust in the surgeon and prematurely seeking a second opinion. When the pain does not respond to conservative measures and becomes chronic, the next step should be an attempt at pain management with blocks by the surgeon or a pain management clinic. The exception would be when the pain can be easily localized and is most likely due to a single tack. Then laparoscopic exploration is indicated. The trigger site is marked just prior to laparoscopic exploration, and at laparoscopy the mark is palpated to guide removal of the corresponding tack or staple.

If conservative measures, medications, and therapeutic injections fail, laparoscopic re-exploration is the next step in treating chronic pain. If CT or ultrasound suggests a recurrence or other pathology as the cause of the pain, treatment is straightforward at the time of re-exploration. Because a positive preoperative study may be the exception, a complete laparoscopic exploration of the previous repair is necessary. The surgeon should approach the previous repair using a transabdominal laparoscopic approach (TAPP). The first step is to look for intraperitoneal pathology that might be causing the patient's pain (Fig. 16.2). The next step is to begin to open the peritoneum above the repair away from the mesh in a virgin area. If the mesh is flat and in the proper position, it should be left alone. Tacks or staples should be removed if they are loose or appear to correspond to the distribution of pain. If the surgeon finds

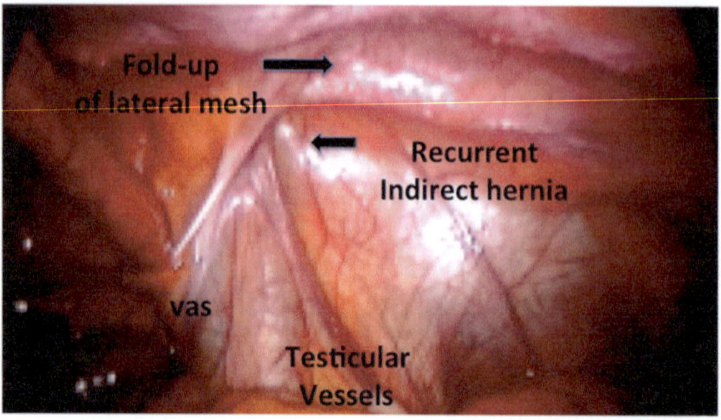

Fig. 16.2. Lateral fold-up of mesh with an indirect recurrent hernia in a patient with groin pain.

mesh rolled up or folded in on itself, it should be removed. Pressure on sensory nerves from hardened mesh can result in chronic groin pain, especially related to activity and change in position. Removal, however, can be difficult. In some cases, the careful use of cautery is required to cut the mesh away from the wall. A careful search for recurrence of the original hernia or a missed hernia completes the exploration (Fig. 16.3). It is not unusual to find a defect that was missed by preoperative radiographic studies. If a recurrence is found to be the cause, it should be repaired with the addition of new mesh. If recurrence is associated with bunched-up, hardened mesh (Fig. 16.4), the surgeon should consider removing the mesh and delaying repair of the recurrence. A recurrent hernia can be repaired from an anterior approach after the patient recovers from the laparoscopic exploration. This staged approach better defines the etiology of the pain and leads to fewer long-term problems and less confusion.

"When should all of the mesh be removed?" is a frequently asked question. There are those who believe it is best to remove all of the mesh [29]. This has not been our approach. Unless the mesh has hardened into a rock-like mass that is potentially irritating or compressing a nerve, it has been our approach to leave it in place. The risk of damaging surrounding structures by unnecessarily removing mesh far outweighs any potential benefit.

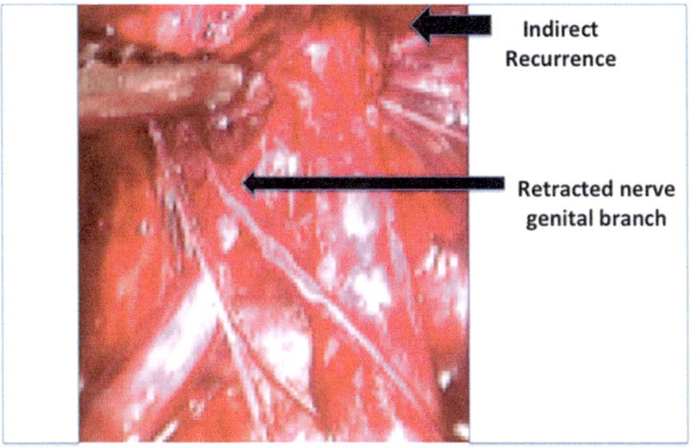

Fig. 16.3. The dissected anatomy of the patient with rolled-up mesh demonstrating the indirect recurrence and the position of the genital branch of the genital–femoral nerve.

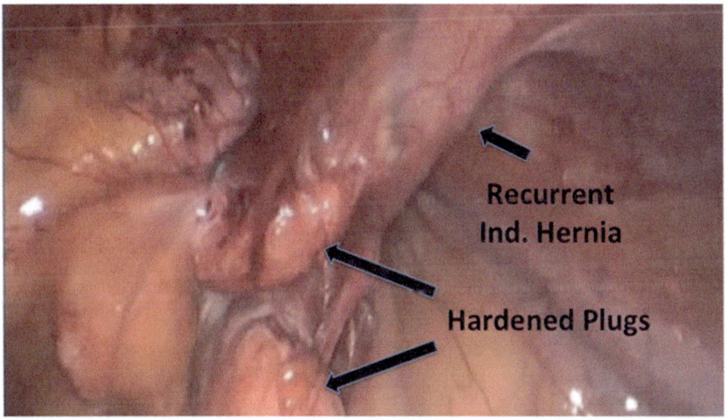

Fig. 16.4. The laparoscopic view in a patient with severe groin pain reveals two hardened plugs with a recurrent indirect hernia. The plugs were removed using cautery and a flat mesh was placed to repair the recurrence.

Preventing Chronic Pain after Posterior Hernia Repair

Decreasing the incidence of chronic pain following posterior inguinal hernia repair is as important as knowing how to treat it. Some approaches have proved to be effective, while others are only potentially effective.

It has been established that using no fixation is an effective way to potentially decrease the incidence of pain following laparoscopic inguinal hernia repair without increasing the incidence of recurrence [30]. Other approaches such as glue fixation [31, 32] and self-adhering meshes [33, 34] have the same potential of decreasing chronic pain by eliminating penetrating fixation, but long-term randomized studies are needed.

As previously discussed, the mesh itself can be the cause of chronic pain following posterior inguinal hernia repair. By far the most common cause of mesh-induced pain is from mesh that bunches or folds in on itself, creating a hard noncompressible mass. The best way of preventing this is to carefully flatten the mesh and make sure that it is not rolled up as the CO_2 is evacuated at the end of a TEP repair, or as the peritoneum is closed at the end of the TAPP repair. Whether using lighter weight larger pore meshes will decrease the incidence of chronic pain is still questionable. Short-term studies have shown that utilizing lightweight meshes decreases acute discomfort. One randomized study has shown a decrease in chronic pain but an increase in recurrence with lightweight mesh [35]. Others have so far failed to show a difference in chronic pain when light and heavyweight meshes are compared [36, 37]. Whether altering mesh composition or using less invasive forms of fixation of mesh can reduce the incidence further is yet to be determined. By far the most important way to decrease the incidence of chronic pain is through the use of proper techniques as previously outlined in this chapter.

Chronic pain after posterior inguinal hernia repair can occur because of improper technique, but also when everything is done perfectly. It is therefore important that all surgeons be able to recognize the causes of such pain, and to be able to reduce the incidence and to treat the pain based upon its root cause. Remember that most postoperative pain will respond to conservative management. Because patients become extremely unhappy and discouraged when they develop chronic pain, it is important to handle it appropriately. It is essential that the pain management plan be discussed with the patient so that they feel like they are participating in their own care. This usually prevents patients from reaching out for a second opinion and possibly making the situation worse. When appropriate, however, an aggressive approach as outlined in this chapter must be followed to eliminate chronic pain after posterior inguinal hernia repair.

References

1. Annandale T. Case in which a reducible oblique and direct inguinal and femoral hernia existed on the same side and were successfully treated by operation. Edinburgh Med J. 1876;27:1087.
2. Cheatle GL. An operation for the radical cure of inguinal and femoral hernia. Br Med J. 1920;2:68.
3. Henry AK. Operation for a femoral hernia by midline extraperitoneal approach. Lancet. 1936;1:531.
4. Nyhus LM, Condon RE, Harkins HN. Clinical experiences hernia repair for all types of hernia of the groin, with particular reference to the importance of transversalis fascia analogues. Am J Surg. 1960;100:234–44.
5. Rignault DP. Properitoneal prosthetic inguinal hernioplasty through a Pfannenstiel approach. Surg Gynecol Obstet. 1986;163(5):465–8.
6. Stoppa RE, Rives JL, Warlaumont CR, Palot JP, Verhaeghe PJ, Delattre JF. The use of Dacron in the repair of hernias of the groin. Surg Clin North Am. 1984;64(2): 269–85.
7. Wantz GE, Fischer E. Unilateral giant prosthesis reinforcement of the visceral sac. In: Fitzgibbons RJ, Greenburg AG, editors. Nyhus and Condon's hernia. 5th ed. Philadelphia, PA: Lippincott Williams and Wilkins; 2002. p. 219.
8. Schultz LS, Graber JN, Peritrafitta J, Hickok DF. Laser laparoscopic herniorrhaphy: a clinical preliminary results. J Laproendosc Surg. 1990;1(1):41–5.
9. Felix EL. Laparoscopic inguinal hernia repair. In: Eubanks WS, Swanstrom LS, Soper NJ, editors. Mastery of endoscopic and laparoscopic surgery. 2nd ed. Philadelphia, PA: Lippincott, Williams and Wilkins; 2009. p. 553.
10. McKernan JB, Laws HL. Laparoscopic repair of inguinal hernias using a totally extraperitoneal prosthetic approach. Surg Endosc. 1993;7(1):26–8.
11. Felix EL, Michas CA, Gonzalez Jr MH. Laparoscopic hernioplasty: why does it work? Surg Endosc. 1997;11(2):36–41.
12. Felix EL, Scott S, Crafton B, Geis P, Sewell R, McKernan B. A Multicenter study: causes of recurrence after laparoscopic hernioplasty. Surg Endosc. 1998;12(3):226–31.
13. Bittner R, Schmedt CG, Swarz J, Kraft K, Leibl BJ. Laparoscopic transperitoneal procedure for routine repair of groin hernia. Br J Surg. 2002;89(8):1062–6.
14. Tamme C, Scheidbach H, Hampe C, Schneider C, Köckerling F. Totally extraperitoneal endoscopic inguinal hernia repair (TEP). Surg Endosc. 2003;17(2):190–5.
15. Eker HH, Langeveld HR, Klitsie PJ, van't Riet M, Stassen LP, Weidema WF, et al. Randomized clinical trial of total extraperitoneal inguinal hernioplasty vs. Lichtenstein repair; along term follow-up study. Arch Surg. 2012;147(3):256–60.
16. Singh AN, Banasai VK, Misra MC, Kumar S, Rajeshwari S, Kumar A, et al. Testicular functions, chronic pain, and life after laparoscopic and open mesh repair of inguinal hernia: a prospective randomized trial. Surg Endosc. 2012;26(5):1304–17.
17. Langeveld HR, van't Riet M, Weideman WF, Stassen LP, Stassen LP, Steyerberg EW, Lange J, et al. Total extraperitoneal inguinal hernia repair compared with Lichtenstein (the LEVEL-TRIAL): a randomized controlled trial. Ann Surg. 2010;251(5):819–24.

18. Pollak R, Nyhus LM. Groin hernia. In: Cameron JL, editor. Current surgical therapy. 2nd ed. St Louis: Mosby; 1986. p. 268.
19. Nienhuijis S, Staal E, Keemers-Geis M, Rosman C, Strobbe L. Pain after open preperitoneal repair versus Lichtenstein randomized trial. World J Surg. 2007;31(9): 1751–7.
20. Sajid MS, Craciunas L, Singh KK, Sains P, Baig MK. Open transinguinal preperitoneal mesh repair of inguinal hernia: a targeted systematic review and meta-analysis of published randomized controlled trials. Gastroenterol Rep. 2013;1(2):127–37.
21. Condon RE. The anatomy of the inguinal region and its relation to groin hernia. In: Nyhus LM, Condon RE, editors. Hernia. 4th ed. Philadelphia: J.B. Lippincott; 1995. p. 16–72.
22. Rosenberger RJ, Loeweneck H, Meyer G. The cutaneous nerves encountered during laparoscopic repair of inguinal hernia: new anatomical findings for the surgeon. Surg Endosc. 2000;14(8):731–5.
23. Spaw AT, Ennis BW, Spaw LP. Laparoscopic hernia repair: the anatomic basis. J Laparoendosc Surg. 1991;1(5):269–77.
24. Aniibali R, Quinn TH, Fitzgibbons Jr RJ. Anatomy of the inguinal region from the laparoscopic perspective: critical areas for laparoscopic hernia repair. In: Bendavid R, editor. Prostheses and abdominal wall hernias (Medical Intelligence Unit). Austin, Texas: R G Landes; 1994. p. 82.
25. Tarply JL, Holzman MD. Groin hernia. In: Cameron JL, editor. Current surgical therapy. 8th ed. Philadelphia, PA: Elsevier Mosby; 2004. p. 545.
26. Koninger J, Redecke J, Butters M. Chronic pain after hernia repair: randomized trial comparing Shouldice, Lichtenstein and Tapp. Langenbecks Arch Surg. 2004;389(5):361–5.
27. Vander Pool AE, Harlarr JJ, den Hoed PT, Weidema WF, van Veen RN. Long-term follow-up evaluation of chronic pain after total extraperitoneal repair of primary and recurrent inguinal hernia. Surg Endosc. 2010;24(7):1707–11.
28. Markos V, Brown EF. CT herniography in the diagnosis of occult hernias. Clin Radiol. 2005;60(2):251–6.
29. Aasvang EK, Kehlet H. The effect of mesh removal and neurectomy on persistent post-hernotomy pain. Ann Surg. 2009;249(2):327–34.
30. Tam KW, Liang HH, Chai CY. Outcomes of staple fixation of mesh versus nonfixation in laparoscopic total extraperitoneal inguinal hernia repair: a meta-analysis of randomized controlled trials. World J Surg. 2010;34(12):3065–74.
31. Horisberger K, Jung MK, Zing U, Schob O. Influence of type of fixation in endoscopic totally extraperitoneal hernia repair (TEP) on long term quality of life. World J Surg. 2012;37(6):1249–57.
32. Berney CR, Yeo AE. Mesh fixation with fibrin sealant during endoscopic totally extraperitoneal hernia approach: a review of 640 repairs. Hernia. 2013; 17(6):709–17.
33. Fumagalli Romario U, Puccetti F, Elmore U, Massaron S, Rosati R. Self-Gripping mesh versus staple fixation in laparoscopic inguinal hernia repair: a prospective comparison. Surg Endosc. 2013;27(5):1798–802.

34. Birk D, Hess S, Garcia-Prado C. Low recurrence and low chronic pain associated with inguinal hernia repair by laparoscopic placement of Parietex ProGrip mesh: clinical outcomes of 220 hernias with a mean follow-up at 23 months. Hernia. 2013; 17(3):313–20.
35. ODwyer PJ, Kingsnorth AN, Molloy RG, Small PK, Lammers B, Horeyseck G. Randomized clinical trial assessing impact of a lightweight or heavyweight mesh on chronic pain after inguinal hernia repair. Br J Surg. 2005;92(2):166–70.
36. Nikkolo C, Murruste M, Vaasna T, Seepter H, Tikk T, Lepner U. Three-year results of a randomized clinical trial comparing lightweight mesh for inguinal hernioplasty. Hernia. 2012;16(5):555–9.
37. Chui LB, Ng WT, Sze YS, Yuen KS, Wong YT, Kong CK. Prospective randomized controlled trial comparing lightweight versus heavyweight mesh in chronic pain incidence after TEP repair of bilateral inguinal hernia. Surg Endosc. 2010; 24(11):2735–8.

17. The Orthopedic Perspective on Groin Pain: The Native and Prosthetic Hip

Calin Stefan Moucha

Editor's Comment (BPJ)

Patients often erroneously associate a complaint of groin pain with the possibility of having an inguinal hernia, and will seek consultation with their primary care physician or with a general surgeon for this reason. It cannot be stated too strongly that even in the setting of finding an obvious inguinal hernia on physical exam, if the chief complaint is groin pain, it is paramount to complete a thorough pain history and exam to assure that the pain itself is not from other etiologies. Often an MRI will help rule out these other etiologies, but the history and physical can help direct the surgeon to order the correct MRI. As this chapter points out, there are a large number of orthopedic injuries that can present as groin pain, and these complaints can lead both the patient and untrained surgeon to incorrectly diagnose an inguinal hernia and possibly even mistreat the complaint with an inguinal hernia repair. All general surgeons who see patients who complain of groin pain should be familiar with the detailed information provided in this excellent chapter written by a distinguished and experienced orthopedic surgeon.

Introduction

Orthopedic and general surgeons commonly see patients with groin pain. As in any other specialty, obtaining an appropriate history from the patient is vital. Young patients generally have completely different causes of groin pain than elderly ones. A history of acute or repetitive

B.P. Jacob et al. (eds.), *The SAGES Manual of Groin Pain*,
DOI 10.1007/978-3-319-21587-7_17

trauma, often seen in younger athletes, will guide the workup differently than pain not associated with a specific injury. A past medical history of extensive alcohol abuse, steroid usage, and certain conditions such as sickle cell anemia or lupus will guide the workup toward specific diagnoses. Duration of symptoms, progression, and alleviating and exacerbating factors need to be properly identified. Exact location and radiating patterns of pain help distinguish intra-articular from extra-articular musculoskeletal causes. Associated symptoms such as weakness, numbness, and paresthesias are also important to identify, as they can be associated with pathology of the spine.

This chapter reviews musculoskeletal causes of groin pain by dividing them into three categories: intra-articular, extra-articular, and groin pain after hip replacement. The focus will be predominantly on diagnosis. Treatment of these conditions will not be discussed in much detail as it is beyond the scope of this manual.

Extra-articular Causes of Groin Pain

A large majority of patients who present with hip pain do not actually have intra-articular hip pathology. Determining whether the cause of pain is intra-articular or extra-articular early on will help streamline the diagnostic process. Patients with extra-articular causes of groin pain will usually complain of pain on the side of the hip and/or in the buttock that radiates into the groin.

Greater Trochanteric Pain Syndrome Greater trochanteric pain syndrome (GTPS) encompasses several diagnoses that include *trochanteric bursitis* and *tendinosis* (or even a degenerative tear) *of the gluteus medius or minimus muscles*. While the direct source of pain is extra-articular, it is not uncommon for patients to present with groin pain. Patients usually have pain while driving or lying down on the affected side.

Gait examination may reveal a *Trendelenburg gait*, indicative of weak abductor muscles, which should not be confused with an *antalgic gait* (painful gait). A Trendelenburg gait is characterized by downward tilting of the contralateral pelvis during stance on the weakened side. On physical examination, we usually see direct tenderness to palpation over the greater trochanter or just proximal to it. The *Trendelenburg sign* is found in patients with weak or paralyzed hip abductor muscles (medius and/or minimus). A positive Trendelenburg sign means that the affected hemipelvis sags during *one-legged stance*. Most importantly, however, intra-

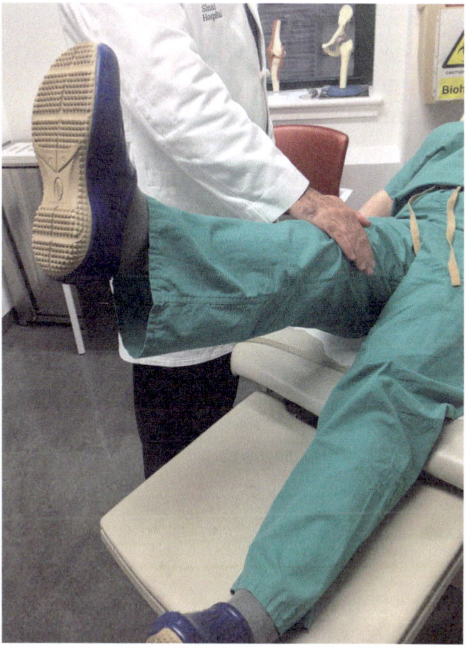

Fig. 17.1. Intra-articular hip pathology is excluded by painless hip rotation and a negative Stinchfield test.

articular hip pathology is excluded by painless hip rotation and a negative Stinchfield test (Fig. 17.1). *The Stinchfield test* is done with the patient in a supine position and the examiner applying downward pressure on the thigh as the patient actively elevates the leg as high as possible; pain during this maneuver signifies a positive test and is indicative of likely intra-articular pathology. An injection with steroid into the trochanteric bursa region is both diagnostic and often therapeutic. On rare occasions, open or endoscopic bursectomy may be considered, but results are not very predictable. Refractory cases warrant obtaining an MRI or an ultrasound to rule out more significant abductor tears or an intra-articular process [1]. Lastly, *coxa saltans externa*, a less common cause of GTPS, refers to snapping of the iliotibial band or the anterior border of the gluteus maximus muscle over the greater trochanter when the hip is brought from a flexed position to an extended position. While it less commonly causes pain, when it does, the pain may radiate from the side of the hip into the groin. After physical therapy trials, surgical treatment consists of a variety of iliotibial band release procedures [2, 3].

Lumbar Spine Disease Spine diseases, such as thoracolumbar discogenic pain at multiple levels or L2–L4 nerve root impingement, can all cause groin and thigh pain, with or without associated back pain [4]. The mechanism is sometimes complex, as the pain can be radicular pain from nerve root compression or pain from nerve endings on the herniated disc itself. The existence of sensory nerve endings in the annulus fibrosus of the human lumbar intervertebral disc has been described and well documented [5]. *Patients sometimes find it hard to understand how their groin pain is originating from their back when they do not present with back pain.* Physical exam of the hip is usually normal. Limited spine flexion, extension, and/or lateral bending are sometimes seen. Many of these patients have very weak core musculature. The femoral nerve traction test is done with the patient in a prone position with the knee flexed to 90° and the hip fully extended; pain in the anterior thigh suggests a L2–L4 nerve root impingement. One of the most common findings, however, appears to be tight hamstring musculature. *Tight hamstrings lead to hip flexion contractures, a subtle crouched gait, and compensatory pressure on the spine*. Physical therapy focusing on core strengthening, hamstring stretching, and lumbar stabilization is the first line of treatment. MRI of the spine is sometimes needed, followed by selective spinal diagnostic and therapeutic injections.

Osteitis Pubis Osteitis pubis is a noninfectious inflammatory process of the pubic symphysis commonly seen in runners and in athletes involved in cutting sports such as soccer and hockey. Previous trauma, overuse, and vaginal delivery are all risk factors. Patients often present with groin pain that is activity related. On physical exam they usually have normal hip range of motion, nontender abductor muscles laterally, and focal tenderness to palpation over the pubic symphysis. Weak core musculature is often noted; pain with resisted hip adduction or passive hip abduction may also be found. These latter physical exam findings are often confirmed by *tendinosis of the rectus abdominis and adductor longus insertions on an MRI.* Plain radiographs of the pelvis are usually helpful as well, as they typically show widening of the symphysis with blurring of the cortical margins and sometimes cysts. Physical therapy is the first line of treatment followed by steroid injections. Surgical bony resection with preservation of the rectus and pubic ligaments is rarely done [6].

Pubic Ramus Fractures Fractures of the superior and/or inferior pubic rami are commonly seen in elderly patients who have sustained a low-energy fall. These patients present with acute groin pain that was not present prior to trauma. They usually have painless hip rotation but focal tenderness over the bony pelvis lateral to the pubic symphysis. It is common for these patients to report having had recent hip radiographs that were normal. Unfortunately, the pelvis has rarely been evaluated. Hip radiographs often do not show rami fractures; it is imperative to obtain an AP of the pelvis when evaluating groin pain. Treatment of these fractures is most commonly nonoperative with unrestricted weight bearing.

Iliopsoas Pathology The two most common iliopsoas pathologies seen by orthopedic surgeons are snapping (coxa saltans interna) and tendinitis. *Internal snapping of the iliopsoas tendon* is actually an extra-articular process. Patients usually present with an audible snap and anterior groin pain. As the hip is extended, the iliopsoas tendon travels from lateral to medial catches at the iliopectineal eminence or on the femoral head. On occasion, the snapping can be palpated directly in the groin. *Dynamic ultrasound* may be useful in the diagnosis. MRI is sometimes indicated, as it can show resultant hip labral tea [7]. *Iliopsoas tendinitis* is a relatively rare entity in patients with native hips and seen most commonly in patients involved in activities that require repetitive hip flexion (rowing, uphill running, and ballet). Patients generally present with anterior hip pain that radiates to the knee and sometimes with knee pain alone. The most common physical exam findings are painless hip rotation, pain with resisted hip flexion, and pain with passive hip extension. Initial treatment of both coxa saltans interna and iliopsoas tendinitis is always stretching (best done in a luge position) and, when necessary, steroid injections [8]. Open and arthroscopic releases of the iliopsoas tendon have been reported, but complications such as symptomatic intra-abdominal fluid extravasation [9] and anterior hip instability [10] have been reported. Exclusion of the other potentially life-threatening pathologies with which abdominal surgeons are very familiar, such as an iliopsoas abscess, is clearly important.

Intra-articular Causes of Groin Pain

Anterior groin pain that radiates deep into the hip and sometimes radiates into the groin should always raise the suspicion of an intra-articular process. In the now classic *C sign* (Fig. 17.2) suggestive of intra-articular hip pathology, patients will place their hand over the affected hip with the

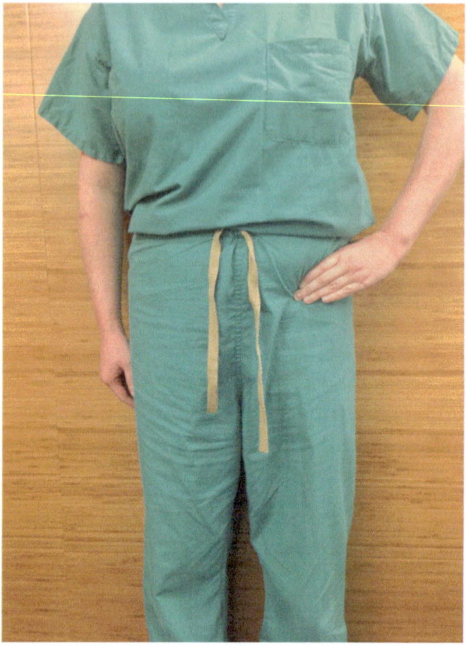

Fig. 17.2. The classic C sign suggestive of intra-articular hip pathology.

index finger in the groin and the thumb placed proximal to the greater trochanter in the shape of the letter C [11, 12]. Patients will also commonly have a positive *Stinchfield test*, described earlier in this chapter. Lastly, *an active straight leg raise* with the supine patient actively raising the heel of the leg by flexing the hip about 30° is also suggestive of intra-articular pathology: during this test, hip flexors produce joint reactive forces up to two times the patient's body weight across the hip joint itself.

Arthritis and Avascular Necrosis More than 21 % of the US population aged 18 or older have arthritis or other rheumatic conditions, and that percentage increases as people age. The number of people in the USA who have arthritis is projected to increase to 67 million, or 25 % of the adult population, by the year 2030. Osteoarthritis is the most common form and the hip is commonly affected. Patients present with pain in the groin, diminished hip motion, difficulty putting on their shoes or socks, and inability to ambulate extensively. Physical exam reveals a positive Stinchfield test and C sign. Nonoperative treatment consists of intra-articular steroid or hyaluronate injections [13]. Surgical intervention is a total hip arthroplasty. Avascular necrosis, commonly seen in the femoral head, is also a common reason for groin pain, especially in patients with risk factors such as steroid use, alcohol abuse, coagulopathies, sickle cell disease, Gaucher's disease, and decompression sickness. When radiographs are normal and suspicion is high, patients should undergo an MRI of the hip. Depending on the stage of avascular necrosis, treatment includes protected weight bearing, bisphosphonate treatment, electrical stimulation, electromagnetic fields, core decompression, bone grafting, autologous mesenchymal cells, osteotomies, and arthroplasty procedures [14].

Hip Synovitis and Septic Arthritis Transient synovitis of the hip is a short-lived acute inflammatory process usually seen in boys aged 2–10 following an upper respiratory tract infection. Generally a diagnosis of exclusion, it must be differentiated from a septic hip, which is also commonly seen in this patient population. Patients present with groin pain and sometimes difficulty putting weight on the limb. In addition to the aforementioned tests for intra-articular pathology, these patients will have pain with log rolling of the hip while in extension. Kocher et al. have provided useful ways of differentiating between these two entities [15]. Patients with transient synovitis require close observation, while those with septic arthritis most commonly require arthroscopic or open irrigation and debridement of the hip joint.

Femoroacetabular Impingement *Femoroacetabular impingement (FAI)* occurs when anatomic variations in hip anatomy lead to impingement between the acetabulum and the femoral head–neck junction. FAI is believed by many to be a common pathway to hip arthritis, especially in younger patients. Impingement is generally classified into Cam impingement and Pincer impingement. *CAM impingement* is due to prominence of the anterosuperior head–neck junction or diminished head–neck offset. *Pincer impingement* is secondary to acetabular overcoverage of the femoral head for a variety of reasons such as coxa profunda or acetabular retroversion. FAI may lead to chronic groin pain, especially in younger adults who often go on to have symptomatic arthritis. FAI is also a probable predisposing factor to *labral tears*, most of which, even in the face of trauma, would probably rarely occur otherwise. Patients with FAI may have some of the classic signs of intra-articular hip pathology and also report pain with tests such as the anterior impingement test or FADIR (flexion adduction internal rotation). Sophisticated radiographs and MR arthrography are some of the methods of choice in further studying these conditions [16]. Diagnostic hip injections, when positive, may lead to a variety of arthroscopic and/or open joint preserving procedures [17].

Femoral Neck Stress Fractures Athletes, especially runners, and military recruits are two groups of patients at risk of femoral neck stress fractures. Delay in diagnosis can be devastating [18]. Female athletes with anorexia and amenorrhea with groin pain and difficulty weight bearing should be aggressively worked up with either a bone scan or an MRI. Concomitant osteoporosis workup and evaluation for vitamin D deficiency are important as well. Fractures on the compression side of the femoral neck may be treated nonoperatively, while those on the tension side generally require surgical fixation.

Hip Instability Chronic groin pain in hyperlax patients, especially female athletes, can be a sign of hip instability. Many of these patients have underlying acetabular dysplasia with decreased femoral head coverage [19]. The exact mechanism of pain in these patients is unclear, but may be due to early cartilage degeneration, labral tears, or synovitis. MRI in patients who have associated dysplasia may show a hypertrophied iliocapsularis muscle in the anterior aspect of the hip [20]. Physical therapy focusing on strengthening hip musculature is the first line of treatment. Refractory cases in patients without dysplasia may require open or arthroscopic capsular plication procedures [21]. When dysplasia is present, correction of bony abnormalities is critical [22].

Groin Pain after Total Hip Replacement

The demand for primary total hip arthroplasties during the period 2005–2030 is estimated to grow by at least 174 % to 572,000 [23, 24]. While this operation is extremely successful, there are still some patients who continue to have pain. Patients who do not have a pain-free interval after surgery and experience pain that is different than they did preoperatively need an infection workup: abnormal C-reactive protein and sedimentation rate should prompt a hip aspiration. Persistence of preoperative pain should raise the question of an improper indication for the hip replacement. Pain in the groin after hip replacement is usually related to the area of the acetabulum; thigh pain often relates to the femoral component. Iliopsoas tendinitis is not uncommon and often missed. Acetabular component malpositioning into retroversion (Fig. 17.3) is usually the cause, and treatment begins with a steroid injection and sometimes leads to component revision [25]. More recently,

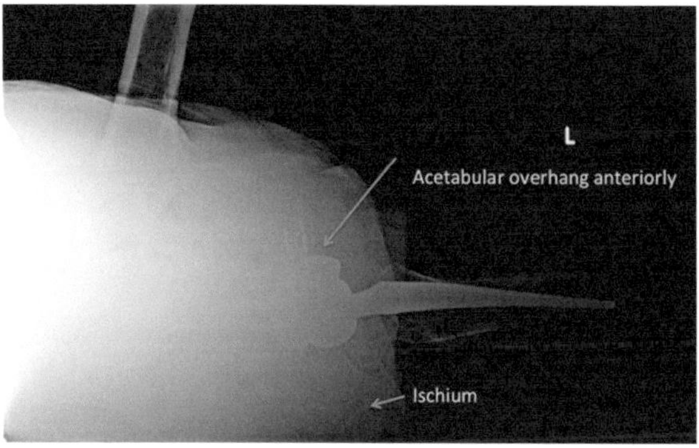

Fig. 17.3. Acetabular component malpositioning into retroversion.

a b

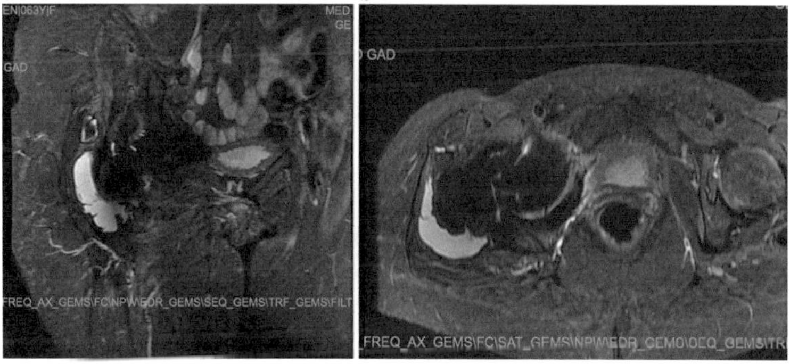

Fig. 17.4. (**a**, **b**) Metal corrosion from metal-on-metal hip implant or modular components can lead to painful adverse local tissue response (ALTR) and so-called "pseudotumors".

cases done through an anterior approach show transient irritation of the iliopsoas tendon that often subsides without aggressive intervention. Stress fractures around cementless acetabular component, hematomas, impingement, loosening, or leg length discrepancy that aggravates preexisting spinal disease can all lead to groin pain. Metal corrosion from metal-on-metal hip implants or modular components can lead to painful effusions and so-called "pseudotumors" (Fig. 17.4) [26, 27]. Lastly, groin pain in a patient with a previous hip fracture treated with a hip replacement may in fact have a partial hip replacement, also known as a hemiarthroplasty. A good number of these patients, deemed "low demand" and given this type of implant, go on to have groin pain due to acetabular degeneration, either preexisting or progressive since surgery.

Conversion to a total hip replacement during which the acetabulum is resurfaced is often curative. Referral to a revision joint replacement surgeon should be considered when the primary surgeon cannot identify the cause of pain after a hip replacement.

Conclusion

Multiple musculoskeletal causes of groin pain exist. While this chapter discussed common and some rare diagnoses, it was not all-inclusive. Oncological causes of musculoskeletal groin pain in particular were not discussed, as the topic is quite broad and beyond the scope of this manual. Orthopedic surgeons evaluate groin pain using a slightly different perspective. Only by cross-pollinating knowledge between different specialties will we gain a better understanding of our own.

References

1. Westacott DJ, Minns JI, Foguet P. The diagnostic accuracy of magnetic resonance imaging and ultrasonography in gluteal tendon tears—a systematic review. Hip Int. 2011;21(6):637–45.
2. Ilizaliturri Jr VM, Martinez-Escalante FA, Chaidez PA, Camacho-Galindo J. Endoscopic iliotibial band release for external snapping hip syndrome. Arthroscopy. 2006;22(5):505–10.
3. White RA, Hughes MS, Burd T, Hamann J, Allen WC. A new operative approach in the correction of external coxa saltans: the snapping hip. Am J Sports Med. 2004; 32(6):1504–8.
4. Oikawa Y, Ohtori S, Koshi T, Takaso M, Inoue G, Orita S, et al. Lumbar disc degeneration induces persistent groin pain. Spine (Phila Pa 1976). 2012;37(2):114–8.
5. Bogduk N, Tynan W, Wilson AS. The nerve supply to the human lumbar intervertebral discs. J Anat. 1981;132(Pt 1):39–56.
6. Hechtman KS, Zvijac JE, Popkin CA, Zych GA, Botto-van BA. A minimally disruptive surgical technique for the treatment of osteitis pubis in athletes. Sports Health. 2010;2(3):211–5.
7. Domb BG, Shindle MK, McArthur B, Voos JE, Magennis EM, Kelly BT. Iliopsoas impingement: a newly identified cause of labral pathology in the hip. HSS J. 2011; 7(2):145–50.
8. Sofka CM, Collins AJ, Adler RS. Use of ultrasonographic guidance in interventional musculoskeletal procedures: a review from a single institution. J Ultrasound Med. 2001;20(1):21–6.
9. Kocher MS, Frank JS, Nasreddine AY, Safran MR, Philippon MJ, Sekiya JK, et al. Intra-abdominal fluid extravasation during hip arthroscopy: a survey of the MAHORN group. Arthroscopy. 2012;28(11):1654–60.e2.

10. Fabricant PD, Bedi A, De La Torre K, Kelly BT. Clinical outcomes after arthroscopic psoas lengthening: the effect of femoral version. Arthroscopy. 2012;28(7):965–71.
11. Martin HD, Shears SA, Palmer IJ. Evaluation of the hip. Sports Med Arthrosc. 2010;18(2):63–75.
12. Plante M, Wallace R, Busconi BD. Clinical diagnosis of hip pain. Clin Sports Med. 2011;30(2):225–38.
13. Migliore A, Bella A, Bisignani M, Calderaro M, De Amicis D, Logroscino G, et al. Total hip replacement rate in a cohort of patients affected by symptomatic hip osteoarthritis following intra-articular sodium hyaluronate (MW 1,500–2,000 kDa) ORTOBRIX study. Clin Rheumatol. 2012;31(8):1187–96.
14. Mont MA, Jones LC, Hungerford DS. Nontraumatic osteonecrosis of the femoral head: ten years later. J Bone Joint Surg Am. 2006;88(5):1117–32.
15. Kocher MS, Zurakowski D, Kasser JR. Differentiating between septic arthritis and transient synovitis of the hip in children: an evidence-based clinical prediction algorithm. J Bone Joint Surg Am. 1999;81(12):1662–70.
16. Smith TO, Hilton G, Toms AP, Donell ST, Hing CB. The diagnostic accuracy of acetabular labral tears using magnetic resonance imaging and magnetic resonance arthrography: a meta-analysis. Eur Radiol. 2011;21(4):863–74.
17. Leunig M, Ganz R. The evolution and concepts of joint-preserving surgery of the hip. Bone Joint J. 2014;96-B(1):5–18.
18. Johansson C, Ekenman I, Törnkvist H, Eriksson E. Stress fractures of the femoral neck in athletes. The consequence of a delay in diagnosis. Am J Sports Med. 1990; 18(5):524–8.
19. Nunley RM, Prather H, Hunt D, Schoenecker PL, Clohisy JC. Clinical presentation of symptomatic acetabular dysplasia in skeletally mature patients. J Bone Joint Surg Am. 2011;93 Suppl 2:17–21.
20. Babst D, Steppacher SD, Ganz R, Siebenrock KA, Tannast M. The iliocapsularis muscle: an important stabilizer in the dysplastic hip. Clin Orthop Relat Res. 2011; 469(6):1728–34.
21. Smith MV, Sekiya JK. Hip instability. Sports Med Arthrosc. 2010;18(2):108–12.
22. van Bergayk AB, Garbuz DS. Quality of life and sports-specific outcomes after bernese periacetabular osteotomy. J Bone Joint Surg (Br). 2002;84(3):339–43.
23. Kurtz S, Ong K, Lau E, Mowat F, Halpern M. Projections of primary and revision hip and knee arthroplasty in the United States from 2005 to 2030. J Bone Joint Surg Am. 2007;89(4):780–5.
24. Kurtz SM, Ong KL, Lau E, Bozic KJ. Impact of the economic downturn on total joint replacement demand in the United States: updated projections to 2021. J Bone Joint Surg Am. 2014;6(8):624–30.
25. Trousdale RT, Cabanela ME, Berry DJ. Anterior iliopsoas impingement after total hip arthroplasty. J Arthroplasty. 1995;10(4):546–9.
26. Gill IP, Webb J, Sloan K, Beaver RJ. Corrosion at the neck-stem junction as a cause of metal ion release and pseudotumour formation. J Bone Joint Surg (Br). 2012;94(7): 895–900.
27. Berend KR, Morris MJ, Adams JB, Lombardi Jr AV. Metal-on-metal hip arthroplasty: going, going, gone…—affirms. J Bone Joint Surg (Br). 2012;94(11 Suppl A):75–7.

18. Algorithmic Approach to the Workup and Management of Chronic Postoperative Inguinal Pain

Johan F.M. Lange Jr.

With the success of tension-free mesh-based inguinal repair, the incidence of chronic postoperative inguinal pain (CPIP) has surpassed that of recurrence after open and laparoscopic herniorrhaphy. Due to different definitions of CPIP, the reported incidence of CPIP ranges from 1 to 63 % [1–7]. Significant CPIP affects daily life for 5–10 % of patients [8, 9]. As quality of life has become the most relevant outcome of inguinal hernia repair, and considering the high prevalence of CPIP, a systematic approach is needed for optimal management.

Etiology of CPIP

CPIP can have several causes. Literature of the past decade has put focus on the possible role of the three inguinal nerves (ilioinguinal, iliohypogastric, and genitofemoral). Although there is no substantial evidence that the inguinal nerves should be identified to preserve them or to cut them purposely, there is consensus that inguinal hernia surgery should be performed with "nerve-awareness." Besides the assumption that iatrogenic nerve damage or interference of the mesh with the nerves plays a key role in the pathogenesis of CPIP, other potential risk factors have been defined: a mesh that is sutured too tight instead of floppy, medial fixation of the mesh to the periosteal rim of the pubic bone, a too narrow neo-annulus, and herniotomy (obsolete) with a twisted peritoneal sac have all been described as potential causes for CPIP [10]. It is also known that preoperative pain is a risk factor for CPIP [11, 12]. In this respect, one should be alert when a patient presents with a sharp groin pain without an obvious inguinal hernia found during physical

B.P. Jacob et al. (eds.), *The SAGES Manual of Groin Pain*,
DOI 10.1007/978-3-319-21587-7_18

examination and only a small hernia detected by ultrasonography. This represents the type of at-risk patient having a predisposition for the development of CPIP. Finally, a wrinkled mesh, also known as meshoma, is a rare but infamous cause for CPIP [13].

Algorithmic Approach

Surgical literature has already paid much attention to risk factors, etiology, and prevention of CPIP. On the other hand, there is only little known about the management of CPIP. What to do when the damage has already been done? What is the policy to follow when a patient presents himself with possible CPIP? It is very hard to answer this question with regard to the heterogenic causes of CPIP, different operation techniques, and lack of scientific evidence. Certainly, there are indications that triple neurectomy can be successful in some cases of patients with CPIP, but it is not the remedy for all patients, and the challenge remains to select the right patient with CPIP for the right treatment. This chapter tries to capture the elements of how to deal with this complex diagnosis in an algorithm (Fig. 18.1).

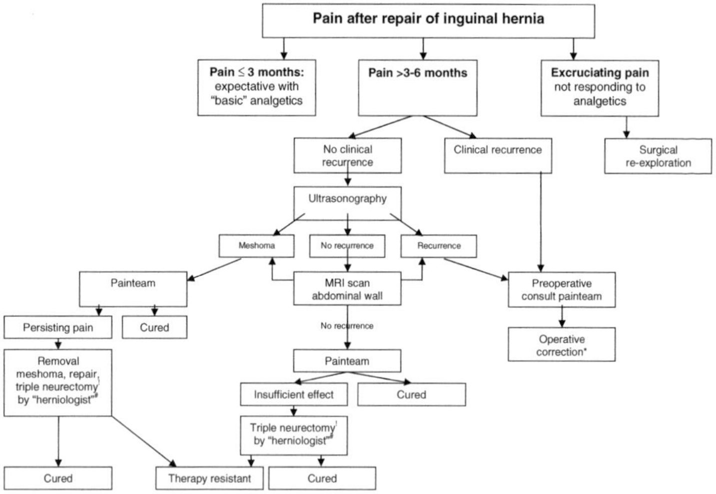

! Including proximal genitofemoral nerve-neurectomy in case of chronic pain after open or laparoscopic preperitoneal mesh technique
Open or endoscopic procedure
* In case of neuropathic pain anterior correction in combination with triple neurectomy is optional

Fig. 18.1. Consensus algorithm for the management and treatment of CPIP (from Lange et al. [14], with kind permission Springer Science+Business Media).

Until recently no such algorithm has existed and current practices were mainly guided by personal opinion and expertise. While it was impossible to include every perspective and address every subtlety in dealing with this complex diagnosis, the proposed algorithm involved many dedicated inguinal hernia surgeons and addressed the general issues that are important in the diagnosis and management. This algorithm approach is not intended as a solid law or rigid guideline, but hopefully will serve as a guide for practicing surgeons, pain physicians, primary doctors and the multidisciplinary services that assist in treating this important group of patients [14].

Timing

The algorithm starts with the two categories of patients after inguinal hernia surgery requiring medical attention: patients with pain immediately after surgery (acute pain) and patients who develop pain later during the postoperative course. This second group is also subdivided in two categories: patients who only complain in the early postoperative phase and those who have persistent pain or develop pain after some months. Acute, excruciating pain is considered an indication for early re-exploration. If postoperative pain develops later during the postoperative course, or if pain persists beyond the normal postoperative recovery period, an expectative phase of 3 months is indicated. During this time, analgesics and other conservative measures are recommended.

Diagnostics

If pain persists after 3 months, inguinal hernia recurrence should be excluded based on physical examination. In case of clinical recurrence, operative correction is indicated, with or without triple neurectomy, depending on the type of pain (neuropathic or nociceptive). If physical examination does not demonstrate recurrence, ultrasonography is recommended as the initial diagnostic procedure to exclude occult recurrence or meshoma. If ultrasonography is unrevealing, cross-sectional imaging with MRI might detect recurrence, meshoma, or other pathologies.

If recurrence is identified and associated with pain, open anterior repair is recommended in conjunction with triple neurectomy if accompanied by neuropathic pain. From the perspective of pain management and remedial surgery for inguinodynia, if the initial hernia operation was an anterior repair (Lichtenstein, Shouldice, Bassini, McVay), laparoscopic correction does not represent the primary recommended modality because positioning of mesh in the preperitoneal space may lead to additional neuropathy (main trunk of genitofemoral nerve and preperitoneal segment of its genital branch). This is contrary to the recommendations for simple recurrence without neuropathic pain, which would favor a laparoscopic approach. If laparoscopic repair of recurrence fails to address the pain, it would not be possible to differentiate whether the source of pain is from neuropathy of nerves in front or behind the transversalis fascia. If the initial hernia operation was a posterior repair (TEP, TAP, PHS, TIPP, or other preperitoneal repair), anterior repair is recommended with open "extended" triple neurectomy, including the genitofemoral nerve trunk if needed. Laparoscopic repair for recurrence may be performed, but neuropathic pain if present must be addressed with retroperitoneal triple neurectomy proximally to the site of neuropathy.

If no anatomical pathology is identified, the surgeon should refer the patient to a pain management team familiar with CPIP. In addition to pharmacologic and behavioral treatment, interventions play a major role in the diagnosis and treatment of CPIP. Nerve blocks of the ilioinguinal, iliohypogastric and genitofemoral nerves are of significant importance, as they serve both a diagnostic and therapeutic role. If conservative or interventional modalities are unsuccessful or not durable, surgical intervention should be offered. If the original operation involves mesh in the preperitoneal space from open or laparoscopic repair, open extended triple neurectomy to resect the genitofemoral trunk or laparoscopic retroperitoneal triple neurectomy is indicated [15].

The International Association for the Study of Pain (IASP) broadly classifies postherniorrhaphy inguinodynia into nociceptive and neuropathic pain [16]. Nociceptive pain is caused by activation of nociceptors by nociceptive molecules. It is caused by tissue injury or inflammatory reaction. Neuropathic pain is caused by direct nerve injury. It is characterized by inguinodynia with radiation to the scrotum/femoral triangle, paresthesia, allodynia, hyperpathia, hyperalgesia, hyperesthesia, hypoesthesia, and positive Tinel's sign. There is no precise demarcation between nociceptive and neuropathic pain and the complexity of diagnosis is increased by social, genetic, patient, and psychological factors.

In-depth knowledge of groin neuroanatomy is of paramount importance to prevent and treat CPIP. Knowledge of the original operative technique and detailed evaluation of the original operative report will help to determine the likely etiologies of CPIP and the nerves at risk.

The diagnosis is also very much dependent on a detailed history and physical examination. Physical exam findings are dependent on the neuroanatomic course of the three inguinal nerves, their respective dermatomes, and the presence of mesh or recurrence. Tools including preoperative dermatomal mapping, quantitative sensory testing, imaging and diagnostic interventions (nerve blocks) help to characterize the etiology and direct treatment [15].

Open and Endoscopic Treatment of Neuropathic Pain

Treatment of the patient with CPIP remains a challenge and several different therapeutic strategies have been proposed. Conservative treatment with pharmacologic, topical, behavioral and expectant measures is advocated in all patients. Interventional techniques, including nerve infiltration, blockade, neuromodulation, and ablative techniques, have all been used in the management of CPIP. Results of selective or triple neurectomy of one or more of the three inguinal nerves and resection of meshoma have been published with practical efficacy. Despite this high volume of information, no consensus on the management of CPIP has been published and high-level evidence on the management of CPIP is lacking. Triple neurectomy described by Amid et al. in 1995 is currently an accepted surgical treatment for neuropathic pain refractory to conservative measures [17].

While some surgeons have had success with selective neurectomy, triple neurectomy is generally recommended due to neuroanatomic and technical considerations [15, 18]. There is significant cross-innervation between the inguinal nerves and reoperating in the scarred field becomes increasingly more difficult and morbid for subsequent remedial operations.

Extensive study of the anatomical variation of these nerves from the retroperitoneum to its terminal branches in the inguinal canal demonstrates significant variation in the number, location, and cross-innervation of these three nerves [19]. Additionally, visual identification of the nerve at the time of reoperation cannot adequately exclude injury. Electron micrography of grossly normal nerves resected at the time of triple neurectomy often demonstrates ultrastructural nerve damage. It is often challenging to identify nerves in the scarred field. Reoperation, especially with concurrent mesh removal, carries the added risk of recurrence, vascular injury, testicular compromise and visceral injury.

Best available evidence suggests that triple neurectomy has higher efficacy than selective neurectomy [15].

Open or endoscopic methods are available to perform triple neurectomy, depending on the type of prior repair, the presence of recurrence or meshoma, and if orchialgia is present. Open triple neurectomy involves re-exploration through the prior operative field and is indicated when recurrence or meshoma are present or for the treatment of patients who originally underwent , anterior repair without preperitoneal placement of mesh. The ilioinguinal nerve is identified laterally to the internal ring, between the ring and the anterior superior iliac spine. The iliohypogastric nerve is identified within the anatomical cleavage between the external and internal oblique aponeurosis. The nerve is then traced proximally within the fibers of the internal oblique muscle to a point laterally to the field of the original hernia repair. Failure to do so may leave the injured intramuscular segment of the nerve behind. The inguinal segment of the genital branch of the genitofemoral nerve can be identified adjacent to the external spermatic vein between the cord and the inguinal ligament and traced proximally to the internal ring where it is severed. Alternatively, the nerve may be visualized within the internal ring through the lateral crus of the ring. Standard triple neurectomy does not address neuropathy of the preperitoneal nerves (main trunk of genitofemoral nerve and preperitoneal segment of its genital branch) after open or laparoscopic preperitoneal repair. In these cases, an "extended" triple neurectomy may be performed, dividing the floor of the inguinal canal to access the genitofemoral trunk in the retroperitoneum directly over the psoas muscle.

Nerves should be resected proximally to the field of the original hernia repair. Although there are no specific data available, ligation of the cut ends ,of the nerves to avoid sprouting and neuroma formation and intramuscular insertion of the proximal cut end to keep the nerve stump away from scarring within the operative field are recommended [17]. Neurolysis, which does not address ultrastructural changes of nerve fibers, is not recommended. Simple removal of entrapping sutures or fixating devices while leaving the injured nerves behind is also not recommended and does not address irreversible damage to the nerve.

Endoscopic retroperitoneal triple neurectomy allows for access proximally to all potential sites of peripheral neuropathy, overcoming many of the limitations of open triple neurectomy after laparoscopic or open preperitoneal repair [15, 20, 21]. Prior preperitoneal laparoscopic or open procedures may damage or entrap the nerve in the preperitoneal position, rendering proximal access to the three nerves a challenge. Endoscopic access to these three nerves in the retroperitoneum allows for definitive identification of the ilioinguinal and iliohypogastric nerves at the L1 nerve root overlying the quadratus lumborum muscle and the genital and femoral branches of the genitofemoral nerve exiting from the psoas muscle. The operative technique is safe and proximal to the field of scarring from all prior inguinal hernia repairs. Complications including deafferentation hypersensitivity are a significant concern. In addition to numbness in the groin region and flank, patients undergoing proximal neurectomy may develop bulging of the lateral abdominal wall because of the additional loss of motor function of the iliohypogastric and ilioinguinal nerve (innervation of transversus abdominus and oblique muscles). In the absence of recurrence or meshoma, endoscopic management may be the preferred technique for definitive operative management of CPIP. Selection of appropriate patients is most important and management is best deferred to experienced hernia specialists.

The consensus recommendation is that reoperation for CPIP should be performed only by experienced surgeons [17].

CPIP caused by recurrence is a less common etiology in these predominantly neuropathic pain syndromes. However, it still represents a cause of CPIP ,largely to be excluded by physical examination and imaging. This contrasts with neuropathic pain, for which there are no reliable tests and can be considered a diagnosis "per exclusionem." Ultrasound is recommended as the first diagnostic test for recurrence (and meshoma) because of costs and facility. It becomes more complicated if there is a possible combined origin for pain: recurrence and neuropathic pain. In case of the combination of recurrence and neuropathic pain, it is important to consider the prior technique of hernia repair and the location of the mesh, as this will dictate the ideal approach for neurectomy and the subsequent repair (Fig. 18.2). In case of initial anterior repair anteriorly to the transversalis fascia (i.e., Lichtenstein procedure, Shouldice, tissue repairs), a standard open triple neurectomy including resection of the intramuscular segment of the iliohypogastric nerve should be performed [22]. Laparoscopic repair of the recurrence can potentially lead to neuropathy of the preperitoneal nerves (main trunk of genitofemoral nerve and preperitoneal segment of its genital branch), and when combined with open triple neurectomy, it would not be possible to differentiate between neuropathy of nerves in front and behind the fascia transversalis as source of pain after this second operation. An alternative is to perform an endoscopic hernia repair and an endoscopic triple neurectomy. In case

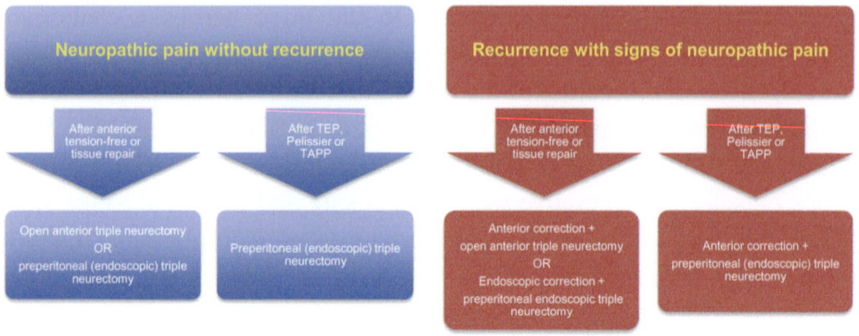

Fig. 18.2. Consensus recommendations for operative approach of CPIP (from Lange et al. [14], with kind permission Springer Science+Business Media).

of recurrence and neuropathic pain after preperitoneal mesh repair [i.e., totally extraperitoneal repair (TEP), transinguinal preperitoneal (TIPP), transabdominal preperitoneal repair (TAPP)], the recurrence should be corrected with an anterior technique (preferably, Lichtenstein procedure) to avoid the prior scarred field. However, a triple neurectomy via this approach would not be useful, as the potentially damaged nerves are located behind the fascia transversalis. As a result, the anterior correction should ideally be combined with an endoscopic triple neurectomy in the "untouched" plane proximally to the preperitoneal mesh. A remedial laparoscopic operation is an alternative approach with proximal neurectomy if indicated.

Mesh Removal

Partial or complete mesh removal is indicated in case of meshoma pain refractory to conservative management. Meshoma as a pathologic entity can present in different gradations from mass-like density to more subtle effects of mesh wrinkling or fibrosis. While meshoma will require surgical intervention if persistent and severe, occasionally patients whose overall pain levels improve can manage without re-exploration and removal. If the pain team is able to decrease the pain with pharmacologic, behavioral, and interventional treatment, this would be preferable. The greatest morbidity in these reoperative surgeries is from removal of the mesh given its apposition to vital structures with the potential for bleeding, testicular loss, visceral injury, and creation of a new hernia. Any potential to spare a patient from surgery is advisable.

Instead of removal of meshoma only or removal of the nerves entrapped by meshoma only, systematic triple neurectomy is recommended because of neuronanatomic and technical considerations [20]. From a neuroanatomic perspective, there is significant cross-innervation of the nerves within the inguinal canal and within the preperitoneal space. Any neuropathic pain cannot be isolated to one specific nerve, and if neurectomy is performed, all potentially damaged nerves within an operative field should be taken. From a technical perspective, reoperative surgery to remove the mesh will likely damage the nerves within the operative field, and the panel advised neurectomy.

Orchialgia

Chronic testicular pain (orchialgia) has been left out of the scope of this algorithm, focusing primarily on inguinal pain. In most cases of orchialgia, the etiology is neuroanatomically and causatively distinct from CPIP. Accordingly, triple neurectomy is typically ineffective for this indication. The management of orchialgia after inguinal herniorrhaphy remains challenging, and it is important to note that it can arise after all variants of inguinal repair. Resection of the paravasal fibers or spermatic cord denervation might be an option for patients with neuropathic testicular pain but must be performed proximally to the level of pathology. Orchiectomy remains an option, but should be reserved only for refractory cases with evidence of nociceptive pain and parenchymal testicular compromise [23].

Conclusion

Since the nature of this algorithm is expert opinion, it should not be considered as a strict guideline. Rather, it should serve as a practical tool for surgeons and clinicians treating the complex problem of CPIP. The algorithm can help direct appropriate management based upon the standard practice of an international group of surgeons considered expert on inguinal hernia surgery. It will also serve as a standard for further research representing the starting point for a developing dynamic algorithm.

In conclusion, with the frequency of inguinal hernia correction as one of the most performed operations worldwide and the high incidence of CPIP, there is need for guidelines with regard to the management of CPIP.

This algorithm will hopefully serve as a guide to the management of these patients and help to improve clinical outcomes. If an expectative phase of a few months has passed without any amelioration of CPIP, a multidisciplinary approach is indicated and a pain management team should be consulted. If conservative measures fail and surgery is considered, triple neurectomy or correction for recurrence with or without neurectomy should be performed. Surgeons less experienced with remedial operations for CPIP should not hesitate to refer their patients to dedicated hernia-surgeons.

Acknowledgements *Special thanks to the following colleagues involved in the emergence of this algorithm:*

M.E. Arregui, C.R. Berney, F. Berrevoet, G. Campanelli, D.C. Chen, R.J. Fitzgibbons Jr., D. van Geldere, C. de Gheldere, H. van Goor, J. Jeekel, A.N. Kingsnorth, J.F. Lange Sr., R. Kauffman, M. Miserez, O. Mjaland, A. Montgommery, S. Morales-Conde, F.E. Muysoms, E.P. Pellisier, J.P.E.N. Pierie, R.J. Ploeg, W.M. Reinpold, M.P. Simons, R.K.J. Simmermacher, F. Ugahary, A.R. Wijsmuller

This chapter is based on an article published in *Hernia* (Lange JF, Kaufmann R, Wijsmuller AR, Pierie JP, Ploeg RJ, Chen DC, Amid PK. An international consensus algorithm for management of chronic postoperative inguinal pain. Hernia. 2015;19(1):33–43), with the kind permission of the editor and Springer Science + Business Media.

References

1. Franneby U, Sandblom G, Nordin O, Nyrén O, Gunnarsson U. Risk factors for long-term pain after hernia surgery. Ann Surg. 2006;244:212–9.
2. Bay-Nielsen M, Nilsson E, Nordin P, Kehlet H, Swedish Hernia Data Base the Danish Hernia Data Base. Chronic pain after open mesh and sutured repair of indirect inguinal hernia in young males. Br J Surg. 2004;91(10):1372–6.
3. Callesen T, Bech K, Kehlet H. Prospective study of chronic pain after groin hernia repair. Br J Surg. 1999;86(12):1528–31.
4. Cunningham J, Temple WJ, Mitchell P, Nixon JA, Preshaw RM, Hagen NA. Cooperative hernia study. Pain in the postrepair patient. Ann Surg. 1996;224(5):598–602.
5. Holzheimer RG. Low recurrence rate in hernia repair—results in 300 patients with open mesh repair of primary inguinal hernia. Eur J Med Res. 2007;12(1):1–5.
6. Mikkelsen T, Werner MU, Lassen B, Kehlet H. Pain and sensory dysfunction 6 to 12 months after inguinal herniotomy. Anesth Analg. 2004;99(1):146–51.
7. Vironen J, Nieminen J, Eklund A, Paavolainen P. Randomized clinical trial of Lichtenstein patch or Prolene Hernia System for inguinal hernia repair. Br J Surg. 2006;93(1):33–9.

8. Kehlet H, Jensen TS, Woolf CJ. Persistent postsurgical pain: risk factors and prevention. Lancet. 2006;367(9522):1618–25.
9. Kehlet H. Chronic pain after groin hernia repair. Br J Surg. 2008;95(2):135–6.
10. Amid PK. Lichtenstein tension-free hernioplasty: its inception, evolution, and principles. Hernia. 2004;8(1):1–7.
11. Poobalan AS, Bruce J, Smith WC, King PM, Krukowski ZH, Chambers WA. A review of chronic pain after inguinal herniorrhaphy. Clin J Pain. 2003;19(1):48–54.
12. Wright D, Paterson C, Scott N, Hair A, O'Dwyer PJ. Five-year follow-up of patients undergoing laparoscopic or open groin hernia repair: a randomized controlled trial. Ann Surg. 2002;235(3):333–7.
13. Amid PK. Radiologic images of meshoma: a new phenomenon causing chronic pain after prosthetic repair of abdominal wall hernias. Arch Surg. 2004;139(12):1297–8.
14. Lange JF, Kaufmann R, Wijsmuller AR, Pierie JP, Ploeg RJ, Chen DC, Amid PK. An international consensus algorithm for management of chronic postoperative inguinal pain. Hernia. 2015;19(1):33–43.
15. Amid PK, Chen DC. Surgical treatment of chronic groin and testicular pain after laparoscopic and open inguinal hernia repair. J Am Coll Surg. 2011;213(4):531–6.
16. Merskey H, Bogduk N. Classification of chronic pain: descriptions of chronic pain syndromes and definitions of pain terms. In: Task Force of the IASP. 2nd ed. Seattle: IASP Press; 1994. p. 209–14.
17. Alfieri S, Amid PK, Campanelli G, Izard G, Kehlet H, Wijsmuller AR, Di Miceli D, Doglietto GB. International guidelines for prevention and management of post-operative chronic pain following inguinal hernia surgery. Hernia. 2011;15(3):239–49.
18. Aasvang E, Kehlet H. Surgical management of chronic pain after inguinal hernia repair. Br J Surg. 2005;92(7):795–801.
19. Wijsmuller AR, Lange JF, Kleinrensink GJ, van Geldere D, Simons MP, Huygen FJ, Jeekel J, Lange JF. Nerve-identifying inguinal hernia repair: a surgical anatomical study. World J Surg. 2007;31(2):414–20. Discussion: 421–2.
20. Chen DC, Hiatt JR, Amid PK. Operative management of refractory neuropathic inguinodynia by a laparoscopic retroperitoneal approach. JAMA Surg. 2013;148(10): 962–7.
21. Song JW, Wolf Jr JS, McGillicuddy JE, Bhangoo S, Yang LJ. Laparoscopic triple neurectomy for intractable groin pain: technical report of 3 cases. Neurosurgery. 2011; 68(2):339–46.
22. Amid PK, Hiatt RH. New understanding of the causes and surgical treatment of postherniorrhaphy inguinodynia and orchalgia. J Am Coll Surg. 2007;205(2):381–5.
23. Chen DC, Amid PK. Persistent orchialgia after inguinal hernia repair: diagnosis, neuroanatomy, and surgical management: Invited comment to: Role of orchiectomy in severe testicular pain and inguinal hernia surgery: audit of Finnish patient insurance centre. Rönka K, Vironen J, Kokki H, Liukkonen T, Paajanen H. Doi 10.1007/s10029-013-1150-3. Hernia. 2015;19(1):61–3.

19. Radiologic Evaluation for Postoperative Groin Pain

Joseph M. Miller, David N. Ishimitsu, and Rola Saouaf

The use of mesh in inguinal hernia repair has substantially decreased hernia recurrence; however, the trade-off appears to have been an increase in the incidence of inguinodynia, or chronic inguinal pain [1]. While pain-generating complications like hematoma or abscess formation are likely to be visible with any modality (ultrasound, computed tomography, magnetic resonance), most other complications are subtle in nature. Barring frank recurrence, mesh abnormalities such as migration, meshoma, or mesh reaction are outside the scope of ultrasound (US) evaluation, and may be of indeterminate significance on computed tomography (CT). Inappropriate nerve division, in particular, requires magnetic resonance (MR) imaging, as neuroma is currently beyond the capability of CT. As such, MR should be considered both the definitive and first-line modality for the most specific evaluation of the postoperative groin.

Fluid Collections

Fluid collections may form anywhere along the path of surgical intervention in the days following herniorrhaphy. They can cause pain and discomfort or lead to subsequent complications. In the case of early postoperative bulge, US can be used to quickly evaluate for bowel content versus the presence of fluid collection. On US imaging, mesh appears isoechoic to surrounding tissues once incorporated [2]. Seromas consist of simple fluid and appear hypoechoic with bright posterior margins, a property known as posterior acoustic enhancement. Hematomas

B.P. Jacob et al. (eds.), *The SAGES Manual of Groin Pain*,
DOI 10.1007/978-3-319-21587-7_19

and abscesses may be indistinguishable by US, as it is impossible to exclude whether a hematoma has been secondarily infected. Traditional ultrasound characteristics of hematoma include a lacy, reticular pattern indicative of clot with the preservation of posterior acoustic enhancement. Definitive abscess reveals thickened walls with posterior acoustic enhancement and complex echotexture; unlike the linear features of organizing hematoma, an abscess may have a mixed appearance of irregular bright and dark foci. All three collections may reveal fluiditic motion when compressed by US probe or on Valsalva, and Doppler US should demonstrate no internal flow pattern in any case.

The primary distinguishing feature of the various fluid collections on CT is attenuation, or density, as measured in Hounsfield Units. On modern multidetector CT machines, simple fluid such as that found in a seroma is considered to range from 0 to 20 HU, though typically in the lower range [3]. Hematoma and abscess both appear more complex than simple fluid; the presence of gas is a strong feature of infected fluid, though no CT feature can guarantee sterility [4]. While the appearance of a thick rim and surrounding soft tissue edema may help to identify abscess, intravenous contrast may be needed for definitive characterization, as abscess walls tend to enhance while hematomas do not.

MR allows for contrast-free characterization of fluid collection. The fluid of a simple seroma should appear dark on T1 and bright on T2 without appreciable wall thickness. While abscess also appears dark on T1 and bright on T2, its T2 signal may be more heterogeneous and its wall thickened, with surrounding edema and tissue reaction apparent on STIR (short tau inversion recovery) sequences, in particular (Fig. 19.1).

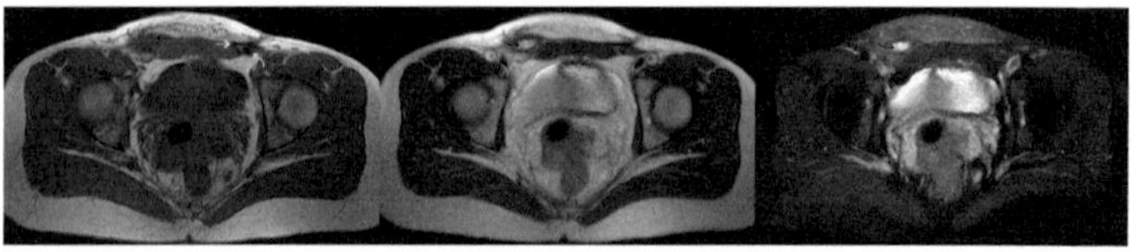

Fig. 19.1. Axial MR T1-weighted, T2-weighted, and fat-saturated T2-weighted images of bilateral flat mesh repairs. The normal mesh on the *left* (*white arrow*) appears smooth and linear, particularly on T1. Dark "blooming" artifacts (*curved white arrow*) are commonly encountered in the postoperative groin and are due to the presence of metal tacks or staples. In comparison, the mesh on the *right* is distorted and difficult to visualize (*black arrow*) due to mesh infection. There is a large fluid collection (*curved black arrows*) just posterior to mesh material, best visualized on the fluid-sensitive T2 sequences.

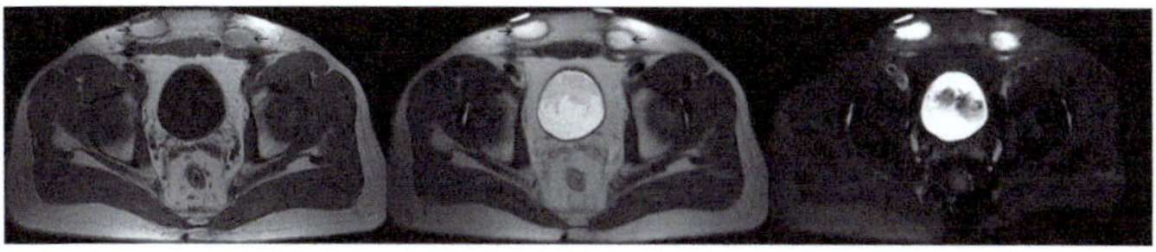

Fig. 19.2. Axial MR T1-weighted, T2-weighted, and fat-saturated T2-weighted images of bilateral flat mesh repairs. Both mesh are intact; however, large superficial fluid collections are seen bilaterally (*black arrows*). While simple seroma would appear bright on T2 sequences only, this fluid is also intermediately bright on T1 sequences, suggesting the presence of significant blood products. Both groins were percutaneously drained, demonstrating hematoma.

Subacute blood products should appear bright on T1, differentiating between hematoma and abscess [5]. If uncertainty still exists, contrast may be utilized to demonstrate peripheral enhancement in abscess. Additionally, diffusion-weighted imaging (DWI) may be used to differentiate the diffusion-restricting abscess (DWI bright, ADC dark) from organizing hematoma, although early hematoma may restrict as well (Fig. 19.2) [6].

Mesh Complications

Mesh complications often present with subtle imaging findings, and may require knowledge of the operative technique utilized to diagnose definitively. US does not reliably identify the mesh, especially if it is folded, balled up, or otherwise complicated (Fig. 19.3). As such, US is not recommended as a first-line imaging modality to evaluate the postoperative groin after mesh implantation when the integrity of the mesh itself is in question. Due to the combination of low material density and minimal profile, normal mesh material is often indistinguishable from surrounding tissue on CT [2], requiring the radiologist to discern a postoperative state from the presence and location of the patient's surgical scars (Fig. 19.4). Even in states of chronic inflammation (e.g., mesh reaction), it may be impossible to specifically identify pathology on the basis of CT alone. On MR, flat mesh materials appear as dark linear bands on T1 sequences, slightly thicker than normal fascial planes, but may be more difficult to identify among their surrounding tissues on fluid-sensitive sequences (Fig. 19.5).

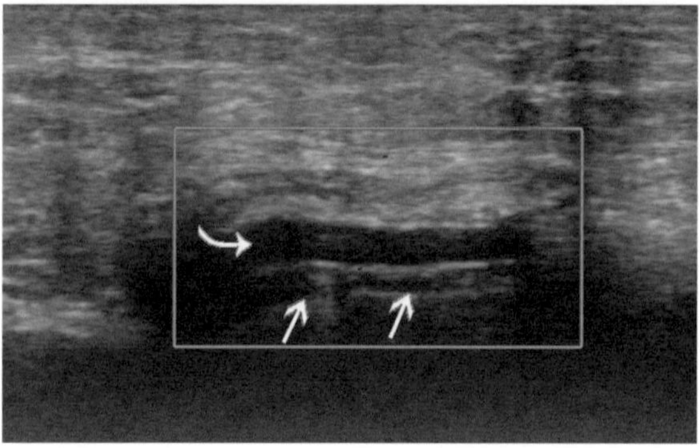

Fig. 19.3. The flat mesh (*white arrows*) shown on this Doppler US is hardly conspicuous, and would be even less so if not for the small fluid collection (*curved white arrow*) overlying it.

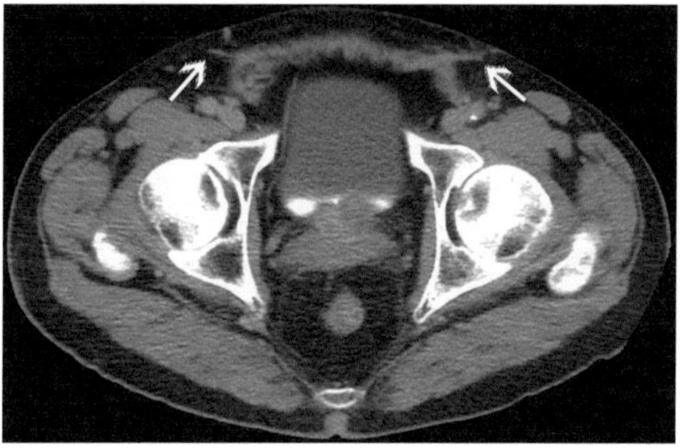

Fig. 19.4. The bilateral flat mesh (*white arrows*) seen in this axial CT of the pelvis look similar to scar tissue, making it difficult to differentiate subtle mesh abnormalities.

Normal mesh should appear smooth, and wrinkling may represent migration with subsequent focal recurrence or inflammatory response. While mesh may be fixed in position with sutures, staples, or tacks, shifting of mesh material may occur. If slippage does occur, even partial mesh migration may result in gross recurrence of bowel protrusion, or may present as subtle herniation of peritoneal or preperitoneal fat. Interposition of even small amounts of fat between mesh material and the abdominal wall may be a cause of significant pain. Dynamic MR

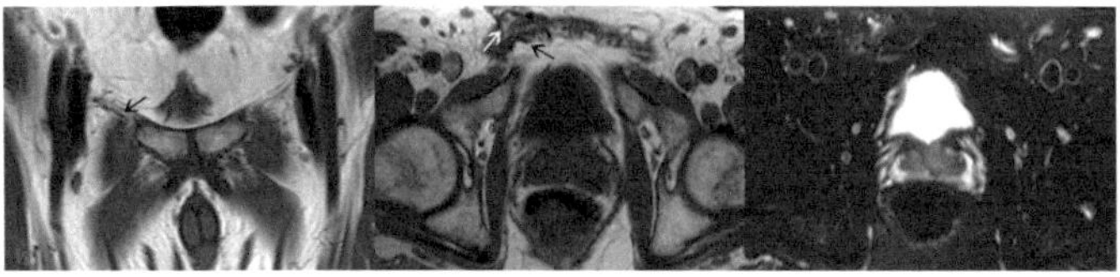

Fig. 19.5. Coronal MR T1-weighted, axial T1-weighted, and axial fat-saturated T2-weighted images of flat mesh material (*black arrows*) closing an indirect defect of the right inguinal canal (*white arrow*). Flat mesh materials typically appear as a *thick hypointense line*. While there is some undulation of this mesh, there is no inappropriate folding and no significant mass effect to suggest meshoma or mesh reaction.

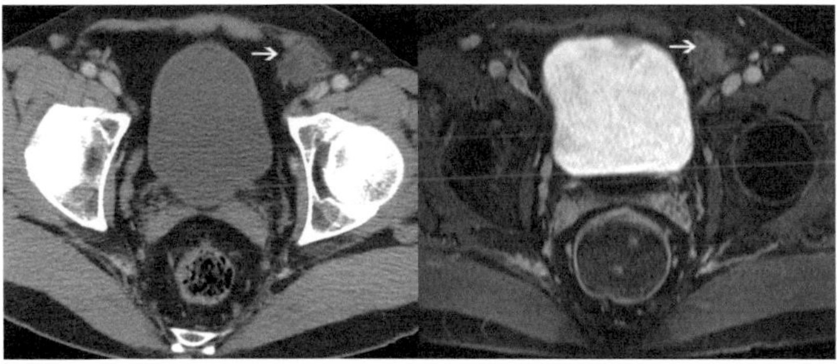

Fig. 19.6. Contrast-enhanced axial CT and corresponding contrast enhanced T1-weighted axial MR images of mesh plug within the left direct space. Both modalities reveal a large "mass" (*white arrows*) with similar density/intensity as soft tissue. This appearance is due to incorporation of soft tissue within the mesh plug's interstices. Though there is mild irregularity within the plug, the lack of contrast enhancement and normal appearance of the surrounding fat confirms lack of mesh pathology.

sequences are particularly capable of identifying such pathology, which may be missed by CT.

Pain associated with more complex mesh materials such as plug or sandwich designs are uniquely challenging to diagnose, as many radiologists are not aware of the existence of such materials and may not be able to recognize normal postoperative appearance without access to detailed operative notes or direct communication by the referring physician. Volume-occupying materials utilized in repair of large, patulous defects often incorporate significant biological material into their interstices, appearing on imaging as large pseudomasses and resulting in misdiagnosis (Fig. 19.6). As an example, our own personal case series includes a post-herniorrhaphy patient with significant pain who underwent percutaneous biopsy of "bilateral inguinal masses" that were in fact mesh Perfix™ plugs that were impressing upon the contents of the inguinal canal, resulting in testicular venous congestion and chronic pain. Most plugs are found within the inguinal canal with a cone appearance, and conversion to a different shape (especially a narrow needle-like one) may be indicative of abnormality, suggesting migration as an etiology of groin pain. When properly positioned, sandwich mesh, such as the Prolene™ Hernia System, should be seen on MR as a T1 dark band superficial to the internal inguinal ring or direct fascial defect, with a second layer positioned preperitoneally: the presence of fat between the layers represents normal appearance.

When part or all of the implanted mesh material migrates or folds into a circumscribed ball and produces mass effect on surrounding tissues, it is referred to as a meshoma and can result in significant pain and discomfort [7]. Similar inflammatory response to mesh materials without apparent deformity is called mesh reaction, and can be characterized by soft tissue edema. Mild tissue reaction has been reported even years after herniorrhaphy without evidence of infection [8]. The identification of both pathologies has led to the development of at least partially bioabsorbable mesh materials in an effort to limit tissue response, although these materials may predispose to recurrent hernia [9]. This is because bioabsorbable devices consist of decreased amounts of synthetic materials that are instead interlaced with collagen for structural integrity. As that collagen is broken down over time, less mesh is left in place. Totally synthetic-free "biologic" mesh is also finding clinical applications and would be expected to be indistinguishable on imaging from surrounding tissues without gross pathology present.

Neurologic Complications

The iliohypogastric, ilioinguinal, and genitofemoral nerves all traverse the areas involved in surgical repair of inguinal hernia, and as such are predisposed to injury during herniorrhaphy, depending on the technique used. As a quick review of the relevant neuroanatomy: [10] the iliohypogastric nerve runs deep to the internal oblique muscle, its cutaneous branch emerging about a centimeter above the external ring of the inguinal canal; the ilioinguinal nerve traverses the inguinal canal and supplies the tissues around and overlying the external inguinal ring; the genital branch of the genitofemoral nerve enters the inguinal canal at the

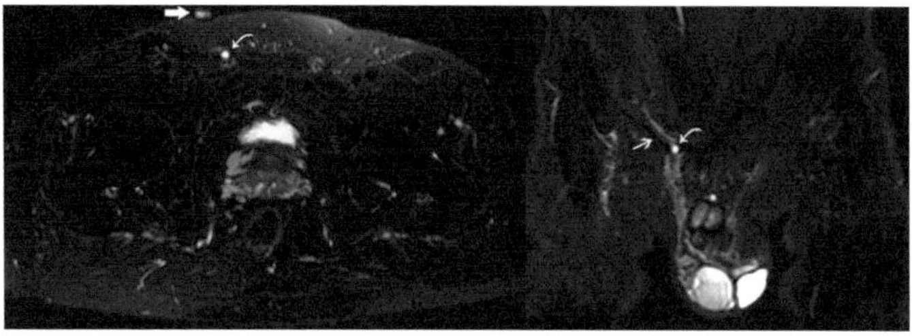

Fig. 19.7. Axial and coronal MR neurogram images. Mesh is present within the right inguinal canal (*white arrow*), its inferior border in contact with the ilioinguinal nerve. An ilioinguinal neuroma (*curved white arrows*) has formed as a result, and is found directly below the patient's pain as indicated by the skin fiducial marker (*large white arrow*).

internal inguinal ring and follows the spermatic cord/round ligament into the scrotum/labia majora; and the femoral branch of the genitofemoral nerve passes under the inguinal ligament alongside the external iliac artery, innervating the femoral triangle.

Whether directly identified or merely implied by the course of the mesh's edge, the presence of sutures, tacks, or staples along the expected path of these nerves should raise the suspicion of neuropathic etiology for chronic post-herniorrhaphy pain. Entrapment, perineural fibrosis, and neuroma are all readily apparent on MR, presenting as T2 hyperintensity within the affected nerve (Fig. 19.7). MR neurograms are specifically protocoled non-contrast MR images that allow for high-resolution evaluation of the peripheral nervous system, but suffer from low signal-to-noise ratios and should ideally be performed with a 3T magnet if available.

Other Complications

The two most common laparoscopic approaches to inguinal hernia repair are known as TAPP (transabdominal preperitoneal) and TEP (totally extraperitoneal), where TAPP perforates the peritoneum twice and TEP remains outside the peritoneal cavity altogether (Fig. 19.8) [11, 12]. The TAPP approach in particular can predispose to scarring and adhesion formation. CT is the preferred method of evaluation for postoperative bowel obstruction, or trocar-associated bowel injury. Inflammatory response to mesh may involve nearby pelvic structures such as the bladder, resulting in intermittent nonspecific pain syndromes

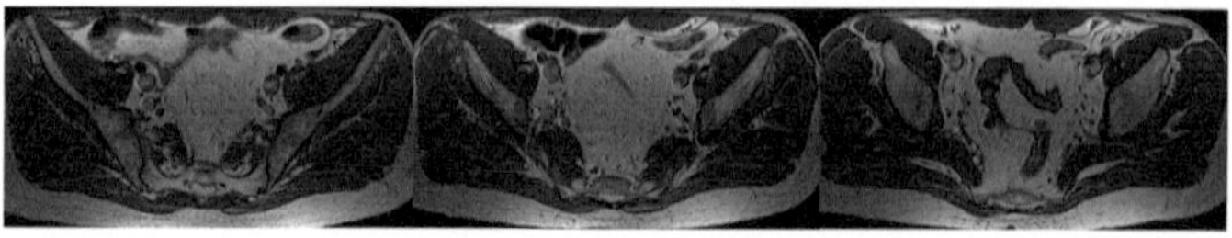

Fig. 19.8. Axial MR T1-weighted images of laparoscopic mesh (*black arrows*) placed with a transabdominal preperitoneal (TAPP) approach.

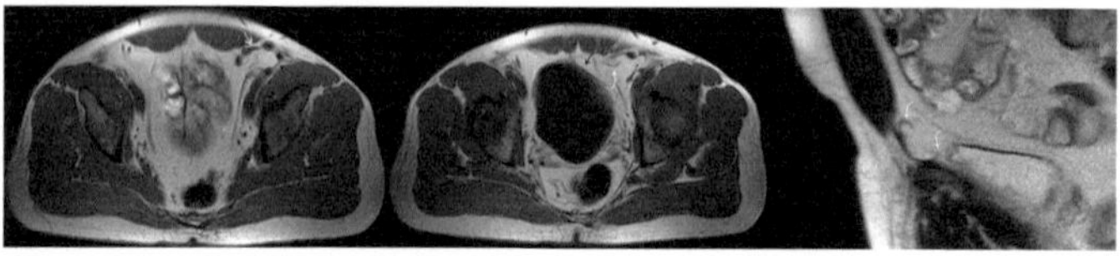

Fig. 19.9. Pair of axial MR T1-weighted images show an example of abnormal laparoscopic mesh (*black arrow*), with areas of irregular central contour (*white arrow*) and numerous tacks. The margins of the mesh are curled and folded (*curved white arrows*). The bladder is distorted (*curved black arrow*), suggestive of local mass effect. Significant adhesions were present at surgical exploration.

that are temporally related to urination or defecation. It is important to identify subtle findings on images related to these repairs, such as bladder asymmetry or its adhesion to a site or segment of the repair (Fig. 19.9). Regardless of approach, attachment of mesh to the pubic tubercle by tack or staple may result in periosteal reaction, another source of chronic pain.

References

1. Fischer JE. Hernia repair: why do we continue to perform mesh repair in the face of the human toll of inguinodynia? Am J Surg. 2013;206(4):619–23.
2. Parra JA, Revuelta S, Gallego T, Bueno J, Berrio JI, Fariñas MC. Prosthetic mesh used for inguinal and ventral hernia repair: normal appearance and complications in ultrasound and CT. Br J Radiol. 2004;77(915):261–5.
3. Israel GM, Bosniak MA. How I do it: evaluating renal masses. Radiology. 2005;236(2):441–50.
4. Allen BC, Barnhart H, Bashir M, Nieman C, Breault S, Jaffe TA. Diagnostic accuracy of intra-abdominal fluid collection characterization in the era of multidetector computed tomography. Am Surg. 2012;78(2):185–9.

5. Noone TC, Semelka RC, Worawattanakul S, Marcos HB. Intraperitoneal abscesses: diagnostic accuracy of and appearances at MR imaging. Radiology. 1998;208(2): 525–8.
6. Attariwala R, Picker W. Whole body MRI: improved lesion detection and characterization with diffusion weighted techniques. J Magn Reson Imaging. 2013;38(2): 253–68.
7. Amid PK. Radiologic images of meshoma: a new phenomenon causing chronic pain after prosthetic repair of abdominal wall hernias. Arch Surg. 2004;139(12):1297–8.
8. Bahçeci T, Nursal GN, Aydın M. Intense FDG uptake around the inguinal surgical mesh 5 years after operation: case report and review of the literature. Mol Imaging Radionucl Ther. 2012;21(1):35–7.
9. Symeonidis D, Efthimiou M, Koukoulis G, Athanasiou E, Mamaloudis I, Tzovaras G. Open inguinal hernia repair with the use of polyglycolic acid/trimethylene carbonate absorbable mesh: a critical update of the long-term results. Hernia. 2013;17(1): 85–7.
10. Apte G, Nelson P, Brismée JM, Dedrick G, Justiz 3rd R, Sizer Jr PS. Chronic female pelvic pain—part 1: clinical pathoanatomy and examination of the pelvic region. Pain Pract. 2012;12(2):88–110.
11. Bittner R, Schmedt CG, Schwarz J, Kraft K, Leibl BJ. Laparoscopic transperitoneal procedure for routine repair of groin hernia. Br J Surg. 2002;89(8):1062–6.
12. Faure JP, Doucet C, Rigouard P, Richer JP, Scépi M. Anatomical pitfalls in the technique for total extra peritoneal laparoscopic repair for inguinal hernias. Surg Radiol Anat. 2006;28(5):486–93.

20. Management of Groin Pain: Interventional and Pharmacologic Approaches

Anuj Malhotra

Editor's Comment (BPJ)

Many hernia surgeons do not have a set algorithm for managing a patient with the chief complaint of chronic groin pain. Thus when surgeons encounter a patient who is in distress from chronic groin pain, all too often the ability to care appropriately for that patient can break down and patient satisfaction can plummet. Written by a dedicated pain management specialist, this chapter provides an amazing overview of the management strategies for chronic groin pain and highlights the importance of having a dedicated pain team to assist in what can sometimes be a very complicated management plan. It is a must read!

Introduction

Pain is defined by the International Association for the Study of Pain as "an unpleasant sensory and emotional experience associated with actual or potential tissue damage" [1]. Postoperative pain is a normal, adaptive phenomenon which prevents further harm such as inadvertent trauma and generally resolves with tissue healing. However, for some individuals this pain persists and becomes chronic. The exact point at which this transition occurs is unknown and, in fact, some patients experience a pain-free interval of weeks to months prior to onset of chronic pain [2, 3]. Regardless, if the pain persists for more than 3 months, it is often classified as chronic [1].

Pain can be further classified as nociceptive or neuropathic. Nociceptive pain is often mediated by inflammation and can represent an

B.P. Jacob et al. (eds.), *The SAGES Manual of Groin Pain*,
DOI 10.1007/978-3-319-21587-7_20

extended variant of normal tissue healing. It often presents with signs and symptoms consistent with an inflammatory process, including localized edema, erythema, excessive scar formation, and mechanical sensitivity without radiation. It may represent a reaction to a foreign body such as stitch, staple, or mesh. Neuropathic pain is more commonly characterized by shooting, burning, or electric pain; evidence of altered sensory perception, often in the distribution of a particular nerve; and proximal or distal pain relief with local anesthetic nerve block. It may result from direct nerve injury, entrapment of a nerve in scar tissue, or neuroma formation. Neuroplasticity of the central and peripheral nervous system likely plays a role in the development of chronic pain, and thus appropriate and timely diagnosis and treatment may influence overall outcome.

Risk Factors

Identifying patients at risk for postoperative groin pain can aid in vigilant monitoring and early recognition of the transition from acute to chronic pain. Studies have identified the following risk factors for the development of persistent pain following groin hernia surgery: age younger than 40 years, preoperative pain, prior groin surgery within 3 years, severe postoperative pain, postoperative complications, preoperative dysesthesia, female gender, and anterior hernia repair [4–13]. In a survey of 2500 Swedish patients 2–3 years removed from groin surgery, 30 % reported residual groin pain with 11–14 % reporting pain severe enough to interfere with daily activities [5].

First-Line Treatments

First-line treatment of postoperative groin pain is generally conservative, to allow for expected resolution and to avoid interventions that may slow tissue healing. Initial treatment is generally with as-needed analgesic dosing with the goal of tapering off as pain improves. Opioids are indicated in the acute postoperative period, exerting their effects in the brain and dorsal horn of the spinal cord to inhibit ascending signals to the somatosensory cortex. These are avoided in the long term, if possible, due to development of tolerance, potential for opioid-induced hyperalgesia, and concerns for addiction or diversion. Nonsteroidal anti-inflammatories inhibit cyclooxygenase, thereby decreasing inflammation and pain related to local tissue injury. They inhibit platelet function to variable degrees, and for this reason are sometimes contraindicated in the immediate postoperative period. Acetaminophen and paracetamol are thought to work centrally through inhibition of prostaglandin synthesis, and are often used as adjuncts due to easy tolerability and good safety profile.

Second-Line Treatments

If pain does not improve or if escalating rather than tapering doses of "prn" medications are observed, antineuropathic agents are often initiated. These include topical agents such as lidocaine 5 % cream or patch, which blocks sodium channels and produces a local anesthetic effect in the underlying skin and superficial soft tissues. This can be of particular use if a superficial injury is expected, such as scar neuroma. Gabapentinoids are often second line due to minimal interaction with other medications and acceptable safety profile. Gabapentin and pregabalin act as modulators of calcium channels, found in abundance on small nerve terminals and in the dorsal horn of the spinal cord. They are dosed 2–3 times daily and require slow titration up to a therapeutic dose to allow for tolerance of sedation side effects. Tricyclic antidepressants (TCAs) are also of use in treating neuropathic pain, with a mechanism of action at multiple sites, including serotonin and norepinephrine reuptake inhibition, sodium channel blockade, and anticholinergic effects. Second-generation TCAs such as nortriptyline and desipramine, the active metabolites of first-generation amitriptyline and imipramine, are often more easily tolerated due to fewer anticholinergic side effects. Selective norepinephrine receptor inhibitors (SNRIs) such as venlafaxine and duloxetine have also been shown to improve neuropathic pain and are more likely to be of benefit than selective serotonin reuptake inhibitors (SSRIs). If these options are exhausted, anticonvulsants with sodium or calcium channel blocking properties such as topiramate, levetiracetam, and carbamazepine may be tried, though these agents require closer monitoring due to potential for rare but serious adverse events.

Interventional Targets

Beyond temporary relief of symptoms, treatment of chronic postoperative groin pain often includes interventions designed to identify and treat the affected area. From least to most invasive, these include diagnostic nerve blocks, selective nerve root blocks/transforaminal epidural steroid injections, therapeutic nerve blocks, pulsed radiofrequency neuromodulation of peripheral nerves or dorsal root ganglia, cryoablation, radiofrequency ablation, chemoneurolysis, and implantable peripheral field stimulation. Identifying the nerve that is most likely to be injured is the first step in pursuing an interventional approach. Groin pain by definition occurs between the abdomen and thigh, and for this reason the nerves supplying the skin and structures in this area are referred to as "border nerves" [14]. These include the ilioinguinal, iliohypogastric, genitofemoral, and lateral femoral cutaneous nerves arising from the T12–L3 anterior rami as they form the upper lumbar plexus.

1. Iliohypogastric nerve—formed by L1 with contributions from T12. Exits the lateral border of the psoas and runs in the retroperitoneum along the quadratus lumborum before emerging through the transversus abdominis plane at the iliac crest. It innervates the infraumbilical skin to the inguinal ligament.
2. Ilioinguinal nerve—formed by L1 with contributions from T12. Follows a similar course as the iliohypogastric nerve before passing lateral to the internal inguinal ring, running anterior to the spermatic cord to innervate the medial thigh, root of penis and upper scrotum in men, and mons pubis and labia majora in women.
3. Genitofemoral nerve—formed by L1 and L2. Pierces the psoas and runs along its anterior surface before splitting into a genital and femoral branch. The femoral branch runs with the external iliac artery under the inguinal ligament and innervates the anterolateral thigh. The genital branch passes through the inguinal canal and runs along with the spermatic cord to innervate the testicle and scrotum. It can run inside or outside the cord itself [14].
4. Lateral femoral cutaneous nerve—formed by L2 and L3. Exits the lateral border of the psoas and runs inferolaterally to the anterior superior iliac spine (ASIS) and then passes under the inguinal ligament and over the sartorius, located between the fascia lata and iliaca, providing innervation to the lateral thigh.

It should be noted that there is significant overlap and communication between nerves in the groin. In particular, the ilioinguinal nerve distribution may overlay that of the iliohypogastric, and, indeed, they are commonly inverse in size. Similarly, the ilioinguinal nerve has been shown to share innervation of the genital branch of the genitofemoral nerve in up to one-third of cadaver dissections [14]. Due to this anatomical variability, landmark-based techniques for nerve blockade are unlikely to be successful or specific, as they rely on large volumes of anesthetic infiltrated over a large area to ensure spread to the involved nerves and tissue planes. For example, four different landmark-based techniques medial to the ASIS and above the inguinal ligament are commonly advocated for ilioinguinal nerve block; however, high failure rates are reported, even in children where the anatomy is generally superficial [15].

Localizing Options

Ultrasound guidance offers several advantages that make it highly suited for diagnosing and treating groin pain. The machines are portable, there is no radiation exposure, the superficial locations of the "border nerves" can be easily visualized, and the lack of surrounding bony structures allows for in-plane needle advancement for accurate, safe, and highly specific diagnostic and therapeutic interventions. Nerve stimulation may be included to ensure close proximity to the involved nerves and can also be of diagnostic value, as stimulating the injured nerve may replicate the patient's usual pain. An initial block of the affected nerve using a low volume of local anesthetic can be performed with confirmation of sensory block in the expected distribution. If this block relieves the patient's usual pain, then neuralgia in this distribution is the likely diagnosis. If pain continues despite appropriate block, then another source for pain should be investigated.

1. Iliohypogastric and ilioinguinal nerve block—A linear high-frequency probe is placed with the lateral end at the ASIS and the probe oriented in the transverse axis to visualize the three layers of the abdominal wall. A needle is advanced in plane from medial to lateral with the target between the transversus abdominis and internal oblique layers, where the nerves can often be visualized 1–2 cm medial to the ASIS [16].
2. Genitofemoral nerve block—A linear high-frequency probe is placed perpendicular to the inguinal ligament with the medial end at the pubic tubercle. The spermatic cord and accompanying nerve are visible in cross section within the inguinal canal, and a needle can be advanced in plane from medial or lateral with the target within the canal and outside the cord [16].
3. Lateral femoral cutaneous nerve block—A linear high-frequency probe is placed with the lateral end at the ASIS and the probe oriented along and inferior to the inguinal ligament. The nerve is

located just below the subcutaneous tissue, deep to the fascia lata and superficial to the fascia iliaca above the sartorius muscle. A needle can be advanced from medial to lateral in this superficial plane.

If there is high suspicion of injury to one of these nerves but there is no improvement with nerve block, neuralgia proximal to the site of blockade should be considered. In this case, diagnostic selective nerve root block/transforaminal epidural steroid injection can be performed. During this procedure, a needle is positioned at the neural foramen of the selected level using fluoroscopic guidance. Contrast is injected to ensure specific spread to the nerve root alone. A small amount of local anesthetic is then injected, and blockade can be confirmed via dermatomal skin testing. If anesthetizing the nerve roots that form the nerve suspected to be injured transiently improves the pain, this can indicate involvement of the nerve in a proximal location. However, it should be noted that given overlapping nerve root innervation of the "border nerves," results may only indicate the presence of neuralgia and not the specific nerve involved.

Therapeutic Options

Once the pain source is identified, several interventional options can be tried for therapeutic benefit. Repeating the block with the addition of a corticosteroid can extend the duration of the block and may also cause mild local tissue or scar atrophy, which can lessen nerve entrapment. Good prognostic signs for this approach include increasing duration and intensity of benefit with subsequent blocks. If good relief is obtained but no extended duration occurs with repeat intervention, pulsed radiofrequency (PRF) can be tried. PRF uses radiofrequency stimulation through an insulated needle with an active 5–10 mm tip that is placed in proximity to the nerve to be treated. Neuromodulation is believed to occur due to inhibition of evoked synaptic activity [17, 18]. PRF has advantages over traditional RF, including limited tissue damage and ability to treat superficial structures due to less reliance on creation of a thermal lesion. PRF can be applied to peripheral nerves or to the dorsal root ganglia of involved nerve roots. Evidence is at the level of case series, but results thus far have been promising [19–21].

Alternatively, true ablation can be performed using a variety of techniques. However, as these techniques cause neurolysis, additional concerns must be considered. First, these techniques should not be performed on any nerves with a significant motor component as weakness will occur. Fortunately, the border nerves are primarily sensory, but given the proximity of DRGs to the spinal nerve, these cannot be safely ablated. Second, injury to adjacent structures may occur, such as to blood vessels which may be coagulated by radiofrequency ablation. Finally, neurolysis can create an area of desensitized skin, which can be bothersome for certain patients and may progress to anesthesia dolorosa, a feared complication manifested by pain in the area despite numbness to stimulation. Nonetheless, these approaches may be of significant benefit in appropriate cases and after full discussion of risks and benefits with the patient. Ablation may be performed with injection of a chemical neurolytic such as phenol or dehydrated alcohol. Radiofrequency ablation creates a thermal lesion at 80 °C along the active needle tip. Cryoablation creates a supercooled −70 °C "ice ball" around the neural sheath, which leads to decreased transmission. Again, case reports have shown promising results with decreased pain and analgesic requirements [22].

For patients refractory to percutaneous interventions, peripheral nerve field stimulation may be attempted. During the trial, stimulating leads are placed either in proximity to the affected nerve or more subcutaneously in the painful area. The leads are then sutured at the skin, and the external lead contacts connected to a patient-controlled pulse generator. The patient can then vary the stimulation intensity and pattern to produce a vibratory sensation that overlaps the affected area. After a trial of 3–5 days, the patient reports on any resulting benefit, and if there is evidence of increased function and decreased need for analgesics, a permanent device can be placed with an implantable pulse generator. Peripheral stimulation is based on the gate control theory proposed by Wall and Melzack, wherein activation of large fibers suppresses painful input from small fibers. This varies from the traditional use of stimulating leads in the dorsal column of the spinal cord and is considered an off-label use. Several case series have shown significant and sustained benefit with peripheral stimulation, which may hold particular promise long term, as there is unlikely to be a recurrence of pain after a pain-free interval as can occur with regeneration of ablated nerves following neurolysis [23–26].

Conclusion

Several recent expert reviews have attempted to quantify the effect of various interventions for chronic postsurgical groin pain [16, 27]. Many promising modalities have been identified, as above; however, differences

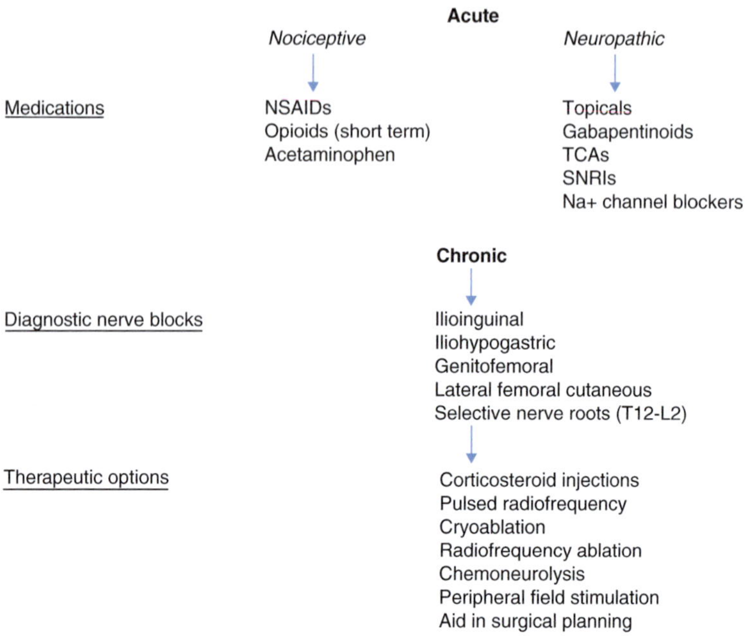

Fig. 20.1. Management of groin pain: interventional and pharmacologic approaches.

in reporting and study methodologies make it difficult to compare outcomes between interventions. In general, a stepwise approach is advocated, beginning with appropriate pharmacotherapy, then progressing to diagnostic nerve blocks, and finally to therapeutic interventions (Fig. 20.1) [27]. Early referral to a pain management physician can help expedite this process and possibly prevent the transition from acute to chronic pain. For certain patients, despite the use of multiple modalities, pain will not improve substantially or for any prolonged duration. For these patients, information from any interventions that at least temporarily lessen their usual pain may be used to help inform more definitive surgical intervention.

References

1. Merskey H, Bogduk N. Classification of chronic pain: descriptions of chronic pain syndromes and definitions of pain terms. Task force on taxonomy of the IASP. 2nd ed. Seattle: IASP Press; 1994.

2. Poobalan AS, Bruce J, Smith WC, King PM, Kurkowski ZH, Chambers WA. A review of chronic pain after inguinal herniorrhaphy. Clin J Pain. 2003;19(1):48–54.
3. Smith WC, Bourne D, Squair J, Phillips DO, Chambers WA. A retrospective cohort study of post mastectomy pain syndrome. Pain. 1999;83(1):91–5.
4. Kehlet H, Jensen TS, Woolf CJ. Persistent postsurgical pain: risk factors and prevention. Lancet. 2006;367(9522):1618–25.
5. Fränneby U, Sandblom G, Nordin P, Nyrén O, Gunnarsson U. Risk factors for long-term pain after hernia surgery. Ann Surg. 2006;244(2):212–9.
6. Reinpold WM, Nehls J, Eggert A. Nerve management and chronic pain after open inguinal hernia repair: a prospective two phase study. Ann Surg. 2011;254(1):163–8.
7. Kalliomäki ML, Meyerson J, Gunnarsson U, Gordh T, Sandblom G. Long-term pain after inguinal hernia repair in a population-based cohort; risk factors and interference with daily activities. Eur J Pain. 2008;12(2):214–25.
8. Nienhuijs SW, Boelens OB, Strobbe LJ. Pain after anterior mesh hernia repair. J Am Coll Surg. 2005;200(6):885–9.
9. Poobalan AS, Bruce J, King PM, Chambers WA, Krukowski ZH, Smith WC. Chronic pain and quality of life following open inguinal hernia repair. Br J Surg. 2001;88(8):1122–6.
10. Liem MS, van Duyn EB, van der Graaf Y, van Vroonhoven TJ, Coala Trial Group. Recurrences after conventional anterior and laparoscopic inguinal hernia repair: a randomized comparison. Ann Surg. 2003;237(1):136–41.
11. Gottrup H, Andersen J, Arendt-Nielsen L, Jensen TS. Psychophysical examination in patients with post-mastectomy pain. Pain. 2000;87(3):275–84.
12. Macrae W, Bruce J. Chronic pain after surgery. In: Wilson PR, Watson PJ, Haythornthwaite JA, Jensen TS, editors. Clinical pain management: chronic pain. London: Hodder Arnold; 2008. p. 405.
13. Aasvang EK, Møhl B, Bay-Nielsen M, Kehlet H. Pain related sexual dysfunction after inguinal herniorrhaphy. Pain. 2006;122(3):258–63.
14. Rab M, Ebmer J, Dellon AL. Anatomic variability of the ilioinguinal and genitofemoral nerve: implications for the treatment of groin pain. Plast Reconstr Surg. 2001;108(6):1618–23.
15. van Schoor AN, Boon JM, Bosenberg AT, Abrahams PH, Meuring JH. Anatomical considerations of the pediatric ilioinguinal/iliohypogastric nerve block. Paediatr Anaesth. 2005;15(5):371–7.
16. Peng P, Tumber P. Ultrasound-guided interventional procedures for patients with chronic pelvic pain—a description of techniques and review of literature. Pain Physician. 2008;11(2):215–24.
17. Mitra R, Zeighami A, Mackey S. Pulsed radiofrequency for the treatment of chronic ilioinguinal neuropathy. Hernia. 2007;11(4):369–71.
18. Van Zundert J, Patijn J, Kessels A, Lamé I, van Suijlekom H, van Kleef M. Pulsed radiofrequency adjacent to the cervical dorsal root ganglion in chronic cervical radicular pain: a double blind sham controlled randomized clinical trial. Pain. 2007;127(1–2):173–82.

19. Cohen SP, Foster A. Pulsed radiofrequency as a treatment for groin pain and orchialgia. Urology. 2003;61(3):645.
20. Rozen D, Parvez U. Pulsed radiofrequency of lumbar nerve roots for treatment of chronic inguinal herniorrhaphy pain. Pain Physician. 2006;9(2):153–6.
21. Rozen D, Ahn J. Pulsed radiofrequency for the treatment of ilioinguinal neuralgia after inguinal herniorrhapy. Mt Sinai J Med. 2006;73(4):716–8.
22. Fanelli RD, DiSiena MR, Lui FY, Gersin KS. Cryoanalgesic ablation for the treatment of chronic postherniorrhaphy neuropathic pain. Surg Endosc. 2003;17(2):196–200.
23. Paicius RM, Bernstein CA, Lempert-Cohen C. Peripheral nerve field stimulation in chronic abdominal pain. Pain Physician. 2006;9(3):261–6.
24. Rauchwerger JJ, Giordano J, Rozen D, Kent JL, Greenspan J, Closson CWF. On the therapeutic viability of peripheral nerve stimulation for ilioinguinal neuralgia: putative mechanisms and possible utility. Pain Pract. 2008;8(2):138–43.
25. Stinson LW, Roderer GT, Cross NE, Davis BE. Peripheral subcutaneous electrostimulation for control of intractable postoperative inguinal pain: a case report series. Neuromodulation. 2001;4(3):99–104.
26. Walter J, Reichart R, Vonderlind C, Kuhn SA, Kalff R. [Neuralgia of the genitofemoral nerve after hernioplasty. Therapy by peripheral nerve stimulation]. Chirurg. 2009;80(8):741–4. *[Article in German].*
27. Thomassen I, van Suijlekom H, van der Gaag A, Nienhuijs S. Intervention techniques for chronic postherniorrhaphy pain. Eur Surg. 2012;44(3):132–7.

21. Dermatome Mapping: Preoperative and Postoperative Assessment

Rigoberto Álvarez

History

During the last two decades, there has been an important decrease in recurrence rates after inguinal hernia repair [1–3]. Consequently, postoperative chronic pain has become the more frequent and important outcome that occupies our attention [4]. The proper study of chronic postoperative pain both prior to its treatment and during subsequent monitoring is of great significance, to the extent that the success or failure of our management depends on proper terminology, characterization, description, and diagnosis. Unfortunately, at present there are no radiological or electromyographic studies that show us with certainty the real scenarios that we face in clinical practice. The available pain evaluation questionnaires are more oriented to determine the degree of limitation and disability than to provide us with reliable guidelines to implement the proper therapies for each individual case [5, 6]. The significance cannot be understated, given the importance of the labor and legal concerns due to the implications and impact of disabling pain after inguinal hernia repair. The generally accepted time frame for defining chronic postoperative pain is after the third postoperative month [7, 8]. There are two main types of postoperative pain that contribute to post-inguinal herniorrhaphy inguinodynia. The first is the nociceptive or non-neuropathic pain, which is the most common and is caused by the rupture of soft tissues such as skin, subcutaneous tissue, and muscles; as a result of tissue trauma, cutting, or cauterization during surgery; as well as pain caused by the presence of orchialgia, meshomas, or granulomas. The second type of pain is neuropathic. It usually appears in the early postoperative period and involves injury of one or several nerve branches, mainly caused by two mechanisms: when these nerves are cut

B.P. Jacob et al. (eds.), *The SAGES Manual of Groin Pain*,
DOI 10.1007/978-3-319-21587-7_21

intentionally or incidentally, or due to entrapment of the nerve by a suture or penetrated when fixating with tacks. This type of pain is the most persistent and severe in intensity.

It is noteworthy that after more than 10 years evaluating patients with chronic postoperative pain, we have not had a single case of lateral femoral cutaneous nerve involvement. The most frequently affected nerve by both open and laparoscopic approaches has been the ilioinguinal nerve, followed by the genital branch in the open technique, especially in patients in whom plugs were placed in the internal ring where this branch emerges, or when the round ligament was severed. The femoral branch of the genitofemoral nerve was involved mainly in the laparoscopic approach [9]. We believe that the mechanism of injury of the femoral branch is caused by traction, pulling, or rupture of this structure when dissecting near to the iliac vessels. This is probably the reason why most of these patients improve with conservative management, since the mechanism of injury is not entrapment. Even so, the recommended treatment for patients with persistent pain involving femoral branch dermatome after 3 months is the lumboscopic approach and truncal genitofemoral nerve resection [10, 11]. The surgical management often involves removal of the mesh [12, 13] and tacks with neuropathy addressed with selective neurectomy [8, 9, 13], triple neurectomy [14, 15], or extended or quadruple truncal neurectomy [10, 11].

Introduction

Dermatome Mapping

Clinical evaluation of the dermatomes involved with each nerve trunk has been a routine part of our general medical practice due to our academic training in routine neurological assessment; therefore, to consider dermatome mapping in a comprehensive evaluation of the patient with chronic postoperative groin pain is of utmost significance when addressing such patients. For this reason, in 1998 we developed and implemented the dermatome mapping test (DMT) as an integral tool in the evaluation of our patients [11, 15]. Since then, DMT has shown us a high sensitivity when matching mapped results with the surgical findings. More importantly, it has demonstrated high sensitivity with histological results [11] and postsurgical evaluation and outcomes (see Clinical Cases below).

Subsequent analysis has allowed us to implement technical modifications such as extending our evaluation to include the upper third of the thigh, because we found a high frequency of injury or involvement of the femoral branch, especially in patients approached laparoscopically. The dermatome mapping aims to determine the specific source of pain with regard to type and intensity in order to establish therapeutic guidelines. Unlike the rest of the proposed assessments that primarily determine the impact on quality of life in the patient [5, 6], we consider the two completely different evaluation techniques yet highly complementary to each other.

The current diagnostic methods for the evaluation of chronic postoperative pain are primarily clinical. Electromyographical studies and evoked potentials do not show a reliable sensitivity or specificity and are difficult to implement and interpret. Imaging in general such as computerized tomography (CT) magnetic resonance imaging (MRI) or ultrasound (US) has demonstrated only limited utility in cases of granulomas or meshomas; however, they are of no use in chronic pain of neuropathic origin.

Technique Description

As a stimulator, a regular ballpoint pen is used to apply the pressure needed to assess the deep sensation of the dermatome evaluated. Three permanent markers black, red, and blue (Fig. 21.1) are used to mark and delineate assessed areas.

In all patients with chronic postoperative pain, dermatome mapping test (DMT) can be implemented using as reference a point one-inch lateral to the umbilicus contralateral to the region to be evaluated. With respect to this point, sequentially go from the superior iliac crest to the midline at a distance of no more than one inch between each point radially, continuing down to the upper third of the scrotum and penis for males, or the labia for women. Continue inferiorly to reach and evaluate the upper third of the thigh, including the anterior, lateral, and medial sides (Figs. 21.2, 21.3, 21.4, 21.5, 21.6, 21.7, 21.8, 21.9, 21.10, and 21.11).

Once the dermatome mapping is completed, proceed to photograph the area and integrate this into the clinical record in order to have an objective view of this event. We can follow up with subsequent mapping to compare with previous DMTs in cases of vague pain scenarios or in preoperative versus postneurectomy pain assessments after quadruple, triple, or selective neurectomy.

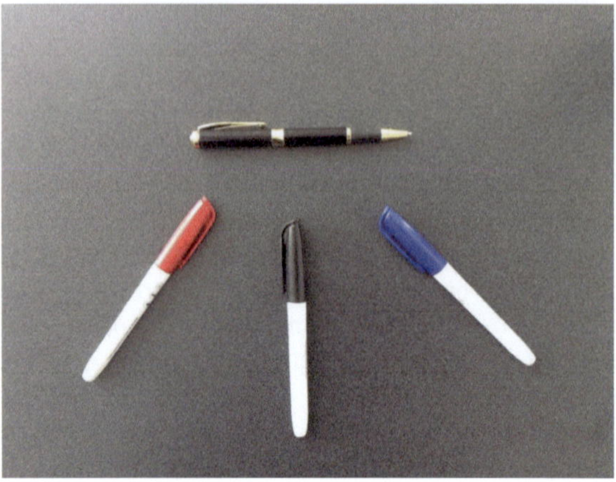

Fig. 21.1. Material: A ballpoint pen and three permanent markers, *black* for normal sensation (isoesthesia), *red* for pain or tenderness (hyperesthesia), and *blue* for anesthesia, numbness, or discomfort (hypoesthesia). Each point is evaluated and marked as follows: (**a**) A *circle* in *black* for those who have a similar sensation to the reference para-umbilical point. (**b**) A *cross* in *red* for those points where the patient feels pain and/or hypersensitivity. (**c**) A *minus* in *blue* for those points of anesthesia, hypoesthesia, or discomfort (superficial burning or numbness).

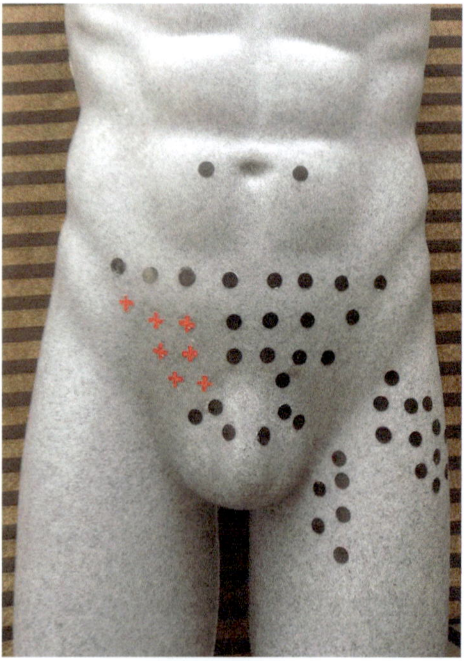

DMC = R-N-1-IX

Fig. 21.2. Dermatome mapping compatible with postoperative pain due to right ilioinguinal nerve involvement. *DMC* Dermatome Mapping Classification.

Fig. 21.3. Dermatome mapping compatible with postoperative pain due to left iliohypogastric nerve involvement.

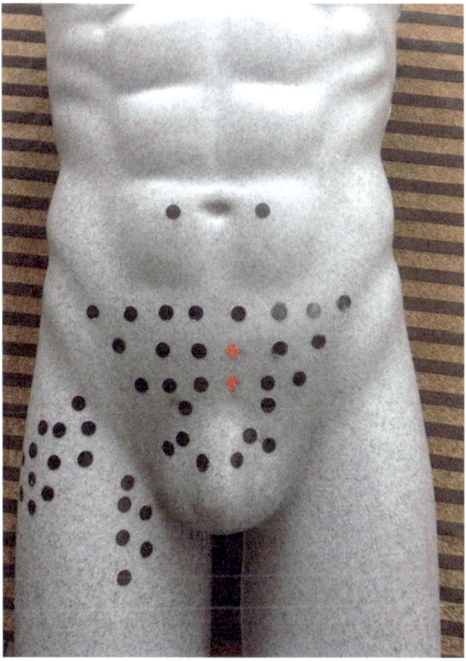

DMC = L-N-2-VIII

Fig. 21.4. Dermatome mapping compatible with postoperative pain due to left genital branch involvement. This injury occurs mainly when the round ligament is severed or plugs are placed in the internal ring where this branch emerges.

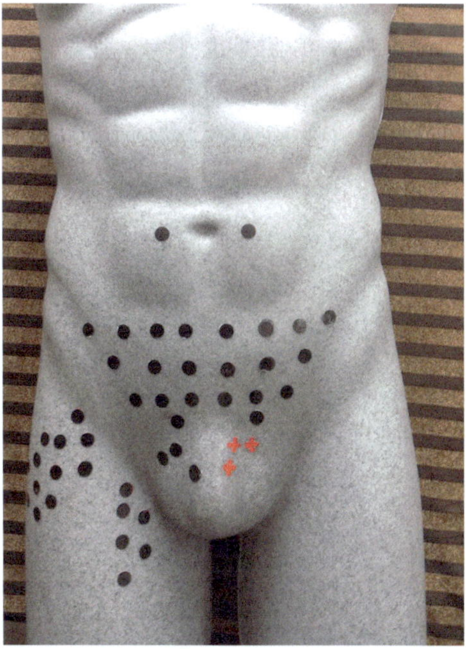

DMC = L-N-3-X

Fig. 21.5. Dermatome mapping compatible with postoperative pain due to left femoral branch involvement. This injury occurs mainly in patients approached laparoscopically.

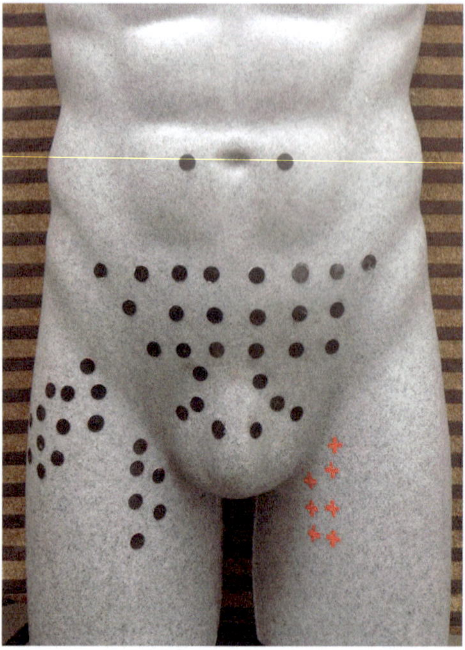

DMC = L-N-4-X

Fig. 21.6. Dermatome mapping compatible with postoperative pain due to left ilioinguinal nerve and left femoral branch involvement. This injury occurs mainly in patients approached laparoscopically.

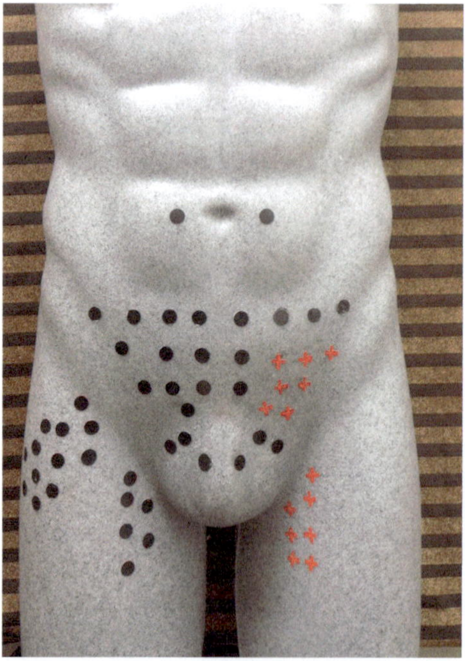

DMC = L-N-1-X / L-N-4-IX

Fig. 21.7. Dermatome mapping compatible with postoperative denervation, numbness, or discomfort after a successful right triple neurectomy. As a follow-up to chronic postoperative pain assessment.

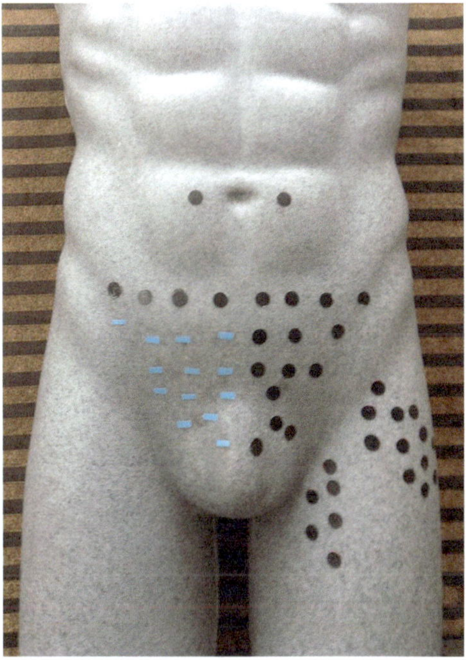

DMC = R-D-1,2,3-II

Fig. 21.8. Dermatome mapping compatible with postoperative denervation, numbness, or discomfort due to successful selective neurectomy of the left ilioinguinal nerve. As a follow-up to chronic postoperative pain assessment.

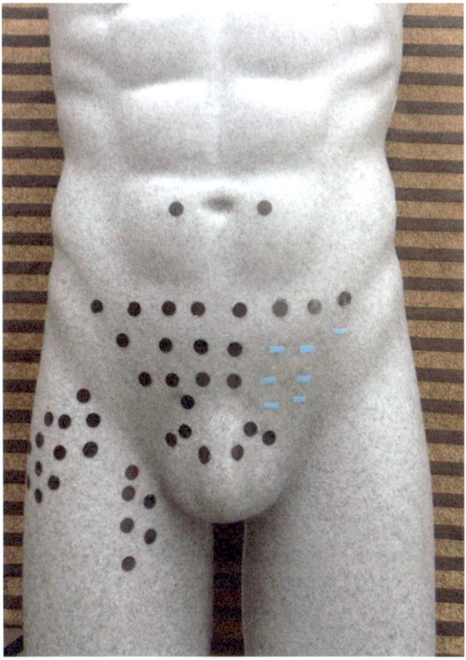

DMC = L-D-1-I

Fig. 21.9. Dermatome mapping compatible with a non-neuropathic or nociceptive *VAGUE* postoperative etiology of pain because it does not follow the pattern of a dermatome. This is probably the case of a simulator due to malingering, conversion, or other non-anatomic pathology. In these scenarios, dermatome mapping may be repeated a week later and photographed, with expectations of a complete different pattern.

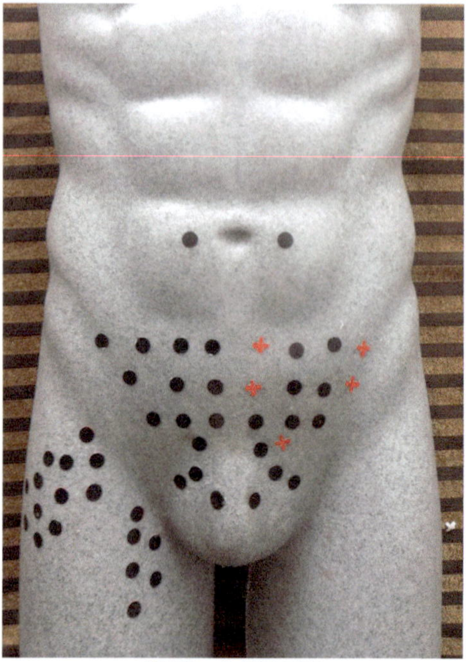

DMC = L-NN-V-X

Fig. 21.10. Dermatome mapping compatible with a non-neuropathic or nociceptive postoperative *PUBALGIA*.

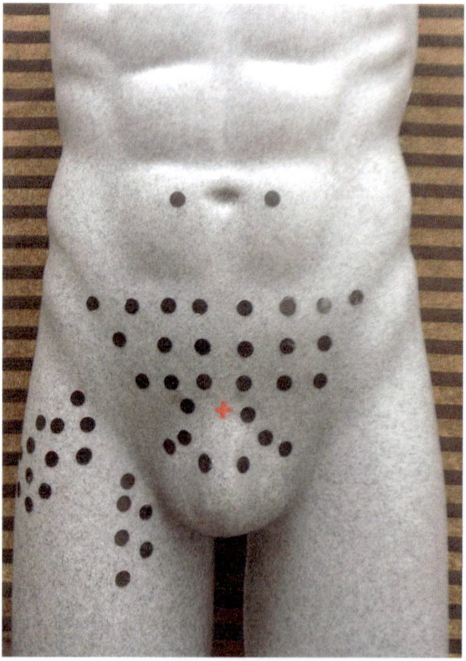

DMC = C-NN-P-VI

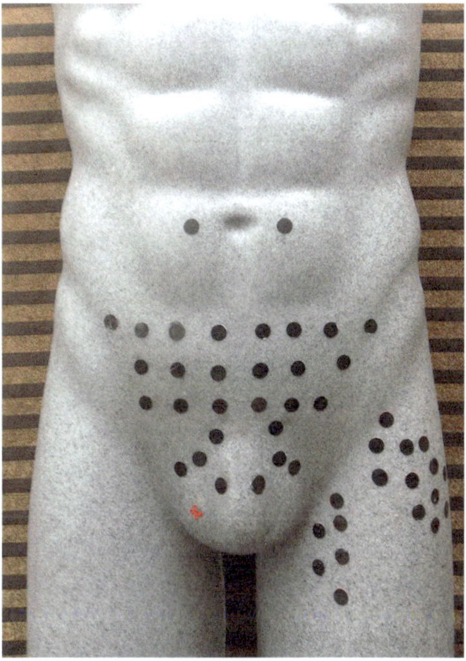

Fig. 21.11. Dermatome mapping compatible with a non-neuropathic or nociceptive postoperative right *ORCHIALGIA*.

Dermatome Mapping Classification

Traditionally, in the era when recurrence was our main concern, chronic postoperative inguinal pain had been referred to using the general term "inguinodynia" for all possible types of inguinal pain manifestations. However, the variation of clinical presentations and etiologies necessitates that we broaden and deepen our understanding of this complex condition [16]. Primarily, it is important to establish the origin of the pain and to classify the *location* and *type* of pain that we are dealing with: neuropathic (follows the pattern of a dermatome of specific nerve trunks), nociceptive (non-neuropathic), or paresthesias (denervation or discomfort due to incidental or therapeutic neurectomy).

Subsequently, we must determine the specific *source* of pain. That is, to establish which nerve or multiple nerve trunks are involved, and at the same time determine if the source of pain is non-neuropathic (meshomas, orchialgia, pubalgia, or vague) or some other mechanism when the nerve trunks are clearly preserved.

Finally, it is imperative to have a record of how the *intensity* of pain is perceived by the patient. For this purpose, the visual analogue scale (VAS) is used in conjunction with this classification system.

We reported the original Dermatome Mapping Classification in 2009 [9]; the updated 2015 DMC is presented in (Fig. 21.12) with respect to the

Location	Type	Source	Intensity VAS
R Right	N Neuropathic	1 Ilio inguinal Nerve	I - X
L Left		2 Ilio hypogastric Nerve	I - X
C Central		3 Genital Branch	I - X
		4 Femoral Branch	I - X
		5 Lateral F Cutaneous Nerve	I - X
		T1-T12 Costal Nerves	I - X
	NN Non Neuropathic	G Granuloma	I - X
		H Hernia	I - X
		L Lipoma	I - X
		M Meshoma	I - X
		O Orchialgia	I - X
		P Pubalgia	I - X
		S Sports Hernia	I - X
		V Vague	I - X
	D Denervation	1 Ilio inguinal Nerve	I - X
	Discomfort	2 Ilio hypogastric Nerve	I - X
		3 Genital Branch	I - X
		4 Femoral Branch	I - X
		5 Lateral F Cutaneous Nerve	I - X
		T1-T12 Costal Nerves	I - X

Fig. 21.12. DMC (Dermatome Mapping Classification). Dermatome Mapping Classification (DMC) is classified evaluating 4 aspects in chronic postoperative pain: location, type, source, and intensity (updated version of the 1999 original [9]).

pre- and postoperative evaluations of patients with chronic postoperative pain. Currently, we are updating the DMC regularly, expanding it, for example, not only to evaluate the inguinal region but also to consider thoracic nerves for cases when postoperative pain occurs in the abdominal region due to injury or entrapment of thoracic nerves during an inguinal laparoscopic surgery, or when using transfascial sutures, for instance, in the Rives or laparoscopic techniques. We have also modified the DMC to include Dr Chen's contribution of the term "vague" for nonspecific situations, subjective inconsistencies, or patients simulating pain that had previously been termed "faking."

The Dermatome Mapping Test (DMT) results are presented and sorted according to the author's classification (DMC), which takes into account 4 main aspects:

1. *Pain location*
   L: Pain or discomfort is located on the left side
   R: Pain or discomfort is located on the right side
   C: Pain or discomfort is located on the center

2. *Type of pain or denervation discomfort*
   N: Neuropathic pain that follows the path of a nerve branch or dermatome
   NN: Non-neuropathic pain that does not follow the pattern of a nerve branch or dermatome
   D: Denervation incidental or secondary to neurectomy, manifested by anesthesia, hypoesthesia, or numbness

3. *Source of neuropathic pain or discomfort and denervation symptoms*
   Source of neuropathic pain or denervation symptoms

   1. For the dermatome of the ilioinguinal nerve
   2. For dermatome of the iliohypogastric nerve
   3. For the dermatome of the genital branch
   4. For the dermatome of the femoral branch
   5. For the dermatome of the lateral femoral cutaneous nerve

   T1 through T12 For each thoracic nerve dermatome

   Source of non-neuropathic pain

   Those symptoms that do not follow a nerve dermatome pattern. The following nomenclature for the source of pain or discomfort is assigned

G: Granuloma
H: Hernia
L: Lipoma
M: Meshoma
O: Orchialgia
P: Pubalgia
S: Sports Hernia
V: Vague

4. *Intensity of pain or discomfort*
   The Visual Analogue Scale (VAS) is reported with Roman numerals (I through X). These data are integrated into the Dermatome Mapping Classification (DMC) for chronic postoperative pain (see Fig. 21.12).

Clinical Cases

See Figs. 21.13, 21.14, and 21.15.

Conclusion

Dermatome Mapping Test (DMT) is a)simple and cost-effective technique that requires only a ballpoint pen and three felt-tipped markers.

This test can be performed in the surgeon's office. A photograph is taken that provides an objective record of a subjective situation such as postoperative chronic pain.

The Dermatome Mapping Classification (DMC) establishes and incorporates a broadened common language that names every specific scenario encountered in post-inguinal herniorrhaphy pain. It allows us to characterize and communicate the multifactorial pain that patients present with and to discuss and form treatment plans in a logical fashion. It additionally provides an excellent tool for postoperative assessment and follow-up to document and communicate the efficacy of our interventions.

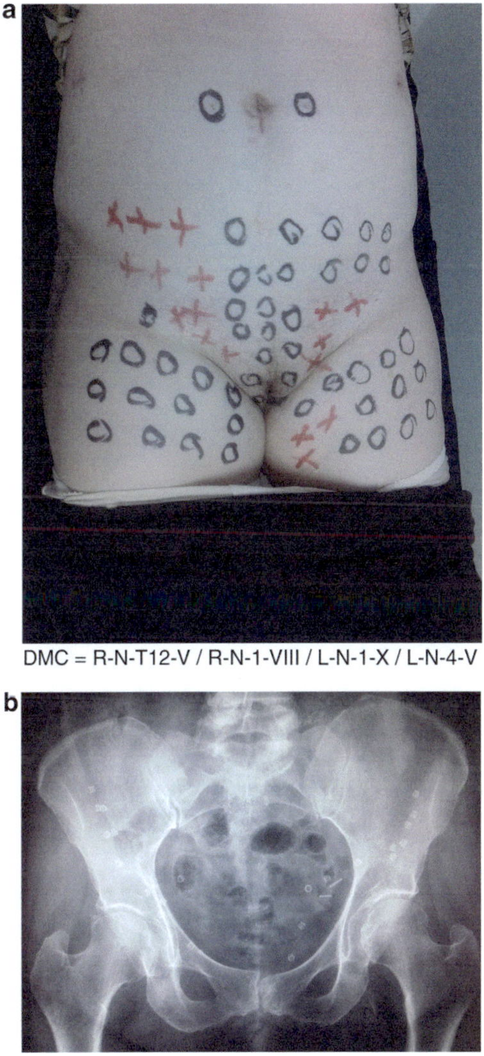

Fig. 21.13. (**a**) Dermatome mapping compatible with LAPAROSCOPIC severe neuropathic postoperative pain due to bilateral ilioinguinal nerve, right T12 nerve, and left femoral branch involvement. (**b**) X-ray shows 11 tackers on each side and two hemoclips.

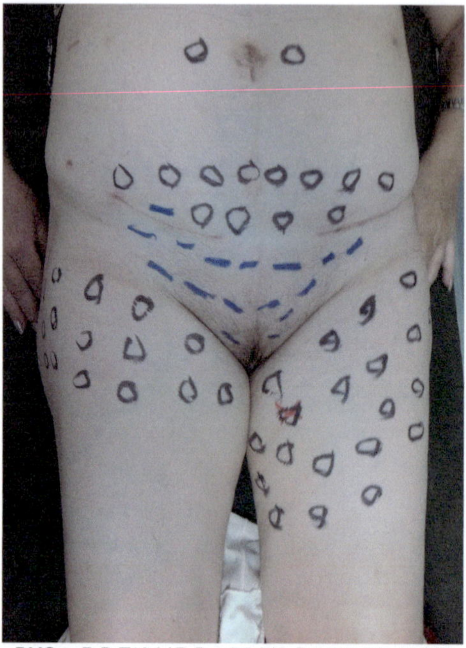

DMC = R-D-T12-I / R-D-1,2,3-I / L-D-1,2,3 -I / L-N-4-III

Fig. 21.14. Same patient as Fig. 21.13 after successful bilateral triple neurectomy and multiple tackers removal. The symptoms improved significantly after 1 year. Left femoral branch pain persists in 1–2 points of the femoral branch dermatome although less intense compared to the preoperative assessment. This patient would have been a candidate for truncal quadruple lumboscopic neurectomy.

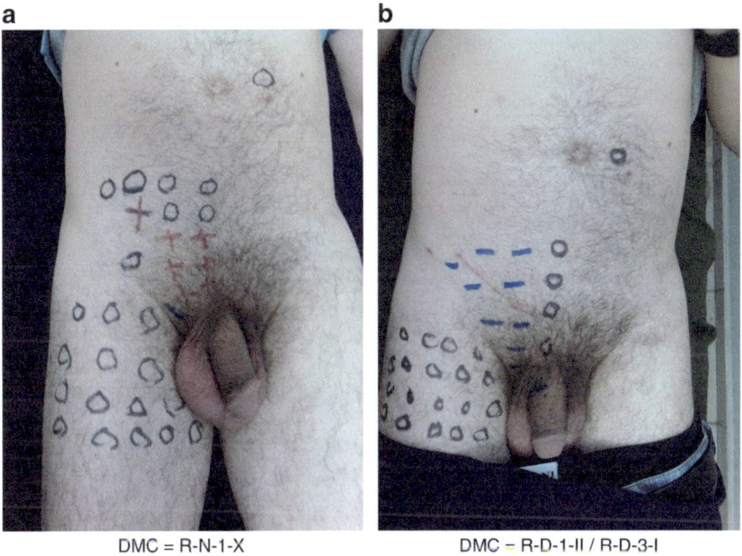

Fig. 21.15. (**a**) Dermatome mapping compatible with neuropathic postoperative chronic pain due to right ilioinguinal nerve involvement after an open mesh plug technique. (**b**) Dermatome mapping of the same patient that shows denervation due to selective neurectomy of the right ilioinguinal nerve and genital branch after plug removal with the iliohypogastric nerve dermatome seen intact. Successful decrease on ilioinguinal pain intensity from X down to II in VAS Visual Analog Scale.

References

1. The Swedish Hernia Register [Internet]. http://www.svensktbrackregister.se/. Accessed 12 Jul 2014.
2. Danielsson P, Isacson S, Hansen MV. Randomised study of Lichtenstein compared with Shouldice inguinal hernia repair by surgeons in training. Eur J Surg. 1999; 165(1):49–53.
3. Beets GL, Oosterhuis KJ, Go PM, Baeten CG, Kootstra G. Longterm followup (12–15 years) of a randomized controlled trial comparing Bassini-Stetten, Shouldice, and high ligation with narrowing of the internal ring for primary inguinal hernia repair. J Am Coll Surg. 1997;185(4):352–7.
4. de Lange DH, Wijsmuller AR, Aufenacker TJ, Rauwerda JA, Simons MP. Neuralgic pain, a significant complication after a Lichtenstein procedure for inguinal hernia repair. Ned Tijdschr Geneeskd. 2008;152(41):2205–9 [Article in Dutch].
5. Staal E, Nienhuijs SW, Keemers-Gels ME, Rosman C, Strobbe LJ. The impact of pain on daily activities following open mesh inguinal hernia repair. Hernia. 2008;12(2):153–7.

6. Lermite E, Arnaud JP. Prospective randomized study comparing quality of life after shoudice or mesh plug repair for inguinal hernia: short-term results. Surg Technol Int. 2012;22:101–6.
7. Classification of chronic pain. Descriptions of chronic pain syndromes and definitions of pain terms. Prepared by the International Association for the Study of Pain, Subcommittee on Taxonomy. Pain Suppl. 1986;3:S1–226.
8. Campanelli G, Bertocchi V, Cavalli M, Bombini G, Biondi A, Tentorio T, et al. Surgical treatment of chronic pain after inguinal hernia repair. Hernia. 2013;17(3): 347–53.
9. Álvarez Quintero R, Mayagoitia González JC. Chapter 37. Dolor inguinal crónico posoperatorio o inguinodinia. In: Mayagoitia González JC, editor. Hernias de la pared abdominal: tratamiento actual. 3rd ed. Editorial Alfi: México; 2015. p. 293–9 [Book in Spanish].
10. Amid PK, Chen DC. Surgical treatment of chronic groin and testicular pain after laparoscopic and open preperitoneal inguinal hernia repair. J Am Coll Surg. 2011; 213(4):531–6.
11. Chen DC, Hiatt JR, Amid PK. Operative management of refractory neuropathic inguinodynia by a laparoscopic retroperitoneal approach. JAMA Surg. 2013;148(10): 962–7.
12. Heise CP, Starling JR. Mesh inguinodynia: a new clinical syndrome after inguinal herniorrhaphy? J Am Coll Surg. 1998;187(5):514–8.
13. Bischoff JM, Enghuus C, Werner MU, Kehlet H. Long-term follow-up after mesh removal and selective neurectomy for persistent inguinal postherniorrhaphy pain. Hernia. 2013;17(3):339–45.
14. Amid PK. Causes, prevention, and surgical treatment of postherniorrhaphy neuropathic inguinodynia: triple neurectomy with proximal end implantation. Hernia. 2004;8(4):343–9.
15. Álvarez Qiomterp R, Anaya Prado R, Malé VE. Inguinodinia: Mapeo por dermatomas como método diagnóstico. Cirujano General. 2004;26(4):265–9 [Article in Spanish].
16. Turk DC, Rudy TE. IASP taxonomy of chronic pain syndromes: preliminary assessment of reliability. Pain. 1987;30(2):177–89.

22. Management of Inguinal Hernia Recurrences (When Pain Is the Primary Symptom)

Keri A. Seymour and Jin S. Yoo

Roughly 27 % of men and 3 % of women in industrialized countries will undergo inguinal hernia repair [1]. Surgical repair of inguinal hernias has progressed from primary repair to mesh repair and now includes minimally invasive techniques. The most common postoperative complications of inguinal hernia repair include recurrence and chronic pain. Incorporating the variety of repair techniques, recurrence is reported in 5–15 % of patients after inguinal hernia repair, and chronic pain is reported in up to 63 % of patients depending on the definition used [2–6]. Management of recurrence and pain after inguinal hernia repair requires a multimodal and multidisciplinary approach. Surgical planning requires a familiarity with the anatomy, realistic patient expectations, understanding of previous repair, and open and laparoscopic options for remedial repair.

Postoperative Pain

Postoperative pain affects quality of life, leads to decreased physical activity, and results in additional medical care and surgical procedures [5]. Postoperative pain is multifactorial; its origin includes visceral, somatic, or neuropathic sources (Table 22.1). Pain experienced in the early postoperative period is frequently nociceptive in origin [4, 7]. It presents as tenderness along the incision, inguinal canal, scrotum, and thigh, and is due to tissue inflammation or irritation from mesh or suture material. Neuropathic pain presents as allodynia, hypoesthesia, paresthesia, or hyperesthesia [8]. The definitions of chronic pain may vary among studies; however, it is frequently defined as pain 3 months after surgery and lasting for more than 2 months.

B.P. Jacob et al. (eds.), *The SAGES Manual of Groin Pain*,
DOI 10.1007/978-3-319-21587-7_22

Table 22.1. Origin of postoperative pain after inguinal hernia repair.

Visceral origin
Vascular injury
Infectious
Urologic, gastrointestinal, or gynecologic disorders
Somatic origin
Scar tissue
Meshoma
Hernia recurrence
Periosteal inflammation
Neuropathic origin
Nerve injury or entrapment
Scar tissue
Mesh or suture irritation

Anterior and Posterior Approach

Neuropathic pain involves damage or injury to the nerves of the inguinal canal, particularly nerves running in the anterior (inguinal) and posterior (preperitoneal) space. There are five sensory and somatic nerves that are susceptible to injury during inguinal hernia repair. These nerves are iliohypogastric, ilioinguinal, genital and femoral branch of the genitofemoral, and the lateral femoral cutaneous nerve.

The anterior approach to repair of inguinal hernia includes the Bassini, McVay, Shouldice, and Lichtenstein repairs. These repairs may expose the nerves running in the inguinal canal (iliohypogastric, ilioinguinal, and genital branch of the genitofemoral) to injury or entrapment. Recurrence after open (anterior) inguinal hernia repair occurs in up to 6 % of patients and pain is associated with 15 % of repairs [9].

Minimally invasive inguinal hernia repair, laparoscopic transabdominal preperitoneal (TAPP) and totally extraperitoneal repair (TEP), is associated with a recurrence rate of 3 % and postoperative pain occurs after 2 % of repairs [9]. These repairs place mesh posterior to the rectus fascia, in the preperitoneal space. The "triangle of pain" is outlined as inferior to the inguinal ligaments; the apex is the internal ring, and anterolateral to the gonadal vessels [4]. The lateral femoral cutaneous nerve, femoral nerve, and femoral and genital branch of the genitofemoral nerves lie in this region [5]. Placement of tacks in the triangle of pain can inadvertently entrap the nerves.

Risk Factors

Pain and recurrence after inguinal hernia repair are related to a combination of patient-related risk factors, technical considerations, and operative approach. Technical errors include inadequate mesh coverage, mesh folding, and mesh migration. Burcarth et al. recommend that females undergo laparoscopic repair of inguinal hernias in order to evaluate an unappreciated femoral hernia. Smoking is associated with impaired wound healing due to hypoxia and decreased collagen formation [10]. Table 22.2 lists the preoperative, perioperative, and postoperative risk factors associated with pain after inguinal hernia repair [1, 4, 5, 11].

Evaluation

Evaluation of recurrence when pain is the presenting symptom after inguinal hernia repair should begin with a thorough history and physical exam. The history should include the frequency, location, and triggers of pain. The physical exam should focus on a bulge, fascial defect, and

Table 22.2. Risk factors associated with pain after inguinal hernia repair.

Preoperative risk factors
Young age
Female sex
Pain prior to surgery
Obesity
Recurrent hernia
Direct inguinal hernia
Smoking
Perioperative risk factors
Surgeon experience
Neurolysis
Fixation with suture or staples
Lightweight mesh
Local anesthesia
Excessive dissection
Postoperative risk factors
Recurrence
Hematoma
Wound infection

reproducible pain. The operative report from the previous surgery should be reviewed, including the type of repair, size of the defect, size and type of mesh, handling of nerves, and type of fixation. Diagnostic imaging—ultrasound, computed tomography scan, or magnetic resonance imaging—supplements the management, excludes recurrence or meshoma, and assists in the diagnosis [4].

Supportive Treatment

Treatment of postoperative pain involves a multidisciplinary approach, including medications, behavior modification, and therapeutic intervention. Courtney et al. found that 30 % of patients have resolution of postoperative inguinal hernia repair pain, 45 % have reduced pain, and 25 % continue to have chronic pain [12]. A period of watchful waiting with symptomatic treatment with a multimodal therapy that includes behavior modification, NSAIDs, and opioid medications is recommended. Additionally, a multidisciplinary group approach that consists of the primary care provider and a dedicated pain specialist (anesthesiologist, neurologist, psychiatrist) is recommended. Adjunctive modalities such as nerve stimulators, steroid injections, or nerve blocks can be both diagnostic and therapeutic. Specific to this subgroup of patients with pain in the presence of a known recurrence, it is important to characterize the potential etiologies of pain so that all contributing factors can be addressed at the time of remedial surgery for both recurrence and pain.

Surgical Options

Reoperation for a recurrent inguinal hernia is considered at the time of identification either by physical exam or imaging studies. A trial of conservative measures, careful diagnostic evaluation, and treatment for the pain component is prudent to help delineate if the pain is primarily due to the recurrence or if neuropathy, meshoma, or other anatomic issues are causative. At the time of repair of a recurrent inguinal hernia, surgery for pain management may include an operative neurectomy or possible removal of mesh, depending on the presentation and suspected etiology of pain.

Neuropathic pain refractory to conservative measures identified from history, physical examination, and adjunctive testing may not improve with recurrent hernia repair alone, and the inguinal and preperitoneal nerves are placed at additional risk during remedial surgery. Neurectomy is advisable at the time of reoperation in these cases. In 2002, Amid described a one-stage, anterior triple neurectomy to treat postherniorrhaphy pain. Triple neurectomy resects the iliohypogastric, ilioinguinal, and genital branch of the genitofemoral nerve. Triple neurectomy successfully treated pain in 80 % of patients. Adequate dissection lateral to the internal ring is necessary to identify the proximal portion of the ilioinguinal nerve. The iliohypogastric is found between the external and internal oblique, and the genital branch of the genitofemoral nerve is identified near the external spermatic vein. The nerves, along with any mesh that incorporates the nerves, are resected with the proximal nerve ends placed in the internal oblique muscle or allowed to retract into the inguinal ring [11]. Excessive scar tissue may distort the anatomy and make a triple neurectomy difficult; however, the nerves have overlapping sensory innervation and triple neurectomy is preferred [4, 7]. Complications include persistent numbness, vascular or bowel injury, testicular atrophy, recurrence of the hernia, wound infection, and pain [2, 7, 8]. Neurectomy must be performed proximal to the site of any nerve injury in order to be effective. A retroperitoneal approach to triple neurectomy is indicated for neuropathic pain after prior preperitoneal posterior repair through either an extended anterior approach or laparoscopic approach.

Recurrence After Anterior Approach

There are different options to manage recurrence of an inguinal hernia combined with chronic pain after anterior repair. A standard open reoperation allows for identification, repair, mesh removal if needed, and neurectomy for neuropathic complaints. The operative repair is more challenging due to reoperation within the scarred inguinal field. Alternatively, Rosen et al. treated chronic neuralgia after anterior inguinal hernia repair with a combined laparoscopic and open approach. A diagnostic laparoscopy was the initial step to evaluate hernia recurrence, mesh contraction, incorporation of vas deferens, and pelvic anatomy. Using a standard TAPP repair, mesh was placed in the preperitoneal space. Then the anterior incision exposed the mesh, which was removed along with all sutures, including the tacking suture at the pubic bone. The operation was completed after ilioinguinal and iliohypogastric neurectomy was performed. Over 90 % of patients reported resolution of pain [2].

Recurrence After Posterior Approach

Chronic pain and recurrence after laparoscopic inguinal hernia repair require additional consideration. Pain with recurrence after posterior (preperitoneal) repair should be treated with anterior repair of the recurrence. A laparoscopic approach to remove tacks, staples, suture, mesh, or a trapped nerve and the addition of an "extended" peritoneal triple neurectomy may be required to treat chronic pain [7]. The extended triple neurectomy involves dissection in the retroperitoneal space to the psoas muscle to locate the genitofemoral trunk along with the ilioinguinal and iliohypogastric nerve adjacent to the quadratus lumborum. Successful relief of pain with extended triple neurectomy is reported in over 85 % of patients [13]. Laparoscopic repair of the recurrence remains an option but carries the risks and challenges of reoperating in the scarred preperitoneal field.

Conclusion

Inguinal hernia repair is a common surgery performed worldwide. Prevention of recurrence and chronic pain rely on preoperative knowledge of risk factors. Recurrence and chronic pain complicate 5–15 % of inguinal hernia repairs. Multimodal intervention should begin 3 months after the original hernia repair. Surgical management includes repair of any recurrence, mesh removal, and/or neurectomy if indicated. Successful treatment can resolve pain in up to 80 % of patients.

References

1. Bjurstrom MF, Nicol AL, Amid PK, Chen DC. Pain control following inguinal herniorrhaphy: current perspectives. J Pain Res. 2014;7:277–90.
2. Rosen MJ, Novitsky YW, Cobb WS, Kercher KW, Heniford BT. Combined open and laparoscopic approach to chronic pain following open inguinal hernia repair. Hernia. 2006;10(1):20–4.
3. Chatzimavroudis G, Papaziogas B, Koutelidakis I, Galanis I, Atmatzidis S, Christopoulos P, et al. Lichtenstein technique for inguinal hernia repair using polypropylene mesh fixed with sutures vs. self-fixating polypropylene mesh: a prospective randomized comparative study. Hernia. 2014;18(2):193–8.
4. Ferzli GS, Edwards ED, Khoury GE. Chronic pain after inguinal herniorrhaphy. J Am Coll Surg. 2007;205(2):333–41.

5. Nienhuijs SW, Rosman C, Strobbe LJ, Wolff A, Bleichrodt RP. An overview of the features influencing pain after inguinal hernia repair. Int J Surg. 2008;6(4):351–6.
6. Neumayer L, Giobbie-Hurder A, Jonasson O, Fitzgibbons Jr R, Dunlop D, Gibbs J, et al. Open mesh versus laparoscopic mesh repair of inguinal hernia. N Engl J Med. 2004;350(18):1819–27.
7. Lange JF, Kaufmann R, Wijsmuller AR, Pierie JP, Ploeg RJ, Chen DC, Amid PK. An international consensus algorithm for management of chronic postoperative inguinal pain. Hernia. 2015;19(1):33–43.
8. Vuilleumier H, Hubner M, Demartines N. Neuropathy after herniorrhaphy: indication for surgical treatment and outcome. World J Surg. 2009;33(4):841–5.
9. Liem MS, van der Graaf Y, van Steensel CJ, Boelhouwer RU, Clevers GJ, Meijer WS, et al. Comparison of conventional anterior surgery and laparoscopic surgery for inguinal-hernia repair. N Engl J Med. 1997;336(22):1541–7.
10. Burcharth J, Pommergaard HC, Bisgaard T, Rosenberg J. Patient-related risk factors for recurrence after inguinal hernia repair: a systematic review and meta-analysis of observational studies. Surg Innov. 2015;22(3):303–17. 2014 Sep 30. pii: 1553350614552731.
11. Amid PK. A 1-stage surgical treatment for postherniorrhaphy neuropathic pain: triple neurectomy and proximal end implantation without mobilization of the cord. Arch Surg. 2002;137(1):100–4.
12. Courtney CA, Duffy K, Serpell MG, O'Dwyer PJ. Outcome of patients with severe chronic pain following repair of groin hernia. Br J Surg. 2002;89(10):1310–4.
13. Amid PK, Chen DC. Surgical treatment of chronic groin and testicular pain after laparoscopic and open preperitoneal inguinal hernia repair. J Am Coll Surg. 2011;213(4):531–6.

23. Mesh Removal for Chronic Pain: A Review of Laparoscopic and Open Techniques

Lisa A. Cunningham and Bruce Ramshaw

Introduction

The incidence of chronic pain or discomfort after inguinal hernia repair is much higher than previously thought, and studies suggest it could be higher than 50 %. Although most of these patients have mild to moderate pain, in a review by Bay-Nielsen et al., the rate of chronic pain after repair that interfered with work or leisure activity was determined to be 11 %, which has the potential to affect many individuals, as there are approximately 800,000 repairs performed each year in the United States [1, 2]. Regardless of the actual incidence, awareness of this problem is increasing in surgeons and other physicians caring for such patients, some of whom are desperate for help.

Fortunately, the vast majority of patients who experience chronic pain symptoms after hernia repair have mild to moderate symptoms and do not require invasive intervention to maintain a good quality of life. For patients in whom chronic pain negatively affects their quality of life, it can threaten the ability to function independently and to work for a living. This degree of chronic pain can also have detrimental effects on family and social relationships. It is not uncommon for some patients with severe chronic pain to verbalize a willingness to commit suicide if their pain cannot be relieved.

Chronic groin pain after hernia repair is a complex problem involving many variables, making it difficult to fully understand and treat. Chronic pain in this patient population can be a result of patient factors, other diagnoses besides inguinal hernia, the surgical technique and quality of the repair, the mesh and fixation materials used, and even the

B.P. Jacob et al. (eds.), *The SAGES Manual of Groin Pain*,
DOI 10.1007/978-3-319-21587-7_23

patient's experience and environment in the pre-, peri-, and postoperative period. One single factor may not be the sole cause for chronic pain, and most often multiple factors play a role. Because of this complexity, there are many treatment options available to patients, which range from noninvasive medications and lifestyle modifications to invasive procedures such as nerve blocks and reoperation. The same treatment option may not benefit each patient, and in some patients, a multimodal approach is necessary to achieve maximal pain relief.

This chapter will focus on the surgical approach for the management of chronic groin pain, including mesh removal, although some other treatment options will be discussed. We will also discuss a team-based model for providing care that attempts to deal with complex problems such as chronic groin pain after inguinal hernia repair.

Types of Pain

In general, pain following an inguinal hernia repair can be divided into two groups, nociceptive and neuropathic [3]. Nociceptive pain is caused by activation or sensitization of peripheral nociceptors, specialized nerve endings that respond to chemical, mechanical, or thermal stimulus. Neuropathic pain is the consequence of injury to peripheral or central nervous structures.

Acute pain from surgery is caused by noxious stimulation due to tissue injury and is usually nociceptive. There are two subtypes of acute pain, somatic and visceral. Visceral pain can occur in the groin when the intestines become involved and may be due to mesh adherence and/or erosion into the bowel. The somatic component of nociceptive pain further subdivides into superficial or deep pain. Superficial pain is sensed by unimodal nociceptors in the skin and subcutaneous tissues that evoke a sharp, pricking type of pain, while deep somatic pain is sensed by polymodal receptors in the muscles, tendons, joints, and bones that bring about a long-lasting dull, aching, or burning pain that is typically less well localized. The ability to localize pain is affected by the intensity and duration of the painful stimulus. In addition, nociceptors display sensitization following repeated stimulation that can manifest as an enhanced response to noxious stimuli or an acquired responsiveness to non-noxious stimuli. Sensitization of nociceptors is proposed as a key component of peripheral pain disorders.

Neuropathic pain is due to partial or complete injury to the nerves. This type of pain is characterized by partial or complete sensory loss or changed sensory function due to damage of the afferent transmission system. Neuropathic pain may be associated with hyperpathia, including allodynia (a painful response to harmless stimuli), hyperalgesia (an exaggerated response to harmful stimuli), and hyperesthesia (increased response to mild stimulation). In contrast to nociceptive pain, neuropathic pain may persist in the absence of noxious stimuli.

Pain that continues after the normal course of healing from injury or surgery is termed chronic pain and varies in presentation in each individual patient. Chronic pain may be purely nociceptive or neuropathic or present on a spectrum between the two.

Chronic pain can cause long-term changes in the nervous system signaling and processing pathways via neural plasticity. Neural plasticity (or neuroplasticity) is a process in which modulations at the synaptic and neurotransmitter level or relearning at the cortical level alters previously established pathways. This mechanism of remodeling is helpful in situations where relearning is needed, such as when a right-hand-dominant individual now has to become left-hand-dominant after amputation but can also contribute to the development of chronic pain and the associated hyperpathia. Neuroplasticity in conjunction with sensitization makes it especially challenging to understand, manage, and treat chronic pain.

Causes of Groin Pain

As mentioned in the introduction, the cause(s) of chronic postoperative pain after inguinal hernia repair are many and complex and may not always be related to the repair itself. Many factors play a role in the development of chronic groin pain after inguinal hernia repair. The patient, the surgeon, and the materials used can all contribute to the development of pain. In fact, preoperative groin pain predicts an increased likelihood of postoperative chronic groin pain, which is supported by the findings of Aasvang et al., who also identified preoperative pain response to heat as a contributing factor to the development of postoperative pain [4]. In the 2014 update to the European Hernia Society (EHS) guidelines, level one evidence showed that preoperative pain and early postoperative pain are independent risk factors for chronic pain and that postoperative chronic pain diminishes over time, with the risk of chronic pain after hernia surgery decreasing in incidence with age [5].

The surgeon may also contribute to the development of pain through their chosen technique. In many studies, the laparoscopic approach has been shown to have a decreased incidence of postoperative chronic groin pain, although in some reports the incidence of chronic groin pain after laparoscopic repair may still be as high as almost 30 %. In a meta-analysis of data comparing the incidence of chronic groin pain after open versus laparoscopic hernia repair, the EHS found no significant difference; however, one small study found that the severity of the pain present was less in the laparoscopic group at 10 years post-operation [6].

The sutures, fixation devices, and mesh may also play a significant role. Chronic groin pain can occur after non-mesh, suture-only inguinal hernia repair; however, when mesh is used, it can significantly contribute to the development of chronic groin pain through an inflammatory response between the mesh and surrounding tissue. The inflammatory reaction can cause nearby nerves to become entrapped in the mesh directly or cause traction injury to nerves as tissues become scarred and contract. The degree of inflammatory response in an individual patient is impossible to predict at this time, although explanted mesh studies that may help address this issue are in progress. In some patients, mesh may be relatively inert, whereas in others it may migrate, fold, or erode through local structures.

When mesh is used, fixation devices such as tacks and staples can contribute to the development of pain, particularly if deployed near or into nerves as they course through the operative field. Sometimes, after surgical repair, a nerve injury is acutely painful and obvious, leading to immediate, excruciating pain and paresthesia in the recovery room as the patient awakens from anesthesia. This is often caused by direct injury to the nerve from a mesh fixation device. In this case, it may be appropriate to return to the operating room immediately to remove the offending fixation device. Familiarities with the course of the nerves at risk in open and laparoscopic inguinal hernia repair have decreased the risk of this complication.

Treatment of Chronic Groin Pain After Inguinal Hernia Repair

There are different levels of pain management that can be used to help control pain after inguinal hernia repair. They range from noninvasive to invasive treatments and may be provided by the surgeon alone or in conjunction with other care providers such as pain specialists.

Noninvasive Options

For nonsevere or acute pain following inguinal hernia repair, the initial treatment is rest, ice, and/or heat to the groin, anti-inflammatory medication, and sometimes a mild narcotic medication. This strategy is appropriate to try for several weeks unless pain is severe or significantly worsens within a short period of time, despite conservative treatment.

Once chronic pain develops, other medications that may be considered include antidepressants, serotonin, and norepinephrine reuptake inhibitors (SNRIs), neuroleptics, antispasmodics, muscle relaxants, corticosteroids, anticonvulsants, topical local anesthetics, alpha-adrenergic agonists, and increased opioid narcotic doses. The use of these medications to attempt to treat chronic pain and allow a patient to return to most normal activities may require activity restrictions while the patient is experiencing chronic pain.

Other noninvasive options include physical therapy and transcutaneous electrical neural stimulation (TENS). A TENS unit works by stimulating large epicritic afferent fibers that sense variations in temperature and touch. Stimulation of these larger fibers outcompetes and potentially causes complete conduction block of the afferent signal from smaller pain fibers.

TENS is thought to produce analgesia by stimulating large afferent fibers. It may have a role for patients with mild to moderate acute pain and those with chronic low back pain, arthritis, and neuropathic pain. The gate theory of pain processing suggests that the afferent input from large epicritic fibers competes with that from the smaller pain fibers.

Although evidence in support of complementary alternative medicine (CAM) for treatment of chronic groin pain is not conclusive, it may be a useful adjunct to traditional approaches in certain patients. CAM may include acupuncture, spinal manipulation, massage therapy, relaxation techniques, tai chi, yoga, and herbal supplements.

To obtain optimal treatment success, psychological, emotional, spiritual, and family counseling may also be required to address the psychosocial factors that may have contributed to the development of chronic pain or may be a result of the enormous toll that chronic pain can exact on a person's life, especially when it has been present for a long period of time.

Invasive Nonsurgical Options

For more severe pain and pain that worsens or persists for more than a few weeks, it is appropriate to offer the patient more aggressive pain management. This can include administering inguinal nerve blocks for diagnostic and potentially therapeutic purposes. If results of the injection are good, but pain returns, additional nerve blocks may be appropriate. Some patients will obtain sufficient pain relief to return to a full quality of life after one or more nerve blocks.

Cryoablation and radiofrequency ablation can also be used and involve placing specialized needles and probes near the affected nerve and causing coagulation at very low or very high temperatures, respectively. These methods work by destroying the nerves at the site of application, thereby providing at least temporary pain relief. It is still possible for nerve regeneration to occur; long-term studies with definitive evidence is lacking.

Surgical Options

If the pain has persisted for more than 3–6 months, and/or the pain is severe or worsening despite other nonsurgical therapies, it is appropriate to consider an operation in an attempt to relieve the pain. Prior to surgery, it is very important to address preoperative, operative, and potential postoperative complications and factors prior to proceeding with surgery.

Preoperative Management

In the preoperative setting, it is important to address goals, to assess the impact of chronic pain on the patient's quality of life and for risk factors for continued pain, and to consider alternatives. It is also important to assess and address any comorbidities and to discuss previous treatment modalities that the patient has tried, if any, as these may help predict their response to operative intervention.

Not everyone will benefit from surgery, and it is important to convey to patients that the pain may stay the same, improve only partially, or could in fact get worse. If neurectomy is to be performed, it is also important to address postoperative numbness in the distribution of the affected nerves.

Operative Management

For surgeons who are experienced with pelvic and groin laparoscopy, a diagnostic laparoscopy is an appropriate first step. A laparoscopic view will identify intra-abdominal adhesions and possibly interstitial and/or recurrent hernias. An interstitial hernia can occur as a defect through the deeper layers of the groin, but not completely through all layers of the groin or through the mesh when placed in an open Lichtenstein-type hernia repair. Sometimes, offending tacks or staples can be viewed and removed without entering the preperitoneal space. After intraperitoneal exploration, the preperitoneal space may be explored laparoscopically to view the cord structures and nerves (genital and femoral branches of the genitofemoral nerve and lateral femoral cutaneous nerves) that course along the psoas muscle with and lateral to the spermatic cord and internal ring and usually posterior to the iliopubic tract. The location and course of the nerves in the preperitoneal space can be variable, especially in patients with a previous groin operation(s). Fixation devices such as sutures, tacks, and/or staples and mesh (placed laparoscopically or through some open techniques) can be identified in the preperitoneal space. The laparoscopic exploration of the preperitoneal space may include repairing an interstitial or recurrent hernia and/or removal of mesh (including plugs that may be visualized laparoscopically) and/or fixation devices. If a hernia is found and thought to be the cause of the pain, the goal of the operation is to provide a durable hernia repair. If there is no hernia, the goal is to eliminate any adhesions from the groin and to clear the groin of all foreign materials (mesh and fixation devices), freeing up the cord structures and nerves. Neurolysis (freeing up the nerves) is frequently possible with a laparoscopic approach; however, a neurectomy may be indicated if a nerve is embedded in scar and/or mesh and cannot be freed and/or if the patient chooses a planned neurectomy during the preoperative shared decision process.

The laparoscopic removal of mesh from the preperitoneal space of the groin can be a difficult and potentially dangerous procedure, especially if the previous mesh had been cut and passed behind the cord structures. Injuries to the cord structures, the iliac vein and artery, the obturator vessels, the inferior epigastric vessels, and the bladder are all possible. Even inadvertent bowel injury is possible, especially if there are bowel adhesions to the groin or mesh. Sometimes it is appropriate to leave a portion of mesh on one or more of these structures to minimize the risk of injury.

If the previous hernia repair was done laparoscopically or if an open approach was performed, during which the mesh was placed completely into the preperitoneal space, it is possible that a laparoscopic approach alone will result in the maximal benefit from an operation. The reason for this is that the nerves in the preperitoneal space are different from those typically involved if the mesh is located in a more superficial tissue plane (as in a Lichtenstein repair). It is important to note that the nerves typically involved in neuropathic pain are different for these two different mesh locations. An open triple neurectomy (addressing the more superficial nerves) will not likely help relieve the pain from mesh or fixation devices that are located in the preperitoneal space (potentially causing neuropathic pain from the deeper nerves). Figures 23.1 and 23.2 illustrate the nerves in the groin and why mesh placed in different locations can result in different nerve injuries. This also explains why a traditional open triple neurectomy may not be effective when mesh is located in the preperitoneal space.

For patients who have had an open inguinal hernia repair with a technique including placing mesh in the preperitoneal space and in more superficial locations (plug and patch, Prolene Hernia System, Ultrapro Hernia System, etc.) or a technique where no mesh is placed in the preperitoneal space, it is likely that an open groin exploration will be required to achieve the maximal benefits from a surgical approach. The open exploration includes removal of mesh and any other material that may be causing pain. Nerves that course in the groin in the intermuscular location (the iliohypogastric, ilioinguinal, and genital branch of the genitofemoral) may be divided and the distal ends implanted into muscle. There is some debate about whether to search for and divide all nerves or only the nerves involved in the scar tissue, mesh, or other fixation devices. Because of the difficulty in finding nerves outside of the field of dissection and the potential to cause complications, it has been our practice to divide only the nerves (neurectomy) or free nerves (neurolysis) that are involved in the scar tissue, mesh, and/or fixation devices, but not to look for additional nerves in otherwise normal-appearing tissue. After the open approach is completed, the groin is closed with three layers of absorbable suture, and then the skin is closed with a subcuticular stitch.

At this point in our experience, we have not placed a permanent synthetic, absorbable synthetic, or biologic mesh after mesh removal for pain, regardless of whether the procedure was laparoscopic only or a laparoscopic and open combined procedure. We have not placed a mesh during this operation in an attempt to minimize the potential of causing

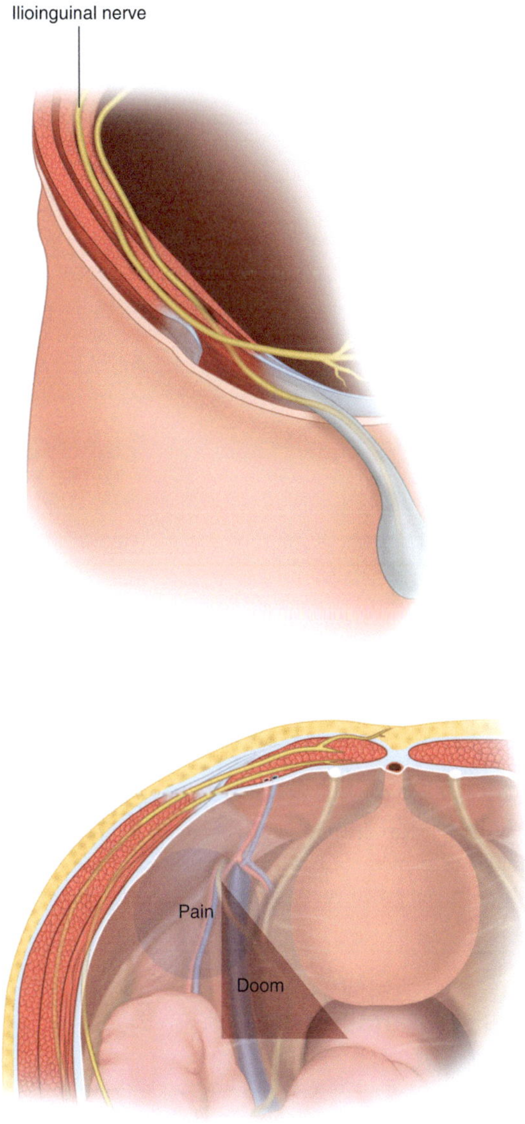

Fig. 23.1. Nerves in the left groin (anterior view and laparoscopic view).

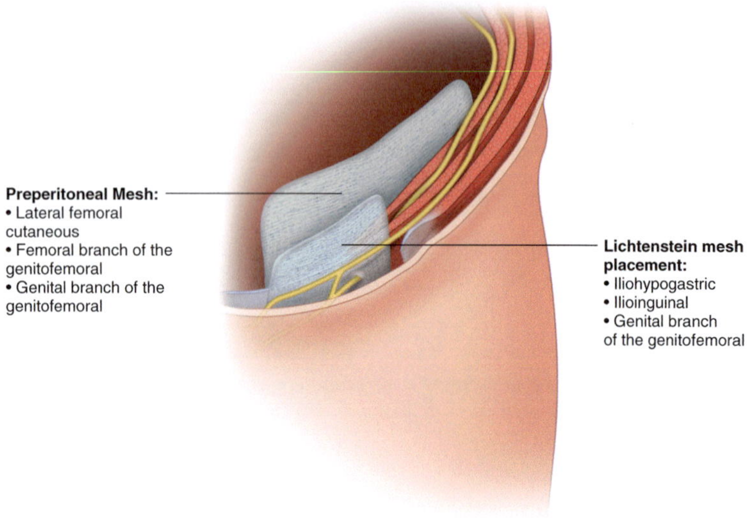

Fig. 23.2. Mesh placement (preperitoneal and Lichtenstein) and the nerves potentially at risk for the left groin.

additional or new pain from a mesh and/or from mesh fixation, when the goal of the operation is to relieve pain. The exception to this is when an interstitial or recurrent hernia is found at laparoscopy. If a hernia defect is identified after a laparoscopic mesh removal, a laparoscopic primary suture repair is performed. For all procedures that include open mesh removal, a three-layer groin reconstruction is performed using absorbable sutures.

Postoperative Management

The patient is often discharged the same day or within 24–48 h of the operation, unless there are complications. However, for patients on high dosages of opioid agonists, a longer hospital stay for pain control and monitoring may be required. In this early postoperative period, the initial treatment of pain is identical to the treatment of nonsevere or acute groin pain and includes rest, ice, and/or heat to the groin, anti-inflammatory medication, and a mild narcotic medication. A bowel regimen to prevent constipation and bloating may also be helpful. It is often helpful to provide the patient with a multitude of pain medication options and then allow them to choose which works best for them, as they may already know from previous experience. Pain control in the immediate postoperative period is essential, as some studies have shown an increased risk for developing chronic pain in patients whose postoperative pain scores are high. Perioperative multimodal pain management inducing transversus abdominis plane (TAP) block and intra-op block with long-acting local anesthetic may help to minimize pain control issues in the immediate postoperative period.

Some patients will have continued difficulty with pain control following surgery, especially if they were taking high doses of opioid medications prior to the operation. All of the pain management techniques previously discussed can be considered for use in the postoperative course. Typically, as the pain and inflammation from the operation resolve, the patient will become increasingly aware of the results from the operation and will report that their improvement levels off 2–4 months after the operation. Following surgery, it is important to track a patient's progress for improvement. It may help to compare a patient's preoperative assessment of pain on a standardized questionnaire to their postoperative pain to examine for objective changes.

Postoperative Complications

In the early postoperative period, complications include wound infection, seroma, and hematoma. As mentioned, postoperative pain control may be difficult.

The long-term complications pertinent to this procedure include hernia recurrence and inadequate resolution of pain. Nonsurgical pain management should be continued and adjusted accordingly for pain that is not resolved in an attempt to improve a patient's quality of life. A continued search for factors that contribute to the development of chronic groin pain after inguinal hernia repair is essential to predict subpopulations at risk for this problem and to potentially alter treatment options based on new knowledge when the concept of predictive analytics and complex systems data management is applied [see Chap. 45, "Value-Based Clinical Quality Improvement (CQI) for Chronic Groin Pain after Inguinal Hernia Repair"].

For the patient who has a hernia recurrence after an operation to relieve pain from a prior hernia repair, the decision to undergo another hernia repair may be a difficult one. If another repair is performed, consideration should be given to the approach (open or laparoscopic) and to the choice of mesh, including options that are not permanent, such as resorbable synthetic and biologic meshes. In this situation, involving the patient in a shared decision process to determine the technique and materials to be used may be helpful to give the patient some control in determining their choice for hernia repair.

Prevention of Chronic Groin Pain After Inguinal Hernia Repair

There have been many attempts to minimize chronic pain over the years, mostly aimed at altering surgical technique. Previous studies have shown mixed results in attempting to prophylactically identify and divide the ilioinguinal, iliohypogastric, and/or genital branch of the genitofemoral nerves during open inguinal hernia repair. In a multicenter prospective study by Alfieri et al., identification and preservation of nerves are directly correlated to the development of chronic pain postoperatively [7].

The laparoscopic approach compared to open inguinal hernia repair has some of the strongest evidence showing a decrease in acute and chronic pain based on several studies. However, some studies have shown a minimal difference in pain after the first 24–48 h, and some studies show increased severity of pain with a laparoscopic repair.

The other most studied factors in the prevention of chronic pain are the mesh and fixation devices used. Several studies have evaluated lightweight mesh to look for a decreased incidence of chronic pain. Some older studies showed inconclusive results or only slight improvement when using lightweight mesh. However, many of these studies also showed a slight increase in hernia recurrence in patients with a lightweight mesh. In two recent studies comparing open and laparoscopic repair, lightweight mesh was associated with a decreased risk of developing chronic groin pain and for the development of other groin symptoms, including stiffness and the sensation of a foreign body, and was not associated with increased risk of hernia recurrence [8, 9]. Decreasing or eliminating tack, staple, and suture fixation, or using glue, has also shown some potential to decrease pain, but with a potential for an increase in recurrence rate.

One other aspect to consider is pain management in the postoperative period as mentioned previously. Some texts suggest that good pain control initially may help to reduce the development of chronic pain by preventing chronic pain pathways from developing through sensitization and neuroplasticity. For this reason, we have developed a multimodal pain treatment protocol in conjunction with our anesthesiologists that includes TAP blocks and controlling nausea to prevent emesis immediately postoperative to reduce strain on newly reconstructed groins. Improvements in preoperative preparation and maximizing bowel function can also help improve outcomes. A summary of perioperative multimodal management strategies is presented in Table 23.1.

Incorporating Systems Science Solutions into the Management of Chronic Groin Pain After Inguinal Hernia Repair

The attempts to isolate and improve one variable and the limited success with this strategy highlight the fact that chronic pain is a complex problem that can rarely be solved with a simple solution. Complex problems require a systems approach that includes identifying and defining processes and variables, including outcomes that measure value (including quality and satisfaction measures as well as costs for the entire cycle of care). It is important to remember that each person is different and that what works for one may have deleterious effects on another; it may have no benefit or cause harm and therefore may be wasteful.

In our approach to treatment using principles of systems and complexity science, we have built a team around hernia disease, with the ultimate goal of creating more teams around other definable patient groups. This places all the focus on the patient process and value-based outcomes generated by the process.

Currently, our hernia team includes surgeons, anesthesiologists, nurses, a patient care manager, care coordinators, a clinical quality improvement manager, a biologic/materials engineer, a mechanical engineer, residents, and other team members. Former patients and their family members participate and help provide care for current patients

Table 23.1. Options for multimodal perioperative management of patient who undergo surgery for the treatment of chronic groin pain after inguinal hernia repair.

	Pre-op	Intra-op	Post-op
Prep/medical/emotional Anesthesia	Weight loss/exercise/nutrition Smoking cessation Counseling	General anesthesia Low pressure Pneumoperitoneum/AirSeal (for lap cases)	Activity as tolerated Many small meals Ice/heat/support
GI function	Bowel cleansing Liquids/Colace/magnesium Entereg		Colace Magnesium Entereg
Medications (up to one week pre-op and post-op 3–5 days)	Lyrica Neurontin		Lyrica Neurontin Opioid agonist (IV/oral)
Medications (immediately pre-op and post-op in hospital)	Ofirmev	Toradol Ibuprofen Ofirmev	Toradol Ibuprofen Ofirmev
Nerve block	TAP: Long-acting local anesthetic Decadron Buprenorphine Consider epidural	Additional local infiltration with long-acting local Xylocaine gel for bladder catheter	
Antiemetic therapy		H2-blocker, Reglan, Zofran (prior to end of case), Emend (for high-risk PONV)	Zofran in PACU

TAP Transversus abdominis plane, *PONV* Postoperative nausea and vomiting, *PACU* Postanesthesia care unit and their family members by providing support and sharing their experiences. By including others outside of the core hernia team, we are able to participate in hernia care communities through face-to-face meetings, video and teleconferencing, and Internet social networking. The clinical portion of our team cares for patients through their dynamic care process and allows for shared decision making with the patient at multiple steps in the care process.

The first step in our process involves interaction with the clinical manager or care coordinator, who begins to develop a caring relationship with the patient and his or her family. From this interaction, we get to develop a sense of the person as well as gather relevant clinical data. Prior to having a clinical visit, it may beneficial for the patient to see other providers first if there are areas of concern identified by the patient care manager. Also, allowing patients to speak with former patients who have suffered with similar types of symptoms can be very beneficial. After the initial clinic visit, more options may be considered or surgery may be offered. The patient's care is followed throughout the hospital and for the entire length of recovery by team members until the patient returns to a good quality of life.

The clinical quality improvement portion of the hernia team is focused on objective outcome measures and identification of anomalies to learn and improve. The analysis of explanted synthetic hernia mesh is the largest clinical quality improvement project currently in progress for our hernia team. Many of these meshes have been explanted from patients who have suffered from chronic groin pain after inguinal hernia repair. By defining dynamic care processes and identifying and measuring quality, satisfaction, and financial outcome measures, the objective of our hernia team is to generate clinical quality improvement data that will help identify ways to improve the value of care delivered.

Summary

Chronic groin pain after inguinal hernia repair is a complex problem that can cause significant impairment to those who are affected as well as for their loved ones. It is a difficult problem to treat and takes a tremendous toll on the individual who suffers. For patients with severe, lasting groin pain or those whose quality of life is impacted despite noninvasive or minimally invasive measures to control pain, it may be appropriate to offer a surgical option for treatment. Surgical treatment includes a diagnostic laparoscopy to look for intra-abdominal adhesions, interstitial and recurrent hernias, and foreign body materials in the preperitoneal space. For previous laparoscopic repairs and for those open repairs that result in mesh placement in the preperitoneal space, this may be all that is required to achieve maximal improvement from an operation. For most open inguinal hernia repair techniques, an additional open groin incision may be necessary to achieve maximal improvement from an operation. Removal of all foreign body materials and division and/or lysis of all involved nerves with a three-layer groin closure using absorbable suture can be accomplished through the open groin incision. Even when the original cause of the pain has been successfully eliminated at operation, complete pain relief may not be achieved and additional pain management may be required. Because of the complexity of this problem and the psychological and emotional impact that can occur, this problem is best addressed by a multidisciplinary team.

References

1. Bay-Nielsen M, Perkins FM, Kehlet H. Pain and functional impairment 1 year after inguinal herniorrhaphy: a nationwide questionnaire study. Ann Surg. 2001;233(1):1–7.
2. Zendejas B, Ramirez T, Jones T, Kuchena A, Ali SM, Hernandez-Irizarry R, et al. Incidence of inguinal hernia repairs in Olmsted County, MN: a population-based study. Ann Surg. 2013;257(3):520–6.
3. Rosenquist RW, Vrooman BM. Chronic pain management. In: Butterworth JF, Mackey DC, Wasnick JD, editors. Morgan and Mikhail's clinical anesthesiology. 5th ed. New York: McGraw-Hill; 2013. p. 1023–86.
4. Aasvang EK, Gmaehle E, Hansen JB, Gmaehle B, Forman JL, Schwarz J, et al. Predictive risk factors for persistent postherniotomy pain. Anesthesiology. 2010;112(4):957–69.
5. Miserez M, Peeters E, Aufenacker T, Bouillot JL, Campanelli G, Conze J, et al. Update with level 1 studies of the European hernia society guidelines on the treatment of inguinal hernia in adult patients. Hernia. 2014;18(2):151–63.
6. Bignell M, Partridge G, Mahon D, Rhodes M. Prospective randomized trial of laparoscopic (transabdominal preperitoneal-TAPP) versus open (mesh) repair for bilateral and recurrent inguinal hernia: incidence of chronic groin pain and impact on quality of life: results of 10 year follow-up. Hernia. 2012;16(6):635–40.
7. Alfieri S, Rotondi F, Di Giorgio A, Fumagalli U, Salzano A, Di Miceli D. Influence of preservation versus division of ilioinguinal, iliohypogastric and genital nerves during open mesh herniorrhaphy: prospective multicentric study of chronic pain. Ann Surg. 2006;243(4):553–8.

8. Sajid MS, Leaver C, Baig MK, Sains P. Systematic review and meta-analysis of the use of lightweight versus heavyweight mesh in open inguinal hernia repair. Br J Surg. 2012;99(3):29–37 (Author reply 446).
9. Sajid MS, Kalra L, Parampalli U, Sains PS, Baig MK. A systematic review and meta-analysis evaluating the effectiveness of lightweight mesh against heavyweight mesh in influencing the incidence of chronic groin pain following laparoscopic inguinal hernia repair. Am J Surg. 2013;205(6):726–36.

24. Open Triple Neurectomy

Ian T. MacQueen, David C. Chen, and Parviz K. Amid

Background

Recurrence rates after inguinal hernia repair have improved since the adoption of tension-free techniques and the routine use of mesh. However, postherniorrhaphy chronic pain still represents a substantial burden of morbidity for patients after inguinal hernia repair. Depending upon definition, the rate of postherniorrhaphy chronic pain reported in the literature varies widely, from 0 % to upward of 60 % [1, 2]. The Swedish Hernia Registry reports that severe or debilitating postherniorrhaphy chronic pain occurs at a rate of between 5 and 7 % [3].

Development of inguinodynia is independent of the method of hernia repair [4–6], but an in-depth understanding of the causes of pain, groin neuroanatomy, and technical aspects of the initial operation is necessary to successfully manage this complication [6–8]. These factors determine the operative options available to address chronic pain after inguinal hernia repair. Effective management is needed, given the person and societal consequences of postherniorrhaphy chronic pain on quality of life, disability, and healthcare utilization.

Pain Classification

Postherniorrhaphy inguinodynia is classically divided into two broad categories, nociceptive pain and neuropathic pain. Nociceptive pain is the result of tissue injury and local inflammatory reaction. It is mediated by endogenous nociceptive molecules and their action on nociceptors. Neuropathic pain, in contrast, results from direct nerve injury. In the postherniorrhaphy patient, neuropathic pain symptoms may include

B.P. Jacob et al. (eds.), *The SAGES Manual of Groin Pain*,
DOI 10.1007/978-3-319-21587-7_24

inguinodynia radiating to the scrotum or femoral triangle, paresthesia, allodynia, hyperpathia, hyperalgesia, hyperesthesia, hypoesthesia, or positive Tinel sign. The mechanisms of nerve injury include indirect or direct structural damage and entrapment injuries, caused by suture, folded mesh or meshoma, or fixating devices. In practice, nociceptive pain and neuropathic pain are not discrete categories but exist on a spectrum with significant overlap between the two. The complexity of diagnosis is increased by social, genetic, patient, and psychological factors.

Anatomy

The neuroanatomy of the groin is complex and highly variable from the retroperitoneal lumbar plexus to the terminal branches exiting through the inguinal canal [9, 10]. Familiarity with this anatomy is central to avoiding nerve injury or entrapment. Evidence suggests that the rate of postherniorrhaphy chronic pain can be reduced to less than 1 % by careful identification and handling of the relevant nerves and by preventing their injury or direct contact with mesh [5]. The three nerves most commonly implicated in postherniorrhaphy chronic pain are the ilioinguinal nerve (IIN), the genital branch of the genitofemoral nerve (GFN), and the iliohypogastric nerve (IHN) (Fig. 24.1) [11]. Additionally, the main trunk of the GFN, the femoral branch of the GFN, and the lateral femoral cutaneous nerve (LFC) may be involved, especially if the original repair was done via a laparoscopic or open preperitoneal approach (Fig. 24.2) [12]. Understanding the potential location of nerve injury based upon mechanism, subjective symptoms, and physical examination (dermatomal mapping, quantitative sensory testing) is crucial to successful operative remediation.

The IIN lies over the anterior surface of the spermatic cord and is covered by the investing fascia of the internal oblique muscle. During inguinal hernia repair, this fascia protects the nerve from direct contact with the mesh and should not be disrupted. Contrary to prior teaching, dissection of the IIN from the cord should be avoided because destruction of this protective fascia increases the risk of perineural scarring or entrapment by the implanted mesh.

The genital branch of the GFN enters the deep inguinal ring and traverses the inguinal canal within the spermatic cord. It is covered by the deep cremasteric fascia, which protects it from contact with the mesh. Its location is most easily identified by its close proximity to the external spermatic vein, which appears as a blue line immediately adja-

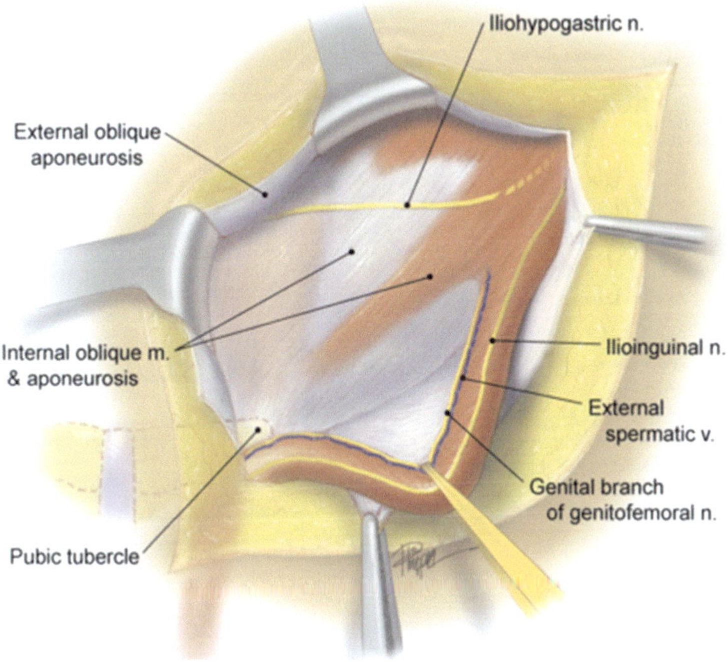

Fig. 24.1. Inguinal neuroanatomy (From Chen et al. [11] with kind permission Springer Science+Business Media).

cent to the nerve. When isolating the cord, care must be taken to visualize the nerve and maintain its position with the other cord structures as the cord is separated from the inguinal floor. The deep cremasteric fascia should be kept intact to avoid perineural scarring or contact between the nerve and mesh.

The IHN lies between the internal and external oblique muscle layers of the abdominal wall. The investing fascia of the internal oblique protects the nerve from contacting the mesh. The IHN can be identified by opening the anatomic cleavage between the internal and oblique layers high enough to expose the internal oblique aponeurosis. This simple maneuver allows for easy identification of the portion of the nerve that lies superficial to the internal oblique aponeurosis. There is an additional, more proximal segment of the nerve that lies within the internal oblique muscle. This intramuscular segment is commonly injured because it is not visible during hernia repair. Suturing the internal

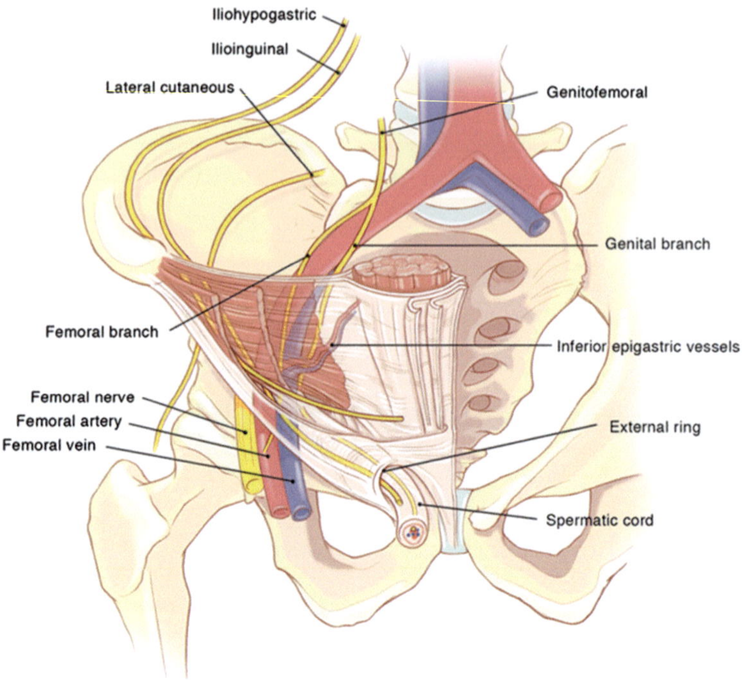

Fig. 24.2. Retroperitoneal neuroanatomy (From Wagner et al. [12], with kind permission ©McGraw-Hill Education).

oblique muscle to the inguinal ligament or mesh implant can potentially result in injury or entrapment of this segment. In approximately 5 % of patients, the IHN runs deep to the internal oblique aponeurosis, passing directly through the internal and external oblique layers simultaneously and is not visible within the inguinal canal at the time of initial repair, placing this nerve at risk [7]. This should be noted by the operating surgeon at the visible point of its exit, and care should be taken to avoid passing suture or fixation material through the internal oblique aponeurosis in the anticipated course of this nerve, risking injury or entrapment of the underlying nerve.

Surgical Management of Neuropathic Pain

Operative management for pain after inguinal surgery has been reported as early as 1942 with Magee describing genitofemoral causalgia as a source for post-inguinal surgery pain [13]. Selective IIN,

IHN, and GFN neurolysis or neurectomy, removal of mesh and fixation material, and revision of the prior herniorrhaphy are common but less effective options for treatment [14–16]. Neurolysis, which does not address ultrastructural changes of nerve fibers, has limited efficacy and is not recommended [5]. Simple removal of entrapping sutures or fixating devices while leaving the injured nerves behind is also inadequate [5]. Selective single or double neurectomy may be effective for some patients, but does not address ultrastructural changes or microscopic neuromas of seemingly normal-appearing nerves during reoperation [14–16]. For patients with high probability of isolated neuropathy based on history, symptoms, mechanism (e.g., IHN entrapment in a Pfannenstiel incision, neuropathy from isolated trocar site, or LFC injury from lateral tack placement), physical examination including dermatomal mapping/distribution, and sensory testing, selective neurectomy is reasonable in experienced hands. However, for most anterior hernia repairs and posterior approaches utilizing fixation, multiple nerves may be involved, and triple neurectomy is more effective.

From a technical perspective, reoperation in the scarred field increases difficulty and morbidity for subsequent remedial operations. Anatomically, the significant variation and cross-innervation of the inguinal nerves in the retroperitoneum and inguinal canal make selective neurectomy less reliable [5]. A recent study by Bischoff et al. described their experience with selective neurectomy in 54 patients with chronic pain after open mesh repair [17]. The IIN, IHN, and GFN were identified in 40 (74 %), 20 (37 %), and 13 (24 %) patients, respectively, illustrating the challenge of reoperative nerve identification in experienced hands. It is difficult to precisely isolate the inguinal nerves involved, and frequently there is more than one nerve implicated in postherniorrhaphy chronic neuropathic pain. Triple neurectomy of the IIN, IHN, and genital branch of the GFN, pioneered in our institute in 1995, is currently a universally accepted surgical treatment for neuropathic pain refractory to conservative measures and is arguably the most effective option [5, 18]. Our experience has included over 700 patients using an open approach with a success rate of over 85 % and 50 selected cases using a laparoscopic retroperitoneal approach with a 90 % success rate. Operative neurectomy in conjunction with removal of meshoma, when present, provides effective relief in the majority of well-selected patients with refractory neuropathic inguinodynia [5].

Timing and Patient Selection

A systematic approach is imperative for proper identification of patients suited for operative intervention. Recommended timing of surgical intervention for postherniorrhaphy pain unresponsive to standard nonsurgical modalities is 6 months to 1 year after the original hernia repair [1, 5]. Failure of conservative measures, in of itself, is not an indication for further surgery. Successful outcomes are entirely dependent upon choosing patients with discrete neuroanatomic problems amenable to surgical correction. A thorough preoperative evaluation should include symptomatology, review of the prior operative report for technique (specifically, the type of repair, type of mesh used, position of the mesh, method of fixation, and nerve handling), imaging to assess for meshoma or other anatomic abnormalities, and response to prior interventions [5, 19]. The patients most likely to benefit from operative neurectomy are those with neuropathic pain isolated to the inguinal distribution that was not present prior to the original operation and that showed improvement with diagnostic and therapeutic nerve blocks.

Risks of Surgery

Operative remediation of inguinodynia carries risk of complications, including refractory pain, exacerbation of underlying pain, deafferentation hypersensitivity, and anticipated permanent numbness involving unilateral labial numbness and potential associated sexual dysfunction in women. Risks related to reoperation in the scarred field include bleeding, disruption of the prior hernia repair, recurrence, vascular injury, and testicular loss. These risks should be discussed with the patient and documented prior to proceeding to operation.

Technique

Triple neurectomy involves resecting segments of the IIN, the genital branch of the GFN, and the IHN from a point proximal to the original surgical field to the most distal accessible point. The main trunk of the GFN over the psoas muscle may also be resected in the case of pain after open or laparoscopic preperitoneal hernia repair, as described below

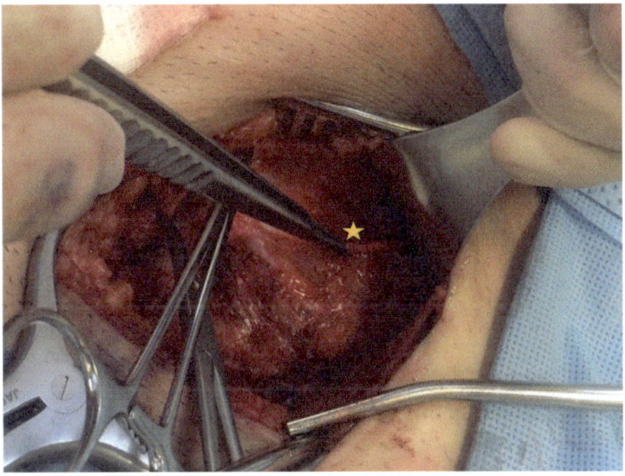

Fig. 24.3. Identification and neurectomy of ilioinguinal nerve.

[20]. This chapter focuses on open triple neurectomy with laparoscopic triple neurectomy described separately. Exposure for open triple neurectomy typically utilizes the same incision as the original anterior repair. If the original repair was done laparoscopically, a standard inguinal incision is used. Extending the incision more cephalad and lateral than typical for a hernia repair facilitates the exposure of the proximal portions of the IIN and IHN. Additionally, this allows for access to the inguinal canal proximal to scarred mesh within the canal.

The IIN is typically identified lateral to the deep inguinal ring and divided as proximally as possible (Fig. 24.3). The IHN is identified in the plane between the internal and external oblique aponeurosis. It is traced proximally to its intramuscular segment and divided proximal to the field of the original repair. Including the intramuscular segment in the resection avoids missing an occult injury in this segment. If the IHN is noted to be one of the subaponeurotic variants described previously, the internal oblique aponeurosis is split proximal to the point where the nerve traverses both internal and external oblique aponeurosis, and this hidden portion of the nerve is divided (Fig. 24.4). The genital branch of the GFN is identified adjacent to the external spermatic vein under the cord or through the lateral crus of the internal ring. In a standard triple neurectomy for pain after anterior repairs, it is ligated and divided at the internal ring (Fig. 24.5).

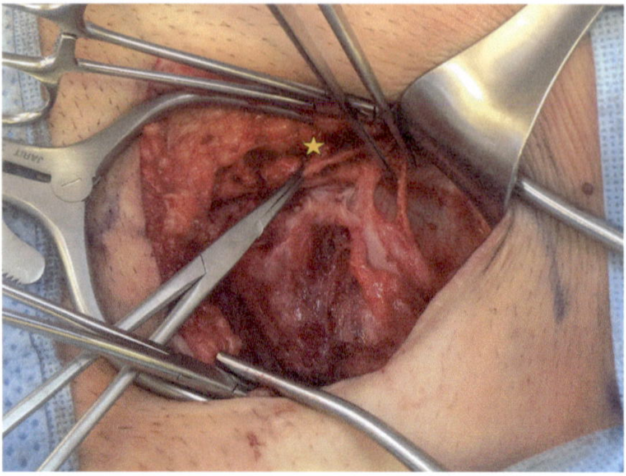

Fig. 24.4. Identification and neurectomy of iliohypogastric nerve.

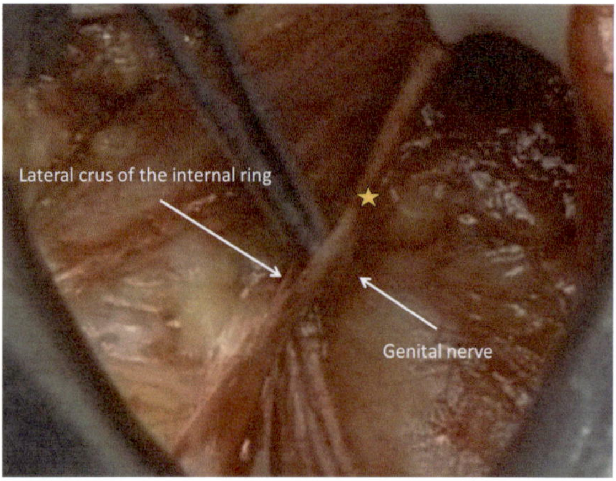

Fig. 24.5. Identification and neurectomy of genital nerve in the inguinal canal.

While there is no scientific consensus for handling the cut ends of the nerves, ligation of the cut nerve ends closes the neurilemma and may reduce neuroma formation. Our standard practice includes burying the proximal nerve stump into surrounding muscle to protect it from the inflammation and scarring of the operative field.

Chronic Pain After Preperitoneal Hernia Repair

The preperitoneal nerves most commonly implicated in postherniorrhaphy neuropathy are the main trunk, femoral branch, and the preperitoneal segment of the genital branch of the GFN. These nerves lack a fascial covering and are therefore at increased risk of injury if allowed to contact mesh. Neuropathic injuries of these nerves can be addressed by laparoscopic triple neurectomy (discussed in Chap. 25) or open extended triple neurectomy, which includes segmental resection of the main genitofemoral trunk in the retroperitoneum [20]. For open extended triple neurectomy, the exposure utilizes the same split made in the internal oblique muscle during resection of the intramuscular segment of the IHN. The underlying transversus abdominis muscle is similarly split, and the parietal peritoneum is swept medially and cephalad to expose the psoas muscle and the main trunk of the GFN as it courses along the body of the muscle (Fig. 24.6). Resecting the GFN at this level additionally addresses neuropathic pain due to injury of the main trunk, the femoral branch, or the preperitoneal segment of the genital branch, which are inaccessible during standard triple neurectomy.

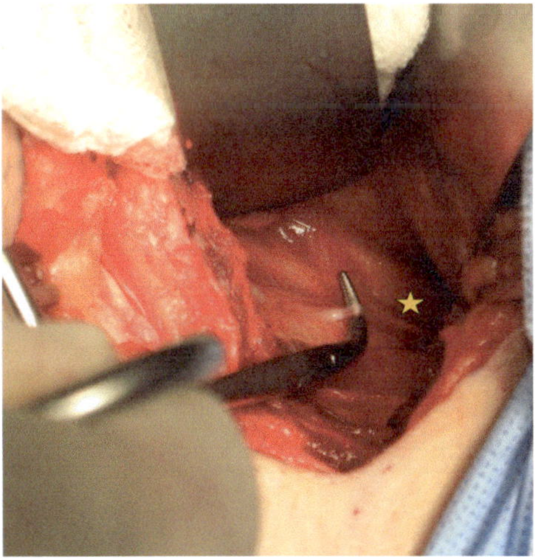

Fig. 24.6. Identification and neurectomy of retroperitoneal genitofemoral nerve trunk over psoas muscles (extended triple neurectomy).

Postherniorrhaphy Orchialgia

Occasionally, postherniorrhaphy orchialgiaaccompanies inguinodynia, and it is important to distinguish testicular pain from the scrotal pain often associated with genital neuralgia in inguinodynia. If true orchialgia exists, it will not respond to triple neurectomy alone. Postherniorrhaphy orchialgia is complex and is likely caused by neuropathy of the paravasal nerve fibers and the autonomic fibers that accompany the cord structures. In patients identified preoperatively, segmental resection of the lamina propria of the vas deferens, which includes the paravasal nerves, may be performed at the time of triple neurectomy. This procedure has been successful, but the results are inconsistent [19]. In orchialgia after preperitoneal mesh repair, vas neurolysis with open triple neurectomy is ineffective, as the injury is proximal to this intervention. Accessing the autonomic nerve plexus proximal to the mesh may be accomplished laparoscopically or robotically in these cases of preperitoneal repair (Fig. 24.7).

Results

Our experience includes over 700 patients who have undergone open triple neurectomy. In patients whose original repair did not enter the preperitoneal space, we now achieve satisfactory resolution of

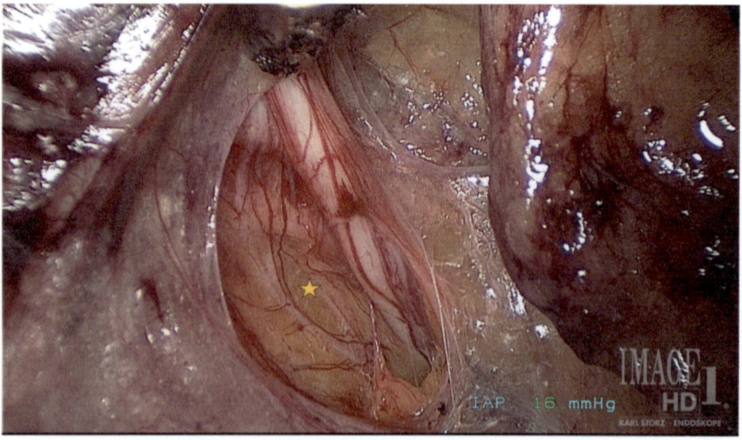

Fig. 24.7. Vas deferens and autonomic nerve plexus.

postherniorrhaphy inguinodynia in 95 % of patients. These results represent patients whose triple neurectomy included resection of the intramuscular segment of the IHN, a technique we have employed since 2004. Prior to this modification, only the extramuscular portion of the IHN was resected, with an associated success rate of 85 % [7].

We have performed open extended triple neurectomy including the main trunk of the GFN in 40 patients with chronic inguinodynia following preperitoneal mesh inguinal hernia repair, with over 90 % of these patients experiencing significant improvement of their pain. We have additionally combined paravasal neurectomy with triple neurectomy for 24 patients with postherniorrhaphy groin pain and orchialgia. The orchialgia was eliminated in 83 % of patients. These limited series suggest that both procedures are safe and effective, though additional study is indicated before they become standard practice.

Conclusion

There is no level 1 or 2 evidence regarding the operative management of inguinodynia, and best available recommendations are derived from case reports, case series, expert opinion, and expert consensus [5, 18]. Our experience with over 750 triple neurectomy operations (700+ open, 50+ laparoscopic) performed by two surgeons (PKA and DCC) is the largest single-institution experience. Since the inception of the Lichtenstein Hernia Institute in 1984, we have additionally evaluated and treated thousands of patients without surgery, with mesh removal, selective neurectomy, quadruple neurectomy, and all other variants of therapy. Triple neurectomy, pioneered in our institute, remains the most definitive and common remedial operation performed. The operative principles of open triple neurectomy involve segmental resection of the IIN, the genital branch of the GFN, and the IHN proximal to the site of injury and resection of the intramuscular portion of the IHN. For patients with a prior preperitoneal hernia repair, open triple neurectomy must be extended to the retroperitoneum to include the main trunk of the GFN, or this nerve can be addressed during laparoscopic triple neurectomy. Patients with concurrent postherniorrhaphy orchialgia may benefit from combining paravasal neurectomy with open triple neurectomy. With success rates of over 90 %, triple neurectomy provides the greatest chance of improving pain and symptoms and is the most definitive option to remediate these problems in an operative field that will ideally never be entered again. Outcomes are highly dependent upon patient selection and experience, and a logical plan of care must be tailored for each patient based upon mechanism, symptoms, anatomy, and technical considerations. Prevention is by far the most important and effective means of preventing inguinodynia and improving patient outcomes.

References

1. Poobalan AS, Bruce J, Smith WC, King PM, Krukowski ZH, Chambers WA. A review of chronic pain after inguinal herniorrhaphy. Clin J Pain. 2003;19(1):48–54.
2. Hakeem A, Shanmugam V. Inguinodynia following Lichtenstein tension-free hernia repair: a review. World J Gastroenterol. 2011;17(14):1791–6.
3. Franneby U, Sandblom G, Nordin O, Nyren O, Gunnarsson U. Risk factors for long-term pain after hernia surgery. Ann Surg. 2006;244(2):212–9.
4. Bay-Nielsen M, Perkins FM, Kehlet H. Pain and functional impairment 1 year after inguinal herniorrhaphy: a nationwide questionnaire study. Ann Surg. 2001;233(1):1–7.
5. Alfieri S, Amid PK, Campanelli G, Izard G, Kehlet H, Wijsmuller AR, Di Miceli D, Doglietto GB. International guidelines for prevention and management of post-operative chronic pain following inguinal hernia surgery. Hernia. 2011;15(3):239–49.
6. Lichtenstein IL, Shulman AG, Amid PK, Montllor MM. Cause and prevention of postherniorrhaphy neuralgia: a proposed protocol for treatment. Am J Surg. 1988;155(6):786–90.
7. Amid PK, Hiatt JR. New understanding of the causes and surgical treatment of postherniorrhaphy inguinodynia and orchalgia. J Am Coll Surg. 2007;205(2):381–5.
8. Aavsang E, Kehlet H. Surgical management of chronic pain after inguinal hernia repair. Br J Surg. 2005;92(7):795–801.
9. Klaassen Z, Marshall E, Tubbs RS, Louis Jr RG, Wartmann CT, Loukas M. Anatomy of the ilioinguinal and iliohypogastric nerves with observations of their spinal nerve contributions. Clin Anat. 2011;24(4):454–61.
10. Rab M, Ebmer J, Dellon AL. Anatomic variability of the ilioinguinal and genitofemoral nerve: implications for the treatment of groin pain. Plast Reconstr Surg. 2001;108(6):1618–23.
11. Chen DC, Amid PK. Technique: Lichtenstein. In: Jacob BP, Ramshaw B, editors. The SAGES manual of hernia repair. New York: Springer; 2013. p. 41–54.
12. Wagner JP, Brunicardi FC, Amid PK, Chen DC. Inguinal hernias. In: Brunicardi FC, Andersen DK, Billiar TR, Dunn DL, Hunter JG, Matthews JB, Pollock RE, editors. Schwartz's principles of surgery. 10th ed. New York: McGraw Hill Medical; 2014. p. 1495–521.
13. Magee RK. Genitofemoral causalgia: (a new syndrome). Can Med Assoc J. 1942;46(4):326–9.
14. Aavsang E, Kehlet H. The effect of mesh removal and selective neurectomy on persistent postherniotomy pain. Ann Surg. 2009;249(2):327–34.

15. Zacest AC, Magill ST, Anderson VC, Burchiel KJ. Long-term outcome following ilioinguinal neurectomy for chronic pain. J Neurosurg. 2010;112(4):784–9.
16. Loos MJ, Scheltinga MR, Roumen RM. Tailored neurectomy for treatment of postherniorrhaphy inguinal neuralgia. Surgery. 2010;147(2):275–81.
17. Bischoff JM, Enghuus C, Werner MU, Kehlet H. Long-term follow-up after mesh removal and selective neurectomy for persistent inguinal postherniorrhaphy pain. Hernia. 2013;17(3):339–45.
18. Lange JF, Kaufmann R, Wijsmuller AR, Pierie JP, Ploeg RJ, Chen DC, Amid PK. An international consensus algorithm for management of chronic postoperative inguinal pain. Hernia. 2015;19(1):33–43.
19. Amid PK. Radiologic images of meshoma: a new phenomenon causing chronic pain after prosthetic repair of abdominal wall hernias. Arch Surg. 2004;139(12):1297–8.
20. Amid PK, Chen DC. Surgical treatment of chronic groin and testicular pain after laparoscopic and open preperitoneal inguinal hernia repair. J Am Coll Surg. 2011;213(4):531–6.

25. Laparoscopic Triple Neurectomy

Stephanie A. Kingman, Parviz K. Amid, and David C. Chen

Introduction

Inguinal hernia repair, with the use of mesh and tension-free techniques, has seen significant improvements in outcomes and decreased recurrence rate. However, postherniorrhaphy chronic pain remains a considerable complication affecting as many as 63 % of patients after surgery [1–3]. Such pain interferes with the physical activity, social interactions, employment, and productivity of 6–8 % of patients after herniorrhaphy, causing notable burden on the individual and society [2–6].

In addition to pain from hernia recurrence, inguinodynia can be caused by factors relating to nociceptive or neuropathic pain [7–12]. Nociceptive pain is induced by tissue injury and inflammation from forceful tissue handling and retraction or from foreign material such as meshoma. Neuropathic pain is provoked by direct nerve injury, perineural scarring, or entrapment injuries by suture, fixating devices, or mesh. Classically, it presents as inguinodynia with radiation to the scrotum/femoral triangle, hyperalgesia, allodynia, hyper- or hypoesthesia, and paresthesias. There is no clear distinction between these two types of pain, and the diagnosis is often complicated by genetic, psychological, social, and economic factors [2–5, 9].

Nonsurgical management, including pharmacologic, interventional, and behavioral therapies, is successful in many patients. Nonetheless, operative intervention is necessary in some cases. The most definitive and effective remedial surgery for refractory neuropathic inguinodynia is triple neurectomy of the ilioinguinal, iliohypogastric, and genitofemo-

B.P. Jacob et al. (eds.), *The SAGES Manual of Groin Pain*,
DOI 10.1007/978-3-319-21587-7_25

ral nerves. This technique, described by us in 1995 with further technical modifications in recent years, has yielded response rates of 85–97 % [13–15].

Triple neurectomy, whereby the three nerves are resected proximal to the area of the initial hernia repair and as distal as possible, is conventionally done via an open anterior approach [13, 16]. The open operation has limitations, as it may be difficult to identify and access the three inguinal nerves in the reoperative field, and there is considerable neuroanatomic variation especially distal to the retroperitoneum within the inguinal canal. Operating in scarred tissues increases the risk of disrupting the previous hernia repair as well as injuring the spermatic cord and testicle. In patients whose initial operation was a preperitoneal (open or laparoscopic) repair, accessing the nerves proximal to the pathology is not always possible from an inguinal approach. These challenges, in addition to causing surgical pain in an already hypersensitive area, make a minimally invasive retroperitoneal approach very desirable.

Preoperative Workup

The recommended timing of surgery for chronic postherniorrhaphy pain not controlled with conservative treatments is 6 months to 1 year after the initial inguinal hernia repair. Prior to surgery, a detailed and methodical preoperative workup is recommended to define the potential causes of a patient's groin pain. This should involve characterization of symptoms, assessment of prior conservative pain management with pharmacologic and interventional therapies, as well as imaging to evaluate for presence of meshoma or other anatomic abnormalities. Previous operative reports should be analyzed for technique such as type of repair, presence, type and position of mesh, method of fixation, and identification and handling of nerves, as these factors would influence the type of intervention and remedial surgery possible. Patients should also have multidisciplinary treatment, including evaluation by a pain specialist. All patients considered for surgery should undergo diagnostic and therapeutic nerve blocks of the ilioinguinal, iliohypogastric, and genitofemoral nerves.

Finally, it is imperative to thoroughly discuss and document possible benefits and risks of remedial surgery with patients, including failure to identify or resect all three nerves, persistent pain despite successful neurectomy due to various etiologies of pain, permanent numbness in the corresponding dermatomal distributions, abdominal wall laxity secondary to partial denervation of the oblique muscles, numbness in the labia in females that may alter sexual sensation, testicular atrophy, and loss of the cremasteric reflex in males. The surgery may cause hypersensitivity from deafferentation that is typically temporary, though its course is unpredictable and may be permanent. This technique does not alter nociceptive pain caused by tissue injury, meshoma, or testicular pain [16].

Surgical Approach

Laparoscopic retroperitoneal triple neurectomy is a 1-stage procedure to access the main trunks of the ilioinguinal, iliohypogastric, and genitofemoral nerves in the lumbar plexus [17]. This access allows the nerves to be resected proximal to any potential site of peripheral neuropathy from the previous surgical field.

The patient is positioned in lateral decubitus position, and the table is flexed to open the space between the iliac crest and costal margin. A 12-mm transverse incision is made in the midaxillary line 3–4 cm above the iliac crest (Fig. 25.1). The external oblique fascia is incised, and the oblique muscles are separated until the retroperitoneum is accessed. An oval dissecting balloon can be placed into this potential space and inflated under direct visualization. This mobilizes the peritoneum in the avascular plane, rotating the viscera medially, and exposes the retroperitoneal space. The dissecting balloon is then exchanged with a 12-mm balloon trocar, and carbon dioxide is used to insufflate to a pressure of 15 mmHg. Another 5-mm port is inserted 2 cm medially under direct visualization. The retroperitoneal fat pad is then dissected medially using laparoscopic cautery or a vessel-sealing device to expose the psoas and quadratus lumborum muscles.

The lumbar plexus should be defined before any neurectomy is performed (Fig. 25.2) [18]. The subcostal nerve can first be identified at the T12 costal margin (Fig. 25.3). The iliohypogastric and ilioinguinal nerves, frequently sharing a common trunk, can then be seen overlying the quadratus muscle at L1 (Fig. 25.4) [19, 20]. The lateral femoral cutaneous nerve originating at L3 is identified lateral to the psoas, crossing the iliacus muscle below the iliac crest (Fig. 25.5). The femoral nerve can also be found lateral and deep to the psoas muscle, but does not require specific dissection. The dissection is then continued toward the groin where the genitofemoral nerve trunk can be noted running over the psoas muscle (Fig. 25.6). Similar to the iliohypogastric and ilioinguinal nerve trunks, the genital and femoral nerve trunks have considerable variability

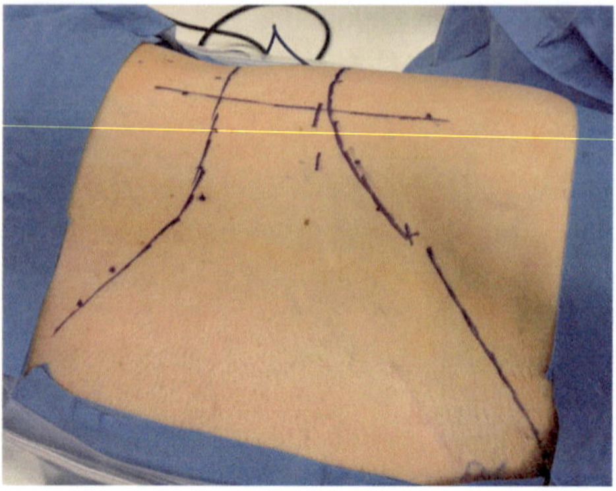

Fig. 25.1. Trocar placement and operative positioning.

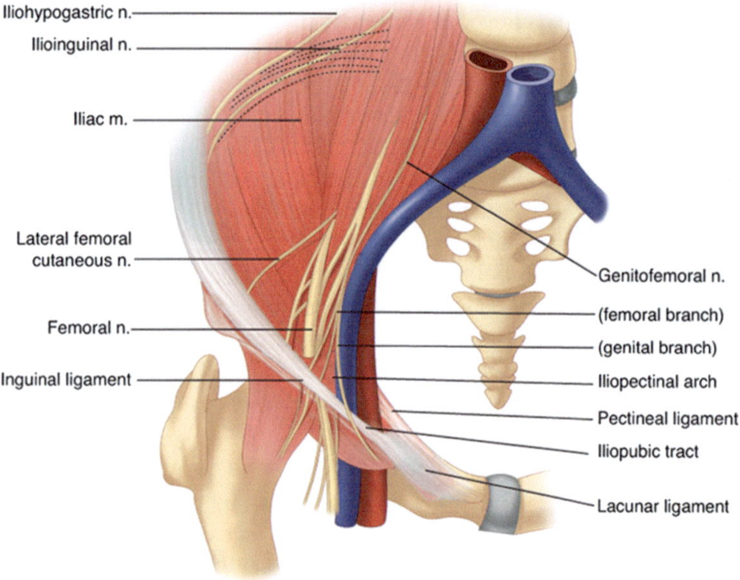

Fig. 25.2. Retroperitoneal lumbar plexus (From Wagner et al. [18], with kind permission ©McGraw-Hill Education).

and often have separate trunks. If the dermatomal distribution of the femoral branch of the GFN is not affected, a separate femoral trunk should be preserved when found. Of note, the ureter and iliac vessels should be identified medial to the psoas and protected (Fig. 25.7).

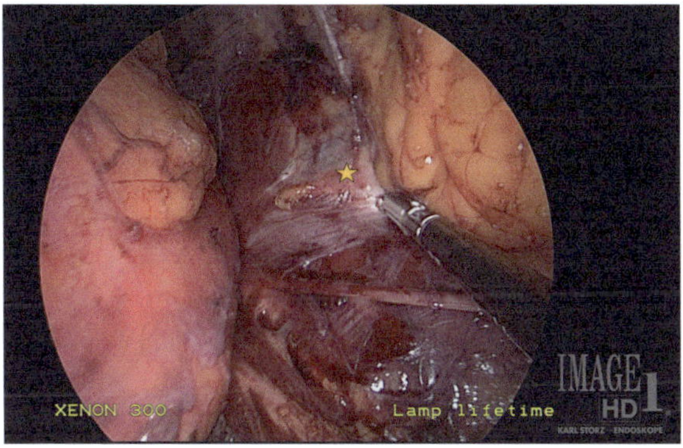

Fig. 25.3. Subcostal nerve trunks and 12th rib at T12 level (*star*). Ilioinguinal/iliohypogastic nerve trunk caudal.

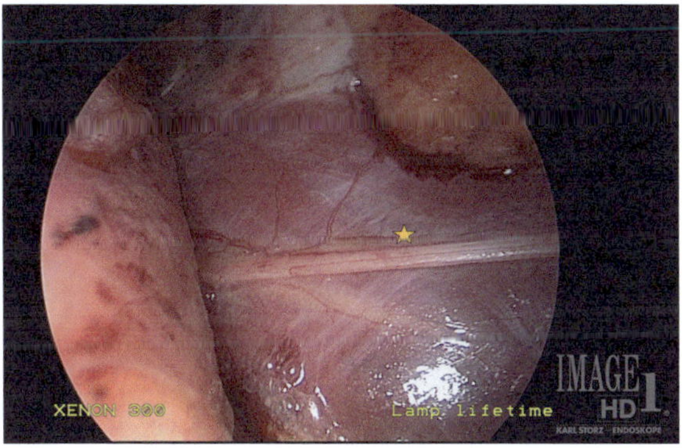

Fig. 25.4. Iliohypogastric and ilioinguinal nerve trunks over quadratus lumborum muscle at L1 level (*star*). Subcostal nerve and 12th rib cephalad.

After all structures have been clearly delineated, the iliohypogastric and ilioinguinal nerves can be resected over the quadratus muscle. With regard to the cut nerve, our preference is to place a clip proximally and distally to close the neurilemma. This theoretically helps to avoid neuroma formation and allows for radiographic identification of the cut nerve if future proximal interventional blocks are needed. The genitofemoral nerve trunk is subsequently clipped and resected over the psoas muscle in a similar fashion. A transabdominal approach may alterna-

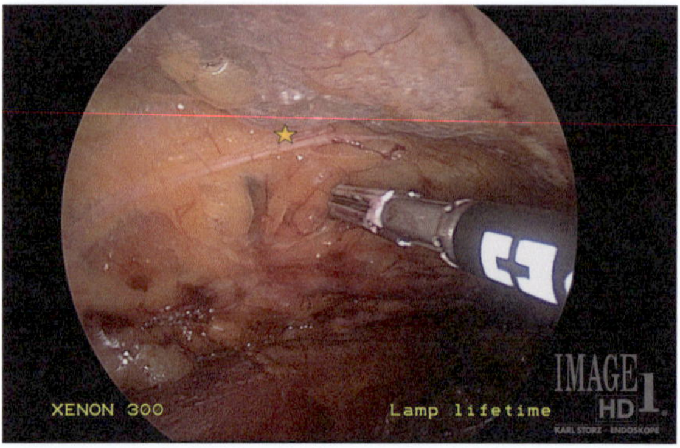

Fig. 25.5. Lateral femoral cutaneous nerve trunk at L3 level (*star*).

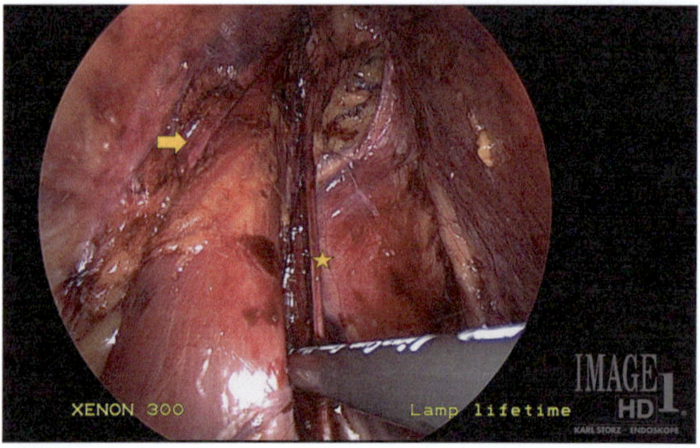

Fig. 25.6. Genitofemoral nerve trunk over psoas muscle (*star*). Femoral nerve lateral to psoas muscle (*arrow*).

tively be used to access the same anatomic planes but requires medial rotation of the viscera and more operative ports.

Outcomes

In our prospective series, 42 patients who presented with chronic inguinodynia not controlled with conservative pain management therapies underwent laparoscopic triple neurectomy (Fig. 25.8). The mean numeric pain scores were significantly reduced (baseline score 8.4) on

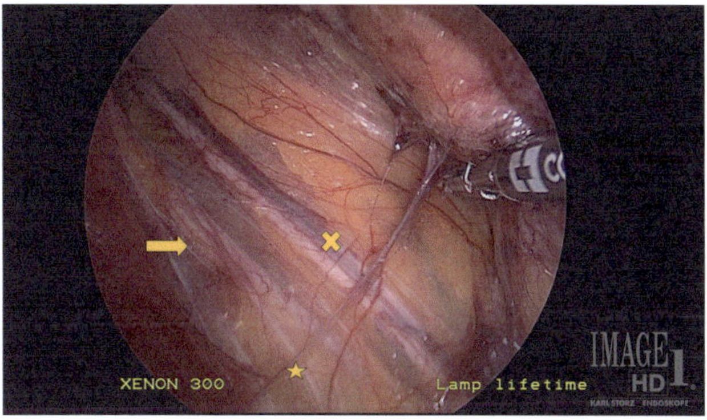

Fig. 25.7. Relationship between ureter (*X*), iliac artery (*arrow*), and genitofemoral nerve trunk over psoas muscle (*star*).

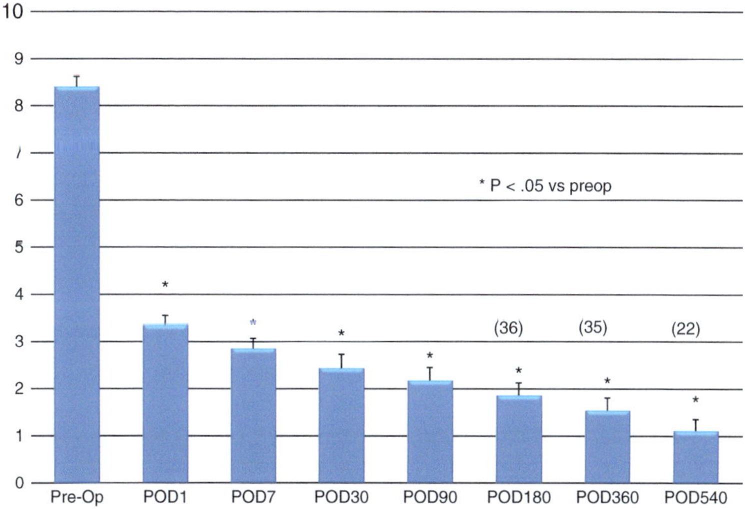

Fig. 25.8. Laparoscopic triple neurectomy outcomes: VAS scores.

postoperative days 1 (score, 3.4; $p<0.001$), 7 (score, 2.8; $p<0.001$), 30 (score, 2.4; $p<0.001$), 90 (score, 2.1; $p<0.001$), and 180 (score, 1.9; $p<0.001$ [17]. Thirty-four patients have been followed to 12 months (pain score 1.5; $p<0.01$), and 20 have been followed over 2 years (pain score 1.1; $p<0.01$). Narcotic dependence was seen to decrease and the activity level of patients increased. All patients reported numbness as anticipated in the distribution of neurectomy. Fourteen (33 %) had transient hypersensitivity consistent with deafferentation, with four patients (9 %) having persistent symptoms greater than 3 months. Seven experienced residual meshoma pain with four of them undergoing a subsequent reoperation for mesh removal. Orchialgia was not improved as expected since paravasal nerves need to be resected to address this problem [16].

Discussion

With the advances in technique of tension-free inguinal hernia repair, chronic groin pain now surpasses recurrence as the most common long-term postoperative complication. This debilitating condition is a result of nociceptive and neuropathic factors. Given the lack of clear discrimination between the two types of pain, confounded with variables such as excitatory coupling between sympathetic and afferent nociceptive fibers, deafferentation hyperalgesia, pain centralization, and neuroplasticity, as well as patient-related factors, prevention of this complication is of key importance [9, 21].

Nociceptive pain can be minimized with gentle handling of tissues and with the use of local anesthetic to decrease the formation of nociceptive molecules. Neuropathic pain can be decreased by meticulous identification and protection of nerves to avoid injury and their direct contact with mesh, which ultimately changes the structure of their fibers. Doing so has been shown to reduce the rate of postherniorrhaphy chronic pain from 5 to 8 % to a fraction of 1 % [20]. Understanding inguinal and preperitoneal groin neuroanatomy as well as the pathophysiology of inguinal pain helps to guide good operative technique in all methods of inguinal hernia repair. Prevention and avoidance of injury at the time of the original operation are of paramount importance.

Numerous operative techniques have been used to address chronic inguinodynia such as revision of the original herniorrhaphy, removal of mesh or fixation device, and selective neurolysis or neurectomy of the ilioinguinal, iliohypogastric, and genitofemoral nerves. These techniques, however, often leave behind injured nerves and do not affect the ultrastructural changes of nerve fibers. Moreover, the considerable variation in anatomy and cross-innervation of the inguinal nerves within the retroperitoneum and inguinal canal can make such procedures unreliable [19, 20, 22–25].

The current most effective therapy for the neuropathic component of inguinal pain is triple neurectomy [12–16, 22–24]. In our ongoing series of laparoscopic retroperitoneal triple neurectomy, we have had a 93 %

success rate in reducing numeric pain scores and narcotic dependence and improving the quality of life and function in daily activities for patients. Numerically, this demonstrates superior results of the laparoscopic retroperitoneal approach to standard open triple neurectomy (80 %) and extended open triple neurectomy, which includes the resection of the genitofemoral nerve trunk (87.5 %). Importantly, it allows us to more effectively treat inguinodynia following posterior repair (i.e., laparoscopic TEP/TAPP), a subset of patients who do not respond as effectively to open anterior triple neurectomy, as the pathology and site of injury typically are proximal to the inguinal operative field.

With the minimally invasive approach, nerve identification, which is often the cause for failure of open neurectomy, is uniformly successful; the anatomy in the retroperitoneal lumbar plexus is less variable, and it obviates the need to reoperate in a scarred field. Risks of remedial surgery are reduced; morbidity and perioperative disability are minimized. Thus, in the absence of recurrence or meshoma, this surgery is our preferred technique for definitive management of chronic postherniorrhaphy inguinodynia due to neuropathic causes.

References

1. Reinpold WM, Nehls J, Eggert A. Nerve management and chronic pain after open inguinal hernia repair: a prospective two phase study. Ann Surg. 2011;254(1):163–8.
2. Kehlet H. Chronic pain after groin hernia repair. Br J Surg. 2008;95(2):135–6.
3. Aasvang E, Kehlet H. Surgical management of chronic pain after inguinal hernia repair. Br J Surg. 2005;92(7):795–801.
4. Aasvang EK, Bay-Nielsen M, Kehlet H. Pain and functional impairment 6 years after inguinal herniorrhaphy. Hernia. 2006;10(4):316–21.
5. Bay-Nielsen M, Perkins FM, Kehlet H, Danish Hernia Database. Pain and functional impairment 1 year after inguinal herniorrhaphy: a nationwide questionnaire study. Ann Surg. 2001;233(1):1–7.
6. Callesen T, Bech K, Kehlet H. Prospective study of chronic pain after groin hernia repair. Br J Surg. 1999;86(12):1528–31.
7. Magee RK. Genitofemoral causalgia: (a new syndrome). Can Med Assoc J. 1942;46(4):326–9.
8. Amid PK. Radiologic images of meshoma: a new phenomenon causing chronic pain after prosthetic repair of abdominal wall hernias. Arch Surg. 2004;139(12):1297–8.
9. Lichtenstein IL, Shulman AG, Amid PK, Montllor MM. Cause and prevention of postherniorrhaphy neuralgia: a proposed protocol for treatment. Am J Surg. 1988;155(6):786–90.

10. Amid PK, Hiatt JR. New understanding of the causes and surgical treatment of postherniorrhaphy inguinodynia and orchalgia. J Am Coll Surg. 2007;205(2):381–5.
11. Heise CP, Starling JR. Mesh inguinodynia: a new clinical syndrome after inguinal herniorrhaphy? J Am Coll Surg. 1998;187(5):514–8.
12. Aasvang EK, Kehlet H. The effect of mesh removal and selective neurectomy on persistent postherniotomy pain. Ann Surg. 2009;249(2):327–34.
13. Amid PK. A 1-stage surgical treatment for postherniorrhaphy neuropathic pain: triple neurectomy and proximal end implantation without mobilization of the cord. Arch Surg. 2002;137(1):100–4.
14. Madura JA, Madura 2nd JA, Copper CM, Worth RM. Inguinal neurectomy for inguinal nerve entrapment: an experience with 100 patients. Am J Surg. 2005;189(3):283–7.
15. Starling JR, Harms BA, Schroeder ME, Eichman PL. Diagnosis and treatment of genitofemoral and ilioinguinal entrapment neuralgia. Surgery. 1987;102(4):581–6.
16. Amid PK, Chen DC. Surgical treatment of chronic groin and testicular pain after laparoscopic and open preperitoneal inguinal hernia repair. J Am Coll Surg. 2011;213(4):531–6.
17. Chen DC, Hiatt JR, Amid PK. Operative management of refractory neuropathic inguinodynia by a laparoscopic retroperitoneal approach. JAMA Surg. 2013;148(10):962–7.
18. Wagner JP, Brunicardi FC, Amid PK, Chen DC. Inguinal hernias. In: Brunicardi FC, Andersen DK, Billiar TR, Dunn DL, Hunter JG, Matthews JB, Pollock RE, editors. Schwartz's principles of surgery. 10th ed. New York: McGraw Hill Medical; 2014. p. 1495–521.
19. Rab M, Ebmer, Dellon AL. Anatomic variability of the ilioinguinal and genitofemoral nerve: implications for the treatment of groin pain. Plast Reconstr Surg. 2001;108(6):1618–23.
20. Klaassen Z, Marshall E, Tubbs RS, Louis Jr RG, Wartmann CT, Loukas M. Anatomy of the ilioinguinal and iliohypogastric nerves with observations of their spinal nerve contributions. Clin Anat. 2011;24(4):454–61.
21. Alfieri S, Amid PK, Campanelli G, Izard G, Kehlet H, Wijsmuller AR, et al. International guidelines for prevention and management of post-operative chronic pain following inguinal hernia surgery. Hernia. 2011;15(3):239–49.
22. Zacest AC, Magill ST, Anderson VC, Burchiel KJ. Long-term outcome following ilioinguinal neurectomy for chronic pain. J Neurosurg. 2010;112(4):784–9.
23. Loos MJ, Scheltinga MR, Roumen RM. Tailored neurectomy for treatment of postherniorrhaphy inguinal neuralgia. Surgery. 2010;147(2):275–81.
24. Kim DH, Murovic JA, Tiel RL, Kline DG. Surgical management of 33 ilioinguinal and iliohypogastric neuralgias at Louisiana State University Health Sciences Center. Neurosurgery. 2005;56(5):1013–20.
25. Giger U, Wente MN, Buchler MW, Krahenbuhl S, Lerut J, Krahenbuhl L. Endoscopic retroperitoneal neurectomy for chronic pain after groin surgery. Br J Surg. 2009;96:1076–81.

26. Chronic Orchialgia: Workup and Management

Jamin V. Brahmbhatt, Ahmet Gudeloglu, and Sijo J. Parekattil

Introduction

Chronic groin or scrotal content (inguinal canal, spermatic cord, testicular, scrotum) pain (CGSCP) is a common presenting problem for several specialists—emergency room, primary care, general surgeon, and urologists. It can be both acute and chronic in nature and can be managed with medical or surgical interventions. The pain can be unilateral or bilateral and intermittent or constant and lasts longer than 3 months [1, 2]. The pain can be idiopathic or caused by nerve irritation or hypersensitivity through vasectomy, hernia repair, sports injury, abdominal surgery, or any intervention that can irritate the genitofemoral or ilioinguinal nerves. Although the exact mechanism for CGSCP is not well understood, one common theme is a two-hit theory. There is a baseline inflammatory or genetic process that leads to Wallerian degeneration of the peripheral nerves. In the groin or scrotum this degeneration may cause hypersensitivty of the ilioinguinal and genitofemoral nerves. A second inciting event—trauma, surgery, or irritation of these nerves—then leads to chronic neuropathic pain in this area (Fig. 26.1) [3].

CGSCP may affect over 100,000 men annually [4, 5]. Prevalence can range up to 33 % of men after vasectomy [6] and 63 % after inguinal hernia repair [7–9]. After hernia repair, the pain can be neuropathic or non-neuropathic secondary to mesh. Even with such a high prevalence after hernia repair, only 1 % of patients who suffer from CGSCP may be referred for further evaluation [10]. In this chapter we will review the current literature and present a structured algorithm for the evaluation and management of CGSCP.

B.P. Jacob et al. (eds.), *The SAGES Manual of Groin Pain*,
DOI 10.1007/978-3-319-21587-7_26

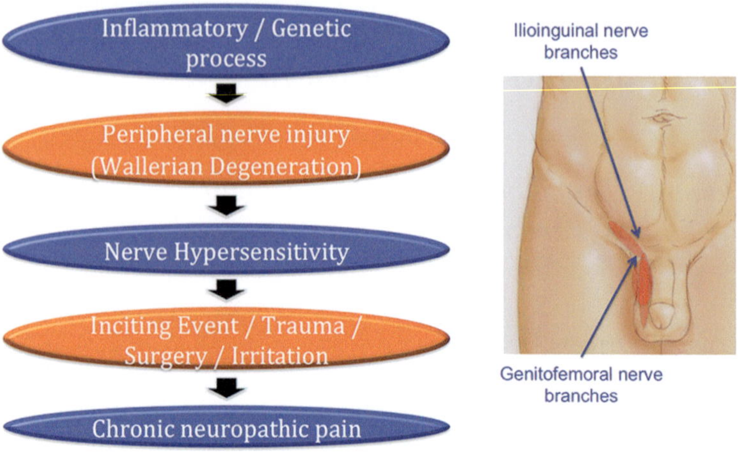

Fig. 26.1. Two-hit theory on cause of chronic groin and scrotal content pain (From Brahmbhatt et al. [3], with kind permission Springer Science+Business Media).

Anatomy and Function

Embryology

At 7–8 weeks gestation, gonads differentiate into testis in the posterior abdominal cavity. After 8 weeks, through the influence of hormones, the testicles begin their descent into the scrotum. Until 7 months, they remain near the inguinal canal. Before birth or within a few weeks after birth, the testicles complete their descent into the scrotum. As the testicle descends from the abdominal cavity, it brings with it layers of the peritoneum and abdominal wall. Figure 26.2 lists the layers from external to internal order. It is important to note that the external spermatic fascia is derived from external oblique fascia, cremasteric muscle from internal oblique fascia, internal spermatic fascia from transversalis fascia, and tunica vaginalis from peritoneum. If the descent does not occur completely, then patients are labeled as having undescended testicles, which may require medical or surgical intervention.

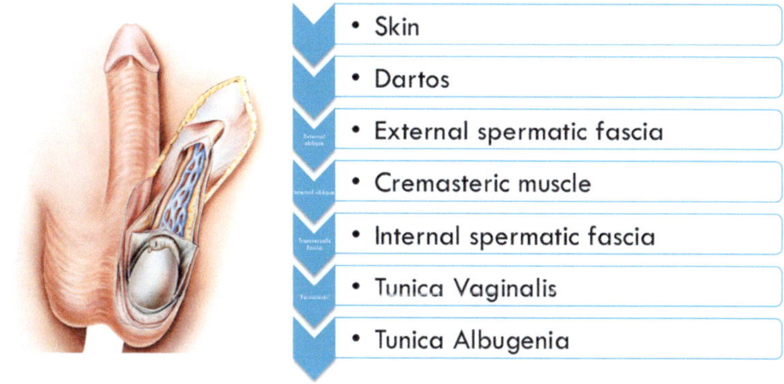

Fig. 26.2. Layers of the scrotal wall.

Anatomy

The testicle is egg shaped with an average length of 5 cm. Testicular function is tightly regulated with signaling from the hypothalamus and anterior pituitary gland. The main function of the testicle is to produce sperm from germ cells and testosterone from Leydig cells. There is another cell type (called Sertoli cells) that is important for support. Sperm travels through the testicle (lobules, seminiferous tubules, rete testis, epididymis) and vas deferens until it mixes with fluid from the seminal vesicles and prostate to form semen. This combined fluid is eventually expelled into the urethra during ejaculation.

The spermatic cord houses the testicular artery, testicular veins (pampiniform plexus), vas deferens, artery of vas deferens, lymphatic vessels, and nerves. Neural innervation to the testicle is via a complex neural network with significant crossover. Afferent innervation of the scrotum originates via somatic nerves in the genital branch of the genitofemoral nerve, ilioinguinal nerves, and autonomic branches from T10-L1 parasympathetic ganglia [11]. The genitofemoral and ilioinguinal nerves provide anterior scrotal wall and thigh innervation. The posterior scrotal wall is innervated via the perineal branch of the pudendal nerve. There is an alternate autonomic pathway between the pelvic plexus and testis via the vas deferens, which explains the positive response to anesthetic injections to the pelvic ganglia [12]. On average, there are 31 small diameter (less than 1 mm) nerve fibers in the spermatic cord. The three primary sites (trifecta nerve complex) of highest nerve density are (in decreasing order): cremasteric muscle fibers, perivasal tissue and vasal sheath, and posterior peri-arterial/lipomatous tissue [13].

Evaluation

Workup of CGSCP begins with a thorough history and physical examination. The characteristics of pain, including onset, duration, and severity, are questioned. Pain is rated using the visual analog scale and externally validated pain impact questionnaire (PIQ-6, Quality-Metrics Inc., Lincoln, RI, USA).

Physical examination focuses on the groin and testicle in the attempt to identify any anatomic causes of the pain, including hernia, varicocele, testicular masses, epididymal cysts, and granulomas from previous vasectomy. All possible causes such as ureteral stones, infection (orchitis or epididymitis), or back problems (lumbar disk hernia) need to be ruled out. Urine analysis, scrotal ultrasonography, abdominal computerized tomography (CT), and spinal magnetic resonance imaging (MRI) should be performed when indicated. Scrotal ultrasound is not necessary when physical examination and urine analyses are normal in patients with chronic scrotal pain. Van Haarst et al. evaluated scrotal ultrasonography imaging of 111 chronic scrotal pain patients with normal physical examination and urine analyses and found 12 epididymal cysts less than 0.5 cm and three subclinical varicocele but no clinical significant abnormalities [14]. Since a significant percentage of CGSCP is idiopathic, patients often have completely negative evaluations. Treatment for these patients is initiated using a structured algorithm (Fig. 26.3).

Medical Treatment

In the absence of any acute findings that require surgical intervention, conservative medical therapy is a first-line treatment [15]. One month of nonsteroidal anti-inflammatory drugs (NSAIDs) is recommended [1]. We usually start with meloxicam 7.5 mg daily or high-dose ibuprofen 600 mg orally three times daily. Newer low-dose NSAIDs such as Zorvolex 35 mg BID-TID can be used to decrease side effect potential.

Sexually transmitted infection with gonorrhea or chlamydia should be considered in men between the ages of 15–35. This is usually treated with azithromycin 1 g orally once (or doxycycline 100 mg orally twice

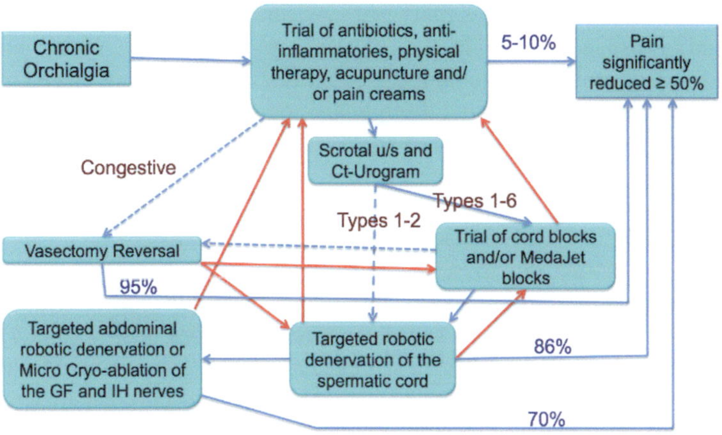

Fig. 26.3. Algorithm for evaluation and management of chronic groin and scrotal content pain.

daily for 10 days) and ceftriaxone 125 mg intramuscularly once. Over the age of 35, *Escherichia coli* is a common urinary pathogen that can cause epididymal infection. *E. coli* infection can be treated with a course of quinolones (ciprofloxacin 500 mg orally twice daily or levofloxacin 500 mg orally once daily for 10 days). A 1-month supply of quinolone therapy can also be considered in combination with NSAIDs for refractory cases.

Antidepressants used at lower doses for chronic pain work by inhibiting the reuptake of norepinephrine. A commonly used class of antidepressants is tricyclics, which include amitriptyline 10–25 mg orally daily and nortriptyline 10–150 mg orally daily. These medications are slowly titrated up to therapeutic levels. The use of these medications can also help the comorbid psychological factors that may contribute to genital pain. In a study of 48 patients with genital pain and no organic findings, psychological disorders were diagnosed commonly, including major depression (27 %), somatization disorder (56 %), and chemical dependency (27 %) [16]. It is critical to taper patients off tricyclic antidepressants—they should never be stopped abruptly.

Anticonvulsants contribute to pain management through their ability to modulate central calcium channels with chronic use via an effect on trafficking [17]. Gabapentin and pregabalin are common anticonvulsants used for neuropathic pain. We start with gabapentin

300 mg orally three times daily and may titrate up as needed. Patients tend to respond well to these medications, but its frequent dosing and side effect profile lead to a high dropout rate. If patients are placed on chronic medications, a multidisciplinary approach to their follow-up is generally recommended.

Spermatic Cord Block

Spermatic cord nerve blocks using local anesthetic with or without steroids can be therapeutic and diagnostic. In our practice, we mix 15 mL 1 % lidocaine, 15 mL 0.25 % marcaine, and dexamethasone 4 mg. The blocks target three areas of high nerve density [13]. First, high-pressure injection of the perivasal tissue is done using a Medi-Jet. Another 5 mL, using a needle, is directly injected directly around the perivasal tissue. We then inject 10 mL medially to the external inguinal ring to target the branches of the ilioinguinal nerve and 10 mL laterally to the external inguinal ring to target genital branches of the genitofemoral nerve.

In patients with no apparent etiology, the blocks can provide temporary relief while helping to predict a positive response to surgical interventions. Benson et al. demonstrated that a positive response to spermatic cord block helps predict a durable and complete resolution of symptoms after microsurgical denervation of the spermatic cord (MDSC) [18]. They achieved an average 89 % decrease in pain with spermatic cord block for median 8 h (1–168 h) in 74 men (77 testicular units). Also improvement from the spermatic cord block was a predictor of overall improvement after MDSC ($p=0.05$).

Microsurgical Targeted Denervation of the Spermatic Cord

Introduction

Microsurgical targeted denervation of the spermatic cord (MDSC) is a minimally invasive surgical option for the management of CGSCP after conservative treatments have failed [2, 19]. An animal study to evaluate the effectiveness of the MDSC procedure showed a significant decrease in median number of nerve fibers remaining around the vas

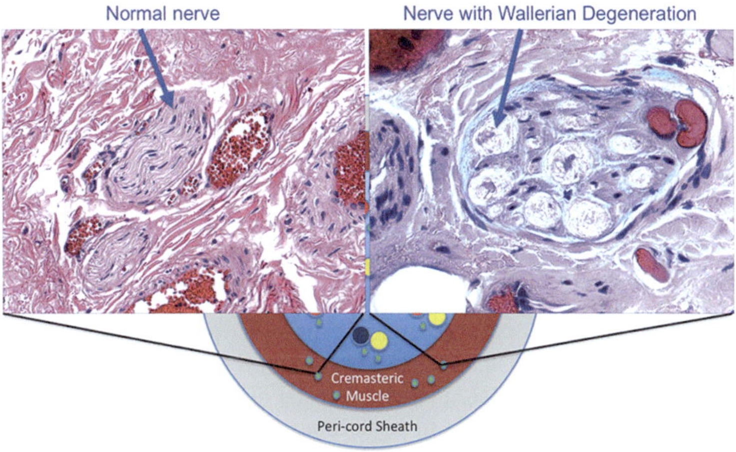

Fig. 26.4. Nerve fiber with and without Wallerian degeneration on H&E staining (From Brahmbhatt et al. [3], with kind permission Springer Science + Business Media).

deferens after MDSC procedure, compared to sham (MDSC = 3.5 nerves, sham = 15.5 nerves, $p = 0.003$) [20]. In this procedure (described in detail below), we target tissue in or around the spermatic cord that carries high-density nerve fibers that contribute to chronic pain. Within these nerve fibers, there are significant anatomic and pathologic differences (Wallerian degeneration) compared to controls (Fig. 26.4) [3, 13]. Targeting these specific areas leads to the preservation of a significant portion of the spermatic cord and potentially fewer complications.

We published our most updated results of the 546 robotic targeted MDSC (RTMDSC) procedures for chronic groin pain [21]. On the last review of our data our total has increased to over 620 patients. Mean preoperative duration of orchialgia in our patients is 2.4 years.The median robotic operative duration was 20 min (range, 10–150 min). Using the externally validated pain assessment tool PIQ-6, we assessed preoperative and postoperative pain. At 6 months there was a 71 % significant reduction in pain and 72 % significant reduction at 1 year. Using the visual analog pain scale, there was an 85 % significant reduction in pain (63 % complete resolution and 22 % greater than 50 %). Complications were limited to one testicular ischemia, two testicular artery injuries (repaired intra-op with no long-term sequelae), one vasal injury (repaired intra-op with no long-term sequelae), 11 hematomas, three seromas, and five wound infections.

There are several advantages to using the robotic platform, which includes improved visualization, decreased tremor, and less dependence on a surgical assistant. Robotic-assisted MDSC seems safe and feasible, and the outcomes appear promising for durable relief.

Technique in Detail

A 1–2 cm transverse subinguinal incision is made. The incision is carried down until the spermatic cord is reached. The spermatic cord is brought up to the surface. Posterior medial and lateral dissection and cauterization are performed to ligate branches of the ilioinguinal and genitofemoral nerves in this area.

The robot is positioned over the patient. A 0° camera lens is utilized. The right, left, and the fourth robot arms are loaded with Black Diamond microforceps, Maryland bipolar grasper, and monopolar curved scissors, respectively (Fig. 26.5) [3]. If a flexible CO_2 laser fiber is used for dis-

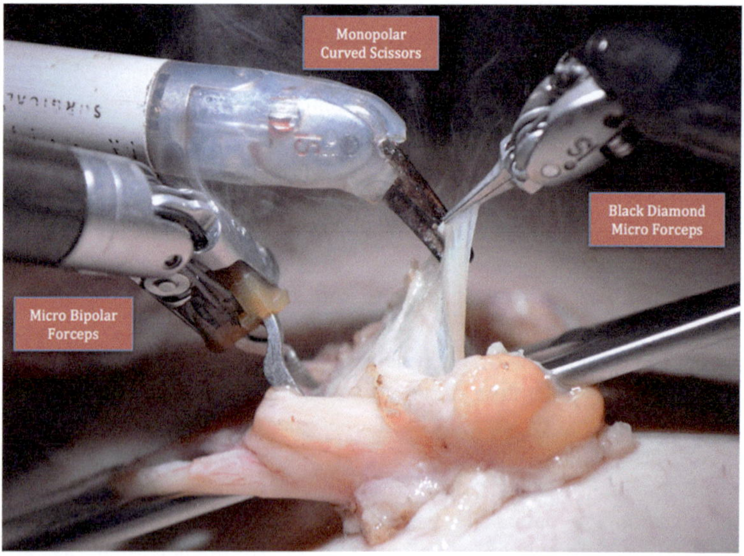

Fig. 26.5. Standard robotic instrumentation for targeted denervation (From Brahmbhatt et al. [3], with kind permission Springer Science+Business Media).

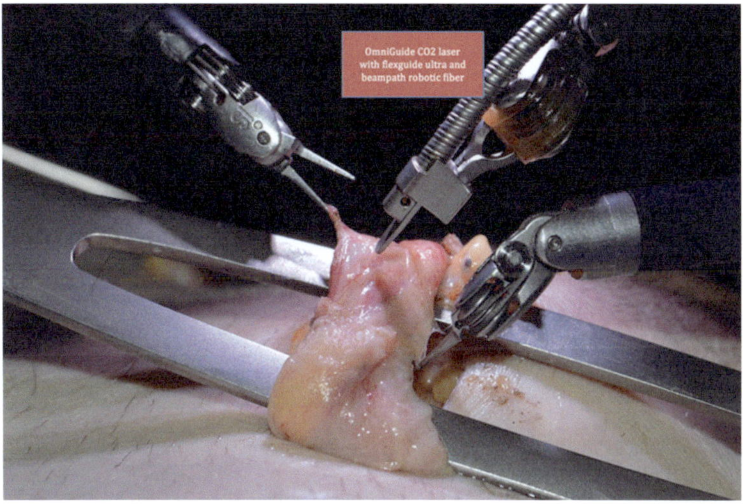

Fig. 26.6. Flexible CO_2 laser instrumentation during targeted denervation (From Brahmbhatt et al. [3], with kind permission Springer Science+Business Media).

section, then the fourth arm is replaced with a Black Diamond microforceps to hold the flexguide laser holder (Fig. 26.6) [3].

The anterior cremasteric muscle is divided. The presence of a testicular artery is confirmed with real-time intraoperative micro-Doppler (Vascular Technology Inc, Nashua, NH). The posterior cremasteric fibers and posterior fat component are ablated. The vas is isolated, and generally the artery and vein to the vas are dissected away from the vas. The perivasal tissue is now ablated. Hydrodissection of the perivasal tissue is now performed (Fig. 26.7) [3], using the ERBEJET 2 hydrodissector (ERBE Inc., Atlanta, GA) to ablate residual nerve fibers.

The cord is now wrapped with AmnioFix (MiMedx, Marietta, GA), which serves as a barrier to reduce scar tissue formation, provide local anti-inflammatory environment, and help with tissue healing. The wrap is loosely secured using 6-0 Prolene or chromic interrupted sutures. The robot is now undocked. The cord is placed back into through the incision, and the deep tissue and skin are closed.

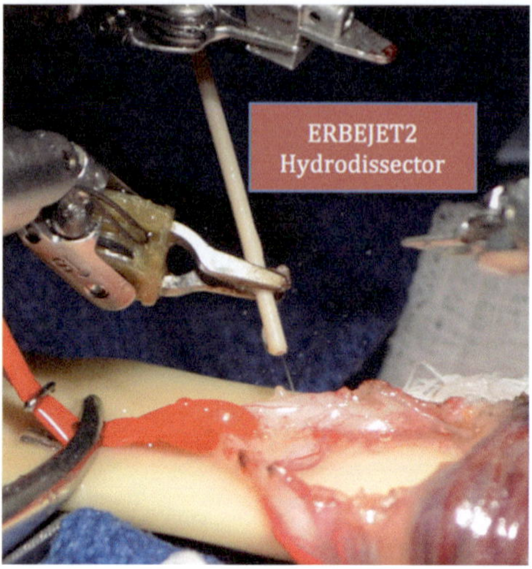

Fig. 26.7. Hydrodissection of residual nerve fibers on perivasal tissue (From Brahmbhatt et al. [3], with kind permission Springer Science+Business Media).

Varicocelectomy for Chronic Scrotal Content Pain

Introduction

Varicoceles are an abnormal enlargement of the pampiniform venous plexus veins within the scrotum. This abnormal enlargement is linked to low sperm counts, decreased sperm quality, infertility, testicle atrophy, and pain. Varicocelectomy (ligation of the enlarged veins) is often performed on its own for fertility, hypogonadism, or in extreme cases, for management of chronic pain. It is performed alongside a targeted denervation if varicoceles are clinically present.

From June 2008 to September 2014, 264 robotic-assisted varicocelectomies were performed in 220 patients. Indications for the procedure were the presence of a grade two or three varicocele and the following conditions: azoospermia in 20 patients, oligospermia in 63 patients, and chronic orchialgia with or without oligospermia in 137 patients. The median duration per side was 25 min (10–80). Median follow-up was 36 months (1–76). 80 % with oligospermia had a significant improvement in sperm count or motility and 30 % with azoospermia converted to oligospermia, and 91 % of the testicular pain patients had a significant reduction in pain (82 % of these patients had targeted denervation of the spermatic cord in addition to varicocelectomy). Two recurrences or persistence of varicocele occurred; one patient developed a small postoperative hydrocele; two patients had postoperative scrotal hematomas; and five patients had wound seroma (treated conservatively).

Technique in Detail

A 1–2 cm subinguinal incision is made over the external inguinal ring. A tongue depressor is placed underneath the cord to keep the cord elevated. The robot is positioned over the patient. A zero-degree camera lens is utilized. The Black Diamond microforceps are used in the right robotic arm, the micro bipolar forceps in the left arm, and the curved monopolar scissors in the fourth arm. The anterior cremasteric sheath of the spermatic cord in now incised to separate the cord structures.

The arteries are identified using real-time micro-Doppler (Vascular Technology Inc, Nashua, NH). All dilated veins are isolated and tied using 3-0 silk (Fig. 26.8) [3]. Vessels are cut with curved monopolar scissors. The cord is placed back into through the incision, and the deep tissue and skin are now closed.

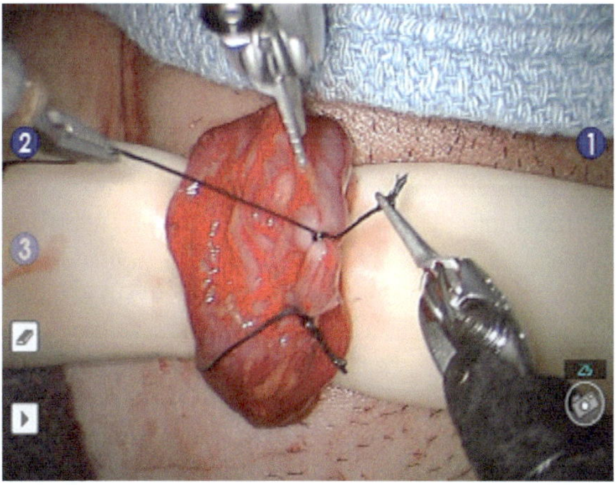

Fig. 26.8. Isolation and ligation of dilated vein (From Brahmbhatt et al. [3], with kind permission Springer Science+Business Media).

Vasectomy Reversal for Post Vasectomy Pain

Introduction

Vasectomy is a common form of contraception. An estimated 40–60 million men worldwide rely on this method of contraception [22]. Complications after vasectomy are rare, but 10–15 % of men may suffer from chronic post vasectomy testicular and groin pain. Congestive epididymitis is one possible mechanism for the chronic discomfort, and therefore, vasectomy reversal is a viable treatment option [19, 23].

Between July 2007 and March 2013, 147 robotic-assisted vasectomy reversals were performed by two fellowship trained microsurgeons. There were 90 robotic-assisted microsurgical vasovasostomy (RAVV) procedures and 57 robotic-assisted microsurgical vasoepididymostomy (RAVE) procedures performed. Twenty of these patients had chronic scrotal pain after vasectomy, and the rest wished to regain fertility. Median patient age was 42 years, and median duration from vasectomy 7 years for RAVV and 11 years for RAVE. Median OR setup duration was 30 min, and median robotic OR duration was 120 min and 150 min for RAVV and RAVE, respectively. After 23 months median follow-up, patency rates (>1 million sperm/ejaculate) were 97 % in the RAVV group and 60 % in the RAVE group. Pain relief occurred in 88 % of the patients who underwent RAVV or RAVE for chronic scrotal pain related to vasectomy.

Technique in Detail: Robotic-Assisted Microsurgical Vasovasostomy

The proximal and distal vas deferens (beyond the previous vasectomy site) is palpated through the scrotal skin. Through the skin, the distal vas is fixed into place with a towel clip. Local anesthetic is infiltrated into this area. A 1–2 cm vertical incision is made over the vas, starting inferiorly from the previously placed towel clip. Using fine electrocautery and sharp dissection, the distal and proximal ends of the vas are dissected free. The distal vas is dissected to allow a tension-free anastomosis to the proximal vas. The proximal vas is carefully transected with a No. 11 blade. Microscopic examination of the proximal vas fluid is performed. If no sperm is present in this proximal fluid, RAVE is performed. If sperm is found, then RAVV is performed. The adventitia from either end of the vasa is now secured together with a 3-0 Prolene suture to allow a tension-free anastomosis.

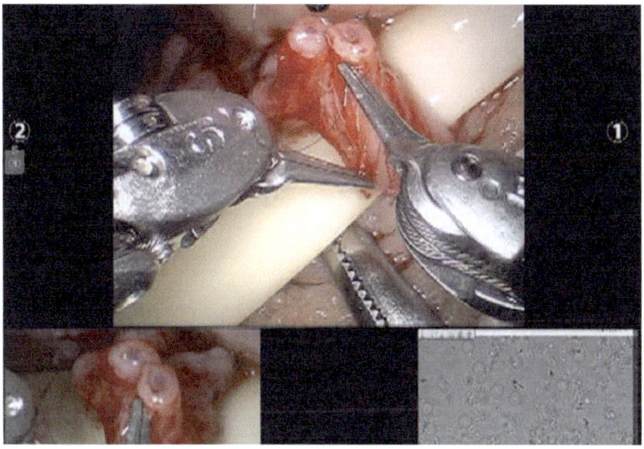

Fig. 26.9. RAVV posterior luminal anastomosis (From Brahmbhatt et al. [3], with kind permission Springer Science+Business Media).

The robot is now positioned over the patient to perform the microsurgical vasovasostomy. Black Diamond microforceps are inserted on the right and left robotic arms. The micro-Potts scissors are inserted onto the fourth robotic arm. The zero degree camera lens is inserted onto the robot camera arm. The two ends of the vas are placed over a 1/4″ Penrose drain. A 9-0 nylon suture is held and manipulated using the Black Diamond forceps in both left and right arms as needle drivers. The posterior muscularis layer of the two ends of the vas is now approximated (Fig. 26.9) [3]. Two or three double-armed 10-0 nylon sutures are now placed inside out to reanastomose the posterior mucosal lumen of the vas. Three double-armed 10-0 nylon sutures are used to close the anterior mucosal lumen of the vas (Fig. 26.10) [3]. Five to six 9-0 nylon sutures are used to approximate the anterior muscularis layer of the vas. The same procedure is now performed on the contralateral side by repositioning the robotic arms. The Penrose drain is gently removed from under the repair. The vas is placed back into the scrotal cavity, and the tissue and skin are closed with absorbable suture.

Technique in Detail: Robotic-Assisted Microsurgical Vasoepididymostomy

The RAVE procedure starts from above when there is no sperm in the fluid from the proximal vas. The scrotal incision is enlarged by 1–2 cm inferiorly. The testicle is delivered, and the tunica is incised to expose

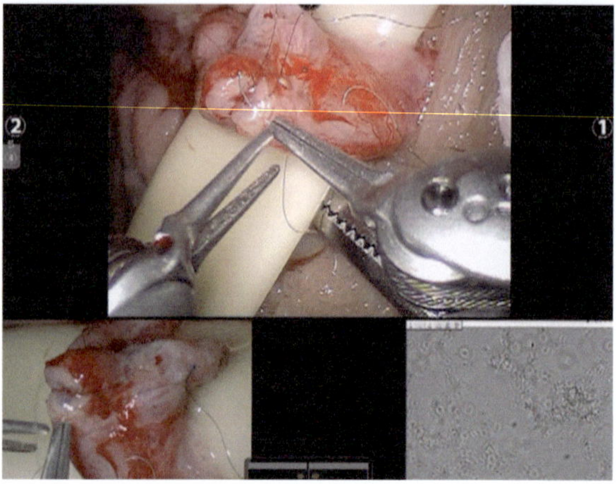

Fig. 26.10. RAVV anterior muscular anastomosis (From Brahmbhatt et al. [3], with kind permission Springer Science+Business Media).

the epididymis. The adventitial layer of the epididymis is incised above the level of epididymal obstruction (blue/gray zone with dilated epididymal tubules above this area). A 3-0 Prolene suture is used to attach the testicle to the adventitia of the vas to prevent tension between the anastomosis. The vas is stripped off the adventitia and flipped toward the epididymal tubules. The robot is now positioned similar to above. Two 10-0 nylon double-armed suture needles are placed longitudinally through a single epididymal tubule to expose the tubule. This tubule is then incised longitudinally using the micro-Potts scissors between the two suture needles to create a lumen in the tubule. The fluid is then aspirated and examined under a separate phase contrast microscope for the presence of sperm.

When sperm is confirmed, two double-armed 10-0 nylon needles in the epididymal tubule are advanced through, and then all four of the needles are brought inside out on the vas mucosal lumen to involute the epididymal tubule lumen into the vas lumen (Fig. 26.11) [3]. Five to six 9-0 nylon sutures are placed circumferentially to approximate the muscularis of the vas to the adventitia of the epididymal tubule (Fig. 26.12) [3]. The testicle and anastomosis are carefully delivered back into the scrotum. The dartos layer and skin are closed.

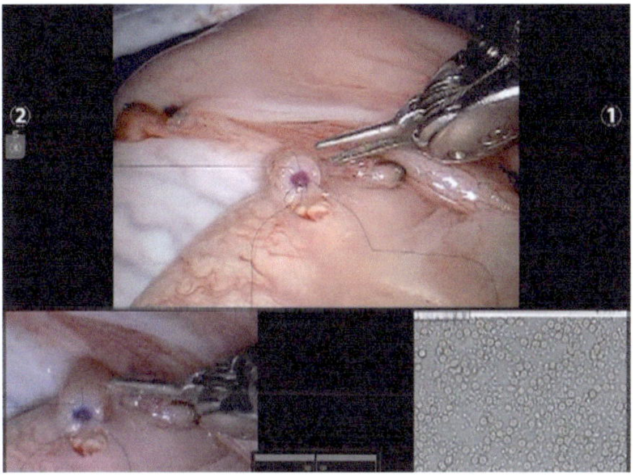

Fig. 26.11. RAVE involution vasoepididymostomy (From Brahmbhatt et al. [3], with kind permission Springer Science+Business Media).

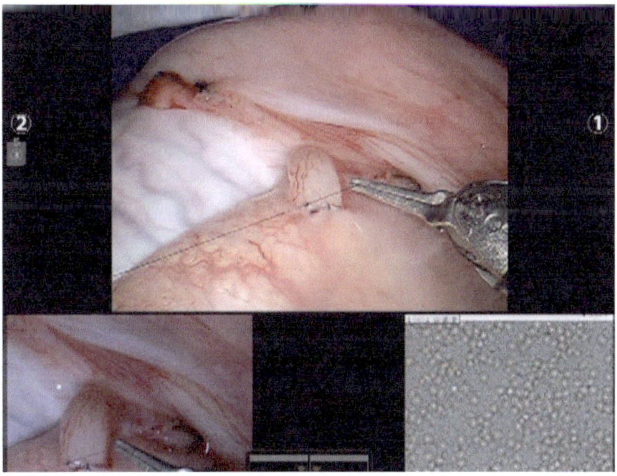

Fig. 26.12. RAVE vas muscularis to epididymal adventitia approximation (From Brahmbhatt et al. [3], with kind permission Springer Science+Business Media).

Salvage Interventions Post Targeted Denervation

Microcryoablation

Approximately, 12–15 % of patients will have persistent pain after denervation [21, 24]. Orchiectomy and epididymectomy could be considered; however, these procedures seem to be controversial due to their low success rates [25–27]. Mirmovich et al. introduced intralesional cryosurgery for the treatment of the pain due to hypertrophic scars and keloids [28]. We adapted their technique and perform microcryoablation of ilioinguinal and genitofemoral nerve fibers for patients with persistent or recurrent CGSCP [29]. A small diameter EndoCare CryoProbe (Healthtronics, Austin, TX) is used to perform cryoablation of the medial and lateral edges of the cord under ultrasound guidance (Fig. 26.13). As of September 2014, we have performed 69 targeted microcryoablations in 60 patients (9 b/l, 22 left, 29 right). At median follow-up of 11 months using the visual analog pain scale, there is a 74 % significant reduction in pain (9 % complete resolution and 65 % greater than 50 % reduction in pain). Using the PIQ6 score, there is 59 % reduction in pain at 6 months. Complications were rare and included one wound infection and one case of increased penile pain.

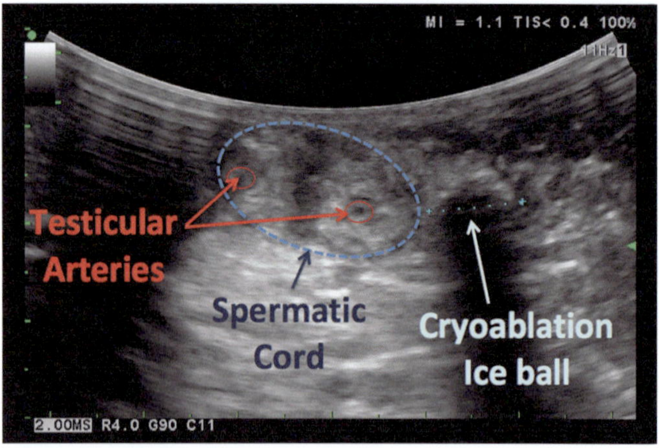

Fig. 26.13. Ultrasound guidance during microcryoablation.

Botox

Another adapted method for the persistent CGSCP patient is spermatic cord block using botulinum toxin. Mori et al. described intracremasteric botulinum-A toxin injection for a patient who had pain due to bilateral cremasteric muscle spasms [30]. We recently began botulinum toxin injection for those persistent CGSCP patients. As of September 2014, we have performed 29 targeted botulinum toxin injections in 25 patients (94 b/l, 11 left, 10 right) at a dose of 100 units. At median follow-up of 8 months, using the visual analog pain scale, there is a 70 % significant reduction in pain (14 % complete resolution and 56 % greater than 50 % reduction in pain). Using the PIQ6 score, there is 40 % reduction in pain at 6 months and 20 % at 1 year. No significant complications have been noted.

Alternative Salvage Interventions

For patients with refractory pain, we may pursue serial nerve blocks, abdominal neurolysis, and, in rare cases, orchiectomy.

Summary

CGSCP is a common problem that is often underdiagnosed. A multidisciplinary approach using a structured algorithm should be used for its evaluation and management.

References

1. Davis BE, Noble MJ, Weigel JW, Foret JD, Mebust WK. Analysis and management of chronic testicular pain. J Urol. 1990;143(5):936–9.
2. Levine L. Chronic orchialgia: evaluation and discussion of treatment options. Ther Adv Urol. 2010;2(5-06):209–14.
3. Brahmbhatt JVGA, Parekattil SJ. Robotic-assisted microsurgery for male infertility and chronic orchialgia. In: Kim KC, editor. Robotics in general surgery. New York: Springer; 2014. p. 365–84.
4. Parekattil SJ, Cohen MS. Robotic surgery in male infertility and chronic orchialgia. Curr Opin Urol. 2010;20(1):75–9.

5. Levine LA. Microsurgical denervation of the spermatic cord. J Sex Med. 2008;5(3):526–9.
6. McMahon AJ, Buckley J, Taylor A, Lloyd SN, Deane RF, Kirk D. Chronic testicular pain following vasectomy. Br J Urol. 1992;69(2):188–91.
7. Alfieri S, Amid PK, Campanelli G, Izard G, Kehlet H, Wijsmuller AR, et al. International guidelines for prevention and management of post-operative chronic pain following inguinal hernia surgery. Hernia. 2011;15(3):239–49.
8. Hakeem A, Shanmugam V. Current trends in the diagnosis and management of post-herniorraphy chronic groin pain. World J Gastrointest Surg. 2011;3(6):73–81.
9. Poobalan AS, Bruce J, King PM, Chambers WA, Krukowski ZH, Smith WC. Chronic pain and quality of life following open inguinal hernia repair. Br J Surg. 2001;88(8):1122–6.
10. Hindmarsh AC, Cheong E, Lewis MP, Rhodes M. Attendance at a pain clinic with severe chronic pain after open and laparoscopic inguinal hernia repairs. Br J Surg. 2003;90(9):1152–4.
11. Kumar P, Mehta V, Nargund VH. Clinical management of chronic testicular pain. Urol Int. 2010;84(2):125–31.
12. Rauchenwald M, Steers WD, Desjardins C. Efferent innervation of the rat testis. Biol Reprod. 1995;52(5):1136–43.
13. Parekattil SJ, Gudeloglu A, Brahmbhatt JV, Priola KB, Vieweg J, Allan RW. Trifecta nerve complex—potential anatomic basis for microsurgical denervation of the spermatic cord for chronic orchialgia. J Urol. 2013;190(1):265–70.
14. van Haarst EP, van Andel G, Rijcken TH, Schlatmann TJ, Taconis WK. Value of diagnostic ultrasound in patients with chronic scrotal pain and normal findings on clinical examination. Urology. 1999;54(6):1068–72.
15. Masarani M, Cox R. The aetiology, pathophysiology and management of chronic orchialgia. BJU Int. 2003;91(5):435–7.
16. Schover LR. Psychological factors in men with genital pain. Cleve Clin J Med. 1990;57(8):697–700.
17. Davies A, Hendrich J, Van Minh AT, Wratten J, Douglas L, Dolphin AC. Functional biology of the alpha(2)delta subunits of voltage-gated calcium channels. Trends Pharmacol Sci. 2007;28(5):220–8.
18. Benson JS, Abern MR, Larsen S, Levine LA. Does a positive response to spermatic cord block predict response to microdenervation of the spermatic cord for chronic scrotal content pain? J Sex Med. 2013;10(3):876–82.
19. Parekattil SJ, Gudeloglu A. Robotic assisted andrological surgery. Asian J Androl. 2013;15(1):67–74.
20. Laudano MA, Osterberg EC, Sheth S, Ramasamy R, Sterling J, Mukherjee S, et al. Microsurgical denervation of rat spermatic cord: safety and efficacy data. BJU Int. 2014;113(5):795–800.
21. Gudeloglu A, Brahmbhatt JV, Parekattil SJ. Robotic microsurgery in male infertility and urology—taking robotics to the next level. Transl Androl Urol. 2014;3(1):102–12.
22. Dohle GR, Diemer T, Kopa Z, Krausz C, Giwercman A, Jungwirth A, et al. European association of urology guidelines on vasectomy. Eur Urol. 2012;61(1):159–63.

23. Adams CE, Wald M. Risks and complications of vasectomy. Urol Clin North Am. 2009;36(3):331–6.
24. Strom KH, Levine LA. Microsurgical denervation of the spermatic cord for chronic orchialgia: long-term results from a single center. J Urol. 2008;180(3):949–53.
25. Costabile RA, Hahn M, McLeod DG. Chronic orchialgia in the pain prone patient: the clinical perspective. J Urol. 1991;146(6):1571–4.
26. Sweeney CA, Oades GM, Fraser M, Palmer M. Does surgery have a role in management of chronic intrascrotal pain? Urology. 2008;71(6):1099–102.
27. Calleary JG, Masood J, Hill JT. Chronic epididymitis: is epididymectomy a valid surgical treatment? Int J Androl. 2009;32(5):468–72.
28. Mirmovich O, Gil T, Goldin I, Lavi I, Mettanes I, Har-Shai Y. Pain evaluation and control during and following the treatment of hypertrophic scars and keloids by contact and intralesional cryosurgery—a preliminary study. J Eur Acad Dermatol Venereol. 2012;26(4):440–7.
29. Gudeloglu A, Brahmbhatt J, Har-Shai Y, Parekattil S. Micro-cryoablation of ilioinguinal and genitofemoral nerve fibers for patients with persistent or recurrent chronic groin or scrotal content pain (abstract). Engineering and urology society 28th annual meeting. San Diego, 4 May 2013.
30. Mori R, Vasavada S, Baker D, Sabanegh Jr E. Treatment of debilitating cremasteric synkinesia with intracremasteric botulinum-A toxin injections. Urology. 2011;78(1):214–6.

Part III
Current Debates

27. The Role of Bioactive Prosthetic Material for the Treatment of Sports Hernias

David S. Edelman

Editor's Comments (ST)

Athletic pubalgia remains one of the most controversial topics in the arena of groin pain. Few surgeons in the world have chosen this specialty for their main focus of practice, and among them, they each have a different approach as to how and why they perform their procedure(s) of choice. Here, the author presents his unique practice of laparoscopic approach to the treatment of athletic muscle tears, with the addition of Surgisis, a porcine tissue derived from small intestine submucosa, and fibrin. The term "BPM" is often applied when discussing Surgisis and other Cook products, whereas the broader term of "biologic" mesh can also be used when discussing human or nonhuman collagen-based grafts. Also, Table 27.1*demonstrates data as provided by each vendor, such as whether the tissue has been intentionally cross-linked. This information may not represent analyses performed independently by scientific laboratories that have shown a special interest in determining the final characteristics of implanted tissue, such as whether the sterilization and other manipulation of the tissue in fact result in unintentional overprocessing or even cross-linking.*

Nevertheless, the technique described is worthy of further discussion and investigation. The author has shown good preliminary results. We look forward to larger population and longer-term data. It would be useful to know if other forms of collagen matrices would also show benefit when implanted in this region and also which specific patients are best candidates to undergo this type of approach for their athletic pubalgia.

B.P. Jacob et al. (eds.), *The SAGES Manual of Groin Pain*,
DOI 10.1007/978-3-319-21587-7_27

Table 27.1. Bioactive prosthetic materials.

Mesh name	Vendor	Source	Cross-linking	Sterilization
AlloDerm®	LifeCell	Human dermis	No	None
AlloMax®	Bard/Davol	Human dermis	No	Gamma irradiation
CollaMend™	Bard/Davol	Porcine dermis	Yes	Ethylene oxide
FlexHD™	Ethicon	Human dermis	No	None
FortaGen®	Organogenesis	Porcine intestine	Yes	Gamma irradiation
MatriStem®	ACell	Porcine bladder	No	E-beam
Peri-Guard®	Synovis	Bovine pericardium	Yes	Liquid alcohol
Permacol™	Covidien	Porcine dermis	Yes	Gamma irradiation
Strattice®	LifeCell	Porcine dermis	No	E-beam
SurgiMend®	TEI Biosciences	Fetal bovine dermis	No	Ethylene oxide
Surgisis®	Cook Medical	Porcine intestine	No	Ethylene oxide
Tutopatch®	Tutogen Medical	Bovine pericardium	No	Gamma irradiation
Veritas®	Synovis	Bovine pericardium	No	E-beam
XenMatrix®	Bard/Davol	Porcine dermis	No	E-beam

Introduction

The use of mesh in the repair of hernias is commonplace. Synthetic meshes such as polypropylene and polyester have been the standard for hernia repairs since the 1980s. Biologic graft material composed of purified porcine small intestinal submucosa was first introduced to the United States in 1998, as an alternative to synthetic biomaterials. These meshes, composed of extracellular matrix (ECM) collagen, fibronectin, associated glycosaminoglycans, and growth factors [1–4], have been extensively investigated in animal models [5–7] and used clinically in many types of surgical procedures. Referred to as bioactive prosthetic materials (BPM), they are considered a scaffold for the binding of growth factors and other cellular elements for the healing response. The subsequent healing response and strength are dependent on ingrowth from the patient's cells and blood vessels into the ECM of the BPM. Fibrin may assist this ingrowth and thus may be added extrinsically as topical fibrin sealant [8]. The balance between ECM synthesis and degradation contributes to the ultimate success of the hernia repair.

Surgisis (Cook Surgical, Bloomington, IN) was the first biologic graft material to be marketed in the United States. I began using it in my practice for hernia repairs and reported my initial results in *Surgical*

Technology International XV in July 2006 [9]. Since then, there have been many more reports using BPM, acellular dermal matrices, and other biologic materials for hernia repair. The aim of this chapter is to review the topic of BPMs and their application to inguinal hernia and sports hernia repairs.

Basic Differences in Bioactive Prosthetic Materials

Table 27.1 summarizes the BPMs currently on the market in the United States. They differ based on their mammalian source (animal or human), tissue of origin (dermal, pericardial, bladder, or intestinal submucosa), as well as their methods of processing (cross-linked or not cross-linked) and sterilization. All of these differences may lead to differences in the healing process and thus clinical outcome. With the exception of AlloDerm®, Surgisis®, and Strattice™, peer-reviewed studies outlining the clinical outcome from implantation of biologic tissue is significantly lacking.

Mammalian source may be considered when choosing among the various BPMs available. Human cadaveric tissues offer the advantage of using allograft (within species) sourcing and thus lacking interspecies rejection risk. The source of such tissues is donor dependent, with variability in composition, health, thickness, and age of the tissue. Additionally, there is risk of disease transmission within species; indeed, there have been reports of disease transmission in human cadaveric allograft products of the dura mater in Japan from the 1990s [10].

Alternatively, animals can be raised to precise specifications to achieve a more consistent product. The risk of allergic response to their ECM is low because of the high homology with similar human proteins. With nonhuman tissues, the risk of tissue rejection remains despite decellularization, as does the rare possibility of disease transmission. The specifics of each biologic material should be known prior to implantation, especially when treating certain populations. For example, an immune-compromised patient may be at higher risk if undergoing implant with a human cadaveric allograft that is not sterilized. Similarly, an atopic patient may be at higher risk of allergic response to a xenograft.

BPMs vary in their tissue of origin. The dermis remains the preferred tissue source, though products made from alternative tissues, such as the pericardium, stomach, bladder, and intestinal submucosa, are also available. There is not enough literature to compare differences in clinical outcomes across different tissues of origin. All tissue sources contain significant amounts of collagen and other ECM proteins. However, some tissues such as the pericardium and intestinal submucosa lack the protein elastin, a significant component of the dermis that gives skin its elasticity. Elastin has been theorized to cause in vivo stretching of the allograft, resulting in diastasis after bridged repair of ventral hernia [11, 12]. At the same time, the elasticity of dermis-based grafts is favorable when implanted in areas where tissue pliability is necessary, such as in breast reconstruction following mastectomy. Thus, the tissue of origin may be a significant factor to consider when choosing the best BPM for the procedure of interest.

Lastly, BPMs are either purposely cross-linked or non-cross-linked at the time of their processing. Cross-linking is a way of stabilizing the graft and making it more resistant to tissue-degrading enzymes and bacteria that break down collagen. While this process may increase the durability of the graft [13], clinical studies have shown that the majority of adverse events associated with hernia repair grafts have occurred with cross-linked products [14]. These complications included acute mechanical failure of the mesh, degradation of the mesh, and poor integration of the mesh. Poor mesh integration is a result of poor angiogenesis into the material, which can lead to encapsulation or prolonged inflammatory response characterized by foreign body giant cell reaction. Recent findings suggest that cross-linking does not necessarily translate to durability, and, while there may yet be a place for cross-linked materials in hernia repair, cross-linked materials need to be used with caution until the optimal degree of cross-linking to overcome these complications can be understood [15].

BPM and Inguinal Hernia Repair

One of the first studies reporting the use of BPM in humans was reported in 2002 [16]. This was a preliminary study on 15 inguinal hernias in 12 patients. The preliminary results were good with no recurrences at 1 year and no chronic pain. Since then, there have been multiple studies demonstrating positive results after inguinal hernia repair with biologic tissue. Fine repaired 51 hernias in 38 patients with

BPM mesh and fibrin sealant [17]. He showed no major complications, one recurrence at 13 months, and chronic pain in three patients (7.9 %). This is lower than what is typically reported with synthetic mesh after open repair, at approximately 12.5 %. We reported our results in 2008, using fibrin sealant alone to laparoscopically secure BPM and polypropylene mesh in comparable groups of 18 patients with 23 repairs [18]. The results were similar in both groups with no long-term chronic pain or hernia recurrences. The biologic group had a few patients with short-term (less than 3 months) groin discomfort. Lastly, in Italy in 2008, Agresta and Bedin reported 11 patients undergoing laparoscopic TAPP hernioplasty with BPM and fibrin sealant [19]. There was one technical error leading to a recurrence at 14.5 months, and there were no reports of chronic pain. He hypothesized for use of BPM in the young patient, where there is a fear of leaving behind a foreign body in the long term.

Acellular extracellular dermal matrix mesh was implanted in 53 patients with 56 hernias using a Lichtenstein repair by Ma et al. in a 2005 report from China [20]. They reported no infections, chronic pain, or discomfort; however, two patients with large direct Gilbert type V and VI hernias had recurrences noted by 18 months. Ansaloni et al. from Italy reported his 2-year follow-up in 2007, on 45 consecutive patients undergoing Lichtenstein repair with BPM [21]. There was a low degree of pain and no recurrences noted. A randomized double-blind trial comparing BPM to polypropylene mesh reported their 3-year follow-up on 70 patients [22]. The incidence of pain was similar, but the degree of pain was less in the biologic mesh group. One recurrence was noted in the polypropylene mesh group. Most recently, Bellows et al. reported their randomized double-blind multicenter trial comparing patients undergoing Lichtenstein hernioplasty with non-cross-linked porcine dermis ($N=84$) to soft polypropylene mesh ($N=88$). Results were equivalent at 3 months [23].

BPM and Sports Hernias

Groin pain in athletes is caused by a wide variety of musculoskeletal disorders. The term "sports hernia" is a poor one and should be replaced by "athletic pubalgia." In patients who have this condition, the insertion site of the rectus muscle and/or adductor longus muscle onto the pubic bone is damaged. When an athlete engages these muscles during activity, there is a lack of function, which can sometimes be perceived as groin pain. Some athletes describe a "weakness." Most athletes describe their inability to plant and turn, or kick, or jump, or perform something similar corresponding to their sport. The physical exam does not reveal a true hernia in most instances; the presence of an inguinal hernia should prompt an inguinal hernia repair.

Many times the examiner can elicit pain at the pubis while the athlete does a Valsalva maneuver. I prefer to perform this exam with the patient lying on my exam table while I place my index finger through the scrotum onto the top of the external ring. I then have the athlete do bilateral straight-leg raises while he lifts his shoulder off the table at the same time. Pain with this maneuver is a diagnostic sign.

An ultrasound with Valsalva may demonstrate an inguinal hernia. If the ultrasound is negative for a true inguinal hernia and the physical exam is equivocal, a noncontrast MRI of the pelvis with attention to the pubis is indicated. With athletic pubalgia, the MRI will show acute or chronic inflammatory changes at the rectus muscle insertion onto the anterior/superior pubis and/or at the adductor longus muscle insertion onto the inferior pubis.

It is recommended initially to offer a trial of rest with NSAIDs, followed by a course of rehabilitative physical therapy. If there is no improvement or continued inability to perform the sport, surgery may be indicated. There is much literature available on this subject.

I prefer the laparoscopic approach, as it has been shown to provide improved recovery over open repair for conventional inguinal hernia repair [24]. I also prefer implantation of biologic repair in this patient population. Clarke et al. first reported the use of biologic mesh, small intestine submucosa (SIS), to repair the abdominal wall in dogs [25]. The resultant repair with SIS was well organized, with dense connective tissue that was well incorporated into the adjacent fascia and skeletal muscle. With this information and my extensive experience with laparoscopic inguinal hernia repair, I have since implanted BPM laparoscopically in 131 athletes. I reported the results of my first ten patients in 2006 [26]. No patient developed a recurrent "sports hernia," and only one patient did not have improvement in symptoms.

The laparoscopic approach offers an excellent visualization of the rectus muscle and conjoined tendon insertion onto the pubis. I use a 10×15 cm biologic mesh, soak it in bupivacaine, and place it in the preperitoneal space. The grasper positions the mesh over the myopectineal space. I use 4–6 absorbable tacks to secure the mesh in place. I spray fibrin sealant on and behind the mesh to secure it to the injured muscle and periosteum. If there is an adductor injury, I make a separate 4 cm skin incision over the affected side, along the inguinal crease. I expose the adductor muscle and make micro-cuts in the tendon. I then use a 4×7 cm biologic mesh and tack it to the inferior pubis and suture it to the adductor muscle using absorbable sutures (3–0 Vicryl). Again, I apply fibrin sealant to the underside of the mesh that is in contact with the tendon.

I have the player and the athletic trainer begin a structured rehabilitation protocol on day 8 after surgery and continue for 3–4 weeks. The results have been excellent, and most of these athletes have returned to full participation of sports within 4–5 weeks.

Lastly, when discussing “biologic” treatments for sports hernias, it is important to mention platelet-rich plasma (PRP). This product has been popularized in the orthopedic literature for treatment of tendinopathies and other orthopedic conditions. PRP is derived from the patient’s own blood and then injected into the site of pain. It is a concentrated source of growth factors and cellular signals that play a role in the biology of healing—not too dissimilar to description of the purpose of biologic mesh. Basic science research infers that PRP may improve tissue healing. There are few clinical studies in humans that confirm the effectiveness of PRP. The controversy is further questioned when one considers that placing a needle into a bone or tendon can stimulate bleeding containing platelets and a healing response occurs. That said, the value of PRP is being proven in certain disease states, demonstrating its role in improving healing, especially for tendinopathies and orthopedic injuries [27].

Conclusions

While there are now several different BPMs available on the market, they differ in their mammalian tissue source, their tissue of origin, and their methods of processing. The existence of these various materials suggests that the ideal mesh is not yet available. Also, the lack of clinical evidence on most of these products prevents surgeons from making evidence-based choices for their use.

From what has been learned over several years of research and published clinical evidence, BPMs can be used clinically in many different surgical procedures with low rates of complications and few reports of mesh rejection. Certain types of hernias seem to be better adapted to the use of BPMs; thus, the pathophysiology of the hernia should be considered when using these materials. BPMs have been successfully implanted in young, healthy individuals who have developed a hernia or weakness due to physical activity of extreme muscular training. As long as the patient does not have a history of recurrent hernias, a large direct inguinal hernia, or a known type of collagen vascular abnormality, BPM can be considered an adjunct to hernia repair. Athletes in particular have fared well by having the biologic material support their tears as they heal their injury.

References

1. McPherson TB, Badylak SF. Characterization of fibronectin derived from porcine small intestinal submucosa. Tissue Eng. 1998;4(1):75–83.
2. Hodde JP, Badylak SF, Brightman AO, Voytik-Harbin SL. Glycosaminoglycan content of small intestinal submucosa: a bioscaffold for tissue replacement. Tissue Eng. 1996;2(3):209–17.
3. Hodde JP, Ernst DMJ, Hiles MC. An investigation of the long-term bioactivity of endogenous growth factor in OASIS Wound Matrix. J Wound Care. 2005;14(1):23–5.
4. McDevitt CA, Wildey GM, Cutrone RM. Transforming growth factor-B1 in a sterilized tissue derived from the pig small intestine submucosa. J Biomed Mater Res. 2003;67(2):637–40.
5. Badylak S. The extracellular matrix as a scaffold for tissue reconstruction. Semin Cell Dev Biol. 2002;13(5):377–83.
6. Badylak S, Kokini K, Tullius B, Whitson B. Strength over time of a resorbable bioscaffold for body wall repair in a dog model. J Surg Res. 2001;99(2):282–7.
7. Hodde J. Naturally-occurring scaffolds for soft tissue repair and regeneration. Tissue Eng. 2002;8(2):295–308.
8. Katkhouda N, Mavor E, Friedlander MH, Mason RJ, Kiyabu M, Grant SW, et al. Use of fibrin sealant for prosthetic mesh fixation in laparoscopic extraperitoneal inguinal hernia repair. Ann Surg. 2001;233(1):18–25.
9. Edelman DS, Hodde JP. Bioactive prosthetic material for treatment of hernias. Surg Tech Int. 2006;15:104–8.
10. Centers for Disease Control and Prevention (CDC). Creutzfeldt-Jakob disease associated with cadaveric dura mater grafts—Japan, 1978–2008. MMWR Morb Mortal Wkly Rep. 2008;57(42):1152–4.
11. Shaikh FM, Kennedy TE, Coyle P, Kavanagh EG, Grace PA. Diastasis as a cause of recurrence in ventral herniorrhaphy with porcine acellular dermal collagen implant. Plast Reconstr Surg. 2011;127(6):167e–9e.
12. Gupta A, Zahriya K, Mullens PL, Salmassi S, Keshishian A. Ventral herniorrhaphy: experience with two different biosynthetic mesh materials, Surgisis and Alloderm. Hernia. 2006;10(5):419–25.
13. Deeken CR, Melman L, Jenkins ED, Greco SC, Frisella MM, Matthews BD. Histologic and biomechanical evaluation of crosslinked and non-crosslinked biologic meshes in a porcine model of ventral incisional hernia repair. J Am Coll Surg. 2011; 212(5):880–8.

14. Harth KC, Rosen MJ. Major complications associated with xenograft biologic mesh implantation in abdominal wall reconstruction. Surg Innov. 2009;16(4):324–9.
15. Cavallo JA, Greco SC, Liu J, Friesella MM, Deeken CR, Matthews BD. Remodeling characteristics and biomechanical properties of a crosslinked versus a non-crosslinked porcine dermis scaffolds in a porcine model of ventral hernia repair. Hernia. 2015; 19(2):207–18.
16. Edelman DS. Laparoscopic herniorrhaphy with porcine small intestinal submucosa: a preliminary study. JSLS. 2002;6(3):203–5.
17. Fine AP. Laparoscopic repair of inguinal hernia using Surgisis mesh and fibrin sealant. JSLS. 2006;10(4):461–5.
18. Edelman DS. Fibrin glue fixation of bioactive extracellular matrix mesh compared with soft prolene mesh for laparoscopic hernia repair. Surg Laparosc Endosc Percutan Tech. 2008;18(6):569–72.
19. Agresta F, Bedin N. Transabdominal laparoscopic inguinal hernia repair: is there a place for biologic mesh? Hernia. 2008;12(6):609–12.
20. Ma SZ, Li XH, Hu J. Acellular extracellular matrix for inguinal hernia repair. Hernia. 2006;10(3):229–31.
21. Ansaloni L, Catena F, Gagliardi S, Gazzotti F, D'Alessandro L, Pinna AD. Hernia repair with porcine small-intestinal submucosa. Hernia. 2007;11(4):321–6.
22. Ansaloni L, Catena F, Coccolini F, Gazzotti F, D'Alessandro L, Pinna AD. Inguinal hernia repair with porcine small intestine submucosa: 3 year follow-up results of a randomized controlled trial of Lichtenstein's repair. Am J Surg. 2009;198(3):303–12.
23. Bellows CF, Shadduck P, Helton WS, Martindale R, Stouch BC, Fitzgibbons R. Early report of a randomized comparative clinical trial of Strattice reconstructive tissue matrix to lightweight synthetic mesh in the repair of inguinal hernias. Hernia. 2014;18(2):221–30.
24. Liem MS, van der Graaf Y, van Steensel CJ, Boelhouwer RU, Clevers GJ, Meijer WS, et al. Comparison of conventional anterior surgery and laparoscopic surgery for inguinal hernia repair. N Engl J Med. 1997;336(22):1541–7.
25. Clarke KM, Lantz GC, Salisbury SK, Badylak SF, Hiles MC, Voytik SL. Intestine submucosa and polypropylene mesh for abdominal wall repair in dogs. J Surg Res. 1996;60(1):107–14.
26. Edelman DS, Selesnick H. "Sports" hernia: treatment with biologic mesh (Surgisis): a preliminary study. Surg Endosc. 2006;20(6):971–3.
27. Hsu WK, Mishra A, Rodeo SR, Fu F, Terry MA, Randelli P, et al. Platelet-rich plasma in orthopaedic applications: evidence-based recommendations for treatment. J Am Acad Orthop Surg. 2013;21(12):739–48.

28. Prevention of Pain: Optimizing the Open Primary Inguinal Hernia Repair Technique

Giampiero Campanelli, Marta Cavalli, Piero Bruni, Andrea Morlacchi, and Gianni Maria Pavoni

The prevention of pain requires that surgeons should take care not only during the entire surgical procedure but also before and after.

This chapter will provide some suggestions based on our experience and up-to-date results from the literature for each of the following steps:

1. Preoperative patient selection
2. Selection of anesthesia
3. Choice of the approach: open anterior versus open preperitoneal
4. Identification and respect of the three nerves
5. Choice of the prosthesis: plug, mesh, lightweight versus heavy-weight, absorbable versus nonabsorbable
6. Choice of fixation
7. Administration of a proper postoperative therapy

Preoperative Patient Selection

Sometimes a patient will complain of apparently inexplicable postoperative pain, even if he has been treated in the same manner as other patients reporting no postoperative pain. Do all patients have the same risk to develop postoperative pain? Aasvang et al. [1] designed a novel prospective study that had the primary end point of assessing the relative contribution of preoperative, intraoperative, and postoperative factors in the development of persistent postoperative pain (PPP) substantially

B.P. Jacob et al. (eds.), *The SAGES Manual of Groin Pain*,
DOI 10.1007/978-3-319-21587-7_28

affecting everyday activities after groin hernia repair. The authors found four factors to be independently correlated to PPP-related impairment: preoperative Activity Assessment Scale (AAS) score, preoperative pain response to heat, intraoperative nerve injury, and early (day 30) postoperative pain intensity. A randomized study by Singh et al. [2] showed that preoperative pain, younger age, open surgery, and 7-day postoperative pain were independent risk factors for chronic pain. In other studies [3, 4] age has again been found to be an independent factor for postoperative pain.

So how should we treat a painful hernia in a young man? All surgeons should choose the operative technique that they know best, is safest in their own hands, and therefore will have the lowest individual risk of postoperative pain. Moreover, looking at our experience, groin pain with a small bulge of posterior inguinal wall is often incorrectly labeled a hernia; however, a proper physical examination and clinical history investigation reveal all the features of the so-called pubic inguinal pain syndrome (PIPS) [5]. Pain in PIPS is usually well localized and tends to be focused on the pubic bone with radiation superiorly to the abdominal rectus insertion and inferiorly to the adductor longus insertion. The site of pain is typically provoked by athletic activities such as kicking, sprinting, and changing directions; the symptoms usually persist the day after; they improve after resting and recur if athletic activities are resumed. Physical examination reveals tenderness or pain over the pubic crest on resisted sit-up ("abdominal crunch test"). Palpation of the internal ring can be painful and a small bulge of the inguinal posterior wall can be detected during coughing, but a palpable lump indicating a classical inguinal hernia is absent. During the adductor test, patients feel a sharp pain in the groin [6].

For these reasons, surgery intended to treat this subset of pain should not be limited to addressing the posterior wall. In order to maximize the chance of relieving preoperative pain, release of the three inguinal nerves in the region and tenotomy of the rectus abdominus rectus and adductor longus should be included.

Selection of Anesthesia

Several randomized studies have compared local anesthesia with general and/or regional anesthesia. They confirm the advantages of local anesthetic, including less postoperative pain [7–13]. For these reasons, it is the favored anesthesia in centers specializing in hernia surgery [14–17]. The administration is technically quite easy but requires training and is successful only if the surgeon handles the tissues gently, has patience, and is competent and facile with the technique. Local anesthesia should be the first option for inguinal hernia repair in the adult, but sedation or general anesthesia with short-acting agents and combined with local infiltration anesthesia may be a valid alternative to local anesthesia alone for surgeons in training or outside of specialized hernia centers [17].

Choice of Approach: Open Anterior Versus Open Preperitoneal

In order to decrease the amount of dissection in the inguinal canal, the manipulation of the inguinal nerves [18], and the interaction between the foreign material of the mesh and the spermatic cord and nerves, placement of the mesh in the preperitoneal space is an option to be considered [19]. By placing the mesh in the preperitoneal space, a more physiologic location for the mesh with intra-abdominal forces on one side and the oblique muscles on the other, fixation becomes less mandatory although may not be completely abandoned [19].

Usher et al. [20] introduced the prosthetic preperitoneal repair performed through an anterior approach using polyethylene mesh (later polypropylene), which was not slit because the spermatic cord was lateralized. Mahorner and Goss [21] used anterior preperitoneal grafts to support the overlying weakened transversalis fascia in two patients with recurrent herniation and destruction of both Poupart's and Cooper's ligaments. Rives [22] pioneered both anterior and posterior preperitoneal prosthetic repairs of groin hernias in France using Mersilene mesh. More recently Kugel (1999) employed this approach through an abdominal gridiron incision, using a fortified patch to reinforce the overlying damaged transversalis fascial floor of the inguinal and femoral canals [23].

With regard to prosthetic preperitoneal repair through a posterior approach, the precursor to modern laparoscopic techniques [totally extraperitoneal repair (TEP), transabdominal preperitoneal repair (TAPP), enhanced or extended view TEP (eTEP)], Stoppa et al. [24] reported on the giant prosthetic reinforcement of the visceral sac (GPRVS) in 1965. Large bilateral Dacron meshes were inserted deep into the weakened transversalis fascia, covering Fruchaud's myopectineal orifice with extensive overlap. This operation provided an alternative to the anterior preperitoneal approach, which, in cases of recurrent herniation, encounters scarring possibly leading to damage of the spermatic cord, nerves, and blood vessels. Wantz [25] proposed a unilateral version of this operation, reaching the preperitoneal spaces of Bogros and Retzius using a transverse incision extending laterally 9 cm from the linea alba and 3 cm below the anterior superior iliac spine.

Gilbert [26] developed a two-layered prosthesis; the superficial portion rests on the transversalis fascial floor of the inguinal canal, while the lower portion lays beneath in the anterior preperitoneal space. The patches were connected by a plug that passed through the internal inguinal ring. A slit in the onlay portion allowed for passage of the spermatic cord to the inguinal canal below the external oblique aponeurosis. Widespread release of the device from its manufacturer (1998) led to the technique being called the Prolene Hernia System (PHS).

More recently the transinguinal preperitoneal repair (TIPP) technique had been proposed by Pellisier [27]: this technique involves a standard anterior approach through the inguinal canal where a patch with a memory ring is placed into the preperitoneal space behind the transversalis fascia. Willaert et al. [19] recently proposed with the Cochrane collaboration a review with the aim to compare the efficacy of an elective open preperitoneal mesh repair via either anterior or posterior approach with the Lichtenstein technique. Efficacy was considered as the absence of chronic pain after at least three months of follow-up. All published and unpublished randomized controlled trials (RCTs) comparing any elective open preperitoneal mesh technique with Lichtenstein repair were considered for inclusion. Strangulated inguinal hernias, bilateral inguinal hernias, and recurrent inguinal hernias were exclusion criteria.

Unfortunately, many studies did not address the primary outcome of the review and only three studies were included. In these three trials, the Lichtenstein technique was compared with Read-Rives technique [28], TIPP [19], and Kugel patch [29]. The last two studies reported less chronic pain after preperitoneal repair; however, the Muldoon study described slightly more chronic pain after preperitoneal repair. Few data are present in the literature about chronic pain after Wantz posterior preperitoneal repair: this technique is usually used in the specialized hernia center for the treatment of very large, incarcerated hernias, recurrent hernias, femoral hernias, or in the treatment of postoperative chronic pain [30].

Identification of and Respect for the Three Nerves

As previously mentioned, intraoperative nerve injury has been shown to correlate independently with postoperative pain-related impairment [1]. Damage to, or entrapment of, one or more of the three inguinal nerves passing through the operative field may cause neuropathic pain. It is not always possible to follow traditional teaching dictating that every effort should be made to identify, preserve, and prevent traumatization or interruption of the inguinal nerves during operation. Inguinal nerves might interfere with placement of mesh or might be traumatized inadvertently during the operation. Several patterns of nerve injury during elective inguinal hernia repair have been described, including inadvertent suture entrapment, partial division, crushing, diathermy burn, or scar encroachment [31]. Identification and routine excision or division of selected inguinal nerves, termed "pragmatic neurectomy," during inguinal hernia repair has been proposed as a method for avoiding postoperative neuralgia [32].

Overall, the systematic review and meta-analysis proposed by Hsu et al. [33], including six RCT studies, indicate that preserving the ilioinguinal nerve during open mesh repair of an inguinal hernia was associated with decreased incidence of sensory loss at 6 and 12 months postoperatively compared with nerve division technique. They found no difference between the two surgical procedures in regard to the occurrence of chronic groin pain or numbness. In the 2014 update of the European Hernia Society Guidelines for the treatment of inguinal hernia in adults [34], all studies with the longest follow-up interval were combined in a new meta-analysis, and they concluded that routine prophylactic resection of the ilioinguinal nerve is not recommended (Grade A). It remains speculative whether this approach would be beneficial in a subset of patients with preoperative risk factors for chronic pain [34].

Importantly, five of the six studies included in the meta-analysis compared the effects of nerve preservation and of prophylactic neurotomy for just the ilioinguinal nerve, ignoring that all three nerves contribute to the sensory innervation of the groin.

Only the study reported by Karakayali et al. [35] focuses on the additional role of the iliohypogastric nerve: in this study patients had been divided into a nerve preservation group, ilioinguinal neurectomy group, iliohypogastric neurectomy group, and a group in which both nerves were transected. The only significant difference between the groups for chronic groin pain at the 1-year postoperative follow-up was between the nerve preservation group and the dual nerve transection group, in favor of the latter.

The majority of surgeons do not routinely detect all three inguinal nerves: the identification of the iliohypogastric nerve ranges between 32 % [36] and 97.5 % [37] of cases and for the genital branch of the genitofemoral nerves ranges between 21.3 % [37] and 36 % [36] of cases. An Italian prospective multicenter study [38] of 973 cases and a French single center study [39] of 1332 cases are the only two published studies reporting the results of the role of the identification of all three inguinal nerves (2305 cases all together) with a long follow-up period (ranging from 1 to 5 years). Both studies concluded that identification and preservation of all three inguinal nerves during open inguinal hernia repairs reduce chronic incapacitating groin pain to less than 1 %: the mean incidence of chronic pain was 0.8 % (range 0–1.6 %). The Italian study [38] also demonstrated that the risk of developing inguinal chronic pain increased with the number of nerves concomitantly undetected. Likewise, the division of nerves was correlated strongly with the presence of chronic pain.

Results from studies reporting data concerning division or neurectomy versus preservation of only the ilioinguinal nerve without giving any data concerning the other two nerves may be distorted because the two nerves not considered could be unintentionally divided or injured during the operation and, for this reason, chronic pain may result [40]. For all these reasons, we strongly suggest the identification and protection of all three inguinal nerves, to avoid removing the nerves from their natural bed as much as possible, and not to remove their covering fascia, as recommended in the international guidelines [41]. In case of a suspected or clearly injured nerve or an "at risk" nerve running in the way of the repair, pragmatic neurectomy is recommended with complete removal and reimplantation of its proximal cut end into the underlying muscle [41].

We also suggest that attention should be given during the placement of mesh to minimize mesh contacting or distorting the course of the inguinal nerves (the medial edge of the mesh sometimes meets and crosses the ilioinguinal or often the iliohypogastric nerve): in this case neurectomy can be done or, preferably, a small window in the edge of the mesh can be cut so that the interaction between mesh and nerve is minimalized.

Choice of the Prosthesis: Plug, Mesh, Lightweight Versus Heavyweight, and Absorbable Versus Nonabsorbable

Polypropylene mesheswere developed in 1959 and have been used commonly since then. Although the use of synthetic mesh substantially reduces the risk of hernia recurrence [42], polypropylene meshes have been found to cause chronic inflammatory reactions that persist for years and can have potentially negative effects, including chronic pain [43].

It has been surmised that the extent of the foreign body reaction with its provoked scar tissue is correlated with the amount of the synthetic material used [44]. This led to the development of so-called lightweight mesh characterized by a reduction in the polypropylene volume, an increase in the pore size, or different web structures [45, 46].

The meta-analysis of RCTs reported by Uzzman et al. [47] shows that lightweight mesh is associated with significantly less chronic groin pain (14.3 %), compared with heavyweight mesh (20.3 %) in Lichtenstein inguinal hernia repair, and less foreign body sensation (15.2 % vs. 26.1 %). These benefits did not appear to be at the expense of an increased rate of hernia recurrence. Smietanski [48] performed a meta-analysis including two more studies in addition to those included in Uzzaman et al. study: results were similar, but there was no difference in the two groups for severe chronic pain. Therefore, while there is clearly a benefit with regard to foreign body sensation, it is still uncertain whether lightweight mesh has a real clinical benefit over heavyweight with regard to severe inguinodynia. No statistical significance was found regarding chronic pain at 5 years in an RCT comparing Lichtenstein with the PHS [49]. We advise lightweight meshes especially in thin patients and in case of small indirect inguinal hernia.

Choice of Fixation

Another factor that should be addressed concerning the prevention of pain is the influence of fixation of the mesh. Penetrating fixating or traumatic devices like sutures, staples, and tacks cause local trauma that may result in nerve injury and chronic pain and should, therefore, be used with caution.

A prospective randomized multicenter trial [50] reported a significant reduction in postoperative pain at 1 and 6 months and a 45 % reduction in incidence of a composite end point regarding chronic disabling complications (pain/numbness/groin discomfort) at 1 year after Lichtenstein repair with fibrin glue (heavyweight) mesh fixation compared to standard suture fixation. The first study on the use of the self-gripping Parietene Progrip mesh (large pore polypropylene with resorbable polylactic acid microgrips) also demonstrated less pain on the first postoperative day versus the use of another large pore polypropylene mesh without gripping capacity [51]. Three other randomized studies comparing atraumatic (cyanoacrylate glue, self-fixating mesh) versus suture fixation in Lichtenstein hernioplasty with a large pore mesh showed no difference in acute or chronic pain [52–54]. Atraumatic mesh fixation (glue, self-fixating mesh) is more expensive than standard fixation, although the operation time was shorter in the majority of the studies. All studies with at least 1-year follow-up showed no differences in recurrence rates. We prefer to adopt a sutureless technique [55] or a fibrin glue technique: the latter is essential when a lightweight mesh is chosen because it allows an immediate fixation and prevents mesh dislocation during closure.

Administration of a Proper Postoperative Therapy

Inadequately treated postoperative pain may be a risk factor for persistent pain after hernia surgery [1]. A systematic review of RCTs up to March 2009 [56] emphasizes the use of a pre- or intraoperative field block (ilioinguinal, iliohypogastric, genitofemoral nerve) with or without local wound infiltration for all patients undergoing open inguinal hernia surgery. The same authors describe a standardized approach to postoperative pain consisting of paracetamol and conventional NSAID or Cox-2-selective inhibitors, followed by opioid administration if needed.

Conclusion

In conclusion, for primary inguinal hernias, surgeons should identify for each individual patient the operative technique they know to be safest in their own hands and with the lowest risk of postoperative pain. It is important to keep in mind the features of patients at risk for postoperative pain and in these cases consider, if available, the laparoscopic approach (and in the case of PIPS the complete approach we suggested previously). The surgeon is always reminded that a proper physical examination and clinical history investigation can avoid mistakes in the choice of optimal technique.

Choosing the operative technique means not only selecting the right approach, but also identifying the right anesthesia, the right mesh, and the correct fixation technique for each patient (tailored approach).

It is usually preferable to use local anesthesia in adult patients (on the condition that the surgeon is experienced in its use), while sedation or general anesthesia in the case of anxious and apprehensive patients or with large scrotal hernias can be used.

The anterior open approach is recommended for primary inguinal hernias [or in case of a first recurrence, "high" lateral (indirect) reducible hernia with a small defect in the thin patient, recurrence R1 according to the Campanelli classification] [57]. The posterior open preperitoneal approach with general anesthesia (Wantz or Stoppa) is reserved for the treatment of very large, incarcerated hernias, femoral hernias, postoperative chronic pain, and recurrent R2 [first recurrence, "low" medial (direct) reducible hernia with a small (<2 cm) defect in the thin patient] and recurrent R3 (multi recurrent hernia and/or not reducible, large wide defect in the posterior inguinal wall, femoral recurrent).

Ideally, nerve identification and preservation with a "no touch" technique for all the three inguinal nerves is recommended. However, if this cannot be achieved, pragmatic neurectomy with division of an injured or at risk nerve is preferred.

The choice of a light- or medium-weight polypropylene flat mesh for "normal" patients is recommended, and we prefer lightweight mesh for thin patients with an indirect inguinal hernia or in the treatment of PIPS. We prefer a sutureless or glue fixation of the mesh to minimize the risk posed by fixation. Always provide appropriate postoperative anti-inflammatory therapy in full dose for 7 days to minimize the risk of converted severe acute pain to chronic pain. These considerations based upon best available data will help to optimize outcomes and limit the incidence and severity of chronic pain with open hernia repair techniques.

References

1. Aasvang EK, Gmaehle E, Hansen JB, Gmaehle B, Forman JL, Schwarz J, et al. Predictive risk factors for persistent postherniotomy pain. Anesthesiology. 2010;112(4):957–69.
2. Singh AN, Bansal VK, Misra MC, Kumar S, Rajeshwari S, Kumar A, Sagar R. Testicular functions, chronic groin pain, and quality of life after laparoscopic and

open mesh repair of inguinal hernia: a prospective randomized controlled trial. Surg Endosc. 2012;26(5):1304–17.
3. Kehlet H, Jensen TS, Woolf CJ. Persistent postsurgical pain: risk factors and prevention. Lancet. 2006;367(9522):1618–25.
4. Macrae WA. Chronic post-surgical pain: 10 years on. Br J Anaesth. 2008;101(1): 77–86.
5. Campanelli G. Pubic inguinal pain syndrome: the so-called sports hernia. Hernia. 2010;14(1):1–4.
6. Cavalli M, Bombini G, Campanelli G. Pubic inguinal pain syndrome: the so-called sports hernia. Surg Technol Int. 2014;24:189–94.
7. Nordin P, Zetterström H, Gunnarsson U, Nilsson E. Local, regional or general anaesthesia in groin hernia repair: multicentre randomised trial. Lancet. 2003;362(9387): 853–8.
8. Ozgün H, Kurt MN, Kurt I, Cevikel MH. Comparison of local, spinal and general anaesthesia for inguinal herniorrhaphy. Eur J Surg. 2002;168(8–9):455–9.
9. van Veen R, Mahabier C, Dowson I, Hop WC, Kok N, Lange J, Jeekel J. Spinal or local anesthesia in Lichtenstein hernia repair: a randomized controlled trial. Ann Surg. 2008;247(3):428–33.
10. Sanjay P, Woodward A. Inguinal hernia repair: local or general anaesthesia? Ann R Coll Surg Engl. 2007;89(5):497–503.
11. Gönüllü NN, Çubukçu A, Alponat A. Comparison of local and general anesthesia in tension-free (Lichtenstein) hernioplasty: a prospective randomized trial. Hernia. 2002; 6(1):29–32.
12. O'Dwyer PJ, Serpell MG, Millar K, Peterson C, Young D, Hair A, et al. Local or general anesthesia for open hernia repair: a randomized trial. Ann Surg. 2003;237(4): 574–9.
13. Gultekin FA, Kuruahvecioglu O, Karamercan A, Ege B, Ersoy E, Tatlicioclu E. A prospective comparison of local and spinal anesthesia for inguinal hernia repair. Hernia. 2007;11(2):153–6.
14. Callesen T, Bech K, Kehlet H. One-thousand consecutive inguinal hernia repairs under unmonitored local anesthesia. Anesth Analg. 2001;93(6):1373–6.
15. Kingsnorth AN, Bowley DMG, Porter C. A prospective study of 1000 hernias: results of the Plymouth Hernia Service. Ann R Coll Surg Engl. 2003;85(1):18–22.
16. Amid PK, Lichtenstein IL. Long-term result and current status of the Lichtenstein open tension-free hernioplasty. Hernia. 1998;2(2):89–94.
17. Simons MP, Aufenacker T, Bay-Nielsen M, Bouillot JL, Campanelli G, Conze J, et al. European hernia society guidelines on the treatment of inguinal hernia in adult patients. Hernia. 2009;13(4):343–403.
18. Nienhuijs S, Staal E, Keemers-Gels M, Rosman C, Strobbe L. Pain after open preperitoneal repair versus Lichtenstein repair: a randomized trial. World J Surg. 2007; 31(9):1751–7.
19. Willaert W, De Bacquer D, Rogiers X, Troisi R, Berrevoet F. Open preperitoneal techniques versus Lichtenstein repair for elective inguinal hernias (Review). Cochrane Database Syst Rev. 2012;7, CD008034.

20. Usher FC, Ochsner J, Tuttle Jr LL. Use of marlex mesh in the repair of incisional hernias. Am Surg. 1958;24(12):969–74.
21. Mahorner H, Goss CM. Herniation following destruction of Poupart's and Cooper's ligaments: a method of repair. Ann Surg. 1962;155:741–8.
22. Rives J. Surgical treatment of the inguinal hernia with Dacron patch. Principles, indications, technique and results. Int Surg. 1967;47(4):360–1.
23. Kugel RD. Minimally invasive, nonlaparoscopic, preperitoneal, and sutureless, inguinal herniorrhaphy. Am J Surg. 1999;178(4):298–302.
24. Stoppa R, Petit J, Abourachid H. [Procédé original de plastie des hernies de l'aine: l'interposition sans fixation d'une prothèse en tulle de dacron par voie médiane sous-péritonéale]. Rev Med Picardie. 1972;1:46–8. (Article in French)
25. Wantz GE. Giant prosthetic reinforcement of the visceral sac. Surg Gynecol Obstet. 1989;169(5):408–17.
26. Gilbert AI. Sutureless repair of inguinal hernia. Am J Surg. 1992;163(3):331–5.
27. Pelissier EP. Inguinal hernia: preperitoneal placement of a memory-ring patch by anterior approach. Preliminary experience. Hernia. 2006;10(3):248–52.
28. Muldoon RL, Marchant K, Johnson DD, Yoder GG, Read RC, Hauer-Jensen M. Lichtenstein vs anterior preperitoneal prosthetic mesh placement in open inguinal hernia repair: a prospective, randomized trial. Hernia. 2004;8(2):98–103.
29. Nienhuijs S, Staal E, Keemers-Gels M, Rosman C, Strobbe L. Pain after open preperitoneal repair versus Lichtenstein repair: a randomized trial. World J Surg. 2007;31(9):1751–7 (Discussion. 1758–9).
30. Campanelli G, Bertocchi V, Cavalli M, Bombini G, Biondi A, Tentorio T, et al. Surgical treatment of chronic pain after inguinal hernia repair. Hernia. 2013;17(3):347–53.
31. Mui WL, Ng CS, Fung TM, Cheung FK, Wong CM, Ma TH, et al. Prophylactic ilioinguinal neurectomy in open inguinal hernia repair: a double-blind randomized controlled trial. Ann Surg. 2006;244(1):27–33.
32. Johner A, Faulds J, Wiseman M. Planned ilioinguinal nerve excision for prevention of chronic pain after inguinal hernia repair: a meta-analysis. Surgery. 2011;150(3):534–41.
33. Hsu W, Chen CS, Lee HC, Liang HH, Kuo LJ, Wei PL, Tam KW. Preservation versus division of ilioinguinal nerve on open mesh repair of inguinal hernia: a meta-analysis of randomized controlled trials. World J Surg. 2012;36(10):2311–9.
34. Miserez M, Peeters E, Aufenacker T, Bouillot JL, Campanelli G, Conze J, et al. Update with level 1 studies of the European hernia society guidelines on the treatment of inguinal hernia in adult patients. Hernia. 2014;18(2):151–63.
35. Karakayali F, Oksuz E, Turk E, Pekmez M, Karabulut Z, Yilmaz T, et al. Effectiveness of multiple neurectomies to prevent chronic groin pain after tension-free hernia repair. Int Surg. 2010;95(1):40–8.
36. Wijsmuller AR, Lange JF, van Geldere D, Simons MP, Kleinrensink GJ, Hop WC, et al. Surgical techniques preventing chronic pain after Lichtenstein hernia repair: state-of-the-art vs daily practice in the Netherlands. Hernia. 2007;11(2):147–51.
37. Bischoff JM, Aasvang EK, Kehlet H, Werner MU. Does nerve identification during open inguinal herniorrhaphy reduce the risk of nerve damage and persistent pain? Hernia. 2012;16(5):573–7.

38. Alfieri S, Rotondi F, Di Giorgio A, Fumagalli U, Salzano A, Di Miceli D, Group Groin Pain Trial, et al. Influence of preservation versus division of ilioinguinal, iliohypogastric, and genital nerves during open mesh herniorrhaphy: prospective multicentric study of chronic pain. Ann Surg. 2006;243(4):553–8.
39. Izard G, Gailleton R, Randrianasolo S, Houry R. Treatment of inguinal hernia by McVay's technique. Apropos of 1332 cases. Ann Chir. 1996;50(9):775–6 (Article in French).
40. Alfieri S, Di Miceli D, Doglietto GB. Prophylactic ilioinguinal neurectomy in open inguinal hernia repair. Ann Surg. 2007;245(4):663.
41. Alfieri S, Amid PK, Campanelli G, Izard G, Kehlet H, Wijsmuller AR, et al. International guidelines for prevention and management of post-operative chronic pain following inguinal hernia surgery. Hernia. 2011;15(3):239–49.
42. Van Veen RN, Wijsmuller AR, Vrijland WW, Hop WC, Lange JF, Jeekel J. Long-term follow-up of a randomized clinical trial of non-mesh versus mesh repair of primary inguinal hernia. Br J Surg. 2007;94(4):506–10.
43. Klinge U, Klosterhalfen B, Muller M, Schumpelick V. Foreign body reaction to meshes used for the repair of abdominal wall hernias. Eur J Surg. 1999;165(7): 665–73.
44. Rutkow IM, Robbins AW. Demographic, classificatory, and socioeconomic aspects of hernia repair in the United States. Surg Clin North Am. 1993;73(3):413–26.
45. Greca FH, de Paula JB, Biondo-Simoes ML, da Costa FD, da Silva AP, Time S, Mansur A. The influence of differing pore sizes on the biocompatibility of two polypropylene meshes in the repair of abdominal defects. Experimental study in dogs. Hernia. 2001;5(2):59–64.
46. Klosterhallfen B, Klinge U, Hermanns B, Schumpelick V. Pathology of traditional surgical nets for hernia repair after long-term implantation in humans. Chirurg. 2000;71(1):43–51 (Article in German).
47. Uzzaman MM, Ratnasingham K, Ashraf N. Meta-analysis of randomized controlled trials comparing lightweight and heavyweight mesh for Lichtenstein inguinal hernia repair. Hernia. 2012;16(5):505–18.
48. Śmietański M, Śmietańska IA, Modrzejewski A, Simons MP, Aufenacker TJ. Systematic review and meta-analysis on heavy and lightweight polypropylene mesh in Lichtenstein inguinal hernioplasty. Hernia. 2012;16(5):519–28.
49. Pierides G, Vironen J. A prospective randomized clinical trial comparing the Prolene Hernia System® and the Lichtenstein patch technique for inguinal hernia repair in long term: 2- and 5-year results. Am J Surg. 2011;202(2):188–93.
50. Campanelli G, Pascual MH, Hoeferlin A, Rosenberg J, Champault G, Kingsnorth A, Miserez M. Randomized, controlled, blinded trial of Tisseel/Tissucol for mesh fixation in patients undergoing Lichtenstein technique for primary inguinal hernia repair: results of the TIMELI trial. Ann Surg. 2012;255(4):650–7.
51. Kapischke M, Schulze H, Caliebe A. Self-fixating mesh for the Lichtenstein procedure—a prestudy. Langenbecks Arch Surg. 2010;395(4):317–22.
52. Paajanen H, Kossi J, Silvasti S, Hulmi T, Hakala T. Randomized clinical trial of tissue glue versus absorbable sutures for mesh fixation in local anaesthetic Lichtenstein hernia repair. Br J Surg. 2011;98(9):1245–51.

53. Pierides G, Scheinin T, Remes V, Hermunen K, Vironen J. Randomized comparison of self-fixating and sutured mesh in open inguinal hernia repair. Br J Surg. 2012;99(5):630–6.
54. Jorgensen LN, Sommer T, Assaadzadeh S, Strand L, Dorfelt A, Hensler M, Rosenberg J, Danish Multicentre DANGRIP Study Group. Randomized clinical trial of self-gripping mesh versus sutured mesh for Lichtenstein hernia repair. Br J Surg. 2013;100(4):474–81.
55. Campanelli GP, Cavagnoli R, Gabrielli F, Pietri P. Trabucco's procedure and local anaesthesia in surgical treatment of inguinal and femoral hernia. Int Surg. 1995; 80(1):29–34.
56. Joshi GP, Rawal N, Kehlet H, Bonnet F, Camu F, Fischer HB, Neugebauer EA, Schug SA, Simanski CJ. Evidence-based management of postoperative pain in adults undergoing open inguinal hernia surgery. Br J Surg. 2012;99(2):168–85.
57. Campanelli G, Pettinari D, Nicolosi FM, Cavalli M, Avesani EC. Inguinal hernia recurrence: classification and approach. Hernia. 2006;10(2):159–61.

29. Prevention of Pain: Optimizing the Laparoscopic TEP and TAPP Techniques

Jorge Daes

Introduction

We have known for a long time that fast and accurate surgery is associated with good postoperative outcomes, including relatively low levels of postoperative pain. However, it is difficult to determine the precise maneuvers that affect outcome. This is especially true in the field of inguinal hernia surgery, where there is a wide variation of techniques among surgeons, and data about different techniques show conflicting results.

It was previously hoped that meta-analyses would answer most of our questions about optimal surgical techniques, but such analyses are limited by lack of methodological rigor in studies and by the fact that biological systems (as well as social and economic systems) are complex and cannot be completely understood by the usual methods of statistical analysis. Experience and common sense have therefore regained their importance for determining optimal surgical techniques.

This chapter describes strategies for optimizing the laparoscopic repair of inguinal hernias to prevent postoperative pain, based on review of the relevant literature, our own experience, and the experience of leading surgeons in the field.

B.P. Jacob et al. (eds.), *The SAGES Manual of Groin Pain*,
DOI 10.1007/978-3-319-21587-7_29

General Aspects of Preventing Pain After TEP and TAPP Hernia Repair

Surgeons should master the laparoscopicanatomy of the inguinal region before performing laparoscopic inguinal hernia repair and should be particularly aware of the anatomy of the inguinal nerves.

Surgeons should be appropriately trained in all techniques for laparoscopic hernia repair, including the totally extraperitoneal (TEP), enhanced or extended view TEP (eTEP), transabdominal preperitoneal (TAPP), and intraperitoneal onlay mesh (IPOM) techniques, in addition to primary closure of defects. Comprehensive training allows surgeons to offer the appropriate procedure according to individual patient characteristics and to convert from one procedure to another if necessary.

One of the best ways to avoid pain after inguinal hernia repair is to avoid operating on patients with unusual preoperative inguinal pain or inguinal pain that is disproportionate to the hernia. Pain is usually not a remarkable symptom of inguinal hernias, except in complex cases. Many patients with disproportionate preoperative pain have a different cause for their pain and develop chronic pain after hernia repair.

We recommend administration of a first-generation cephalosporin during the induction of anesthesia. We do not routinely use prophylactic antithrombotic medication, but use pneumatic compression devices in all patients.

Patients should be reexamined while standing immediately before surgery, and the physical examination findings should be compared with the laparoscopic findings. This is an excellent method for ensuring that hernias are not missed.

We prepare the skin, drape the patient, and set up the equipment while the patient is still awake (but sedated) so that surgery starts almost immediately after the induction of anesthesia, thereby reducing costs and facilitating a faster recovery. Optimal muscle relaxation is important to ensure a fast and easy procedure, and the anesthesiologist should be asked to provide a short period of full relaxation before the start of the operation.

Technical Aspects of Preventing Pain After TEP and TAP Hernia Repair

We inject long-acting local anesthetic into the skin before incision to reduce postoperative pain [1]. Surgeons should be able to perform surgery both in a triangulated position and with the camera lateral to the working ports, so that they can adapt to different setups.

When using the TEP technique, the use of a balloon trocar to create the surgical space makes the procedure easier and faster and reduces blood loss, which may result in less postoperative pain [2].

The eTEP technique creates a larger surgical space and allows a more versatile distribution of ports than the TEP technique. The eTEP technique takes advantage of the fact that the preperitoneal space can be reached from almost any part of the anterior abdominal wall [3]. A video showing this technique can be found online [4].

The creation of large peritoneal flaps during the TAPP procedure facilitates complete dissection of the myopectineal orifice of Fruchaud, placement of a large mesh, and perfect apposition of the peritoneal edges at the end of the procedure. This is a faster, less expensive, and less painful alternative to closing the peritoneum using tacks, glue, or sutures. The peritoneal edges come together as CO_2 is carefully released, and the wound heals quickly. The findings of an experimental study support this approach [5]. We have never closed the peritoneum, and other groups have also reported that they do not close the peritoneum. We have not experienced any cases of bowel obstruction or fistula using this approach. For the same reason, we do not close accidental tears created during the TEP or eTEP procedures. A video showing peritoneal apposition at the end of a TAPP procedure can be found online [6]. However, not physically closing the peritoneal edges is controversial, and most consensus statements recommend some type of peritoneal closure, with suture closure being the most frequently used. The recent development of barbed sutures has increased the ease of peritoneal closure.

We advise a stepwise approach to dissection. In the TEP procedure, we dissect Cooper's ligament (both ligaments in cases of direct hernias), free the lax transversalis fascia from preperitoneal structures in cases of direct hernias, dissect the space of Bogros, divide the posterior transversalis fascia that usually overlaps the indirect sac and peritoneum at the level of the internal ring, and finally identify the indirect sac. A video showing the seldom-mentioned posterior transversalis fascia is available online [7]. Cauterization should be performed with care and avoided at the "triangle of doom" and the electrical hazard zone or "triangle of pain." The use of fine, low-voltage instruments and bipolar cautery helps to avoid damage to sensitive structures and to prevent residual hematoma, which is one of the most commonly cited causes of postoperative pain.

One rarely mentioned cause of pain after hernia surgery is grasping and traction of the cord structures, which is common during open surgery. This pain may be caused by injury to the vasa nervorum. In laparoscopic repair, traction may occur during separation of an indirect sac from the cord structures, as some surgeons grasp the cord structures to dissect them from the sac. We advise pulling the sac medially while dissecting the fibrous and fatty tissues next to the cord structures, using fine Maryland forceps without directly touching the cord structures. As dissection progresses, the sac can be grasped more laterally and rotated medially. This process is continued until the hernia sac is separated from the vas deferens and the spermatic vessels by a bluish transparency. Videos of these maneuvers are available online [8, 9]. It is then possible to deal with the sac in two ways. If the sac does not extend deeply into the scrotum, it can be reduced completely. In cases of large inguinoscrotal hernias, attempting to completely reduce the sac risks damage to the cord structures and the development of orchitis. Failure to deal with the distal sac, however, carries the risk of formation of large and sometimes cumbersome hematomas, seromas, or pseudohydroceles. Repeated drainage and occasionally surgery may be necessary in such cases.

We previously described a technique for managing the distal sac in large inguinoscrotal hernias [10]. After ligating the sac and dividing it distally, at the level of the internal ring, we reduce the distal sac by pulling it out of the scrotum and fixing it high and laterally to the posterior inguinal wall with tacks or sutures. Using this maneuver, we were able to reduce the incidence and severity of seromas, with no cases of postoperative orchitis, testicular pain, or neuralgia [10]. A video showing this maneuver is available online [11].

The next step is parietalization of structures, which consists of proximal dissection of the sac and peritoneum to allow proper placement of the mesh over the cord structures. Extensive proximaldissection helps to prevent recurrence by rolling of the mesh or a sac sliding under the mesh. Parietalization is complete when upward traction of the sac does not move the cord structures. A video showing parietalization is available online [12].

Three important lipomatousstructures should be properly identified during the laparoscopic repair of hernias: lipomas, a fatty spermatic cord, and lymph nodes. Only the first of these should routinely be removed to avoid "hernia" recurrence. Lipomas lie anterolateral to the cord structures, are light yellow, have a thin capsule, are usually devoid of accompanying vessels, and are easily dissected out of the internal ring. A fatty cord, often seen in obese patients or in patients who have undergone bariatric surgery, could be confused with a lipoma and partially divided. A fatty structure with vessels running toward or from the internal ring is usually a cord structure. Lymph nodes are usually positioned posterior and lateral to the cord structures, are dark yellow, are not easily displaced, and "bounce" when pushed with the dissector. Lymph nodes should generally not be resected, to avoid bleeding and nerve damage. The nerves usually run posterior to the lymph nodes. A video showing these structures can be found online [13].

One meta-analysis concluded that the use of a low-weight mesh (less material, large pores, some elasticity) lessens the risk of postoperative pain, groin stiffness, and foreign body sensation, especially during the first few months after laparoscopic inguinal hernia repair [14]. However, another meta-analysis did not have the same findings [15]. Some studies found that the use of a very-low-weight mesh increased the risk of recurrence, probably because of the difficulty of fixing the mesh and the tendency of the mesh to roll up. We use a mid-weight (45 g/m^2), large-pore, polyester mesh. A low- or mid-weight polypropylene mesh is also suitable. The mesh should be at least 15–17 × 10–12 cm in size to completely cover the myopectineal orifice of Fruchaud.

Many studies have reported that lack of fixation does not increase the recurrence rate, reduces the cost of the procedure, and is associated with less postoperative pain, but most of those studies focused on small hernias (defects of <3 cm) or included patients with unmeasured defects [16–20]. The mesh should be fixed according to the learning curve and preference of the surgeon, especially for direct or large hernias.

Tacks are known to cause postoperative pain, but this can be avoided by careful placement. Two tacks placed over Cooper's ligament and one placed high on each of the superomedial and superolateral corners of the mesh are all that are needed. Some studies have reported that the use of fibrin glue causes less acute and chronic pain than stapling, with no difference in recurrence rate [21–25]. Other studies have not found clear differences in chronic pain rates according to the method of mesh fixation [26–28]. Other options include the use of self-fixing meshes [29] and cyanoacrylate.

At the end of the TEP procedure, the inferolateral corner of the mesh should be held in position with a closed grasper (if not fixed by glue or a carefully placed suture) to ensure that it does not roll up anteriorly and medially while the CO_2 is slowly released. We instill diluted bupivacaine into the space at the end of the procedure, which we believe improves recovery, but we do not have definitive evidence to prove this hypothesis. The final maneuvers can be observed in a video posted online [30].

Associations Between Surgical Techniques and Types of Postoperative Pain

Most of the strategies described in this chapter help to prevent nociceptive-type postoperative pain, and some help to prevent visceral-type postoperative pain. The most troublesome postoperative pain is chronic neuralgic pain, which has specific causes. Álvarez studied a series of cases with well-documented dermatome mapping and surgical exploration and found that the ilioinguinal nerve was the most commonly injured nerve (60 %) during laparoscopic inguinal hernia repair (see Chap. 21). This nerve descends from the lumbar plexus, penetrates the abdominal wall muscles, and turns abruptly at the anterior superior iliac spine to run transversely, eventually lying anterior to the cord structures. The nerve is not exposed during the laparoscopic approach but may be injured when a tack penetrates deep into the abdominal wall along the imaginary line between the anterior superior iliac spines, which is an unacceptably low place for fixation. This may occur when using a small mesh or with low placement of the mesh. Tacks should therefore be placed very high on the posterior inguinal wall. The femoral branch of the genitofemoral nerve was the next most commonly injured nerve (40 %) and was usually injured during dissection. The study did not include any cases of injury to the lateral femoral cutaneous nerve, and the occurrence of this injury is generally overstated. It is also useful to know that an intact epineurium prevents nerve damage from adjacent mesh.

References

1. Hon SF, Poon CM, Leong HT, Tang YC. Pre-emptive infiltration of Bupivacaine in laparoscopic total extraperitoneal hernioplasty: a randomized controlled trial. Hernia. 2009;13(1):53–6.
2. Bringman S, Ek A, Haglind E, Heikkinen T, Kald A, Kylberg F, et al. Is a dissection balloon beneficial in totally extraperitoneal endoscopic hernioplasty (TEP)? A randomized prospective multicenter study. Surg Endosc. 2001;15(3):266–70.
3. Daes J. The enhanced view—totally extraperitoneal technique for repair of inguinal hernia. Surg Endosc. 2012;26(4):1187–9.
4. Daes J. Large inguinal hernia repair with E-TEP technique.wmv. 15 Dec 2010. YouTube.com [Internet]. https://www.youtube.com/watch?v=XUyB_WKWV_k. Accessed 2 Feb 2015.
5. Durstein-Decker C, Brick WG, Gadacz TR, Crist DW, Ivey RK, Windom KW. Comparison of adhesion formation in transperitoneal laparoscopic herniorrhaphy techniques. Am Surg. 1994;60(3):157–9.

6. Daes J. Peritoneal apposition after TAPP. Facebook.com [Internet]. https://www.facebook.com/photo.php?v=10201118167996955&l=7043899692607058045. Accessed 2 Feb 2015
7. Daes J. Anatomia de la fascia posterior. 28 Mar 2013. YouTube.com [Internet]. https://www.youtube.com/watch?v=71lDLL1BTpM. Accessed 2 Feb 2015.
8. Daes J. Video 4 de la serie educativa tecnica e-TEP. 28 Mar 2013. YouTube.com [Internet]. http://www.youtube.com/watch?v=Sam3XbYheDk. Accessed 2 Feb 2015.
9. Daes J. Video 5 de la serie educativa tecnica e-TEP. 28 Mar 2013. YouTube.com [Internet]. http://www.youtube.com/watch?v=_xsV2cDmyXU. Accessed 2 Feb 2015.
10. Daes J. Endoscopic repair of large inguinoscrotal hernias: management of the distal sac to avoid seroma formation. Hernia. 2014;18(1):119–22.
11. Daes J. Video 9 de la serie educativa tecnica e-TEP. 28 Mar 2013. YouTube.com [Internet]. https://www.youtube.com/watch?v=Ez19DKCcYdU. Accessed 2 Feb 2015.
12. Daes J. Video 6 de la serie educativa tecnica e-TEP. 28 Mar 2013. YouTube.com [Internet]. https://www.youtube.com/watch?v=5mQc2F_PCOc. Accessed 2 Feb 2015.
13. Daes J. Lipomatous masses—identification and management. Apr 2014. Facebook.com [Internet]. https://www.facebook.com/photo.php?v=10200883828138605&l=6015477704179870708. Accessed 2 Feb 2015.
14. Sajid MS, Kalra L, Parampalli U, Sains PS, Baig MK. A systematic review and meta-analysis evaluating the effectiveness of lightweight mesh against heavyweight mesh in influencing the incidence of chronic groin pain following laparoscopic inguinal hernia repair. Am J Surg. 2013;205(6):726–36.
15. Currie A, Andrew H, Tonsi A, Hurley PR, Taribagil S. Lightweight versus heavyweight mesh in laparoscopic inguinal hernia repair: a meta-analysis. Surg Endosc. 2012;26(8):2126–33.
16. Taylor C, Layani L, Liew V, Ghusn M, Crampton N, White S. Laparoscopic inguinal hernia repair without mesh fixation, early results of a large randomised clinical trial. Surg Endosc. 2008;22(3):757–62.
17. Koch CA, Greenlee SM, Larson DR, Harrington JR, Farley DR. Randomized prospective study of totally extraperitoneal inguinal hernia repair: fixation versus no fixation of mesh. JSLS. 2006;10(4):457–60.
18. Moreno-Egea A, Torralba Martinez JA, Morales Cuenca G, Aguayo Albasini JL. Randomized clinical trial of fixation vs nonfixation of mesh in total extraperitoneal inguinal hernioplasty. Arch Surg. 2004;139(12):1376–9.
19. Smith AI, Royston CM, Sedman PC. Stapled and nonstapled laparoscopic transabdominal preperitoneal (TAPP) inguinal hernia repair. A prospective randomized trial. Surg Endosc. 1999;13(8):804–6.
20. Ferzli GS, Frezza EE, Pecoraro Jr AM, Ahern KD. Prospective randomized study of stapled versus unstapled mesh in a laparoscopic preperitoneal inguinal hernia repair. J Am Coll Surg. 1999;188(5):461–5.
21. Lovisetto F, Zonta S, Rota E, Mazzilli M, Bardone M, Bottero L, et al. Use of human fibrin glue (Tissucol) versus staples for mesh fixation in laparoscopic transabdominal preperitoneal hernioplasty: a prospective, randomized study. Ann Surg. 2007;245(2):222–31.
22. Schwab R, Willms A, Kroger A, Becker HP. Less chronic pain following mesh fixation using a fibrin sealant in TEP inguinal hernia repair. Hernia. 2006;10(3):272–7.

23. Topart P, Vanderbroucke F, Lozac'h P. Tisseel versus tack staples as mesh fixation in totally extraperitoneal laparoscopic repair of groin hernias: a retrospective analysis. Surg Endosc. 2005;19(5):724–7.
24. Brügger L, Bloesch M, Ipaktchi R, Kurmann A, Candinas D, Beldi G. Objective hypoesthesia and pain after transabdominal preperitoneal hernioplasty: a prospective, randomized study comparing tissue adhesive versus spiral tacks. Surg Endosc. 2012;26(4):1079–85.
25. Tolver MA, Rosenberg J, Juul P, Bisgaard T. Randomized clinical trial of fibrin glue versus tacked fixation in laparoscopic groin hernia repair. Surg Endosc. 2013; 27(8):2727–33.
26. Ceccarelli G, Casciola L, Pisanelli MC, Bartoli A, Di Zitti L, Spaziani A, et al. Comparing fibrin sealant with staples for mesh fixation in laparoscopic transabdominal hernia repair: a case control-study. Surg Endosc. 2008;22(3):668–73.
27. Boldo E, Armelles A, Perez de Lucia G, Martin F, Aracil JP, Miralles JM, et al. Pain after laparoscopic bilateral hernioplasty: early results of a prospective randomized double-blinded study comparing fibrin versus staples. Surg Endosc. 2008;22(5): 1206–9.
28. Lau H, Patil NG. Acute pain after endoscopic totally extraperitoneal (TEP) inguinal hernioplasty: multivariate analysis of predictive factors. Surg Endosc. 2004;18(1): 92–6.
29. Wu A, Reiner M, Jacob B. An ongoing prospective study evaluating self-gripping mesh (Parietex Progrip) without additional fixation during laparoscopic total extraperitoneal (TEP) inguinal hernia repair: interim analysis at one year (Abstract FP-5584). Hernia. 2013;17 Suppl 1:S39.
30. Daes J. Video 11 de la serie educativa tecnica e-TEP. 28 Mar 2013. YouTube.com. [Internet]. https://www.youtube.com/watch?v=3n1FlEzq4fA. Accessed 2 Feb 2015.

30. Prophylactic Neurectomy Versus Pragmatic Neurectomy

Ryan Berg and Matthew I. Goldblatt

Editor's Comment (DCC)

The data presented in this chapter represent the best available high-quality studies to date on this topic. The counterargument to the proposition of prophylactic neurectomy is that the incidence of significant chronic pain may be reduced to less than 1 % with three-nerve identification and meticulous operative technique (Alfieri S et al. International guidelines for prevention and management of post-operative chronic pain following inguinal hernia surgery. Hernia. 2011;15(3):239–49). This is lower than the rates of pain reported in both the control and prophylactic neurectomy groups in each of these cited studies. If pain rates can be reduced with good technique, intentionally causing the sensory disturbances or risk of deafferentation pain with neurectomy in all patients may be considered unnecessary. "Pragmatic neurectomy" was coined in response to the concept of prophylactic neurectomy and simply refers to the logical practice of performing a neurectomy at the time of hernia repair whenever a nerve is recognizably injured or is at risk for injury due to its neuroanatomic location, course, or operative factors. Neurectomy in these cases is absolutely recommended. The author's conclusion that no patient in each of these studies developed severe pain is significant, and "prophylactic neurectomy" in at-risk individuals is likely prudent. Mild sensory disturbances and numbness are vastly preferable to chronic pain in high-risk patients. Using preoperative risk calculators such as the Carolinas Equation for Quality of Life (CeQOL; www.carolinashealthcare.org/ceqol*) and understanding high-risk populations for the development of chronic pain will help surgeons to decide who will benefit from prophylactic neurectomy. Tailoring the right operation for each patient in conjunction with a good informed consent makes for good practice.*

B.P. Jacob et al. (eds.), *The SAGES Manual of Groin Pain*,
DOI 10.1007/978-3-319-21587-7_30

Introduction

Chronic inguinal neuralgia is one of the most common and significant complications following open inguinal hernia repair. Incidence of long-term (≥1 year) postoperative neuralgia following Lichtenstein repair ranges from 6 to 29 % [1]. Subsequent patient disability can be debilitating and require multiple further interventions for treatment. Further, while many cases result in out-of-court settlement, it is worth noting that 5–7 % of patients with postoperative inguinal neuralgia will sue their surgeon [2].

The ilioinguinal nerve is a sensory nerve that innervates the skin over the groin, the medial aspect of the thigh, the upper part of the scrotum, and the penile root [3]. Routine ilioinguinal neurectomy has been adopted by many as a means of minimizing the troubling pain that can result from inguinal dissection and hernia repair. It is proposed that excision of the nerve would eliminate the possibility of nerve entrapment, inflammation, neuroma, and fibrosis. The counterargument to this practice is that routine nerve excision may not only decrease the incidence of chronic groin pain, but it may also cause disturbing and potentially disabling neurologic deficits in the aforementioned distribution, including both decreased touch and pain sensations. Examining these arguments is certainly challenging, in large part owing to the significant subjectivity and variability that is inherent to a patient rating his or her severity of pain and loss of sensation. That said, the issue of chronic groin pain after inguinal surgery is by its very nature a subjective complaint, and as such subjective data are necessary and valuable in its study.

Pragmatic Neurectomy

Routine neurectomy is a concept that is not unique to inguinal surgery and is commonly practiced in other general surgical procedures. In 1998, Abdullah et al. performed a randomized, controlled trial studying routine division versus preservation of the intercostobrachial nerve in patients undergoing axillary dissection for breast cancer [4]. This study was performed in an intention-to-treat fashion and as such was essentially a comparison between routine and pragmatic neurectomies. The study demonstrated that there is increased incidence of sensory loss at hospital discharge in the routine neurectomy group (78 vs. 60 %, $p<0.05$), as well as pain (30 vs. 16 %, $p<0.05$). However, differences in pain, diminished sensation, sensory loss, and paresthesia were lost by 3-month follow-up [4]. While the authors suggest that nerve preservation may be favored given the early symptomatic differences, the longer-term follow-up results suggest nerve division does not portend significant functional or sensory deficits. In addition, nerve pain during inguinal hernia repair is typically due to nerve entrapment within suture or mesh, which are not used during axillary dissection.

Ravichandran et al. were among the first to perform a randomized trial on the topic of ilioinguinal neurectomy [5]. This study was, and still is, unique in this body of literature in that it is self-controlled. The authors enlisted patients who were planned for bilateral inguinal hernia repair and randomized the patients' right or left side to undergo routine neurectomy, while the other side had nerve preservation. In comparing the neurectomy side to that of nerve preservation, there was no difference in pain rated on a 10-point scale noted on postoperative day 1 (2.9 vs. 2.5, $p=0.98$). At 6 months, just two patients complained of minor wound discomfort, one on the divided side and one on the preserved side. Physical examination on patients 6 months postoperatively revealed an increased incidence of diminished touch sensation on the divided side (9 patients vs. 1 patient; no *p*-value given) as well as increased incidence of diminished pain sensation (8 vs. 5; no *p*-value given). However, it is important to note that just two of the 20 patients reported any symptoms of numbness at their 6-month follow-up, including one complaining of lateral thigh numbness, an area not supplied by the ilioinguinal nerve [5]. As such, it is reasonable to conclude from this study that patients undergoing routine neurectomy do not have increased incidence of immediate postoperative nor chronic pain, nor do they have increased incidence of symptomatic sensory loss. The study is underpowered to provide statistically significant differences in these groups and does not provide statistical analysis of all its data; however, it is a landmark and otherwise well-designed study of the debate.

As routine neurectomy in inguinal hernia repair increased in popularity, so did the size of the studies. Tsakayannis et al. prospectively observed a cohort of 191 patients, all of whom underwent routine, elective resection of both the ilioinguinal and iliohypogastric nerves [6]. These patients were followed up at 1 month, 6 months, and 1 year to determine their pain rating and degree of sensory loss. At no point postoperatively did any patient report moderate or severe pain. 9.4 % of patients reported subjective numbness at 1 month, while at both 6 and 12 months postoperatively, 6.3 % of patients noted numbness. Patients who complained subjectively of sensory loss were subjected to a detailed physical exam. Just two patients (1 %) had sensory loss confirmed in the distribution of the excised nerves on exam, of which neither deficit was found to be disabling [6]. This study was certainly limited by its observational nature with lack of a control group; however, it serves to demonstrate in a large cohort that neurectomy may be safely performed in the inguinal region without disabling consequences.

Perhaps the largest study on the topic of neurectomy during inguinal hernia repair came from Picchio and colleagues. In 2004, they enlisted 813 patients in a double-blind study and randomized them into routine ilioinguinal nerve transection versus preservation [7]. They followed these patients at 1 month, 6 months, and 1 year postoperatively. Utilizing a survey with a 4-point pain scale (none, mild, moderate, severe), the study did not find any significant differences in patient pain rating between the two groups at any endpoint. The study did, however, find differences in touch and pain sensation between the groups. These sensory deficits were tested with focused, detailed physical examination on follow-up visits. There was an increased incidence in loss of touch sensation for those undergoing routine neurectomy at 1 month (49 vs. 21 %, $p<0.001$), 6 months (29 vs. 6 %, $p<0.001$), and 1 year (11 vs. 4 %, $p=0.002$). There was also an increased incidence in loss of pain sensation for neurectomy at 1 month (56 vs. 45 %, $p=0.004$) and 6 months (33 vs. 25 %, $p=0.04$). There was no difference in loss of pain sensation at 1 year (9 vs. 8 %, $p=0.89$) [7]. Given the size of enrollment and study design, these were the strongest data to date suggesting that there is increased incidence of sensory deficits for patients undergoing routine ilioinguinal nerve excision. At the same time, the study did not address the question of whether these deficits were disabling or disturbing to the patient. As demonstrated previously by Ravichandran et al., it is possible for patients to have little to no subjective complaints of sensory loss despite objective physical exam findings to suggest that a deficit is present. Again, given that the endpoints of chronic groin pain and troubling or disabling neurologic deficits are primarily subjective in nature, to disregard the patients' subjective neurologic complaints is a shortcoming of this otherwise strong evidence.

Prophylactic Neurectomy

As studies on this controversialtopic have continued, evidence has become increasingly suggestive of a potential benefit of routine neurectomy in combating chronic inguinal pain after hernia repair. Dittrick et al. conducted a small retrospective chart review in 90 patients who underwent open inguinal hernia repair. These patients had either routine nerve preservation or nerve excision, primarily owing to differing practice patterns of two surgeons whose patients were included. Data were obtained through patient interview, which in most cases included asking the patient to recall the severity and duration of pain at different time points. The endpoints assessed included 1 month, 6 months, 1 year, and 3 years postoperatively. The study found that patients who underwent routine ilioinguinal neurectomy reported decreased incidence of neuralgia at 1 month (5 vs. 21 %, $p=0.016$), at 6 months (3 vs. 26 %, $p<0.001$), and at 1 year (3 vs. 25 %, $p=0.003$). No difference was observed at 3 years (6 vs. 8 %, $p=0.75$). Additionally, patients were asked to report postoperative paresthesia, and comparison between the groups revealed no differences at any endpoint in the reported incidence of sensory loss [1]. This study was clearly limited in a number of ways, not the least of which includes a likely significant recall bias. The study was also not randomized, had small enrollment, and may have been influenced by the other surgical technical differences between surgeons, and patients may not all have been blinded to the details of their procedure. That said, this was one of the first studies to suggest an improvement in outcomes in patients undergoing routine neurectomy during this procedure and laid the foundation for more convincing evidence to follow.

Among the more recent studies to examine this topic, Malekpour and associates recently conducted a double-blind randomized controlled trial comparing routine ilioinguinal nerve preservation to excision. One hundred twenty one patients were studied, and their pain rated on a 10-point visual analog scale (VAS). Their analysis revealed that the incidence of chronic inguinodynia, defined as the presence of pain at 3 months postoperatively, was lower in the nerve excision group (6 vs. 21 %, $p=0.033$). Further, the mean severity score on the VAS was noted to be lower in the neurectomy group 1 day after surgery (2.2 vs. 2.8, $p<0.001$) as well as at 1 month (0.7 vs. 1.5, $p<0.001$). However, at both 6 months and 1 year, these differences had been eliminated, with median pain scores of 0 in both groups. Patient rating of hypoesthesia at all time points was also equivalent between the groups, with no patients complaining of loss of sensation at 1 year in either treatment arm [2]. As with much of the research on this topic, this study is limited in part by the subjectivity of the data it analyzes. Given this, it is noted that cultural differences may factor into these data, as this Middle Eastern study population provided lower overall VAS scores when compared to their

Western counterparts. This only serves to reinforce that analyzing this topic is quite challenging, as its subjectivity provides such variation.

Among the stronger evidence to date in support of routine ilioinguinal neurectomy in patients undergoing open inguinal hernia repair is that from Mui et al. [3]. These authors also conducted a double-blind controlled trial randomizing 100 patients to undergo either prophylactic ilioinguinal neurectomy or nerve preservation. In follow-up, patients were not only asked to rate their pain generally but also were asked to rate their pain while performing various common tasks. These included coughing 10 times, walking up three flights of stairs, and riding a bicycle for 10 min. Groin sensation was also tested using the Semmes-Weinstein monofilament test. Patients were followed up to 6 months postoperatively. At 1 month, there was not a significant difference between the groups in groin pain or sensation. However, at 6 months, the overall incidence of chronic groin pain was significantly lower in the routine neurectomy group (8 vs. 29 %, $p=0.008$). This incidence of pain at 6 months was also noted to be lower when climbing three flights of stairs (2 vs. 14 %, $p=0.03$) and cycling 10 min (4 vs. 20 %, $p=0.015$). There was no significant difference in pain noted between groups at rest, during normal daily activities, or with coughing. There were also no objective differences in groin numbness or loss of sensation. This study is also notable for including patient overall quality of life in its analysis, and it found no difference between groups at any time point, including at baseline, 1 month, and 6 months [3]. This study is commendable in that it studied pain with multiple tasks as well as included a quality of life measure in its analysis and notable in that it demonstrated improved pain ratings at 6 months in patients undergoing neurectomy. It is important to note, however, that follow-up was fairly short compared to other studies, and while routine nerve excision was noted to decrease pain with various activities, it also showed no significant reduction in the incidence of pain noted at rest, pain with normal activities of daily living, or in patient-reported improved quality of life.

Despite an ever-growing set of data examining the practice of routine ilioinguinal neurectomy in patients undergoing inguinal hernia repair, there is lack of consensus regarding its efficacy and potential morbidity. Again, this may be attributed in large part to the inherent subjectivity of the endpoints in question, namely, the incidence of chronic pain and disabling sensory deficit. While there are data that demonstrate an increased incidence of objective sensory loss on detailed, focused physical exam after neurectomy, there are no strong data that support the assertion that nerve excision increases the incidence of disabling or even perceptible neurologic symptoms. Additionally, the data are also mixed in regard to the efficacy of routine neurectomy in decreasing chronic groin pain. That said, there are certainly no data to suggest that the incidence of chronic groin pain is increased with routine neurectomy. Further, the data do imply that there is at least equivalence in pain outcomes with neurectomy, and there is increasing evidence in well-designed trials that the incidence of chronic pain months after inguinal hernia repair is decreased with routine neurectomy. Given this, it is these authors' opinion that routine ilioinguinal neurectomy is a reasonable treatment option in patients undergoing open inguinal surgery as it does not increase patient morbidity and there is an increasing body of evidence that it may decrease incidence of chronic groin pain.

References

1. Dittrick G, Ridl K, Kuhn J, McCarty T. Routine ilioinguinal nerve excision in inguinal hernia repairs. Am J Surg. 2004;188(6):736–40.
2. Malekpour F, Mirhashemia S, Hajinasrolah E, Salehi N, Khoshkar A, Kolahi A. Ilioinguinal nerve excision in open mesh repair of inguinal hernia: results of a randomized clinical trial: simple solution for a difficult problem? Am J Surg. 2008;95(6):35–40.
3. Mui W, Ng C, Fung T, Cheung F, Wong C, Ma T, et al. Prophylactic ilioinguinal neurectomy in open inguinal hernia repair: a double-blind randomized controlled trial. Ann Surg. 2006;244(1):7–33.
4. Abdullah T, Iddon J, Barr L, Baildam A, Bundred N. Prospective randomized controlled trial of preservation of the intercostobrachial nerve during axillary node clearance for breast cancer. Br J Surg. 1998;85(10):1443–5.
5. Ravichandran D, Kalambe B, Pain J. Pilot randomized controlled study of preservation or division of ilioinguinal nerve in open mesh repair of inguinal hernia. Br J Surg. 2000;87(9):1166–7.
6. Tsakayannis D, Kiriakopoulos A, Linos D. Elective neurectomy during open, "tension free" inguinal hernia repair. Hernia. 2004;8(1):67–9.
7. Picchio M, Palmento D, Attanasio U, Malarazzo P, Bambini C, Caliendo A. Randomized controlled trial of preservation or elective division of ilioinguinal nerve on open inguinal hernia repair with polypropylene mesh. Arch Surg. 2004;139(7):755–8.

31. Triple Neurectomy Versus Selective Neurectomy

Wolfgang M.J. Reinpold and Alexander D. Schroeder

Introduction

While recurrences after groin hernia repair have decreased after the introduction of open and laparoscopic mesh techniques, today chronic pain figures among the most frequent postherniorrhaphy complications [1–6]. Despite the fact that the use of mesh does not lead to an increase of chronic pain, the surgeon's focus of interest has shifted toward the prevention and treatment of chronic inguinodynia. Persistent postoperative pain affects everyday activities in 5–8 % of patients [1, 2, 7, 8] and may cause long-term disability. It is now widely accepted that surgery with inguinal nerve neurectomy is the last treatment option for persistent postherniorrhaphy pain. Before neurectomy, multidisciplinary diagnostics with local and paravertebral infiltrations, MRI of the lower abdomen and spine, and a multimodal nonsurgical treatment of at least 6 months should be performed. The pain management should include a pain specialist, and hernia recurrence should be excluded. Details of the workup and management of chronic postherniorrhaphy pain are described in Chap. 18.

Neurectomy: What Do We Know?

Currently, there are 19 reports on selective neurectomy and four publications on triple neurectomy available. Studies with less than 10 patients were not considered in this report. Tables 31.1 and 31.2 [2, 8–23] summarize the publications on selective neurectomy, and Table 31.3 [5, 6, 24, 25] shows the available data on triple neurectomy. All studies except one have reported significant pain relief after neurectomy. After considering the cumulative data of three multiphase studies [2–5, 7, 16, 26], there

B.P. Jacob et al. (eds.), *The SAGES Manual of Groin Pain*,
DOI 10.1007/978-3-319-21587-7_31

Table 31.1. Selective neurectomy with follow-up <24 months.

Authors	*N*	Primary surgery	Pain surgery	Neurectomy	Mesh removal	New mesh	Pain relief	Follow-up
Loos et al. [8]	54	Suture 24 Mesh 30 Lap 10	Open	IIN 44 GFN 25 IHN 9	Partly 19/54	No	Pain-free 25/49 Improved 12/49 Worsened 8/49	18 months PE and Q
Starling et al. [9]	13	Open	Open	GB, IIN, or IHN	No	No	10/13	Not reported
Starling et al. [10]	17	Open	Open	GB, IIN, or IHN	No	No	12/17	Not reported
Heise et al. [11]	20	Open 17 Lap 3	Open	IIN or IHN	Yes	No	Improved 12/20	Phone-Q, PE
Lee et al. [12]	11	Open 8 Lap 3	Open	IHN, IIN, GFN, or LFCN	[a]	[a]	Nine improved	6–18 months, phone-Q
Deysine et al. [13]	22	Open	Open	IHN	No	No	Twenty-two pain-free	Not reported
Madura et al. [14]	100	Open 87 Lap 13	Open	IHN, IIN, or GFN	Yes	No	Seventy-two pain-free	1 month, PE (no data)
Kim et al. [15]	16	Open	Open	IIN or IIN + IHN	[a]	[a]	14/16	12–46 months, phone-Q
Keller et al. [16]	21	Suture 2 Mesh 14 Lap 6	Comb.	IIN 2, IHN + IIN 9, TNE 7	Yes	Yes	Pain-free 15/21 Improved 5/21 Pain recur 1/21	6 weeks, PE (no data)
Giger et al. [17]	39[b]	Suture 20 Mesh 9 Lap 4	Lapar. ERN	GFN 19 GFN + IIH 20	No	No	Improved 27/39	12 months PE
Vuilleumir et al. [18]	43	Open 31 Lap 12	Open	IIN 35, IHN 11	Yes	Yes	Pain-free 41/43 Improved 2/43	12 months PE

BMR biologic mesh repair, *Sr* suture repair, *Q* questionnaire, *PE* physical examination, *LFCN* lateral femoral cutaneous nerve
[a]Not reported, *QL* quality of life, *Lapar ERN* retroperitoneoscopic neurectomy
[b]Includes noninguinal hernia surgery

Table 31.2. Long-term results after selective neurectomy (>24 months).

Authors	N	Primary surgery	Pain surgery	Neurectomy	Mesh removal	New mesh	Pain relief	Follow-up
Bischoff et al. [2]	54	Open: mesh 46 Plug 8	Open	Selective NE IIN 37 IHN 19 GFN 10	Yes	No	Improved 16/25 Unchanged 6/25 Worsened 3/25	36 months, PE 3 months, Q pain, physical + sexual. Function
Bower et al. [19]	12	Open	Open	IHN, IIN, GFN, LFCN	No	No	Improved 9/12	66 months PE (no data)
Ducic et al. [20]	19	Not reported	Open	Selective neurectomy	No	No	Improvement in pain and quality of life 16/19	24 months PE (no data)
Zacest et al. [21]	18[a]	Not reported	Open	Selective IIN 25, GFN 2	No	No	Pain-free 5/18 Improved 7/18	35 months (3–108 months) Q
Koopmann et al. [22]	67	Open 54 Lap 13	Open	Selective neurectomy	Yes	35 BMR 32 SR	Improved: BMR 29/35 SR 23/32	32 bis 80 months, 56 % phone-Q
Valvekens et al. [23]	12	Open 2 Lap 10	Comb.	Unspecified neurectomy	Yes 6/12	No	Improved 4/12 Unchanged 7/12 Worsened 1/12	32 months (11–118 months) Q

BMR biologic mesh repair, *Sr* suture repair, *Q* questionnaire, *PE* physical examination
[a]Includes noninguinal hernia surgery

Table 31.3. Triple neurectomy studies.

Authors	*N*	Primary surgery	Pain surgery	Neurectomy	Mesh removal	New mesh	Pain relief	Follow-up
Amid and Hiatt [5]	415	Sr 212 Mesh 195 Lap 8	Open	TNE	No	No	Pain-free 85 % Improved 14 %	1 month 100% 6 months>90 % PE and phone-Q (no data)
Amid et al. [6]	16	Open Lap	Open	Extended TNE	Part. Me	Part. Me	Improved 14/16	4–6 weeks, PE (no data)
Campanelli et al. [24]	46	Open	Open ant + post mesh	TNE 44 IHN 2	Yes 40/46	Yes 42/46	Pain-free 40/46 Not improved 6/46	12 months (12–66 months) PE 100 %
Chen et al. [25]	20	Open 10 Lap 10	Lap	Lapar. ERTNE	No 18 Yes 2 (Me)	No	Improved 20/20	22 weeks (16–40 weeks) PE

Sr suture repair, *Q* questionnaire, *PE* physical examination, *QL* quality of life, *TNE* triple neurectomy, *Lapar ERTNE* retroperitoneoscopic triple neurectomy, *Me* meshoma remained 21 studies with 1035 patients. Ninety-three percent of neurectomies were performed after open inguinal hernia repair and 7 % after laparoscopic procedures [transabdominal preperitoneal repair (TAPP); totally extraperitoneal repair (TEP)]. Four studies with 497 patients reported on triple neurectomy [5, 6, 24, 25] and 17 studies with 538 patients on selective neurectomy [2, 8–23]. The overall success rate (patients pain-free or pain improved) of neurectomy is 87 %, 77 % after selective neurectomy, and 98 % after triple neurectomy. The interpretation and comparison of the studies are limited due to different pre- and postoperative pain assessments; different type, duration, and percentage of follow-up; and limited reports on surgical complications. Eight studies did not report on early complications at all [4–6, 9, 13, 19, 23, 24]. Four trials assessed pain-related physical disabilities and restrictions of daily activities [2, 7, 12, 17]. Three trials reported on pain during sexual activity [2, 7, 8].

Only seven studies included a workup of a multidisciplinary pain team [3–6, 10, 17, 25], and 12 publications integrated preoperative peripheral or paravertebral blocks in their study [8, 10, 11, 13, 16–18, 20, 21, 24–26]. Fifty-nine (5.6 %) of the neurectomies were performed retroperitoneoscopically with a success rate of 80 % [17, 25]. Patients do not seem to benefit from a general meshectomy. Table 31.4 [2, 5, 6, 8–25] shows the success rates of neurectomies with or without mesh removal.

The Rationale of Selective Neurectomy

Starling et al. [9, 10] were the first to publish on neurectomy for the treatment of disabling chronic pain after open inguinal hernia repair. After a multidisciplinary approach with conservative pain treatment, as well as local blocks of the inguinal nerves and paravertebral blocks of L1 and L2, ilioinguinal and genitofemoral neuralgia were diagnosed in 19 and 17 patients, respectively. After a selective neurectomy of the entrapped portion of the ilioinguinal nerve (IIN), 17 patients were reported to be completely pain-free. Twelve patients improved after neurectomy of the entrapped genitofemoral nerve (GFN). The authors did not report on mesh removal, and there are no follow-up data in their publications.

The selective neurectomy approach today is commonly used as an alternative to the triple neurectomy concept coined by Amid and colleagues in 2004 [4]. The aim of a selective neurectomy is to resect only painful and damaged nerves and to preserve those nerves that are intact

Table 31.4. Improvement of pain after neurectomy with or without mesh removal [2, 5, 6, 8–25].

	Pain-free or improved (*n*)	**Pain-free or improved (%)**	**Not improved (*n*)**	**Not improved (%)**
Neurectomy without mesh removal	516	95	30	5
Neurectomy with mesh removal	273	77	82	23
Neurectomy + mesh removal + new mesh implantation	154	87	23	13
Neurectomy + mesh removal without new mesh implant	117	66	61	34

and unlikely to cause chronic pain. The concept of selective neurectomy is supported by the fact that nerve resection may lead to neuropathic pain and pain relapse in some cases even after several years [27, 28]. Thus, in this paradigm only definitely damaged nerves should be resected. A nerve can be damaged by suture material, tacks, wadded mesh, or fibrotic nerve encasing scar tissue. The problem consists in diagnosing and differentiating damaged nerves from intact nerves before and during a surgical exploration [4]. Especially after open hernia repair, the spermatic cord and the inguinal nerves are surrounded by scar tissue. During the primary operation or previous operations, inguinal nerves might have been resected, which makes it sometimes very difficult or even impossible to identify the nerves. The difficulty in performing a selective neurectomy is clearly demonstrated in the well-designed recent study by Bischoff et al. [2] that included 54 patients with chronic pain after open mesh repair. The IIN, iliohypogastric nerve (IHN), and GFN were identified in 40 (74 %), 20 (37 %), and 13 (24 %) patients, respectively. Neurectomies of the IIN, IHN, and GFN were performed in 37 (69 %), 19 (35 %), and 10 (19 %) patients, respectively. The data show that only 7 of 73 (10 %) identified nerves were preserved. Despite the fact that all neurectomies in this study were performed by one experienced surgeon, a triple neurectomy would have been possible in at most 10 (19 %) patients. Noteworthy is the fact that 3 years after neurectomy, 3 (12 %) out of 25 patients suffered from an increase of pain compared to their preoperative pain intensity. Moreover, the study demonstrates that even in a country with an excellent structured health-care system, long-term data (36 months) could be obtained from only 25 patients (46 %). The selective neurectomy studies include all single, double, or multiple neurectomies that were not planned as a triple neurectomy before the operation [2, 8–23]. These studies might also include operations where the surgeon planned a triple neurectomy but could not identify all of the three inguinal nerves.

After selective neurectomy, 77 % of the patients reported less pain or were pain-free [2, 8–23]. In six studies the follow-up was 24 months or longer (see Table 31.2) [2, 19–23]. In these studies 60 % of the patients improved. Twelve patients suffered from worse pain after selective neurectomy [2, 8, 23]. The following early postoperative complications have been reported: six wound infections, six hematomas, one wound dehiscence, one seroma, orchiectomy due to impingement in scar tissue, one deep venous thrombosis, and one pulmonary thromboembolism [2, 8–23]. There were five publications of seven cases of ischemic orchitis leading to testicular atrophy [2, 7, 8, 11, 21]. Two studies [2, 7] out of six with mesh removal and without mesh replacement [2, 7, 8, 11, 14, 23] reported six recurrences in 233 patients. In three studies of 88 patients with mesh removal and mesh replacement, there were eight recurrences, seven of these after acellular human dermis repair [16, 18, 22].

The Rationale of Triple Neurectomy

The open triple neurectomy was essentially developed by Amid [3–6], who published on 431 open cases that he performed himself. There is only one triple neurectomy publication from another institution [24].

According to Amid [4], it is extremely difficult to pinpoint the inguinal nerves involved in the pain pathology because:

1. The innervation fields of the three inguinal nerves overlap [9, 29].
2. The IIN, IHN, and genital branch (GB) of the GFN often have peripheral communications, which results in an additional overlap of their sensory innervation [29].
3. At the spinal level the IIN and IHN derive from T12 and L1, and the GFN and IIN receive communication from the first lumbar nerve [9, 29].
4. Frequently there is more than one nerve involved in the postherniorrhaphy neuropathic pain complex syndrome.

Often peripheral nerve blocks and differential paravertebral root blocks cannot precisely discern between damaged and intact nerves. This finding led Amid [3, 4] to develop the concept of triple neurectomy. By the triple neurectomy approach, all three inguinal nerves (IHN, IIN, and GB) that potentially might be involved in the chronic pain pathology are resected.

The concept of triple neurectomy was first published in 2002 [3, 4]. The entire length of the IHN, IIN, and GB of the GFN should be resected as far proximally and distally as possible to be sure that the pain-triggering segment of the nerve is included and intercommunications between the nerves are removed [3, 4]. Later Amid et al. [5] focused their attention on the additional resection of the intramuscular segment of the IHN.

After open or laparoscopic preperitoneal groin hernia mesh repair, the involved segment of the GFN is often located proximal to the internal ring. In these cases the resection of the GB at the level of the inguinal ring is not sufficient. In these cases Amid and Chen [6] recommend the extension of the triple neurectomy to the preperitoneal space and additional resection of the main trunk of the GFN, which can always be identified on the ventral surface of the psoas muscle. According to Amid, the triple neurectomy should be performed without mobilization of the spermatic cord [3, 4]. Only plugs and wrinkled or wadded pieces of mesh (meshoma) should be removed. Recently, Campanelli et al. [24] reported on 40 cases of triple neurectomy with mesh removal and new mesh placement. The open triple neurectomy and extended open triple neurectomy are described in detail in Chap. 24.

Recently, Chen et al. [25] published a series of 20 retroperitoneoscopic triple neurectomies. After a medium follow-up of 22 weeks (16–40 weeks), all patients were pain-free or their pain had improved. According to our recent anatomic study of the retroperitoneal inguinal nerves on 30 fixed cadavers, the retroperitoneoscopic approach allows for reproducible identification of the proximal portion of the IHN and IIN on the surface of the quadratus lumborum muscle and the GFN on the ventral surface of the psoas muscle [30]. The minimally invasive approach allows for reliable proximal nerve identification and triple neurectomy in a territory of virtually untouched tissue.

The results of the open and laparoscopic triple neurectomy are excellent. Ninety-eight percent of the patients are either pain-free, or their pain improved after surgery. There are no reports on pain relapse. However, there are no long-term follow-up data available. Except for one minor wound healing problem after open triple neurectomy and one small lesion of the diaphragm during laparoscopic triple neurectomy (which was intraoperatively fixed), there were neither surgical nor general complications reported, especially no testicular or visceral complications [3].

Summary

Today neurectomy is the last treatment option for patients with disabling persistent postherniorrhaphy pain. Selective or triple neurectomy can be performed open or laparoscopically and give good results with low morbidity. Wrinkled or wadded mesh and plugs should be removed concomitantly. Patients do not seem to benefit from the removal of well-incorporated mesh. According to the available data, triple neurectomy seems to have an edge over selective neurectomy. However, more than 90 % of the published triple neurectomy data derive from a single institution with one dedicated surgeon [3–6, 24, 25]. There are no reports on long-term follow-up after triple neurectomy and scarce long-term follow-up data after selective neurectomy. Fortunately, postneurectomy impairment of pain seems to be a rare complication. Moreover, interpretation and comparison of the studies are limited due to heterogenous demographics, different pre- and postoperative pain assessments, different kinds and duration of follow-up, and limited reports on surgical complications.

There is a strong need for high-quality randomized multicenter trials with uniform pre- and postoperative assessment of pain and pain-related functional and psychologic impairment, as well as long-term follow-up.

References

1. Werner MU. Management of persistent postsurgical inguinal pain. Langenbecks Arch Surg. 2014;399(5):559–69.
2. Bischoff JM, Enghuus C, Werner MU, Kehlet H. Long-term follow-up after mesh removal and selective neurectomy for persistent inguinal postherniorrhaphy pain. Hernia. 2013;17(3):339–45.
3. Amid PK. A 1-stage surgical treatment for postherniorrhaphy neuropathic pain. Arch Surg. 2002;137(1):100–4.
4. Amid PK. Causes, prevention, and surgical treatment of postherniorrhaphy neuropathic inguinodynia: triple neurectomy with proximal end implantation. Hernia. 2004;8(4):343–9.
5. Amid PK, Hiatt JR. New understanding of the causes and surgical treatment of postherniorrhaphy inguinodynia and orchalgia. J Am Coll Surg. 2007;205(2):381–5.
6. Amid PK, Chen DC. Surgical treatment of chronic groin and testicular pain after laparoscopic and open preperitoneal inguinal hernia repair. J Am Coll Surg. 2011;213(4):531–6.
7. Aasvang EK, Kehlet H. The effect of mesh removal and selective neurectomy on persistent postherniotomy pain. Ann Surg. 2009;249(2):327–34.
8. Loos MJ, Scheltiga M, Roumen RM. Tailored neurectomy for treatment of postherniorrhaphy inguinal neuralgia. Surgery. 2010;147(2):275–81.
9. Starling JR, Harms BA, Schroeder ME, Eichman PL. Diagnosis and treatment of genitofemoral and ilioinguinal entrapment neuralgia. Surgery. 1987;102(4):581–6.
10. Starling JR, Harms BA. Diagnosis and treatment of genitofemoral and ilioinguinal neuralgia. World J Surg. 1989;13(5):586–91.
11. Heise CP, Starling JR. Mesh inguinodynia: a new clinical syndrome after inguinal herniorrhaphy? J Am Coll Surg. 1998;187(5):514–8.
12. Lee CH, Dellon AL. Surgical management of groin pain of neural origin. J Am Coll Surg. 2000;191(2):137–42.
13. Deysine M, Deysine GR, Reed Jr WP. Groin pain in the absence of hernia: a new syndrome. Hernia. 2002;6(2):64–7.
14. Madura JA, Madura JA, Copper CM, Worth RM. Inguinal neurectomy for inguinal nerve entrapment: an experience with 100 patients. Am J Surg. 2005;189(3):283–7.

15. Kim D, Murovic J, Tiel R, Kline D. Surgical management of 33 ilioinguinal and iliohypogastric neuralgias at Louisiana State University Health Sciences Center. Neurosurgery. 2005;56(5):1013–20.
16. Keller JE, Stefanidis D, Dolce CJ, Iannitti D, Kercher KW, Heniford BT. Combined open and laparoscopic approach to chronic pain after inguinal hernia repair. Am Surg. 2008;74(8):695–700.
17. Giger U, Wente MN, Buchler MW, Krahenbuhl S, Lerut J, Krahenbuhl L. Endoscopic retroperitoneal neurectomy for chronic pain after groin surgery. Br J Surg. 2009;96(9):1076–81.
18. Vuilleumier H, Hubner M, Demartines N. Neuropathy after herniorrhaphy: indication for surgical for surgical treatment and outcome. World J Surg. 2009;33(4):841–5.
19. Bower S, Moore BB, Weiss SM. Neuralgia after inguinal hernia repair. Am Surg. 1996;62(8):664–7.
20. Ducic I, West J, Maxted W. Management of chronic postoperative groin pain. Ann Plast Surg. 2008;60(3):294–8.
21. Zacest AC, Magill ST, Andersen VC, Burchiel K. Long-term outcome following inguinal neurectomy for chronic pain. J Neurosurg. 2010;112(4):784–9.
22. Koopmann MC, Yamane BH, Starling JR. Long-term follow-up after meshectomy with acellular human dermis repair for postherniorrhaphy inguinodynia. Arch Surg. 2011;146(4):427–31.
23. Valvekens E, Nijs Y, Miserez M. Long-term outcome of surgical treatment of chronic postoperative groin pain: a word of caution. Hernia. 2015;19(4):587–94. 2013 Jun 19 [Epub ahead of print] doi:10.1007/s10029-013-1125-4.
24. Campanelli G, Bertocchi V, Cavalli M, Bombini G, Biondi A, Tentorio T, et al. Surgical treatment of chronic pain after inguinal hernia repair. Hernia. 2013;17(3): 347–53.
25. Chen D, Hiatt JR, Amid PK. Operative management of refractory neuropathic inguinodynia by a laparoscopic retroperitoneal approach. JAMA Surg. 2013;148(10): 962–7.
26. Rosen M, Novitsky Y, Cobb W, Kercher K, Heniford BT. Combined open and laparoscopic approach to chronic pain following open inguinal hernia repair. Hernia. 2006;10(1):20–4.
27. Borsook D, Kussmann BD, George E, Becerra L, Burke DW. Surgically induced neuropathic pain: understanding the perioperative process. Ann Surg. 2013;257(3): 403–12.
28. Schott GD. Delayed onset and resolution of pain: some observations and implications. Brain. 2001;124(Pt 6):1067–76.
29. Moosman DA, Oelrich TM. Prevention of accidental trauma to the ilioinguinal nerve during inguinal herniorrhaphy. Am J Surg. 1977;133(2):146–8.
30. Reinpold W, Schroeder AD, Schroeder M, Berger C, Rohr M, Wehrenberg U. Retroperitoneal anatomy of the iliohypogastric, ilioinguinal, genitofemoral, and lateral femoral cutaneous nerve: consequences for prevention and treatment of chronic inguinodynia. Hernia. 2015;19(4):539–48. doi:10.1007/s10029-015-1396-z. Epub 2015 Jun 17.

32. Chronic Groin Pain: Mesh or No Mesh

Nathaniel F. Stoikes, David Webb, and Guy R. Voeller

Introduction

Inguinal hernia repair techniques have evolved over time. In the days when tissue repairs were more prevalent, recurrence rates were as high as the 10–20 % range. For this reason, the primary outcome of importance was reduction of recurrence. Now that mesh-based repairs of inguinal hernias have reduced recurrence rates, the outcome of postoperative chronic groin pain (CGP) has gained importance. The concern for postoperative CGP has increased in direct correlation with the increased use of synthetic mesh for inguinal hernia repair; thus, many have thought the relationship was a causal one.

However, there are a multitude of risk factors and variables that influence CGP after inguinal hernia repair. The exact role that mesh and its various forms of fixation play in the development of postoperative CGP remains to be determined. In addition, it remains unclear whether the incidence of CGP has actually increased due to the use of mesh for inguinal hernia repairs. It may be that groin pain had been an issue with tissue repairs and it was overlooked due to the main focus on recurrence as an outcome measure.

It is the focus of this chapter to specifically evaluate not only the objective data but also the perceptions surrounding the role that synthetic mesh may play related to inguinal hernia repair and postoperative CGP.

Risk Factors

Regardless of the use of mesh, there are many factors that affect the risk of development of CGP in inguinal hernia repair. The presence of preoperative pain, psychosocial issues, and aberrant nerve anatomy can

B.P. Jacob et al. (eds.), *The SAGES Manual of Groin Pain*,
DOI 10.1007/978-3-319-21587-7_32

all contribute to postoperative CGP development. Furthermore, perioperative factors such as tissue and nerve trauma, seroma, hematoma, and infection can all contribute to chronic pain. Some or all of these factors can be present regardless of a mesh or non-mesh inguinal hernia repair [1]. One must consider all of these risk factors when trying to determine the reason for the development of CGP, but the focus of this chapter is the *role synthetic mesh may or may not play in the development of postoperative CGP*.

Mesh as a Foreign Body

There is no question that all synthetic mesh elicits an inflammatory response. Whether that response is clinically significant is debatable. Animal studies have shown that mesh in contact with nerves does cause inflammatory changes characterized by an increase in fiber diameter and increased nerve demyelination [2]. However, the clinical significance of these findings in animals alone cannot be substantiated.

A translational study, "Mesh-Related SIN Syndrome. A Surreptitious Irreversible Neuralgia and its Morphologic Background in the Etiology of Post-Herniorrhaphy Pain" by Bendavid et al., has recently been published. Given the title alone, one can surmise that the mesh is *perceived* as the sole instigator of chronic pain, though perhaps not in the traditional manner of being a foreign body that causes inflammation. In this study, a scientific model was implemented comparing 10 explants of virgin tissue of the posterior inguinal wall, 10 explants of scar tissue from tissue repairs, and 10 explants from mesh repairs. Mesh was not found to significantly inhibit or promote nerve growth in scar. However, deformation of mesh was found to provide potential compartments for entrapment of nerves and to create more surfaces for random nerve ingrowth into the mesh. These issues can be further potentiated by contraction and migration of mesh, which can occur after it is implanted [3].

Objective scientific findings from implanted mesh provide data to support a convincing case for mesh-related chronic pain. However, based on these findings, one should expect CGP to be an even larger problem than it is currently. In fact, much of the existing clinical data support the contrary. Therefore, despite evidence of the foreign body reaction seen after synthetic mesh implantation, one must understand the contemporary history and clinical data surrounding CGP to gain a full perspective of this complex and multifactorial problem.

Perceptions

Perceptions about mesh use for inguinal hernia repair can vary greatly; trying to understand the thoughts and biases of surgeons, patients, and research data can be challenging. Some believe the use of mesh in and of itself is the cause for the apparent increase in CGP. Others believe there has not been an objective increase in CGP due to mesh, but at the same time they recognize that mesh can play a role in the development of CGP postoperatively.

Fischer recently wrote a commentary on the continued use of mesh for inguinal hernia repair despite the "human toll of inguinodynia" [4]. He comments that "conventional" tissue repairs had sound results, including acceptable recurrence rates of 4–6 % and CGP in 2–4 % of patients. Along the way, mesh repairs became more popular and with it his personal perception of increased incidence of inguinodynia. These complications were superimposed with issues of pending litigation, potential malingering by patients for secondary gain, and "ruined lives." After evaluating the data, including mesh use and nerve management, he concludes that there has been little gained by the use of mesh in inguinal hernia repair due to the risk of chronic debilitating pain and really no improvement of recurrence rates. He contributes the etiology of CGP to the inflammatory response of mesh as it involves the three inguinal nerves (ilioinguinal, iliohypogastric, and genital branch of the genitofemoral nerve). His recommendation is that it would be better to learn to do tissue repairs, similar to the Shouldice repair, so as to not "create" inguinodynia in patients, as it has significant societal and economic implications. He also notes that the U.S. Food and Drug Administration has become increasingly concerned about the issue of CGP and the use of mesh.

The opposite view was expressed by Gilbert, a hernia surgeon specialist and originator of a commonly used mesh prosthesis for inguinal hernia repair. He wrote a response to Fischer's article with a perception that was strikingly different, starting with the issue of inflammation due to a foreign body [5]. He states, "Ordinarily reactions to inert mesh are minimal and short lived." He goes on to interpret the existing data that incriminate mesh to have bias, as they are not the result of randomized controlled trials. His personal experience includes both Shouldice tissue-based repairs and thousands of mesh-based repairs. His perceptions of CGP were that it occurred in his patients with recurrences and not necessarily in those with mesh. He further states that the mesh repair decreases the incidence of recurrence. His feeling is that CGP is due to inadequate knowledge of the groin and is directly related to surgical technique and not necessarily due to the use of mesh. He states the cause of CGP is due to the "absence of careful technique."

There is a difference of opinion between two expert and well-respected surgeons regarding the issue of CGP after inguinal hernia repair. As with most controversial issues, the answer lies somewhere between these two extremes.

Studies Evaluating CGP and Mesh

One of the first papers to describe inguinodynia after mesh repair as a clinical syndrome was from Heise and Starling [6]. They reviewed 117 patients with inguinodynia, 20 of whom had mesh removal, neurectomy (when involved), and tissue repair (modified Bassini or McVay). Sixty percent of patients had favorable results with their technique. Most importantly, the authors did a review of CGP in non-mesh repairs and reported that it was as high as 10.6 % with certain tissue repairs (McVay), and that the etiology most commonly was *entrapment* of nerves. They concluded by noting, "We strongly believe that mesh inguinodynia does occur, [and] will occur more frequently than anticipated now that mesh is used with impunity." This conclusion about mesh in the study is interesting, given the historical data provided about certain tissue repair techniques having a predicted rate of chronic pain of 10.6 %. It also points out the fact that since mesh is now used much more often than autogenous repairs, it is only natural that we speak about CGP in relation to the use of mesh.

Since Starling's work there have been other reviews of CGP and the use of mesh in inguinal hernia repair. Poobalan et al. reviewed CGP and hernia repair in 2001 [7]. They defined chronic pain as pain that persists for greater than 3 months. Forty studies were reviewed and they found that the incidence of chronic pain ranged from 0 to 53 %. Moderate to severe pain was experienced by up to 10 % of patients. Within the review, they found three studies that looked at mesh versus non-mesh repairs and the development of CGP. They found that two of the three studies evaluated reported less CGP with the mesh-based repairs.

Aasvang and Kehlet also reviewed chronic postoperative pain in inguinal hernia in 2004 [8]. In their review, they specifically looked at studies comparing mesh versus non-mesh repairs, and they showed no increase in the incidence of CGP with the use of mesh.

A more recent randomized clinical trial of mesh versus non-mesh methods of inguinal hernia repair was done by van Veen et al. [9]. Three hundred patients were reviewed with follow-up in 153 of the patients out to a median of 129 months. None of the patients in either group had pain as defined by persistent pain or pain interfering with daily activities. Autogenous repairs including Bassini, McVay, and Shouldice techniques were compared to the Lichtenstein repair. At all time points, pain was similar with mesh repair trending toward less pain compared to non-mesh repairs, except at 10 years when neither group had chronic pain. This agrees with most comparative studies that have found the incidence of CGP to be similar between the open mesh repairs versus the autogenous repairs. The Hernia Trialists reviewed 20 trials and over 5000 repairs comparing mesh-based and non-mesh-based repairs for inguinal hernia [10]. The incidence of CGP was equal in both groups. Nordin et al. found similar results when comparing the mesh-based Lichtenstein repair versus the autogenous Shouldice repair [11]. At 3-year follow-up, the incidence of CGP was 4.2 % in the Shouldice repair and 5.6 % in the Lichtenstein group.

The differences between laparoscopic inguinal hernia repair with mesh and the open autogenous Shouldice repair are even more pronounced. Bittner et al. completed a meta-analysis showing CGP in 2.2 % of laparoscopic repairs and 5.4 % of Shouldice repairs [12]. The SMIL study reviewed laparoscopic transabdominal pre-peritoneal (TAPP) repair versus open autogenous Shouldice repair and found CGP to be similar between the two groups (8.5 % TAPP vs. 11.4 % Shouldice) [13]. Koninger et al. looked at 280 patients at 52 months follow-up and compared the incidence of CGP in those having a Shouldice, Lichtenstein, or TAPP repair [14]. CGP was found in 36 % of the Shouldice repairs, 31 % of the Lichtenstein repairs, and 15 % of those with TAPP.

Looking at the collection of studies including autogenous and mesh repairs, it becomes evident that mesh use may not be the sole cause of chronic pain. An objective review of the data actually indicates that the laparoscopic repair, which is a mesh repair, has the lowest incidence of CGP. These findings indicate the complex nature of inguinal hernia repair. Also of importance is the technical detail of each approach.

Mesh Weight

To complicate matters and to reinforce the concept of mesh and the foreign body response, mesh weight has been speculated to contribute to CGP. The general principle is that less synthetic mesh implies less foreign body, less inflammatory response, and therefore less pain.

Numerous studies have evaluated the differences between conventional heavyweight (or normal weight) mesh and lightweight mesh and the development of CGP. Bringman et al. evaluated 600 patients who underwent hernia repair at 3 years [15]. Patients were randomized to have implantation of polypropylene mesh of 80 g/m^2 or a 30 g/m^2. The lightweight mesh group was found to have less pain and less sensation of mesh. The lightweight mesh group was also found to have less "minor" groin problems. Paajanen et al. reviewed 228 patients who were randomized to various lightweight and heavyweight mesh options, and these patients were followed up at 2 years [16]. There was no difference in pain, quality of life, sensation of mesh, or hernia recurrences. Page and O'Dwyer also found no difference in pain scores at one year between patient groups ($N=300$) who underwent repair with either partially absorbable mesh or nonabsorbable mesh [17]. They did find a significantly higher recurrence rate among patients who had repairs using the partially absorbable mesh (5.6 vs. 0.7 %). Currently, there are no strong data to confirm that mesh weight is a contributor to CGP in inguinal hernia repair.

Fixation

Another cause of chronic pain in inguinal hernia repair may be the type of fixation used to secure the mesh. There are a wide variety of options, including sutures (absorbable and permanent), tacks (absorbable and permanent), and adhesives. These various options apply to both laparoscopic and open techniques. Referring to open mesh repair, the TIMELI trial by Campanelli et al. included 319 patients and compared the use of fibrin sealant for fixation versus sutures [18]. At 1 year, there were less disabling complications among patients in the adhesive group, with less pain at 1 month and 6 months. Meta-analysis by Colvin et al. also found a reduction in CGP with adhesive use in open inguinal hernia repair with mesh [19]. Comparisons of suture material in open inguinal hernia have been done as well. Paajanen randomized 162 patients to absorbable (Dexon™, polyglycolic acid) versus permanent (polypropylene) suture fixation with Lichtenstein hernia repair [20]. At 2 years, there was no difference between the two groups. Twenty-four percent described "some" pain in follow-up, but over 90 % of patients were satisfied with their result. In contrast, Jeroukhimov et al. conducted a single-blinded randomized controlled trial comparing Vicryl® (polyglactin 910) and polypropylene fixation with a Lichtenstein approach [21].

There were 100 patients in each arm. Chronic pain rate and time to pain disappearance were higher among patients in the permanent suture group. Similar comparisons have been done comparing adhesive and tacks in laparoscopy. Lovisetto et al. reviewed 197 patients with TAPP repair randomized to fibrin glue or tacks and followed them out to 2 years [22]. Patients who had fixation with fibrin glue had significantly less acute and chronic postoperative pain. Topart et al. evaluated 168 patients undergoing totally extraperitoneal (TEP) technique hernia repair [23]. Chronic pain occurred in 14.7 % of patients who had tacks for mesh fixation versus 4.5 % of patients with fibrin glue.

Basic science studies evaluating different fixation methods and their effects on mesh and CGP are lacking. A recent study by Stoikes et al. compared fibrin glue fixation of lightweight mesh with permanent suture fixation in an animal model [24]. Though sutures were stronger than fibrin glue at 24 h, fibrin glue fixation was found to be adequate at 24 h. At 1 week postoperatively, the fixation strength was equal between the groups. A secondary outcome was evaluation of mesh contraction between the two groups. The contraction rate was consistently greater in the suture group compared to the glue group, although not statistically significant. Possibilities affecting mesh contraction may be that the adhesive group fixates the entire surface of the mesh, thereby preventing folding and wrinkling. This ultimately allows the full area of the mesh to be fixed in granulation tissue. Such a finding links to the previously mentioned study by Bendavid, which showed that disfigured mesh created potential compartments for nerve entrapment, leading to CGP [3].

Within the spectrum of mesh repairs, one can see that fixation choices and careful application of fixation can play a role affecting CGP, independent of the actual type of mesh used. The difference found with fixation alone is an example of the multitude of factors that can affect CGP independent of mesh or mesh type.

Discussion

There is a full spectrum of opinions about the use of synthetic mesh in inguinal hernia repair. There are valid points from both sides of the controversy, but the data show that in reality CGP exists with both tissue repairs and mesh repairs. Their etiologies are likely different. With tissue repairs, CGP may be due to entrapment of nerves by layers of sutures; with mesh, it may be due to nerve entrapment from mesh deformation or a foreign body response causing nerve demyelination [2, 3].

However, the clinical data are not consistent and do not seem to correspond with the objective findings found in the basic science. Overall, autogenous and mesh repairs have been found to have similar outcomes of CGP. Further complicating the landscape are the different outcomes found with different techniques of either autogenous or mesh repairs. It has been suggested that there is more CGP with the McVay repair compared to the Shouldice repair [4]. There are the same issues found with mesh repairs. Whether it is laparoscopic versus open, or fixation with fibrin glue, tacks, or sutures, they have all been evaluated and found to have different outcomes independent of the mesh. In fact, studies have supported that the laparoscopic approach appears to have the best results out of all autogenous and mesh repairs combined. Given this, mesh is clearly not the sole cause of CGP in inguinal hernia repair.

CGP is a multifactorial process that is influenced by the innate complexity of groin anatomy, psychosocial issues, and various technique options requiring different anatomic knowledge for each approach. In spite of the difference of opinion between Fischer and Gilbert, they both indicated in their commentaries that the performance of excellent surgical technique—regardless of actual technique choice—was one of the most important factors in preventing CGP [4, 5]. Therefore, the best approach for an inguinal hernia repair lies in the hands of the surgeon to select a technique in which the surgeon has complete knowledge of all the potential technical pitfalls and is the most comfortable performing.

References

1. Kehlet H, Aasvang E. Chronic pain after inguinal hernia repair. In: Schumpelick V, Fitzgibbons RJ, editors. Hernia repair sequelae. Berlin: Springer; 2010. p. 163–7.
2. Demirer S, Kepenekci I, Evirgen O, Birsen O, Tuzuner A, Karahuseyinoglu S, et al. The effect of polypropylene mesh on ilioinguinal nerve in open mesh repair of groin hernia. J Res Surg. 2006;131(2):175–81.
3. Bendavid R, Lou W, Koch A, Iakovlev V. Mesh-related SIN syndrome. A surreptitious irreversible neuralgia and its morphologic background in the etiology of post-herniorrhaphy pain. Int J Clin Med. 2014;5:799–810.
4. Fischer JE. Hernia repair: why do we continue to perform mesh repair in the face of the human toll of inguinodynia. Am J Surg. 2013;206(4):619–23.
5. Gilbert AI. Hernia repair: do you know your own results? Am J Surg. 2013;207(6):1002–3.
6. Heise CP, Starling JR. Mesh inguinodynia: a new clinical syndrome after inguinal herniorrhaphy? J Am Coll Surg. 1998;187(5):514–8.
7. Poobalan AS, Bruce J, Smith WC, King PM, Krukowski ZH, Chambers WA. A review of chronic pain after inguinal herniorrhaphy. Clin J Pain. 2003;19(1):48–54.

8. Aasvang E, Kehlet H. Chronic postoperative pain: the case of inguinal herniorrhaphy. Br J Anaesth. 2004;95(1):69–76.
9. van Veen RM, Wijsmuller AR, Vrijland WW, Hop WC, Lange JF, Jeekel J. Randomized clinical trial of mesh versus non-mesh primary inguinal hernia repair: long term chronic pain at 10 years. Surgery. 2007;142(5):695–8.
10. EU Hernia Trialists Collaboration. Mesh compared with non-mesh methods of open groin hernia repair: systematic review of randomized controlled trials. Br J Surg. 2000;87(7):854–9.
11. Nordin P, Bartelmess P, Jansson C, Svensson C, Edlund G. Randomized trial of Lichtenstein versus Shouldice hernia repair in general surgical practice. Br J Surg. 2002;89(1):45–9.
12. Bittner R, Sauerland S, Schmedt CG. Comparison of endoscopic techniques versus Shouldice and other open nonmesh techniques for inguinal hernia repair: a meta-analysis of randomized controlled trials. Surg Endosc. 2005;19(5):605–15.
13. Eklund A, Montgomery A, Bergkvist L, Rudberg C, Swedish Multicentre Trial of Inguinal Hernia Repair by Laparoscopy (SMIL) Study Group. Chronic pain 5 years after randomized comparison of laparoscopic and Lichtenstein inguinal hernia repair. Br J Surg. 2010;97(4):600–8.
14. Koninger J, Redecke J, Butters M. Chronic pain after hernia repair: a randomized trial comparing Shouldice, Lichtenstein and TAPP. Langenbecks Arch Surg. 2004;389(5):361–5.
15. Bringman S, Ramel S, Heikkinen TJ, Englund T, Westman B, Anderberg B. Tension-free inguinal hernia repair: TEP versus mesh-plug versus Lichtenstein: a prospective randomized controlled trial. Ann Surg. 2003;237(1):142–7.
16. Paajanen H. A single-surgeon randomized trial comparing three composite meshes on chronic pain after Lichtenstein hernia repair in local anesthesia. Hernia. 2007;11(4):335–9.
17. Page B, O'Dwyer P. Does the choice of prosthetic mesh type make a difference in postherniorrhaphy pain? In: Schumpelick V, Fitzgibbons RJ, editors. Hernia repair sequelae. Berlin: Springer; 2010. p. 279–86.
18. Campanelli G, Pascual MH, Hoeferlin A, Rosenberg J, Champault G, Kingsnorth A, Miserez M. Randomized, controlled, blinded trial of Tisseel/Tissucol for mesh fixation in patients undergoing Lichtenstein technique for primary inguinal hernia repair: results of the TIMELI trial. Ann Surg. 2012;255(4):650–7.
19. Colvin HS, Rao A, Cavali M, Campanelli G, Amin AI. Glue versus suture fixation of mesh during open repair of inguinal hernias: a systematic review and meta-analysis. World J Surg. 2013;37(10):2282–92.
20. Paajanen H. Do absorbable mesh sutures cause less chronic pain than nonabsorbable sutures after Lichtenstein inguinal herniorraphy? Hernia. 2002;6(1):26–8.
21. Jeroukhimov I, Wiser I, Karasic E, Nesterenko V, Poluksht N, Lavy R, Halevy A. Reduced postoperative chronic pain after tension-free inguinal hernia repair using absorbable sutures: a single blind randomized clinical trial. J Am Coll Surg. 2014;218(1):102–7.
22. Lovisetto F, Zonta S, Rota E, Mazilli M, Bardone M, Bottero L, et al. Use of human fibrin glue (Tissucol) versus staples for mesh fixation in laparoscopic transabdominal

preperitoneal hernioplasty: a prospective, randomized study. Ann Surg. 2007;245(2): 222–31.
23. Topart P, Vandenbroucke F, Lozac'h P. Tisseel versus tack staples as mesh fixation in totally extraperitoneal laparoscopic repair of groin hernias: a retrospective analysis. Surg Endosc. 2005;19(5):724–7.
24. Stoikes N, Sharpe J, Tasneem H, Roan E, Paulus E, Powell B, et al. Biomechanical evaluation of fixation properties of fibrin glue for ventral incisional hernia repair. Hernia. 2015;19(1):161–6.

Part IV
Case Reports and Patients' Perspectives

33. Foreign Body Reaction, Fibromyalgia, and Autoimmune Disorders

Shirin Towfigh

Chief Complaint

There is chronic pain after inguinal hernia repair.

History

The patient is a 30-year-old female, BMI 22 kg/m^2, status post-routine laparoscopic bilateral inguinal hernia repair with mesh. Preoperatively, she had presented to her medical doctor with bilateral lower pelvic pain and was diagnosed with presumed inguinal hernias. She was evaluated by a general surgeon and found to have tenderness along her groin areas bilaterally. Her pain was constant, worse with straining during bowel movements, and worse with her menses. She was offered exploratory laparoscopy and inguinal hernia repair. Laparoscopy was normal and TEP inguinal hernia repair was performed with polypropylene mesh.

Postoperatively, the patient's health worsened to the point of debilitation. Her preoperative pain did not resolve. In addition, she progressively worsened in health. She developed chronic pelvic pain; cramping of the lower abdomen and pelvis; urinary frequency; pain with full bladder; bloating; nausea; inability to tolerate normal meals; weight loss; hair loss; subjective feeling of "hotness" without fevers, especially at lower abdomen; thigh numbness and tingling; and feeling of swelling of the upper thighs. She also had vaginal burning and pain. Her menses

B.P. Jacob et al. (eds.), *The SAGES Manual of Groin Pain*,
DOI 10.1007/978-3-319-21587-7_33

were irregular and sometimes missed. Sexual intercourse was painful. She was chronically fatigued. She was unable to work and uses a wheelchair, as she cannot tolerate walking long distances. She is found lying in bed most of the day.

Physical Exam

The patient was found in fetal position, shivering, unable to be examined comprehensively due to severe pain. Temperature and other vital signs were normal. Abdomen was mildly distended but soft. Incisions were well healed. She had 4+ tenderness to light touch along bilateral lower groin and upper thigh areas, without specific dermatomal distribution and no skin changes.

Nonoperative Management Options

The patient was first admitted to the hospital for pain control. This included epidural catheter placement, which helped relieve many of her symptoms. This allowed us to perform imaging and gynecologic and GI evaluation to help determine the plan of care. She had already had intolerances to many medications as well as certain tapes as noted from her prior operation. During this hospitalization, she also showed intolerances (nausea, rashes, swelling) from more medications, tapes, and even IV needles. She was evaluated for autoimmune disorders as well as endocrine abnormalities and complement and nutritional deficiencies, all of which were normal.

Imaging

Abdominal x-ray showed normal pelvis with normal bowel gas pattern. Also, there was a normal number and placement of spiral titanium tacks (i.e., 3–4 on each side, and none placed laterally). Magnetic resonance imaging (MRI) of the anterior pelvis, non-contrast, with valsalva and dynamic views demonstrated intact flat extraperitoneal mesh with no hernia recurrence and no evidence of mesh-related fluid collection or inflammation (Fig. 33.1). Pelvic ultrasound was concerning for adhesive disease and endometriosis.

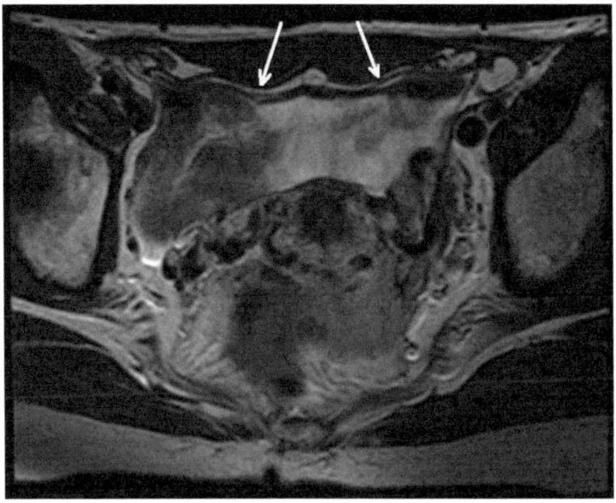

Fig. 33.1. MRI pelvis T2 axial image. Bilateral inguinal mesh found to be in appropriate position (*white arrows*).

Operative Treatment

Based on extensive evaluation by gynecology and general surgery, the patient consented to undergo laparoscopy for diagnoses of endometriosis and mesh-related chronic pain reaction. Laparoscopy demonstrated severe endometriosis, and she underwent extensive adhesiolysis and painstaking endometriosis excision, which involved her rectum, uterus, adnexa, and pelvic side walls. The mesh was confirmed to be flat and in appropriate position. No attempt was made at mesh removal.

Her postoperative recovery was difficult; she required a lot of assistance from the pain management specialists to develop a combination therapy of opioids, neuromodulating medications, muscle relaxants, and antidepressants to help control her pain. She was also maintained on hormonal therapy for her endometriosis.

Upon follow-up, much of her chronic symptoms remained. She continued to have chronic pelvic pain, fatigue, lower abdominal bloating, pain with full bladder, swelling and tingling of the upper thighs, feeling of "hotness," and weakness of the extremities. She was losing her hair. She could not maintain her weight. She remained in bed most of the day and could not function to perform her normal daily activities. She had weaned herself off of most of her medications, as they were ineffective in addressing her symptoms or she developed intolerances to them, such as nausea, vomiting, and dizziness. Evaluation by gynecology and further imaging demonstrated no suggestion of recurrence of her endometriosis.

Due to the direct relationship between the hernia repair with mesh and her debilitated state at such a young age, she was offered laparoscopic mesh removal. She understood that this might not cure her of her problem and that indeed there was no concrete diagnosis. Also, she understood the risks of the procedure, which included the risk of nerve injury and vessel injury at the time of mesh removal. She underwent uneventful laparoscopic mesh removal bilaterally. No hernias were noted after mesh removal.

Postoperative Course and Outcomes

In anticipation of a difficult postoperative course, she had an epidural placed preoperatively. This allowed for smooth recovery postoperatively. She had shown sensitivity to many different pain medications and was able to tolerate pain control with ice and acetaminophen. Pathology of the mesh demonstrated dense fibrosis and chronic inflammation with foreign body giant cell reaction. Over a span of 1 year, she was able to recuperate toward a more normal life. Repeat MRI confirmed no hernia recurrence. She is now eating and gaining weight. Her hair loss has stopped. She is regaining her conditioning with physical therapy.

Discussion

It is unpredictable which patients may develop a mesh reaction. A true mesh allergy is notable as an erythematous blotch on the skin, usually demarcating the exact dimensions of the mesh itself. There may be associated edema or systemic reaction such as fever. Such a mesh allergy is rare and few surgeons have witnessed it.

However, there does seem to be another reaction to mesh, specifically to synthetic mesh, which is a foreign body reaction. To date, there is no literature to support such a clinical problem and its presentation; however, we know that histologically this reaction does occur [1]. Also, clinically, it has been shown very nicely that positron emission tomography (PET) scan may be positive in patients with mesh implantation, demonstrating the inflammatory response to synthetic mesh [2].

In my experience, patients with a mesh foreign body reaction present in very nonspecific constitutional manner; however, many share similar complaints. These include pain in the general area of the mesh implantation, bloating, edema, and feeling that their waist or pants size has increased. It is not uncommon to have other gastrointestinal symptoms, most commonly nausea and sometimes pain with defecation. Some have intermittent diarrhea or constipation. Due to the predominance of the bloating, many are diagnosed with irritable bowel syndrome (IBS). Most get minimal to no response to IBS medications or changes in diet (e.g., gluten-free). Some have urinary symptoms, especially pain with full bladder and urinary frequency and urgency. Many are diagnosed with interstitial cystitis (IC) as a result. The feeling of temperature changes, especially feeling hot in the area, is also a common complaint. Body temperature is rarely over 99° Fahrenheit. Also, numbness, tingling, and swelling of the lower body, mons/base of the penis, and upper thigh are commonly seen symptoms. These symptoms may be in a general pattern of the ilioinguinal or genital nerves, thus leading some to provide nerve blocks, sympathetic blocks, or even neurectomies. In most situations, there is no true neuralgia. Also, most fail treatment with anti-inflammatories, steroids, neurological modulating agents (e.g., gabapentin, pregabalin), and antidepressants.

Most patients with mesh-related foreign body reactions are female. They are also often young and have demonstrated sensitivities and intolerances to a wide range of environmental allergens, medications, and even tapes, the plastic of IV needles, and certain plastics and sutures.

Some patients are already diagnosed with fibromyalgia, which is recognized as a diagnosable disorder by both the National Institutes of Health and the American College of Rheumatology. Unfortunately, not much is known about the cause of fibromyalgia, and there are likely multiple subtypes that are yet to be determined. That said, patients with fibromyalgia tend to be hypersensitized to pain and live in a heightened inflammatory state. Thus, the use of inflammatory agents, such as synthetic implants for hernia repair, can spiral the disease in these patients out of control. Interestingly, as a manifestation of their disease, many of these patients also have bowel and bladder derangements, thus the prevalent diagnoses of IBD and IC among them. Endometriosis and chronic pelvic pain are also notably higher in this patient population. There also seems to be an overlap of this disease with autoimmune disorders that affect inflammation at the tissue level, such as systemic lupus erythematosus and rheumatoid arthritis [3, 4].

Conclusion

We have yet to prove a direct cause and effect of synthetic mesh and chronic debilitating pain. However, some of us—those with surgical practices that treat high volumes of patients with chronic pain after mesh implantation—have noticed a commonality in a subset of these patients. The details of their clinical presentation are noted above. Patients with such a clinical presentation should be evaluated for a possible adverse mesh foreign body reaction, and mesh removal should be a consideration in their treatment plan. Many of these patients are already diagnosed with fibromyalgia or have an autoimmune disorder. Using the same logic, I strongly recommend that patients who present with a known diagnosis of fibromyalgia or autoimmune disorder should not undergo implantation of any inflammatory agent, such as a synthetic mesh, for their hernia repair. If a tissue repair cannot be performed, then a biologic allograft may be considered. Perhaps in the future we will have an understanding of the disease of fibromyalgia and a better understanding of the body's reaction to mesh. And perhaps we will have an objective diagnostic test for either fibromyalgia or mesh reaction prior to implantation of such a foreign body.

References

1. Klinge U, Klosterhalfen B, Müller M, Schumpelick V. Foreign body reaction to meshes used for the repair of abdominal wall hernias. Eur J Surg. 1999;165(7):665–73.
2. Aide N, Deux J-F, Peretti I, Mabille L, Mandet J, Callard P, et al. Persistent foreign body reaction around inguinal mesh prostheses: a potential pitfall of FDG-PET. Am J Roentgenol. 2005;184(4):1172–7.
3. Hawkins RA. Fibromyalgia: a clinical update. J Am Osteopath Assoc. 2013;113(9):l680–9.
4. Yunus MB. Role of central sensitization in symptoms beyond muscle pain, and the evaluation of a patient with widespread pain. Best Pract Res Clin Rheumatol. 2007;21(3):481–97.

34. Patient with Groin Pain After an Athletic Event

Kent W. Kercher

Chief Complaint

Left groin pain.

History

A healthy 20-year-old college scholarship football player presented with a 6-month history of left groin pain. He was referred for possible sports hernia related to pain that developed relatively acutely during summer workouts prior to his most recent season. The pain was initially localized to the pubic tubercle on the left and had improved somewhat over the past 6 months with rest, physical therapy, and alterations in his exercise routine, although he continued to have a deep, gnawing pain localized to the left groin. There was no radiation of pain to the testicle or into the left anterior thigh. He was able to play football during the past season, but had significant ongoing discomfort with sit-ups, lunges, and running, particularly when it involved quick changes in direction. Pain was relieved with rest and oral anti-inflammatory medications. He had been evaluated by an orthopedic surgeon who diagnosed athletic pubalgia and recommended general surgical consultation. The patient wanted to begin spring practice in the upcoming months and preferred to pursue surgical intervention as soon as possible in order to facilitate return to competitive athletics.

B.P. Jacob et al. (eds.), *The SAGES Manual of Groin Pain*,
DOI 10.1007/978-3-319-21587-7_34

Physical Examination

Well-developed male. 5′10″ 246 lbs BMI: 35.

Abdomen: soft, non-tender. No masses.

Focused inguinal examination: Moderate tenderness to palpation at the pubic symphysis, extending laterally along the pubic tubercles to both sides of midline. Focal tenderness to palpation over the external rings and inguinal canals. No palpable hernia defect on either side. Increased discomfort with resisted sit-up. Internal and external hip rotation negative for pain. Mild pain with adduction of the hips against resistance.

Imaging

Magnetic resonance imaging (MRI) pelvis: Bilateral rectus abdominis and adductor longus aponeurosis pubic osteotendinous junction avulsion injuries (Figs. 34.1, 34.2, 34.3, and 34.4). MRI findings of a "secondary cleft" are visible on fluid-sensitive sequences as a curvilinear fluid-signal interface that is continuous with the symphysis pubis and undermines the inserting structures at the pubis.

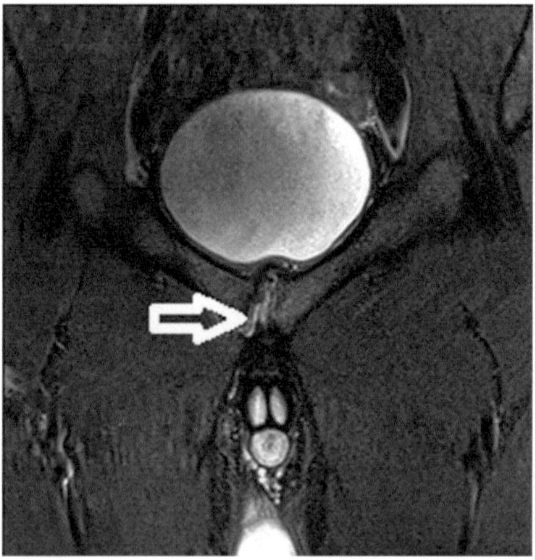

Fig. 34.1. Right adductor tear with secondary cleft sign (fluid in pubic symphysis).

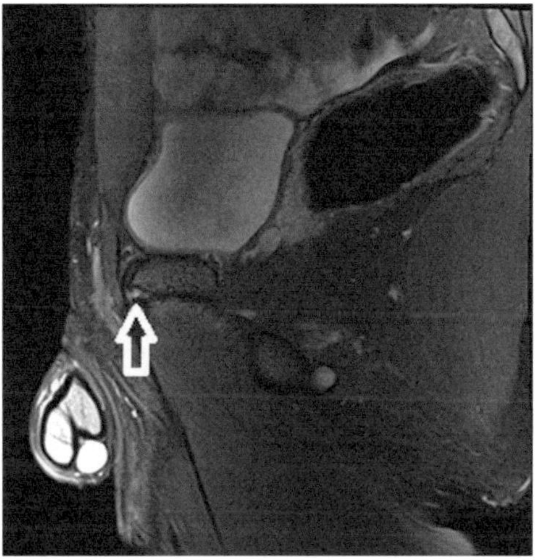

Fig. 34.2. Right rectus avulsion injury with secondary cleft (sagittal view).

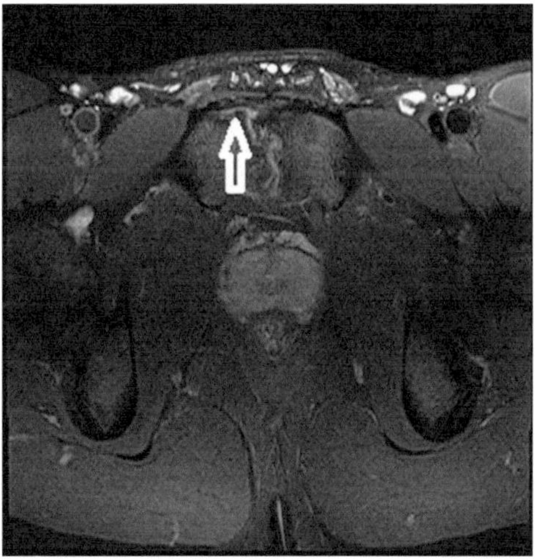

Fig. 34.3. Right rectus avulsion with secondary cleft (axial view).

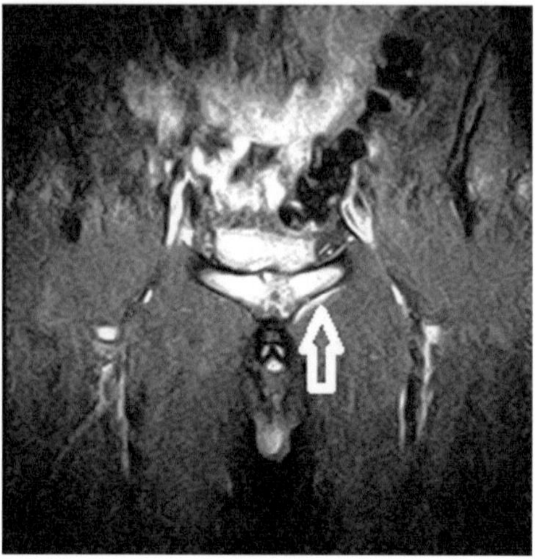

Fig. 34.4. Left adductor tear with adductor edema and secondary cleft.

Diagnosis

Bilateral athletic pubalgia with MRI demonstrating bilateral rectus abdominis and adductor avulsions from their insertions on the pubis.

Nonoperative Management Options

Conservative treatment is the first-line therapy for musculoskeletal strains of the groin. Nonsurgical strategies include anti-inflammatory medications, deep massage, heat or ice, and prolonged rest followed by gradual return to activity. Physical therapy may be effective and should focus on core strengthening to allow for resolution of hip and pelvic muscular imbalance. In patients with radiographic and clinical evidence of osteitis pubis and/or adductor tendinopathy, fluoroscopically guided injection of the symphysis pubis and adductor origin with local anesthetic and/or steroids may be effective.

Operative intervention is generally indicated for chronic pain of greater than 2–3 months duration that is refractory to conservative management, including prolonged rest, physical therapy, nonsteroidal anti-inflammatory drugs (NSAIDS), nerve blocks, and/or steroid injections.

For this patient, nonoperative treatment strategies had been employed for 6 months without significant improvement in symptoms. Further attempts at rehabilitation were felt unlikely to provide relief. Surgical intervention was offered.

Operative Treatment

Laparoscopic Bilateral Transabdominal Pre-peritoneal (TAPP) Repair The technique utilized for sports hernia repair is identical to the standard repair of inguinal hernias and is well described in the literature. Briefly, a three-port technique is used, with a Hasson cannula at the umbilicus and one 5-mm port lateral to the rectus on either side of the umbilicus. The peritoneum is sharply opened at the medial umbilical ligament in a curvilinear fashion extending laterally. The pre-peritoneal space of Retzius is entered medially and the bladder bluntly dissected away from the pubis and Cooper's ligaments. The inferior peritoneal flap is retracted and the cord structures are dissected away from the peritoneum. Any direct or indirect inguinal hernia defects are reduced. The posterior aspect of the rectus insertion is inspected to confirm evidence of attenuation or avulsion injuries of the rectus insertion onto the pubis.

After development of a wide pre-peritoneal pocket bilaterally, a large polypropylene mesh (minimum 12×15 cm) is used to reinforce the entire myopectineal orifice on each side. Bilateral mesh prosthetics are confirmed to overlap in the midline in order to provide for complete reinforcement of the entire myopectineal orifice, Cooper's ligaments, and the pubic tubercle (Figs. 34.5 and 34.6). The mesh is secured with several tacks or staples to Cooper's ligament and then further secured circumferentially with fibrin glue. No tacks or staples are placed into the abdominal wall musculature and no mechanical fixation is utilized below the iliopubic tract. The peritoneum is re-approximated. The same procedure is performed for the contralateral groin in order to allow for wide coverage of all potential inguinal defects on both sides.

Postoperative Course

The patient was discharged to home following surgery and seen in follow-up at 2 weeks and at 6 weeks postoperatively. His postoperative course was uncomplicated and he returned to activity following our

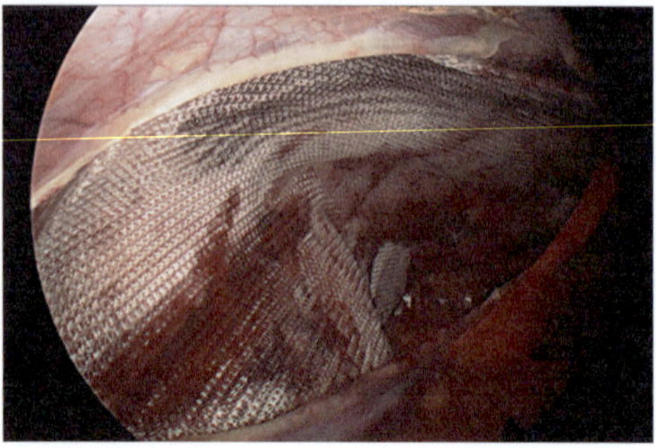

Fig. 34.5. After development of a wide pre-peritoneal pocket, a large polypropylene mesh (minimum 12×15 cm) is used to reinforce the entire myopectineal orifice (*left side*).

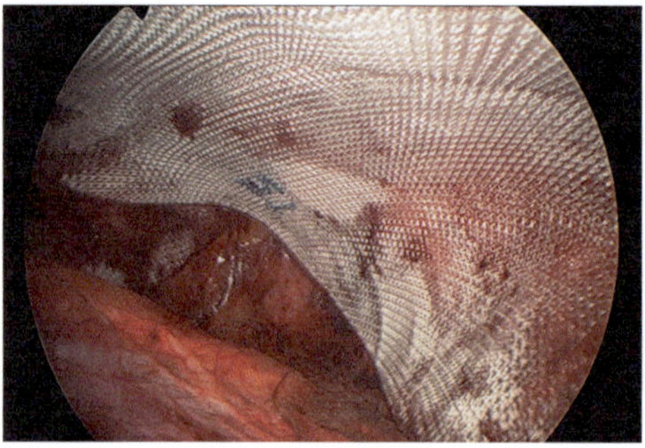

Fig. 34.6. Polypropylene mesh reinforcement of the myopectineal orifice (*right side*). Note mesh overlap of pubic tubercle in the midline.

sports hernia physical therapy protocol (Table 34.1). He began aerobic activity 2 weeks following surgery and gradually progressed to range of motion exercises and resumption of dynamic core training over the following month. He returned to light sporting activity at 6 weeks and was allowed to resume full activity including football 2 months postoperatively. He was able to compete at the collegiate level during the fall football season several months later. He had a successful senior season as a tight end and on special teams as a long snapper.

Table 34.1. Sports hernia repair postoperative protocol.

Phase I: Immediate postoperative phase (weeks 0–2)
Pain and edema control, gentle stretching, walking
Phase II: Intermediate postoperative phase (weeks 2–3)
Gentle strengthening/pelvic stabilization, light exercise (pool, bike)
Phase III: (Weeks 3–4)
Range of Motion, strengthening, dynamic core training, straight plane jogging
Phase IV: (Weeks 4–5)
Light sport specific activity, plyometrics, interval bike training
Phase V: (Weeks 5–6)
Return to full sport activity
General recommendations
Avoid trunk hyperextension for first 2 weeks
Avoid hip extension past 0° for first 2 weeks
Avoid crunch and lifting activities for first 6 weeks
Pain and edema control, ice 3–4 times per day as needed for first week, as needed thereafter
Return to work and sport to be determined on an individual basis by the physician and physical therapist

Outcomes and Discussion

Chronic groin pain in the athlete can be disabling and in some cases career-ending. The exact cause is debated but is theorized to be due to repetitive loading of the pubic symphysis, leading to symphyseal degeneration and loss of mechanical stability. Shearing forces across the pubic symphysis are more prominent in athletes with an imbalance between the strong adductor muscles of the thigh and the relatively weaker lower abdominal wall musculature. These factors are believed to cause weakness and attenuation of the transversalis fascia portion of the posterior wall of the inguinal canal. Weakness in the inguinal floor can lead to localized bulging and compression of the genital branch of the genitofemoral nerve, which is believed to be a source of pain in these patients. Other proposed pathophysiologic mechanisms of injury remain an area of debate and include disruption of the conjoined tendon as well as tears of the rectus abdominis and adductor longus aponeurosis at the point of insertion on the pubis [1].

A number of surgical approaches have been utilized to address the above anatomic abnormalities. All focus on repair or reinforcement of the posterior wall of the inguinal canal by employing the traditional repairs used to treat "true" inguinal hernias. These include open tension repairs (Bassini and Shouldice), anterior tension-free repairs (Lichtenstein), and laparoscopic repairs trans-abdominal preperitoneal repair (TAPP) and total extraperitoneal (TEP)] with mesh [2–6]. Specific "sports hernia" repairs have also been applied and include the open "pelvic floor repair" described by Meyers and the "minimal repair" technique popularized by Muschaweck [7, 8]. The open anterior repairs selectively include decompression and/or division of the genitofemoral nerve with or without an adductor release.

While results vary based upon technique and patient selection, most series report that >90 % of athletes return to full activity within 2–4 months of surgery. In the only randomized control study to date, nonoperative management consisting of physical therapy and NSAIDs was compared with open repair and neurolysis in 66 soccer players with chronic groin pain of more than 3 months duration. In this study, only the surgically treated group demonstrated substantial, statistically significant improvement in symptoms during a 6-month follow-up interval [9].

When chronic groin pain is related to an isolated tear at the adductor longus origin, conservative treatment with rest, NAIDS, physical therapy, and local injection is usually effective, with few patients requiring surgical intervention in the form an adductor tenotomy. In a prospective randomized trial by Holmich and colleagues, incorporation of an active training and core-strengthening program was found to be superior to physical therapy alone in athletes with adductor-related groin pain [10]. For those who fail conservative management of osteitis pubis and/or adductor tendinopathy, periodic injections of the pubic cleft and adductor origin with dextrose and lidocaine have been effective for resolution of chronic pain [11, 12]. In a small subset of athletes who fail these nonoperative measures, adductor tenotomy (surgical release of the adductor longus tendon at its origin from the pubis) has provided encouraging long-term symptomatic and functional results [13].

In our patient, a laparoscopic bilateral TAPP repair with polypropylene mesh was utilized. His recovery was uncomplicated, and he was able to return to full activity within 2 months of surgery. Given the presence of bilateral radiographic findings and clinical symptoms, the laparoscopic approach was felt to offer the ideal solution for mesh reinforcement of both groins. The primary disadvantage of the laparoscopic technique is that it does not afford the opportunity for neurolysis of the sensory nerves, which some authors believe is important in achieving pain relief.

A number of prospective non-randomized series have evaluated the efficacy of laparoscopic TAPP and TEP repairs for athletic pubalgia. Most have included small numbers of patients ($n = 14–131$) and have utilized mesh reinforcement of the myopectineal orifice. Over an average follow-up of 12 months (range 3–48 months), 87–100 % of athletes were able to return to full activity within 3 months of surgery, with many patients resuming full competitive athletics within 3–4 weeks [4, 5, 16, 17]. One potential advantage of the laparoscopic repair is the ability to treat patients with bilateral injuries simultaneously through a three port approach.

In our patient, preoperative imaging demonstrated evidence of bilateral rectus abdominis and adductor longus avulsion injuries, while clinical symptoms and examination findings were localized primarily to the inguinal canals and pubic tubercle, consistent with musculotendinous disruptions of the posterior inguinal wall, transversalis fascia, and the rectus insertions onto the pubis. In our experience, these patients are best served by initial reinforcement of the groin with mesh, as the large majority will experience symptom resolution without any further intervention. For those patients in whom adductor symptoms predominate, adductor release can be performed either as an isolated procedure or can be combined with TAPP, TEP, or open repair using mesh. Since many surgical options exist, patient selection is critical and the approach must be tailored to the specific diagnosis, based upon preoperative clinical and radiographic localization of the anatomic site of injury. As shown in Table 34.2, there is no single procedure that can be universally applied to all patients with athletic pubalgia; the specific intervention must be applied to the pathology being treated. Results have been excellent in general, with the large majority of athletes returning to full sporting activity following surgical intervention [4, 12–17]. Due to small numbers of patients being treated in most series, however, these techniques cannot be compared from a statistical standpoint, nor has there been a controlled trial to determine whether one surgical technique is superior.

Conclusion

Athletic pubalgia (frequently referred to as "sports hernia") is a common entity among athletes, though the specific anatomic and physiologic mechanisms for chronic groin pain remain poorly understood. As a result, no one surgical solution can be applied to all patients. In athletes

Table 34.2. Surgical technique and outcomes.

Author	Number	Return to activity (%)	Surgical technique
Hackney [14]	15	87	Open inguinal floor repair
Canonico [15]	16	100	Lichtenstein
Ingoldby [4]	28	96	50 % Open mesh 50 % Laparoscopic mesh
Paajanen [16]	41	95	TEP
Genitsaris [17]	131	97	TAPP
Topol [12]	24	92	Pubic symphysis/adductor injection
Akermark [13]	16	62	Adductor tenotomy

who present with insidious, deep groin pain in the absence of an inguinal hernia, a sports hernia should be considered. Initial treatment is conservative and should involve a multidisciplinary approach (orthopedist, sports medicine, athletic trainer, and/or physical therapist). Imaging (MRI) and surgical referral may be indicated after a failure of conservative management. While small series of laparoscopic and open repair have provided encouraging results, a multicenter prospective randomized controlled trial is needed.

References

1. Swan Jr KG, Wolcott M. The athletic hernia: a systematic review. Clin Ortho Relat Res. 2007;455:78–87.
2. Polglase AL, Frydman GM, Farmer KC. Inguinal surgery for debilitating chronic groin pain in athletes. Med J Aust. 1991;155(10):674–7.
3. Gilmore OJ. Gilmore's groin: ten years experience of groin disruption—a previously unsolved problem in sportsmen. Sports Med Soft Tissue Trauma. 1991;3:12–4.
4. Ingoldby CJ. Laparoscopic and conventional repair of groin disruption in sportsmen. Br J Surg. 1997;84(2):213–5.
5. van Veen RN, de Baat P, Heijboer MP, Kazemier G, Punt BJ, Dwarkasing RS, et al. Successful endoscopic treatment of chronic groin pain in athletes. Surg Endosc. 2007;21(2):189–93.
6. Economopoulos KJ, Milewski MD, Hanks JB, Hart JM, Diduch DR. Sports hernia treatment: modified Bassini versus minimal repair. Sports Health. 2013;5(5):463–9.
7. Meyers WC, Foley DP, Garrett WE, Lohnes JH, Mandelbaum BR. Management of severe lower abdominal or inguinal pain in high-performance athletes. PAIN (Performing Athletes with Abdominal or Inguinal Neuromuscular Pain Study Group. Am J Sports Med. 2000;28(1):2–8.

8. Muschaweck U, Berger L. Minimal repair technique of sportsmen's groin: an innovative open-suture repair to treat chronic inguinal pain. Hernia. 2010;14(1):27–33.
9. Ekstrand J, Ringborg S. Surgery versus conservative treatment in soccer players with chronic groin pain: a prospective randomized study in soccer players. Eur J Sports Traumatol Rel Res. 2001;23:141–5.
10. Hölmich P, Uhrskou P, Ulnits L, Kanstrup IL, Nielsen MB, Bjerg AM, Krogsgaard K. Effectiveness of active physical training as treatment for long-standing adductor-related groin pain in athletes: randomised trial. Lancet. 1999;353(9151):439–43.
11. Schilders E, Bismil Q, Robinson P, O'Connor PJ, Gibbon WW, Talbot JC. Adductor-related groin pain in competitive athletes. Role of adductor enthesis, magnetic resonance imaging, and entheseal pubic cleft injections. J Bone Joint Surg. 2007;89(10): 2173–8.
12. Topol GA, Reeves KD, Hassanein KW. Efficacy of dextrose prolotherapy in elite male kicking-sport athletes with chronic groin pain. Arch Phys Med Rehabil. 2005; 86(4):697–702.
13. Akermark C, Johansson C. Tenotomy of the adductor longus tendon in the treatment of chronic groin pain in athletes. Am J Sports Med. 1992;20(6):640–3.
14. Hackney RG. The sports hernia: a cause of chronic groin pain. Br J Sports Med. 1993;27(1):58–62.
15. Canonico S, Benevento R, Della Corte A, Fattopace A, Canonico R. Sutureless tension-free hernia repair with fibrin glue (tissucol) in soccer players with chronic inguinal pain: initial experience. Int J Sports Med. 2007;28(10):873–6.
16. Paajanen H, Syvahuoko I, Airo I. Totally extraperitoneal endoscopic (TEP) treatment of sportsman's hernia. Surg Laparosc Endosc Percutan Tech. 2004;14(4):215–8.
17. Genitsaris M, Goulimaris I, Sikas N. Laparoscopic repair of groin pain in athletes. Am J Sports Med. 2004;32(5):1238–42.

35. Chronic Post-inguinal Herniorrhaphy Pain: A Patient's Perspective

David C. Chen and Brian P. Jacob

Introduction

Patient ZP is an acclaimed writer and columnist. We have asked her to share her story to illustrate the experience, frustrations, hopes, and fears from a patient's perspective. Her story also demonstrates the importance of referral patterns and how physicians themselves should pay more attention to the skill set of their trusted referrals. These patients navigate a sea of doctors, studies, interventions, and surgeries often without enough guidance or expertise. Their difficulties finding adequate care turn lives upside down, and while we can usually ultimately make patients better, the road is long, and some lives never return to normal. It is a reminder for all of us to try always to do better.

Background

I am a writer by vocation but a dancer by avocation. As someone whose body is very important to her—not for vanity but for health reasons and a daily sense of well-being—any interruption in that routine has repercussions beyond the norm. Every day I do something—running, walking, hiking, Pilates, core conditioning, or dancing—to keep my body and my brain in shape. Once upon a time I was a Balanchine baby, but most recently, in the last six years, have been studying flamenco.

About seven years ago, I noticed a tiny bump on my lower right abdomen, barely visible, close to my pelvic region. I was not in any pain

B.P. Jacob et al. (eds.), *The SAGES Manual of Groin Pain*,
DOI 10.1007/978-3-319-21587-7_35

or discomfort, but after I noticed it grew ever so slightly about a year later, I pointed it out to my gynecologist in case it was a tumor. She checked it out visibly and said, "I think it's probably a hernia so eventually you should go see a surgeon," and she gave me some referrals. I waited almost another year. Every once in a while after I danced it would pop out slightly further.

My Preoperative Experience

Finally I decided to go see a surgeon. But instead of following the gynecologist's suggestions, I asked my internist for a name or names. I wanted a female surgeon. My doctor gave me the name of a colleague, Dr. A (who, I learned later, was also a trusted friend). Dr. A examined me, ordered a scan, and reported to me shortly after that I indeed had a femoral type hernia and that it eventually would have to be removed. Because it had "infarcted" already, she said I would not be a candidate for laparoscopic surgery (which she did not perform). Dr. A said eventually it was possible that my hernia could incarcerate at any time, when I could be traveling or away from home, and it was best to nip it in the bud. It was left to me to decide if and when to do this open surgery, but I should not wait too long. Mistake number one.

I decided to get a second opinion, as I still had no discomfort but just the occasional tiny bulge. I got a second opinion from the suggestion of my friend who was a prominent vascular surgeon. I again asked for a female surgeon. My friend asked around and got the name of another respected female general surgeon. I went to see Dr. B, who looked at the imaging and agreed that it was a femoral hernia and that it eventually should probably be removed, but that there wasn't any great urgency and I should do it within the year. Regarding technique, at some point, Dr. B told me that if I wanted to pursue laparoscopic surgery, a colleague, Dr. C, would be more experienced at that technique. When I called back to Dr. A, this surgeon opined that I was not eligible for a laparoscopic repair. My second opinion, Dr. B, did not push me toward a consult with Dr. C, an experienced laparoscopic hernia surgeon. Mistake number two.

I eventually decided to do the surgery over the summer when things were quieter at work and made a date with Dr. B to do the surgery. My internist thought I was making a mistake and told me if it were she, she would do it with the original surgeon, Dr. A, at the outpatient clinic. I was told this was better than having surgery at the university, which was much busier and where I was apt to also have medical students involved. So at the last minute, I called Dr. A, who was able to schedule me a few days later in the same time frame away from work I had already allowed for. Mistake number three:

…big mistake not to have consulted the Internet.

I felt that with these two recommendations I had done enough homework, but alas it was an error (mistake number four) not to have consulted the Internet. Though some doctors may complain of this system of checking them out, in fact, it is essential for patients to trade information. It's another important step in making a decision about surgery.

The surgery was short, a little over an hour, and seemed to go well enough. I was sent home to recover with instructions about icepacks and rest. After about a week I was able to get around at home and then eventually left the house after a few weeks. I stayed in touch with my surgeon Dr. A; however, because the healing seemed to be going slower than anticipated, I was still not pain-free and was taking Percocet after a month. I saw the surgeon before my scheduled post-op; actually, it was a partner, as my surgeon was out of town, and I told this partner of my concerns about pain:

I could not sit, I could not lie down…

Over the next weeks and months, I continued to inform the surgeon of this pain. It felt like the pain I remembered having with my IUD in the 1960s, a Dalkon Shield, which was eventually removed from the market. It felt like uterine pain, high up inside, not really near the inguinal nerves. I was unable to exercise or to walk very much without pain and to sit comfortably, and because I am a writer, this totally inhibited my work. I could not do chores or cook or anything domestic either. At this point, I became totally obsessed with reading on the Internet—where I learned much to my dismay about the issues with mesh and this surgery. I read and read and began to compile a list of doctors who might have the answer for me. Finally, after a number of calls, my surgeon apologized to me and said there must indeed be a problem and that I should see another surgeon if I did not want to come back.

I could not sit. I could not lie down. I could not walk. I could not dance. I could not exercise. I could not jog. I could not concentrate on work. I could not interact with family members. I could not care for my mother, who was dying. I could not read. I could not attend work events or any social events. I could not travel. I had sexual dysfunction, bladder dysfunction, and gastritis from the Advil, which Dr. A told me I could take and did not warn me that high doses could corrupt my stomach.

Though I consulted by e-mail and phone with a number of offices all over the country, I had already learned of two surgeons who specialized in revisional surgery after hernia repair. I very much liked Dr. C. Dr. C said after viewing new imaging that my mesh was indeed corrupted, that it had formed a ball—that I had a meshoma—and that in order to excise it, the procedure would likely include cutting my three important sensory nerves, called a triple neurectomy. Dr. D made the same diagnosis but said that the nerves might not need to be cut. Though I preferred Dr. C due to this doctor's patience and accessibility, I waited the requisite six months they had both urged as the "wait-and-see" period for the pain to go away:

Frightened and depressed…

During that time I also consulted pain doctors. Dr. E, a pain doctor, decided to do an exploratory block and scheduled me in the OR, as I was in such discomfort. I was crying all the time, miserable, frightened, and depressed. I was diagnosed with post-traumatic stress disorder. The block did not do much, and so I went ahead and scheduled with Dr. D, as I was hoping to save my nerves, particularly the genital nerve.

In the meantime, my insurer broke off relations with the Dr. C's hospital. Dr. C was most accessible and did not push me in one direction or another, but gave me information as I requested it. Again at the last minute, I became afraid and canceled the surgery with Dr. D and rescheduled with Dr. C. I had to do it within a month, as my insurer would only cover up to three months.

I had numerous bad side effects from the pain medication. Percocet was the only thing that helped, but its side effects of constipation eventually took me back to the doctor, to Dr. F, a gastroenterologist who recommended a diagnostic colonoscopy and some antidepressants. I have always been afraid of antidepressants, and though I filled the prescription, I could not get myself to take them. I did not proceed with the colonoscopy. I felt sure that if I could get off the pain meds, I would return to normal.

Revisional Surgery

I had to move mountains to get the surgery approved with my insurer; it made everything even more painful and complicated. Finally in late March, seven months after the original surgery, Dr. C eventually did the fix. Right before I was wheeled in, I implored Dr. C to save my nerve if possible and begged Dr. C not to put back any more mesh. Dr. C knew my surgery might require more mesh, so no promises were made. But during the five-hour surgery, Dr. C discovered that two of my nerves were on the same trunk and cut those, but that my genital nerve seemed unaffected and decided to risk saving it, since I had asked for this if possible. Dr. C took out the old plug mesh, which was adherent to my insides and my blood vessels. Dr. C advised that the mesh was accomplishing its normal role, but that a 3-dimensional mesh was entirely wrong for my thin pelvic region and femoral canal. It took nearly five hours in the operating room to remove it, and Dr. C felt it best to replace it with a different, more flexible mesh.

Recovery

Right afterward I felt better, but it took another three months for me to feel real relief. I was therefore still on Percocet and terribly frightened that the surgery had not worked. But suddenly, at the end of June, things improved. For the entire summer through September, I was in better shape. I did have pain, but I could manage to travel and did so, and in September, I felt I was well enough to begin dancing again. I was thrilled and relieved. I had my life back:

Profound depression…plunged me into despair.

I gradually built up to dancing again at least three nights a week. I made mistake number five in succumbing to my love for a very special but extremely challenging footwork. One night in mid-October, I knew I had pushed too much. The pain returned and with it a profound depression that plunged me into despair. This time, I could blame nobody but myself.

I went back on the physician rounds, via Dr. C, to a new gynecologist who specialized in pelvic floor repair, a new pain consult, and who was in constant touch with Dr. C, who basically kept me alive. I was back on Percocet, back with the debilitating constipation even with nightly Miralax. I developed a bout of transverse colon diverticulitis from the narcotics. I had to do enemas at least once a week, often more. It was a living hell.

My mother died around the same time that the pain returned, and other family and work complications ensued. Finally, by January, I had a breakdown and returned to Los Angeles. I was bedridden for a month, but with the help of a new psychiatrist was finally able to go on antidepressants. I tried three, and finally the third (Lexapro) made an enormous gradual difference. Not only did it help me emotionally, it masked the pain. I knew the pain was still there, but I was able to resume something of a normal life, to exercise, and to travel. I talked to Dr. C constantly about another surgery, this time to remove the mesh altogether. He said he would ask me to wait again and try everything else short of surgery.

In addition, after the surgery, my right leg and ankle became swollen, and my ankle has never gone back to its original size. I consulted with another vascular surgeon, who thought the lymphatic drainage might have been damaged during reoperation. We never got to the bottom of this.

By then I had been through every type of pain medication possible. But the Lexapro mostly took care of it all. Since then, over two years from the original surgery and a year and a half from the fix, I have had relief. Recently, the pain returned, but I am trying to presume that it is a temporary issue or that I possibly have to up the Lexapro dose. Alas, my hair has fallen out, which is possibly connected with the Lexapro, which has also had the side effect of a little weight gain and a few other side effects, but has so far been worth the trade-off.

Conclusions

In summation I would counsel every doctor performing hernia surgery to really look closely at the statistics upon which this surgery is based. There are very few femoral surgeries to begin with, so the statistics are from a tiny sample. Do not recommend surgery if the patient has no pain. Think about mesh and what it is doing to people's insides and think about going back to plain tissue repair. Think about laparoscopic surgery and see if more people are candidates.

I wish everyone with hernias could have it fixed right the first time. It has totally changed my life, much for the worse, and I will never forgive Dr. A for what she has done to me. Never. She was not qualified to do this surgery. She was not up to date with medical literature and the pros and cons of various meshes. She was too out of date to be doing this surgery and perhaps others. Doctors need to be reevaluated to make sure they keep up with current thinking. Retraining should be mandatory. And doctors should know themselves: if they have any doubts about performing a surgery or about their qualifications, they should pass the patient off to another doctor. It's not worth the money to ruin a life.

36. Sports Hernia with Adductor Tendonitis

Fredrick J. Brody and Jeffrey Harr

Chief Complaint

Pain in the left groin that extends to the thigh.

History

A 21-year-old lacrosse player presents with an 8-month history of pain in his left groin that radiates to his left upper thigh. The patient first developed these symptoms after an arduous lacrosse practice. He continued to practice in discomfort over the next several days, and was evaluated subsequently by his team doctor who treated him with ice and anti-inflammatories. Despite these treatments, his pain persisted and he was unable to perform explosive lateral movements. Ultimately he underwent two cortisone injections in the groin, which enabled him to complete the season in variable degrees of pain. After the season he avoided all athletic activities for 4 weeks, which resolved his symptoms completely. However, he returned for summer league lacrosse, and his groin and thigh pain returned immediately with strenuous activities.

Electronic supplementary material: The online version of this chapter (doi:10.1007/978-3-319-21587-7_36) contains supplementary material, which is available to authorized users. Videos can also be accessed at http://link.springer.com/chapter/10.1007/978-3-319-21587-7_36.

B.P. Jacob et al. (eds.), *The SAGES Manual of Groin Pain*,
DOI 10.1007/978-3-319-21587-7_36

Physical Exam

On physical examination, he was exceedingly fit with minimal body fat, pronounced abdominal muscles, and large quadriceps. His abdomen was flat without overt hernia defects. Upon palpation, the left groin was tender, particularly cephalad to the pubic bone at the insertion of the left rectus abdominis. His tenderness extended toward the pubic tubercle and laterally for 2 cm into the inguinal crease. Active sit-ups, with or without resistance, replicated his groin and abdominal pain. He was also tender along the left adductor longus tendon starting at the inferior aspect of the pubic tubercle, and extending along the tendon for 10 cm. The adductor tendon pain was exacerbated with hip adduction and abduction. He had full range of motion of his hip, and there were no clinically evident inguinal hernias even after multiple Valsalva maneuvers.

Workup

Plain radiographs did not show evidence of femoroacetabular impingement, hip dysplasia, or lumbar or sacroiliac degenerative changes. There was also no evidence of bone resorption or sclerosis. *Magnetic resonance imaging (MRI)* of the abdomen and pelvis revealed high signal uptake on T1- and T2-weighted images along the pubic bone consistent with pubic osteitis. There was also increased uptake along the left rectus abdominis insertion at the pubic bone consistent with edema and a possible tear. A cleft sign was also visible along the inferior portion of the pubic bone at the insertion of the left adductor longus, signifying a tear of the tendon (Fig. 36.1a, b). There was no evidence of any associated intra-articular hip pathology.

Diagnosis

The history, physical exam, and radiologic findings were consistent with the diagnosis of a sports hernia with *adductor tendonitis*. Other differential diagnoses should include iliopsoas strains or bursitis, avulsion injuries of the pubic bone, nerve entrapment syndromes, stress fractures of the femoral neck or pubic rami, vertebral body pathology, and associated hip injuries [1]. The most common hip pathology associ-

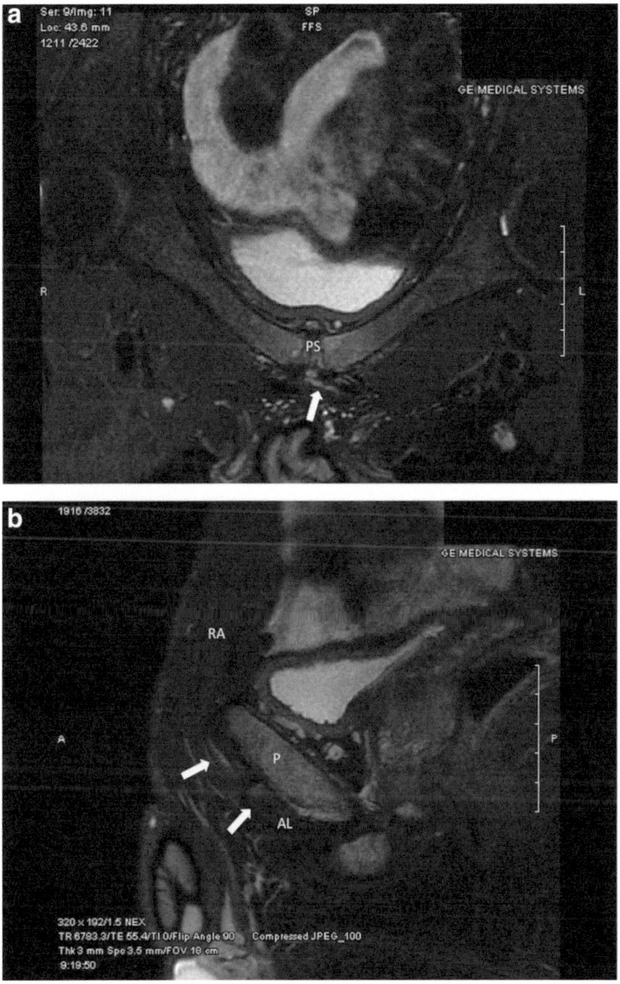

Fig. 36.1. (**a**) T2-weighted axial oblique image of the pelvisdemonstrating a cleft sign (*arrow*) just inferior to the pubic symphysis (PS); (**b**) T2-weighted sagittal image of the pelvis demonstrating hyperintense aponeurotic tears (*arrows*) of the rectus abdominis (RA) and adductor longus (AL) tendon insertions into the pubic bone (P).

ated with a sports hernia is a *labral tear*, which is diagnosed with MRI [2]. Other diagnoses that are not part of the musculoskeletal system include appendicitis, urinary tract infections, testicular pain, varicoceles, round ligament entrapment, endometriosis, and ovarian cysts.

Nonoperative Management

Nonoperative management is initially provided for adductor tendonitis associated with groin pain and may include *3–9 months of rehabilitation with a physical therapist*. These modalities may involve extensive stretching, deep tissue massage, electrical stimulation, and local injections. These injections employ mixtures of steroids with local anesthetics, and are injected directly into the pubic symphysis, pubic tubercle, or associated aponeuroses. Mixtures commonly include methylprednisolone mixed with 0.5 % bupivacaine, but other combinations of methylprednisolone, dexamethasone, lidocaine, and bupivacaine have been used as well. Overall, the majority of adductor tendon injuries should resolve with inactivity and nonoperative therapy. The goal of nonoperative therapy should restore a full range of motion, while maintaining muscle strength and preventing contracture of the adductor tendon or rectus sheath. Ultimately, the patient should regain strength, flexibility, and endurance quickly if nonoperative management is successful. However, chronic injuries located along the musculotendinous junction usually require operative intervention if athletic activities cannot be halted.

Operative Management

Our technique entails an inguinal incision along the skin crease located directly above the superficial ring. The dissection extends through Scarpa's fascia to the external oblique aponeurosis. *Tears in the external obliqueaponeurosis* are repaired with 4-0 nonabsorbable sutures since these tears may cause nerve impingements. The external oblique aponeurosis is then opened superiorly and inferiorly through the superficial ring. The spermatic cord is dissected at the level of the pubic tubercle, retracted laterally, and skeletonized to exclude a hernia sac. Once a hernia sac is excluded, a *relaxing incision is made along the fascia of the conjoined tendon starting at the level of the deep ring*, and extended inferiorly along the pubic bone (Fig. 36.2). The *anterior rectus sheath is then released from the pubic bone* by incising the aponeurosis approximately one cm superior to the pubic bone. This relaxing incision is then extended laterally until the transversalis fascia of the inguinal floor is encountered. The undersurface of the relaxing incision is undermined in an avascular plane. The sports hernia repair is performed by *plicating the previously released fascia of the conjoined tendon to the iliopubic tract* with a continuous 2-0 Prolene suture (Fig. 36.3). The first

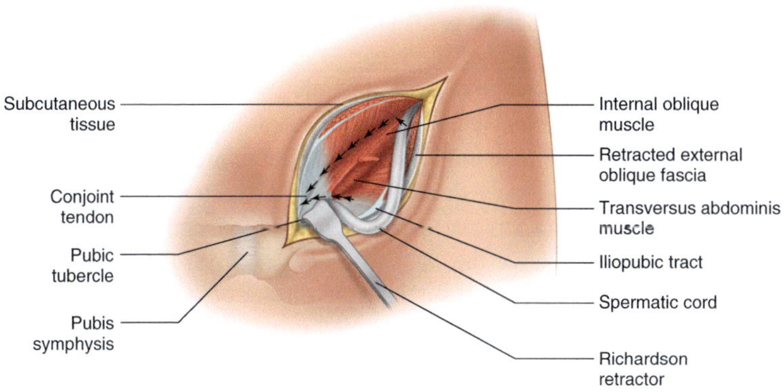

Fig. 36.2. After undermining the conjoined tendon beneath the relaxing the incision, the flap is secured with a running 2-0 Prolene suture to the iliopubic tract. A second 2-0 Prolene is used to imbricate the first suture line.

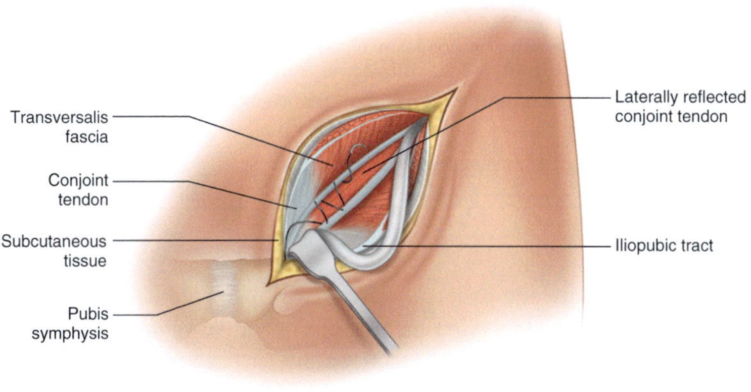

Fig. 36.3. Relaxing incision starts at the pubic tubercle and extends along the conjoined tendon to the level of the deep ring. The relaxing incision also extends medially along the conjoined tendon to release the aponeurosis from the pubic symphysis for approximately 2 cm.

layer starts at the pubic tubercle and extends superiorly to the deep ring. A second layer of 2-0 Prolene imbricates the initial layer. At this point, 10 cc of 0.25 % bupivacaine is injected into the pubic tubercle. Mesh is not placed since the intent of the repair is to lateralize the vector force away from the pubic symphysis and tubercle.

The *adductor tenotomy*is performed to release the vector force that extends inferiorly from the pubic bone to the large muscles of the thigh. Conceptually, the tenotomy divides the vertical vector or the common aponeurosis that incorporates the adductor tendon and rectus sheath. This maneuver dissipates the distracting forces from the pubic bone. Technically, a Deaver retractor is inserted through the inguinal wound, and the subcutaneous tissues are retracted in an avascular plane above the fascia of the adductor longus muscle (Fig. 36.4a). The tendon is easily palpated as a strong band extending from the pubic bone. Starting at the 12 o'clock position and extending medially and inferiorly, the tenotomy is performed 2 cm from the pubic tubercle (Fig. 36.4b). The tenotomy extends toward the 7 o'clock position and is performed in a superficial manner, utilizing a right angle dissector to avoid injury to the underlying muscle. The tenotomy separates the overlying tendon and fascia from the underlying adductor longus muscle. Hemostasis is typically well controlled with the electrocautery as long as the muscle is not divided. Associated nerve fibers are encountered rarely in this plane along the upper medial aspect of the adductor musculature. Once the tenotomy is completed, hemostasis is verified. From the inferior aspect, the pubic tubercle and the proximally divided tendons of the adductor longus are injected with 10 cc of 0.25 % bupivacaine. After closing the incision in layers, the patient is extubated and taken to recovery room in stable condition (Videos 36.1, 36.2 and 36.3).

Outcome

Upon discharge, the patient's activity was restricted to walking and activities of daily living. After his first postoperative visit (14 days after surgery), he was allowed to increase his activities slowly, which involved cycling, jogging, and swimming. At 4 weeks, he resumed core activities, including sit-ups and running an under 8-min mile. At 6 weeks, he initiated activities involving lateralizing movements, but full-speed activities were restricted. After approximately 8 weeks, the patient resumed his normal activities, and ultimately returned for his final sea-

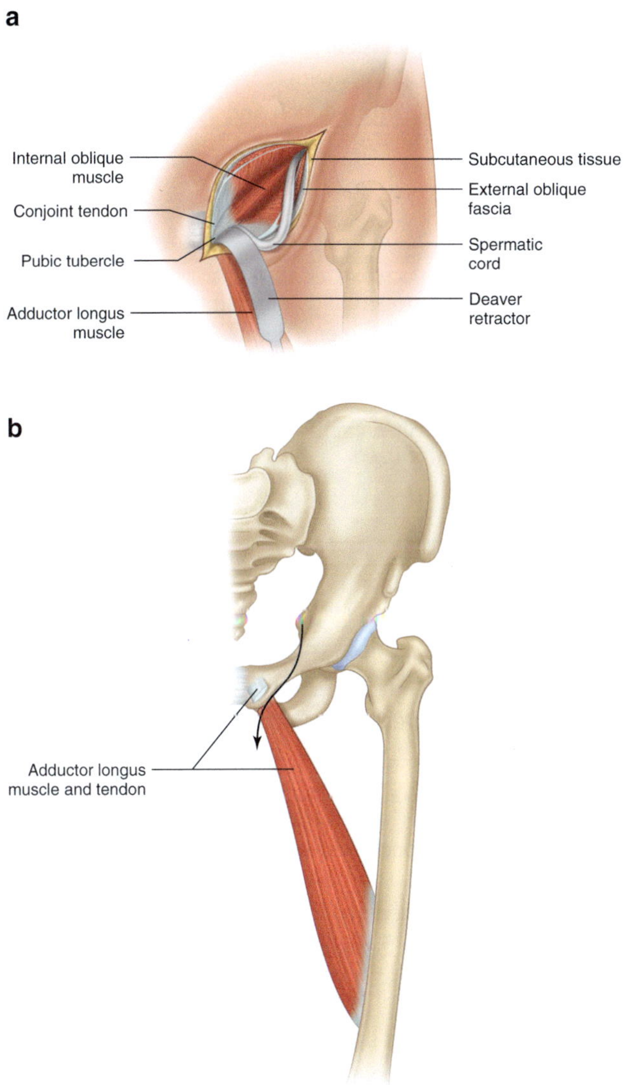

Fig. 36.4. (**a**) A Deaver retractor is placed in the wound and the adductor tendon is visualized with inferior retraction; (**b**) The tenotomy begins at the 12 o'clock position on the adductor tendon and extends to the 7 o'clock position.

son of lacrosse without restrictions. With nonoperative management, approximately 50 % of patients are able to resume full athletic activities, free of pain, within a year [3]. As with this patient who failed nonoperative management, surgery engenders a full recovery in 80–100 % of patients [4].

Complications

The most common postoperative complaints include minor bruising or edema involving the abdomen, thighs, genitalia, and perineum. Postoperative complications are rare and entail seromas, hematomas, dysesthesias, surgical site infections, and penile vein thrombosis [5]. The majority of these adverse events are self-limiting and resolve nonoperatively within 4–6 weeks.

Conclusion

Sports hernias, also referred to as athletic pubalgia, encompass injuries to the tendons and muscles of the rectus abdominis and adductor longus at their respective insertions along the pubic bone. Typically, there is a common aponeurosis that connects both the rectus and adductor mechanism. However, the rectus and adductor muscles exert their forces in competing vectors, which lead to injuries of the common aponeurosis. These injuries, or micro-tears, may induce a degree of compartment syndrome, and concurrent injuries to both aponeurotic regions may be found in up to 23 % of patients [6]. Therefore, both injuries should be treated simultaneously.

It was not until 1991 that Taylor et al. reported their initial experience with an open inguinal approach to address groin pain in athletes [7]. This study was followed quickly in 1992 by a paper from Malycha and Lovell, in which they coined the term "sportsman's hernia" to describe a complex of chronic groin pain in athletes [8]. Despite being a misnomer, this term was adopted quickly to describe chronic groin pain in athletes and has been popularized over the last two decades. The etiology of this pain encompasses injuries to the tendons and muscles of the rectus abdominis and adductor longus at their respective insertions at the pubic bone. Magnetic resonance imaging (MRI) is the optimal test to diagnose this entity, discover concurrent injuries, and rule out other causes of groin pain. Initially, a trial of nonoperative management including physical therapy is recommended. If unsuccessful, surgical intervention is offered, and has a high success rate. An overlying tenotomy (fasciotomy) of both aponeurotic sheaths and tendons usually resolves the underlying edema and inflammation. Consequently, this alleviates the scarring induced from chronic tendonitis, and allows the region to appropriately relax and regain its original propensity for high impact activities.

References

1. Nam A, Brody F. Management and therapy for sports hernia. J Am Coll Surg. 2008;206(1):154–64.
2. Morelli V, Espinoza L. Groin injuries and groin pain in athletes: part 2. Prim Care. 2005;32(1):185–200.
3. Paajanen H, Brinck T, Hermunen H, Airo I. Laparoscopic surgery for chronic groin pain in athletes is more effective than nonoperative treatment: a randomized clinical trial with magnetic resonance imaging of 60 patients with sportsman's hernia (athletic pubalgia). Surgery. 2011;150(1):99–107.
4. Larson CM. Sports hernia/athletic pubalgia: evaluation and management. Sports Health. 2014;6(2):139–44.
5. Meyers WC, McKechnie A, Philippon MJ, Horner MA, Zoga AC, Devon ON. Experience with "sports hernia" spanning two decades. Ann Surg. 2008;248(4): 656–65.
6. Lovell G. The diagnosis of chronic groin pain in athletes: a review of 189 cases. Aust J Sci Med Sport. 1995;27(3):76–9.
7. Taylor DC, Meyers WC, Moylan JA, Lohnes J, Bassett FH, Garrett Jr WE. Abdominal musculature abnormalities as a cause of groin pain in athletes. Inguinal hernias and pubalgia. Am J Sports Med. 1991;19(3):239–42.
8. Malycha P, Lovell G. Inguinal surgery in athletes with chronic groin pain: the 'sportsman's' hernia. Aust N Z J Surg. 1992;62(2):123–5.

37. Patient with Groin Pain After a Plug and Patch Hernia Repair

Christopher G. DuCoin and Garth R. Jacobsen

Postoperative groin pain following hernia repair is common. For most, the pain is acute and resolves with time; however, some patients develop chronic inguinal pain, termed inguinodynia. Inguinodynia may occur after any type of hernia repair. One of the most common inguinal hernia repair techniques is the plug and patch mesh herniorrhaphy. Though extremely popular, this procedure has its own set of potential complications that may lead to inguinodynia.

Chief Complaint

Extreme pain when pressing bump in groin

History

A 60-year-old male presents with a history of a prior open right inguinal hernia repair with the plug and patch technique 4 years earlier. His past medical and surgical history is otherwise not significant except for a history of chronic narcotic use secondary to his inguinal pain. He reports that he can press on the mass and make it disappear, but the maneuver is extremely painful, with pain score increasing to 10/10 in severity. The pain has been present since the operation. It worsens throughout the day, starting with fairly little pain in the morning and crescendoing toward the evening. It radiates into his scrotum. Physical activity makes the pain unbearable. He denies any electrical shock or burning sensation in the groin or skin.

B.P. Jacob et al. (eds.), *The SAGES Manual of Groin Pain*,
DOI 10.1007/978-3-319-21587-7_37

Physical Exam and Workup

On physical exam, he was found to have a firm palpable mass in the right groin that was quite solid and not consistent with that of herniated tissue. He had hyperesthesia in the region of the mass and full deep examination was not feasible due to his pain. His abdomen was benign. He was able to walk without a limp, but this caused an extreme amount of pain. We could not delineate any specific neuropathic distribution of the pain. Targeted injections by our anesthesia pain colleagues provided little relief and therefore did not provide guidance as to which nerve might be definitively involved.

Diagnosis and Management Options (Nonoperative vs. Operative)

We had a discussion regarding his diagnosis and possible treatment options. Nonoperative management was discussed; however, both the patient and surgeon felt that this would be of little merit, as the pain was chronic and conservative therapies had failed over prior years. Indeed, he had resorted to chronic narcotic use as the only modality that helps his pain.

The reducible painful nature of the mass represented either a recurrence of his hernia, poor integration of the plug, or both. Also, nerve entrapment or other involvement could not be definitively ruled out. His surgical treatment options were discussed. The final plan consisted of diagnostic laparoscopy for evaluation of the groin area for recurrence, possible recurrent inguinal hernia repair with mesh, possible laparoscopic versus open explant of mesh plug, and possible triple neurectomy, depending on operative findings. The risk of injury to the testicular vessels and subsequent ischemic orchitis was discussed with the patient, as was the possible need for vasectomy or orchiectomy. It is important in the reoperative patient to discuss these potential complications and their implications, as this may sway some patients to continue with observation.

Operative Management

The operation began with a diagnostic laparoscopy via the umbilicus. There was no inguinal hernia on the left. On the right, the mesh plug could be seen to be freely floating through the deep inguinal ring and was almost completely enveloping the spermatic cord structures. It was felt that the mesh migration and entrapment of the cord structures were creating the patient's inguinal pain with radiation into the scrotum.

At this point a transabdominal preperitoneal (TAPP) hernia repair was initiated with the goal of mesh explantation and placement of a macroporous lightweight polypropylene mesh to cover the myopectineal orifice. Two additional 5 mm trocars were placed in the standard fashion and position, and the preperitoneal space was developed, starting at the anterior superior iliac spine and working medial to the umbilical ligament. Here the mesh was found to encompass the cord structures. This can be seen deep and lateral to the plug (Fig. 37.1). Note the testicular artery medial to the vas deferens, as its normal course lateral to the structure is deviated due to adherence to the mesh plug. Here we weighed the feasibility of removal of the meshoma and the risk of injury to the surrounding structures such as the vas deferens, testicular artery, testicular veins, and the iliac vessels. With slow and meticulous sharp dissection, the mesh was successfully dissected free of the cord structures without injury. Next a standard TAPP recurrent inguinal hernia repair with macroporous lightweight polypropylene mesh was completed. The mesh plug was removed through the umbilical port site, which was enlarged to allow extraction.

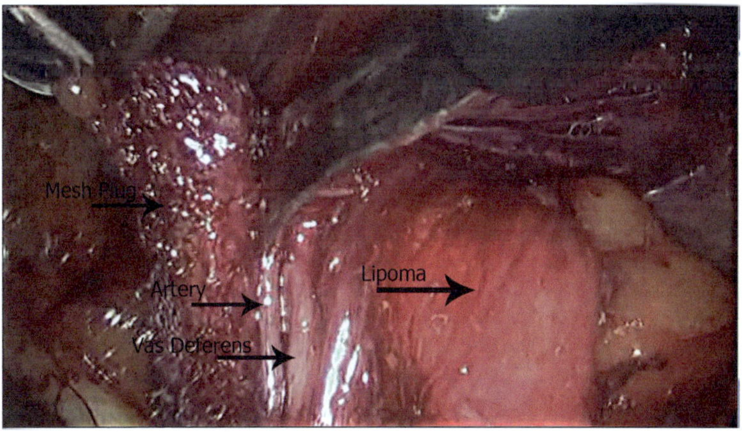

Fig. 37.1. Laparoscopic preperitoneal view of right groin. Mesh plug found deep to internal ring, overlying and densely encompassing the spermatic cord structures. Mesh plug is medial; spermatic cord is lateral. Spermatic cord lipoma noted laterally.

Operative time was 65 min. The patient was discharged home from the postoperative unit. He reported near-immediate improvement of his preoperative chronic pain and was off from all narcotic pain medications by his 2-week follow-up appointment.

Outcomes and Complications

There is at least a single case report of mesh migration for every type of hernia repair, both open and laparoscopic. However, there is an exceedingly greater number of mesh migrations reported for the mesh plug repairs than for flat mesh repairs [1–6]. The location of migration and the organs involved determine the symptoms produced. Reported locations of mesh migration include the scrotum, bladder, and hollow viscous structures such as the cecum and small bowel [1–6]. The symptoms produced include chronic pain, recurrent urinary tract infections, intestinal obstructions, volvulus, and even intestinal perforation [1–6]. One underlying similarity of these reports is that of lack of mesh fixation or fixation of original mesh with absorbable suture [7]. In the end, this patient's pain was relieved by removal of the plug and reinforcement of the myopectineal orifice with a flat sheet of mesh. The authors remain skeptical that the addition of a mesh plug to a standard Lichtenstein herniorrhaphy provides any benefit and, at worst, can result in the above complication.

References

1. Chuback JA, Singh RS, Sills C, Dick LS. Small bowel obstruction resulting from mesh plug migration after open inguinal hernia repair. Surgery. 2000;127(4):475–6.
2. Benedetti M, Albertario S, Niebel T, Bianchi C, Tinozzi FP, Moglia P, et al. Intestinal perforation as a long-term complication of plug and mesh inguinal hernioplasty: case report. Hernia. 2005;9(1):93–5.
3. LeBlanc KA. Complications associated with the plug-and-patch method of inguinal herniorrhaphy. Hernia. 2001;5(3):135–8.
4. Tokunaga Y, Tokuka A, Ohsumi K. Sigmoid colon diverticulosis adherent to mesh plug migration after open inguinal hernia repair. Curr Surg. 2001;58(5):493–4.
5. Moorman ML, Price DP. Migrating mesh plug: complication of a well-established hernia repair technique. Am Surg. 2004;70(4):298–9.
6. Nowak DD, Chin AC, Singer MA, Helton WS. Large scrotal hernia: a complicated case of mesh migration, ascites, and bowel strangulation. Hernia. 2005;9(1):96–9.
7. Jeans S, William G, Stephenson B. Migration after open mesh plug inguinal hernioplasty: a review of the literature. Am Surg. 2007;73(3):207–9.

38. Patient with Groin Pain After Open Inguinal Hernia Repair with Mesh

Jeffrey A. Blatnik and Ajita S. Prabhu

Chief Complaint

Right lower quadrant pain, status post six prior hernia repairs

History

The patient is a 51-year-old thin male with multiple prior inguinal hernia operations with right lower quadrant abdominal pain described as "burning" and at times "stabbing" and "dull." Previously, he has had six hernia operations on the same side. It began with an open inguinal hernia repair with mesh at age 20 years to address preoperative groin pain radiating to his right leg, across his back, and down to his groin area. Approximately 2 years later, he developed stabbing pain in the same location upon routine lifting of objects that he did not consider to be heavy. He underwent a second open repair with mesh at that time and is uncertain as to whether his first mesh was removed. He again had resolution of his symptoms until 2 years later, when he had recurrence of the same symptoms and underwent a third open exploration. He is unsure as to whether mesh was placed at that time. Approximately 1 year afterward, he developed an acutely incarcerated right inguinal hernia with obstructive symptoms. He was taken emergently for a fourth operation and does not know if mesh was placed at that time. Due to the time period during which these operations took place, operative reports were not available for review. Three years later in 2010, the patient developed recurrent pain radiating to his right leg, groin, and back. He had his fifth exploration via an open incision. At that operation, he was noted to have an onlay polypropylene flat mesh on top of the external oblique aponeurosis, as well as

B.P. Jacob et al. (eds.), *The SAGES Manual of Groin Pain*,
DOI 10.1007/978-3-319-21587-7_38

a polypropylene flat mesh on the floor of the inguinal canal, with a small direct hernia recurrence. This was repaired primarily with permanent suture. The onlay mesh over the external oblique aponeurosis was resected, and the mesh on the floor of the canal was left in place. Finally, in 2011 he had painful recurrence of his symptoms and was taken for open right inguinal exploration with resection of all previously placed mesh and permanent suture. The ilioinguinal nerve was not seen or identified. Notably, there was no hernia identified at the time of that operation and therefore no new sutures were placed. This operation was complicated by an immediate postoperative expanding hematoma for which he was emergently explored. He had a brief reprieve from his pain postoperatively. He presented 2 years later with recurrence of his symptoms.

Physical Exam

He had multiple surgical scars in the right lower quadrant, a small recurrence of his hernia, and palpable mesh in the subcutaneous space with tenderness over the area. There was no hypesthesia or allodynia noted.

Imaging

Computed tomography scan of the abdomen and pelvis was obtained that did not show an obvious recurrence.

Nonoperative Management Options

As part of the workup for this problem, the patient was referred to a pain management specialist who performed a comprehensive neurologic exam. He did not feel that the pain was neuropathic in nature. The patient was offered a transcutaneous electrical nerve stimulation unit as a noninvasive measure for pain control. The patient did not pursue this due to the cost and also the desire for a more definitive cure for his pain.

Diagnosis

The patient was considered to have a hernia recurrence (most likely, again,) of his direct hernia. In addition, it was unclear if the mesh alone was contributing to his symptoms. We did not feel that he suffered from any nerve injury or spermatic cord injury. Nor was there any evidence for infection or balling up of the mesh, i.e., meshoma.

Operative Treatment

Once workup was completed, the patient was offered laparoscopic exploration to address the recurrence of his inguinal hernia. Chronic groin pain is not an uncommon complication after open inguinal herniorrhaphy, with an incidence as high as 62.9 % described in some series [1]. A generally accepted definition for the term *chronic groin pain* is the presence of pain in the groin region for greater than 3 months after surgery. This may be further divided into neuropathic pain versus non-neuropathic pain. Neuropathic pain may be related to injury to the ilioinguinal nerve, the iliohypogastric nerve, the genitofemoral nerve, or (rarely) the lateral femoral cutaneous nerve. Nerve injury may be mechanical in nature or otherwise may be related to an adjacent inflammatory process such as granuloma or excess fibrotic reaction or mesh encasement of the nerve structures [2]. For the patient discussed in this scenario, an extensive workup by a pain management physician suggested a non-neuropathic source of pain, hence the decision to take the patient to surgery. There was no role for nonoperative intervention, as the pain was felt to be non-neuropathic in etiology.

We began with a transabdominal laparoscopic evaluation. This identified multiple loops of small intestine densely adhered to the hernia mesh (Fig. 38.1). This finding was despite the fact that all of the patient's previous hernia repairs had been in an open fashion and presumably as an onlay, and per report, all mesh had been removed. These adhesions were taken down sharply, and to avoid injury to the small bowel, a portion of mesh was left adherent to the bowel. This dissection exposed what appeared to be a plug mesh in his internal ring.

Due to the chronic, non-neuropathic nature of his groin pain, it was felt that all previous mesh would need to be removed at this operation. To facilitate subsequent hernia repair following mesh removal, we began by creating a large, extraperitoneal flap. During this portion, we encountered multiple pieces of prior mesh, all of which were removed with a combination of sharp dissection and harmonic scalpel. Great care was taken to avoid injury to the overlying skin, as the patient was very thin, and there was not a significant amount of subcutaneous tissue. Cooper's ligament was identified and served as our inferomedial landmark, and dissection was continued laterally. The plug mesh was identified adherent to the vas deferens, as can often be expected. In the

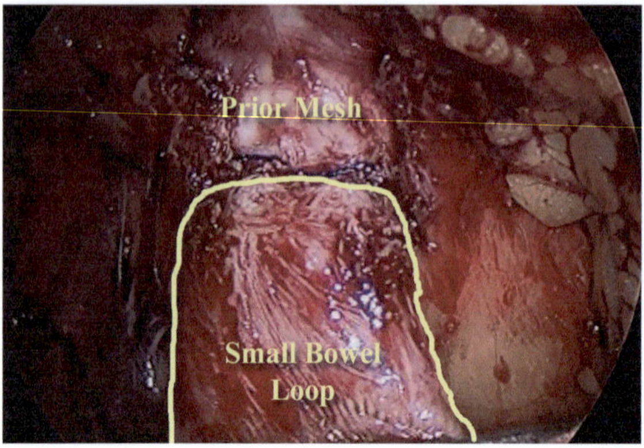

Fig. 38.1. Densely adherent small intestine to prior hernia mesh in retroperitoneal space.

complex re-operative setting, we discuss with the patient preoperatively the potential for division of the vas deferens, as often this cannot be preserved. When completing the inferior dissection of the plug, great caution must be used, as it is in close proximity to the iliac artery and vein (Fig. 38.2). The plug mesh was then removed through the umbilical incision. We then turned our attention to removing the remainder of the prior mesh. This was noted to involve the inferior epigastric vessels; to prevent bleeding during this stage, the vessels were prophylactically ligated with a clip. After removal of all of the prior mesh from the abdominal wall, they were removed from the abdomen in a specimen removal bag.

The patient was clearly left with a defect in the groin. We elected to repair this with an extra-large piece of 3DMax Light Mesh (Bard, Warwick, Rhode Island). This is a preformed, lightweight polypropylene mesh with large pores. This was positioned in our extraperitoneal plane and secured in place with selectively placed tacks into Cooper's ligament. In addition, it is our practice to secure the inferior portion of the hernia mesh with fibrin glue in an effort to prevent recurrence below the mesh. To prevent the mesh from contacting the viscera, it was covered with our previously created peritoneal flap. Prior to completing the procedure, the portion of small bowel that was densely adhered to the mesh was externalized through the umbilical incision to evaluate for any potential injury and was ultimately oversewn.

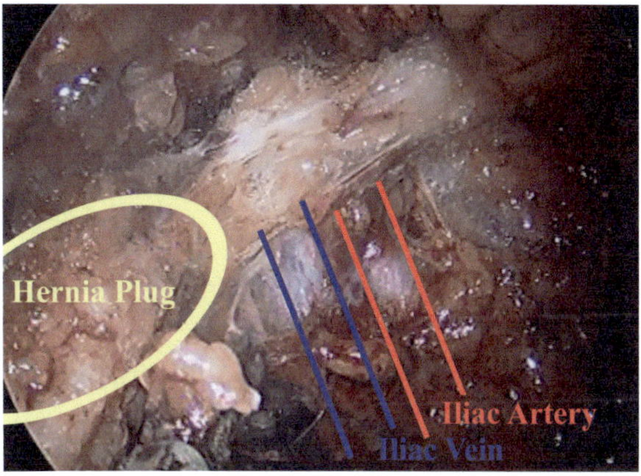

Fig. 38.2. The prior plug mesh is retracted medially to demonstrate its close proximity to the underlying right iliac artery and vein.

Postoperative Course

The patient was admitted to the hospital for observation overnight. He noted a marked improvement in his chronic pain symptoms. He was discharged home, doing well, on postoperative day number one. At 1-month follow-up, our patient reported that his groin pain had completely resolved with no evidence of hernia recurrence on exam.

Outcomes and Discussion

This outcome is similar with two other small case series evaluating laparoscopic management of groin pain following inguinal hernia repair [3, 4]. These studies report that all patients had some improvement in pain symptoms, with the majority having complete resolution. Although this patient had a successful outcome with resolution of his pain symptoms, this is not always the case, as often some patients will have some minor ongoing complaints. Known complications of this operation include recurrent pain or hernia, injury to the adjacent vasculature (inferior epigastric or iliac artery/vein), injury to the vas deferens or other spermatic cord contents (resulting in testicular pain or need for vasectomy or orchiectomy), and injury to the genitofemoral or lateral femoral cutaneous nerves with resultant neuralgia. In conclusion, laparoscopic management of chronic groin pain following open inguinal hernia repair can be technically challenging. However, it often results in symptom improvement for the patient.

References

1. Poobalan AS, Bruce J, King PM, Chambers WA, Krukowski ZH, Smith WC. Chronic pain and quality of life following open inguinal hernia repair. Br J Surg. 2001;88(8): 1122–6.
2. Cunningham J, Temple WJ, Mitchell P, Nixon JA, Preshaw RM, Hagen NA. Cooperative hernia study. Pain in the postrepair patient. Ann Surg. 1996;224(5):598–602.
3. Rosen MJ, Novitsky YW, Cobb WS, Kercher KW, Heniford BT. Combined open and laparoscopic approach to chronic pain following open inguinal hernia repair. Hernia. 2006;10(1):20–4.
4. Keller JE, Stefanidis D, Dolce CJ, Iannitti DA, Kercher KW, Heniford BT. Combined open and laparoscopic approach to chronic pain after inguinal hernia repair. Am Surg. 2008;74(8):695–700.

39. Patient with Groin Pain After a Lichtenstein Hernia Repair

Shirin Towfigh

Chief Complaint

Right groin pain radiating to the base of the penis

History

The patient is a 65-year-old male, status post open onlay mesh repair of his right inguinal hernia using polypropylene mesh in Lichtenstein tension-free onlay method. He reports noting groin pain within the first several weeks after surgery, with no improvement since then. He presents with 1 year of chronic right groin pain and 5 months of severe debilitating pain. The pain is 10/10, ranging from 2/10 to 10/10. It is a sharp, stabbing, hot pain like a "big knife" or "hot poker." It is always in the same area at the lateral edge of his groin wound and with time has radiated farther and farther down his groin. He is now hypersensitive at the right scrotum. He wears restrictive underwear to prevent tugging by or swaying of the scrotum. He fidgets when he sits and does not wear jeans, as the wrinkling of the stiff fabric causes pressure and pain when he sits. He cannot sit on the toilet seat without pain. Walking is now an ordeal, as is getting up to stand and bending. He cannot pick up a bar of soap from the ground. He cannot raise his leg, such as to step over the little bottom lip of his shower door, as this causes pain. He is best when lying flat. He used to be an avid cyclist, but he can no longer cycle.

According to the operative report, the patient had an indirect inguinal hernia. The onlay patch from a medium size plug and patch kit was used as a keyhole mesh. It was sutured with 0 Ethibond sutures. The ilioinguinal nerve was identified throughout its course and protected.

B.P. Jacob et al. (eds.), *The SAGES Manual of Groin Pain*,
DOI 10.1007/978-3-319-21587-7_39

Physical Exam

The patient was in discomfort while sitting at the edge of the chair. He had a healed groin scar and no visible bulge. Palpation elicited 3+ tenderness at the internal ring and along the spermatic cord. A mass of mesh was palpable laterally. He had 3+ hypesthesia and allodynia at the right groin scar and scrotal skin. The testis was descended and without associated tenderness or mass.

Imaging

Magnetic resonance imaging of the anterior pelvis, non-contrast, with Valsalva and dynamic views demonstrated intact flat onlay mesh with no hernia recurrence (Fig. 39.1). He has a significant varicocele on the right. There is no inflammatory reaction noted around the mesh.

Diagnosis

The patient was diagnosed with ilioinguinal neuralgia. This was due to direct injury at the time of his operation versus entrapment due to scar or mesh. He had no other obvious causes for his postoperative pain, including no evidence of hernia recurrence, infection, inflammation, or meshoma. He was offered nonsurgical treatment as the initial modality

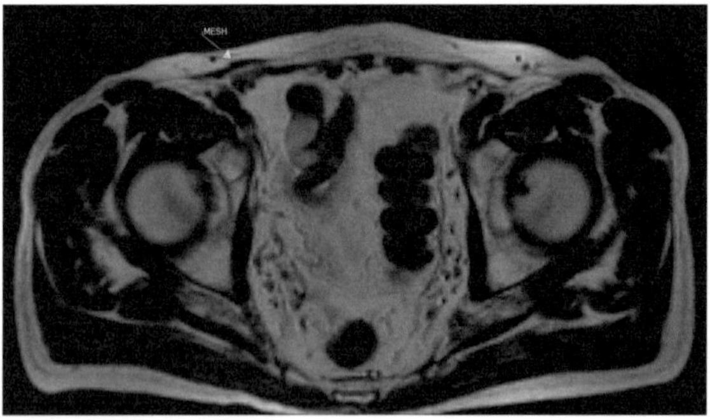

Fig. 39.1. MRI anterior pelvis, non-contrast, with Valsalva and dynamic views demonstrated intact right inguinal hernia repair with no hernia recurrence. T2 axial view here shows intact flat onlay mesh (*yellow arrow*).

for cure. This includes a combination of nerve blocks and neuromodulating medications and anti-inflammatories. If these provide short-term success, they are continued with the goals of long-term success. In our experience, most patients who respond to local injections require 3–5 cycles of nerve blocks and tend to have weeks to months of pain-free episodes after each block as well as overall reduction in their pain score. Only those that have short-term response but no long-term cure are offered surgical options.

Nonoperative Management Options

I first offered the patient a diagnostic nerve block in the office, which was performed with 0.5 % bupivacaine, injected medial and inferior to the anterior superior iliac spine. This resulted in near-complete resolution of the patient's pain. Wiping the area clean after the injection resulted in no pain. As per our protocol among patients with purely neuropathic pain, if they respond positively to local nerve blocks, they are offered serial blocks, no more than every 2 weeks, as their primary mode of treatment. The therapeutic nerve blocks include steroids (10 mg Kenalog). The patient indeed had a very clear improvement with the blocks. These were continued and resulted in reduction of his pain such that he was able to return to work, which involved sitting and standing. The scrotal sensitivity resolved. He did have continued pain along his groin and would even pass out at times due to the pain.

The patient already was under the care of a pain management specialist. He did not tolerate duloxetine, due to rash, and did not tolerate gabapentin due to its side effects. He was using Traumeel topically and 5 % lidocaine patch with no major improvement in symptoms.

The patient did not wish to undergo an operation unless absolutely necessary. After five cycles, the patient was agreeable to surgical exploration.

Operative Treatment

The patient was offered targeted ilioinguinal neurectomy. This was performed in open fashion, anteriorly, with identification of the nerve as it coursed anteriorly and just proximal to the lateral edge of the mesh.

The patient's localized areas of pain were marked in the preoperative area. This often helps intraoperatively with correlation of the pain with the operative findings and also to help guide the procedure. The operation was performed under local anesthesia with sedation. The prior incision was reincised along its lateral portion. Once the external oblique aponeurosis was identified, the lateral edge of the mesh was noted, as was the greenish hue of Ethibond sutures. The fascia was incised and lifted off of the mesh using a combination of blunt, sharp, and cautery dissection. There were two sutures of Ethibond with multiple knots each at the superolateral edge of the mesh. These were both removed. The ilioinguinal nerve was identified entering this area on top of the internal oblique muscle and under the mesh. This was dissected out proximally and distally. The nerve seems to have tracked in the same region as one of these sutures. Thus, most likely the patient had nerve entrapment of the right ilioinguinal nerve with a laterally placed suture within the muscle. The right ilioinguinal nerve was skeletonized proximally and distally. It was tied off and transected distally at the level of the mesh edge. It was then dissected proximally and injected with local anesthetic proximally. A 3-0 Chromic tie was used to tie its end to reduce bleeding from the neurovascular bundle. It was transected and sent to pathology for identification. The stump of the nerve was further dissected and implanted into a pocket of internal oblique muscle just deep to it. The purpose of this is to help reduce the risk of postoperative neuroma. The wound was then closed in layers.

Postoperative Course

The patient had complete resolution of his pain postoperatively. He was followed up for 2 years and has not had any recurrence of his symptoms.

Outcomes and Discussion

Nerve injury at the time of Lichtenstein hernia repair is either due to a technical error (e.g., direct injury, suture entrapment, manipulation, and dissection of the nerve) or due to mesh folding or scar tissue with resultant entrapment of the nerve nearby. In this case, inappropriately placed suture, placed laterally along the mesh, entrapped the ilioinguinal nerve as it coursed through the muscle layers.

The timing of onset of the pain (usually less than 6 weeks vs. after 6 weeks, respectively) can help determine the cause. In the case of direct nerve injury at the time of procedure, such as due to manipulation, cautery, etc., the application of an inflammatory mesh on top of the injured nerve can prevent natural healing and result in progression toward neuroma and/or chronic pain. Imaging can help rule out meshoma and other causes such as hernia recurrence and infection.

Nonsurgical options for neuropathic pain include the use of neuromodulating medications (e.g., gabapentin, duloxetine, tricyclic antidepressants), local anesthetics (topical creams, patches, direct nerve blocks), and anti-inflammatory medications [nonsteroidal anti-inflammatory drugs (NSAIDs), steroids] or a combination therapy. If patients are not cured from these modalities, then the injury is more severe and/or is mechanical in nature, and more invasive therapies are indicated. This includes nerve ablation (e.g., with alcohol, radiofrequency, cryoablation) or surgical neurectomy.

In my experience, patients who respond to nerve blocks with more than 6 h of complete pain relief are best suited for nonoperative management. These patients are provided 3–5 cycles of blocks with bupivacaine and Kenalog steroids, injected in the vicinity of the nerve proximal to the site of injury. If the pain relief duration increases with each injection cycle and the overall pain score reduces with each visit, then this is considered a successful plan of care. At least 20 % of such patients will never require surgical neurectomy.

Conclusion

Patients who undergo Lichtenstein hernia repair with mesh are at risk for chronic pain; the cause of many of these occurrences is technical in nature and thus preventable (Fig. 39.2) [1]. The surgeon evaluating the patient should have a grasp of all the different technical errors that could lead to such complications and rule them out as part of the workup.

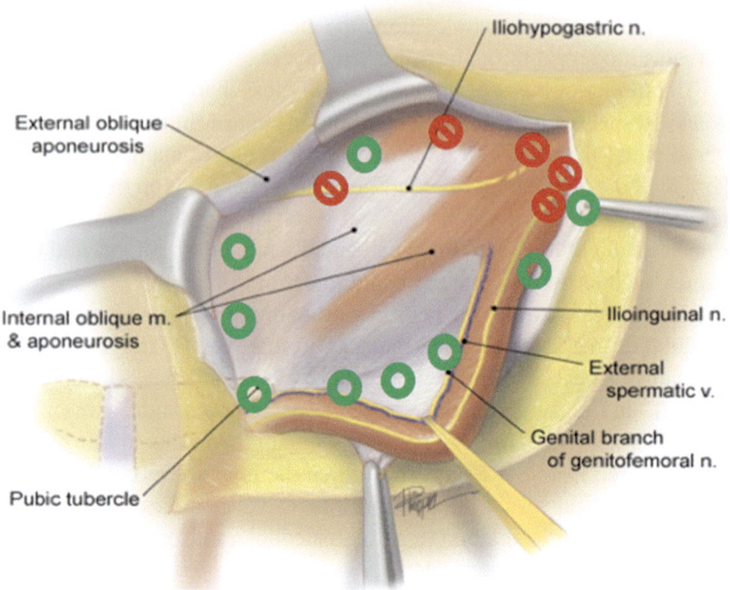

Fig. 39.2. Diagram of left open inguinal dissection prior to onlay repair via Lichtenstein technique. *Green circles* demonstrate areas of safe suture placement, i.e., at or medial to the level of the internal ring. *Red circles* demonstrate areas of unsafe suture placement, including no sutures at areas of visible nerves and in muscle, as the ilioinguinal and iliohypogastric nerves may be traveling within the muscle layers of the internal oblique prior to their emergence onto the conjoint tendon. In general, it is safest not to place any sutures lateral to the internal ring (Adapted from Chen and Amid [1], with kind permission of Springer Science+Business Media).

Reference

1. Chen DC, Amid PK. Technique: Lichtenstein. In: Jacob BP, Ramshaw B, editors. The SAGES manual of hernia repair. New York: Springer; 2013. p. 41–54.

40. Patient with Groin Pain After Tissue Repair, Anterior Approach

Shirin Towfigh

Chief Complaint

Right groin pain after tissue repair

History

The patient is a 73-year-old male, status post classic Shouldice repair of his right inguinal hernia 3 months earlier. Preoperatively he had a bulging hernia with scrotal extension, without significant pain. He now complains of daily groin pain, 7/10, ranging from 2/10 to 9/10. This began 3 weeks postoperatively after an otherwise uneventful early recovery period when he was feeling "amazing." He now reports a feeling of tightness, like a "rubber band" across his lower abdomen at the level of the repair. He feels like he wants to "pop out." The pain is at times burning, sharp, shooting, or a dull constant pain at baseline. The pain radiates to the upper inner thigh as a "minor but irritating" burning stinging pain. The pain also radiates to his flank and he feels pain at his hip bone. He denies testicular pain. He has swelling of the right groin that comes and goes. He also has bloating and feels "filled with gas." He has changed his diet, removed all dairy, and takes daily probiotics, stool softeners, and anti-gas medication, with no improvement. He denies constipation or straining.

Physical Exam

The patient gets up from sitting position with mild distress. His entire lower abdomen seems a bit edematous and bloated. The right groin has a healed incision and is edematous along the wound and its

B.P. Jacob et al. (eds.), *The SAGES Manual of Groin Pain*,
DOI 10.1007/978-3-319-21587-7_40

periphery. He is 2+ tender along the entire groin area, nonspecifically. There is no palpable mass or hernia recurrence. He has no hypesthesia or allodynia in the region.

Imaging

Computed tomography scan performed 2 months postoperatively demonstrated an intact repair without hernia recurrence. There were marked edematous changes without fluid collection.

Diagnosis

The patient was diagnosed with postoperative pain from anterior tissue repair, without evidence of neuropathy. He also had no evidence of spermatic cord injury, which can also at times be seen with this repair. His symptoms of increased pain and swelling are concerning for tearing of the repair, with associated edema and pain. As this is a tension repair, and the patient notably is an elderly male with a relatively large hernia, it is possible that the Shouldice repair was too tight and his tissue is not supportive of such a repair. There is no evidence of infection or hernia recurrence at this time, so conservative management alone is indicated.

Nonoperative Management Options

To address his inflammatory state and possible underlying tissue tearing from a tight repair, the patient was recommended to begin local treatment with ice as well as systemic treatment with nonsteroidal anti-inflammatory drugs (NSAIDs). He was also educated to rest the groin, similar to treatment for a sports hernia, until the area is healed. Once healed, he should begin a physical therapy regimen to regain abdominal core muscle strength and mobility at his hip joint.

Operative Treatment

If the patient does indeed prove to have a hernia recurrence in the future, then he is eligible for repair, which I recommend be performed with mesh. The patient initially did not wish to have mesh repair, as he was concerned about the risk for postoperative groin pain.

Outcomes and Discussion

Tissue repairs are tension repairs by definition. The complications associated with them may be due to direct injury at the time of operation (e.g., nerve injury due to cautery, traction), missed hernia (e.g., femoral hernia), too tight of a repair, or hernia recurrence. In the case of too tight of a repair, as the repair is purely muscle/fascia, and without mesh involvement, it is likely that this will gradually loosen with time. Local therapies to reduce risk factors (e.g., constipation, chronic cough, weight gain) for hernia recurrence can also help reduce pain and allow for improved recovery from such a repair. Also, physical therapy and exercise to improve mobility of the groin area (e.g., cycling) can help with the rehabilitation. Surgical options should be expectant only and address any hernia recurrence or irreversible causes for pain. It is notable that nerve mobilization is necessary in most situations for tissue repair to reduce risk of entrapment in the repair. This may theoretically increase the risk of chronic pain of neuropathic nature. In most cases, however, as there is no additional foreign body or inflammatory status, the patients improve with time and there is no need for intervention [1].

Meta-analysis of 16 trials showed no major difference in chronic pain with the Shouldice repair versus other open techniques [2]. It does have a higher recurrence rate compared to mesh techniques, by a factor of 3.80, but is considered to have a lower recurrence rate than other non-mesh techniques. That said, the chronic pain after Shouldice repair is significant, with over one third of patients having significant pain more than 3 months postoperatively [1, 3, 4]. This is typically higher than most mesh repairs, and so the clinician should evaluate these patients and their recovery with different standards than he does with patients who undergo open or laparoscopic repair with mesh. That said, the incidence of chronic pain rapidly decreases with time and is infrequently debilitating [1].

Conclusion

Patients who undergo tissue hernia repair are at risk for chronic pain, but this is mostly due to a tight repair, missed hernia (e.g., femoral hernia), and/or hernia recurrence. Minor instances of nerve injury may heal on their own, as may too tight of a repair.

References

1. Matthia W, Reinpold J, Nehls J, Eggert A. Nerve management and chronic pain after open inguinal hernia repair. Ann Surg. 2011;254(1):163–8.
2. Amato B, Moja L, Panico S, Persico G, Rispoli C, Rocco N, et al. Shouldice technique versus other open techniques for inguinal repair. Cochrane Database Syst Rev. 2012;4:CD001543.
3. Köninger J, Redecke J, Butters M. Chronic pain after hernia repair: a randomized trial comparing Shouldice, Lichtenstein and TAPP. Langenbecks Arch Surg. 2004;389(5):361–5.
4. Poobalan AS, Bruce J, Smith WC, King PM, Krukowski ZH, Chambers WA. A review of chronic pain after inguinal herniorrhaphy. Clin J Pain. 2003;19(1):48–54.

41. Right Inguinal Hernia with Osteitis Pubis: A Case Report of Osteitis Pubis and Ipsilateral Inguinal Hernia

Naif A. Al-Enazi and Brian P. Jacob

Chief Complaint

Right groin discomfort

History

A 46-year-old male, otherwise healthy, had right groin pain, which started 2–3 years previously and was not that bad at first, but became slowly more pronounced over the years. Pain was described as discomfort, not radiated anywhere localized to the groin area. The pain started suddenly after heavy exercise. It was intermittent, lasted a few seconds, and then disappeared. It was aggravated by exercise and spontaneously disappeared. No history of nausea, vomiting, diarrhea, or constipation. No history of lifting heavy objects or trauma. He was nonathletic.

Allergy history—unremarkable
Past surgical history—lipoma of back
Drug history—none
Past medical history—unremarkable
Social history—nonsmoker, nonathletic

Review of Systems

- Cardiovascular—normal
- Respiratory—breathing well
- Gastrointestinal—unremarkable

B.P. Jacob et al. (eds.), *The SAGES Manual of Groin Pain*,
DOI 10.1007/978-3-319-21587-7_41

- Neurological—unremarkable
- Genitourinary/renal—unremarkable
- Musculoskeletal—unremarkable
- Psychiatric—normal behavior

Focused Examination

BMI: 32 mg/kg^2.

Abdomen: soft and non-tender.

Umbilicus: no hernias.

Left groin: no obvious hernia on exam.

Right groin: patient was examined in standing and lying position, with and without Valsalva maneuver. A small reducible right inguinal hernia was palpable when standing and coughing, and the patient did not have any discomfort over this reducible bulge.

Right pubic bone and tubercle: mildly tender to palpation.

Left pubic tubercle: non-tender.

Symphysis: non-tender and stable to manipulation.

Rectus muscle: when flexed during sit-up maneuver, mildly uncomfortable over right tubercle insertion site only.

RIGHT Hip rotation: non-tender. RIGHT Leg elevation against resistance to evaluate the flexors (flexion at hip joint): mildly uncomfortable over the right tubercle with this maneuver. Reproducible symptoms. RIGHT Leg extension (extension at the hip joint): non-tender and no symptoms.RIGHT Adductor: mildly tender over the right tubercle inseriton with right leg adduction, reproducible symptoms.right groin pain with standing and squatting: NONE. RIGHT GROIN Sensory exam: no numbness, tingling, or hypersensitivity along any of the nerve distribution in the groins. rotation: non-tender.

Leg elevation against resistance to evaluate the flexors: mildly uncomfortable over the right tubercle with this maneuver.

Leg extension: non-tender.

Adductor: mildly tender over the right tubercle with right leg adduction.

Standing and squatting: non-tender.

Sensory exam: no numbness, tingling, or hypersensitivity along any of the nerve distribution in the groins.

Workup

Magnetic resonance imaging (MRI) of the osseous pelvis with attention to the pubic symphysis showed that there was mild marrow edema and subchondral remodeling at the pubic symphysis compatible with osteitis pubis (Fig. 41.1). The adductor tendons were intact. The rectus abdominis aponeurosis was intact as well. There was also a fat-containing right-sided inguinal hernia with direct and indirect component measuring 3 cm (Fig. 41.2).

Diagnosis

While the reducible right inguinal hernia in this patient was obvious, the patient's history and physical examination gave high suspicion for the clinical diagnosis of osteitis pubis as well. The symptoms of osteitis pubis can be presented as any complaint in the groin or lower abdomen [1]. Pain generally is localized over the symphysis and may radiate to the groin, scrotum, perineum, medial thigh, hip, or abdomen [2].

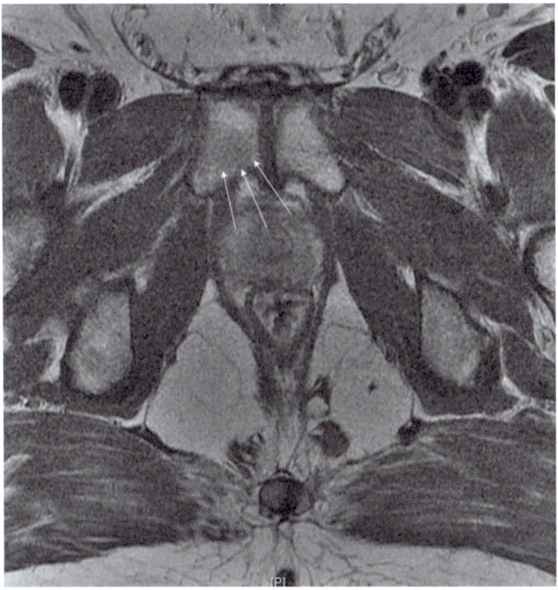

Fig. 41.1. Osteitis pubalgia on MRI.

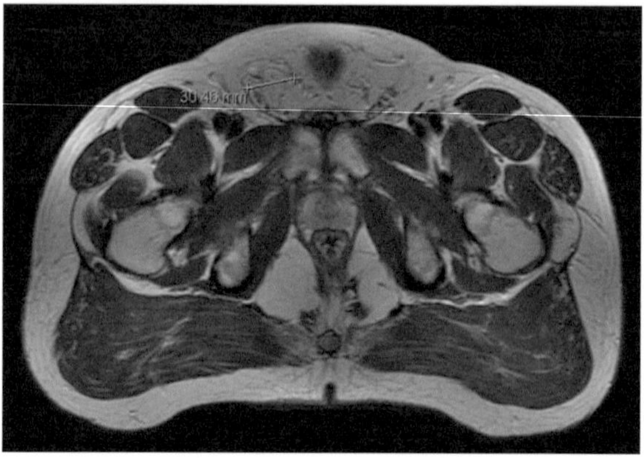

Fig. 41.2. Right inguinal hernia on MRI.

Physical findings for osteitis pubis can vary. It is important always to consider any sport involved, as well as the chronicity of the conditions [3]. Gradual onset of pelvic pain, with pubic symphysis tenderness, is one of the features of osteitis pubis [4]. One great modality for the diagnostic investigation in detecting the osteitis pubalgia is MRI. The differential diagnoses of chronic groin pain that can present similarly to osteitis pubis include musculotendinous strain, osteomyelitis, inguinal hernia, referred low back pain, intra-articular hip disease and genitourinary disease, and adductor tendinopathy. Osteomyelitis of the pubic symphysis is one of the diseases most commonly confused with osteitis pubis [5], so clinical evaluation requires a careful synthesis of history taking, physical examination, and appropriately directed investigation.

Discussion

Osteitis pubis, a rare condition, is characterized by pelvic pain localized over the symphysis pubis, in the lower abdominal muscles or in the perineum. The pain may radiate to the adductor region of the thigh, and patients may describe painful adductor muscle spasms. Aggravating factors are walking and standing from a seated position.

Osteitis pubis, also known as pubalgia or sometimes mislabeled as just athletic pubalgia, is one of the most chronic and debilitating syndromes affecting athletes [6]. It is described as the pubic bone stress injury occurring usually as a result of chronic overload or impaction trauma [7].

Osteitis pubis is also considered as noninfectious, inflammatory damage to the pubic symphysis and its supporting structures [8] anatomically. The pubic symphysis is mainly composed of fibrocartilage and is a nonsynovial, nonvascular joint. The pubic symphysis is reliant on four ligaments to maintain its supportive integrity. Most of the strength and support arise from the superior and inferior ligaments, whereas the anterior and posterior ligaments are of less supportive importance. The pelvic floor musculature, composed of the levator ani and coccygeus, inserts posteriorly at the pubic symphysis. The pectineus, rectus abdominis, and oblique externus muscles, as well as the inguinal ligament, insert near the superior portion of the pubic symphysis. The pubic rami give rise to several muscle origins: adductor magnus, adductor longus, adductor brevis, and gracilis. These muscles make up the adductors of the hip [1].

The prevalence of osteitis pubis among the general population of athletes ranges from 0.5 to 6.2 % [8, 9]. Although many different sports may be associated with osteitis pubis, sports with a higher risk include soccer, football, ice hockey, and rugby [9].

The etiology of osteitis pubis is unknown; repetitive trauma alone or in conjunction with opposing shearing forces across the pubic symphysis is likely the main contributing factor in many athletes [10, 11]. This may be due to different types of movements, including rapid acceleration-deceleration, kicking, and changes in direction.

Management

Nonoperativemanagement of osteitis pubis is similar to that of other causes of chronic groin pain and consists mainly of rest, ice or heat, and nonsteroidal anti-inflammatory drugs (NSAIDs) or other oral medication if patient can not take an anti-inflammatory medication. If these initial modalities are not helpful after a defined trial period (3 weeks for example), then the next intervention to consider would be glucocorticoid injections directly into the pubic symphysis or oral glucocorticoids [12, 13]. Additional nonoperative interventions can then also include physiotherapy focusing on core stability, muscle balance, and rotational hip range of movement; and activity modification. The most important lesson here is that if a patient has osteitis pubalgia and an inguinal hernia, just repairing the hernia alone will not alleviate the pubalgia pain. We therefore recommend initiating treatment for pubalgia first and delaying a hernia repair until the pubalgia pain is better controlled and the patient understands the difference between the hernia pain and the pubalgia pains.

The main goal of physical rehabilitation is to strengthen and stabilize the pelvis and pubic symphysis. Physical rehabilitation, usually for 6–8 weeks, has been found in multiple observational studies to be effective in reducing pain among patients with osteitis pubis [14]. Conservative measures have been shown to be effective in treating osteitis pubis, and the expected duration of treatment before resolution should be around 2–3 months. A majority of the patients with osteitis pubis respond very well to this modality of treatment. After the osteitis is improved or resolved, the hernia repair can then be sought.

Conclusion

Osteitis pubis can be easily missed when patients complain of groin pain and present with a simultaneous inguinal hernia. History and physical examination should include a high index of suspicion. If the diagnosis is suspected, the potential to make the diagnosis will increase, and thus the surgeon and the patient may be able to prevent a case of chronic groin pain that would otherwise be at risk for being associated with an inguinal hernia repair. Usually, the treatment of choice for patients with osteitis pubis is conservative, with rest, painkillers, and physiotherapy or steroid injection. The healing process for most patients is 6–8 weeks. In patients who have an obvious inguinal hernia and osteitis, it is important to educate the patients about the osteitis and its associated symptoms, treatment options, and outcomes before repairing the inguinal hernia.

References

1. Hölmich P. Adductor-related groin pain in athletes. Sports Med Arthrosc Rev. 1997;5:285–91.
2. Fricker PA. Osteitis pubis. Sports Med Arthrosc Rev. 1997;5:305–12.
3. Ruane JJ, Rossi TA. When groin pain is more than 'just a strain': navigating a broad differential. Phys Sportsmed. 1998;26(4):78–103.
4. Holt MA, Keene JS, Graf BK, Helwig DC. Treatment of osteitis pubis in athletes. Results of corticosteroid injections. Am J Sports Med. 1995;23(5):601–6.

5. Pauli S, Willemsen P, Declerck K, Chappel R, Vanderveken M. Osteomyelitis pubis versus osteitis pubis: a case presentation and review of the literature. Br J Sports Med. 2002;36(1):71–3.
6. Rodriguez C, Miguel A, Lima H, Heinrichs K. Osteitis pubis syndrome in the professional soccer athlete: a case report. J Athl Train. 2001;36(4):437–40.
7. Gilmore J. Groin pain in the soccer athlete: fact, fiction and treatment. Clin Sports Med. 1998;17(4):787–93.
8. Batt ME, McShane JM, Dillingham MF. Osteitis pubis in collegiate football players. Med Sci Sports Exerc. 1995;27(5):629–33.
9. Johnson R. Osteitis pubis. Curr Sports Med Rep. 2003;2(2):98–102.
10. Radic R, Annear P. Use of pubic symphysis curettage for treatment-resistant osteitis pubis in athletes. Am J Sports Med. 2008;36(1):122–8.
11. Meyers WC, Foley DP, Garrett WE, Lohnes JH, Mandelbaum BR. Management of severe lower abdominal or inguinal pain in high-performance athletes. PAIN (Performing athletes with abdominal or inguinal neuromuscular pain study group). Am J Sports Med. 2000;28(1):2–8.
12. O'Connell MJ, Powell T, McCaffrey NM, O'Connell D, Eustace SJ. Symphyseal cleft injection in the diagnosis and treatment of osteitis pubis in athletes. AJR Am J Roentgenol. 2002;179(4):955–9.
13. King JB. Treatment of osteitis pubis in athletes: results of corticosteroid injections. Am J Sports Med. 1996;24(2):248.
14. Choi H, McCartney M, Best TM. Treatment of osteitis pubis and osteomyelitis of the pubic symphysis in athletes: a systematic review. Br J Sports Med. 2011;45(1):57–64.

42. Patient with Chronic Pelvic Pain

Shirin Towfigh

Chief Complaint

Left groin pain and pelvic pain

History

The patient is a 41-year-old female with left groin and pelvic pain for 8 months. It first was felt when moving crates of files at work. The pain started in the left groin and radiates up to the umbilicus and around her back. She also has pain that radiates down her leg, mostly anterior thigh. The pain is currently 10/10 and ranges from 6/10 to 10/10. It is a pinching, "so sharp," shooting pain that occurs daily. Any pressure on the area causes pain. This includes cuddling from her children, as she is also very sensitive in the area. She wears skirts and dresses to work, as formal pants and belts cause too much pain in the area. The pain is worse with prolonged standing, sitting, coughing, laughing, sneezing, climbing stairs, getting out of a car or bed, bending, and with crossing legs. Sexual intercourse is painful. The pain is worse during her menses and at the end of the day. She is best when lying flat. She has nausea when the pain is at its worst.

The pain is severe and activity limiting. She has been to the emergency room twice due to pain. She has been evaluated by her gynecologist as well as gastroenterologist and colorectal surgeon. Colonoscopy was normal. She was sent to a pain management specialist, as she was told she has muscle spasm. She underwent local injection, which increased her level of pain.

B.P. Jacob et al. (eds.), *The SAGES Manual of Groin Pain*,
DOI 10.1007/978-3-319-21587-7_42

Physical Exam

The patient was in no discomfort. She has a healed Pfannenstiel incision from her prior Cesarean section. There is 3+ tenderness with associated fullness at the internal ring. There is no visible bulge or reducible mass. She has no hypesthesia and allodynia in the area.

Imaging

Pelvic ultrasound and abdominal ultrasounds were both nondiagnostic. CT scan showed a small left inguinal hernia with fat content.

Diagnosis

The patient has an occult inguinal hernia based on history and physical examination that are suggestive but not diagnostic of a hernia and then imaging which is diagnostic of a hernia. Her symptoms are not suggestive of a gynecologic or gastroenterologic disorder, as she has point tenderness at the internal ring and pain with activity that involves engaging the abdominal muscles.

Operative Treatment

The patient was offered open versus laparoscopic repair. There was a hint of possible femoral hernia on computed tomography (CT) scan, and so laparoscopic repair was considered the best option. Operative findings were of an indirect and femoral hernia. This was repaired with mesh, with fixation to Cooper's ligament, using TEP technique.

Postoperative Course

The patient had complete resolution of her pain as early as in the postoperative recovery unit. She was followed up for 2 years and has not had any recurrence of her symptoms.

Outcomes and Discussion

Occult inguinal hernias are more common among women but are often overlooked as a cause for pelvic pain, as inguinal hernias are considered a male disease. Symptoms are variable and may include pain radiating from the groin to the labia, vagina, down the leg, and around the back. The physical examination finding that is highly specific for such a hernia is point tenderness at the internal ring. In fact, early studies show this to be found in 98–100 % of all occult hernias, and ilioinguinal neuralgia symptoms were noted in 63 % (Tables 42.1 and 42.2) [1]. In our study, the finding of point tenderness at the internal ring meant inguinal hernia was between 13% and 25 % more likely to be the correct diagnosis in women with chronic pelvic pain, when correcting for BMI, age, dysmenorrhea, and radiating pain [2]. The overall positive predictive value of occult hernia when tenderness was elicited on groin examination was 74 %. The sensitivity was 60 % and specificity was 88 %. All other typical findings, such as a visible bulge, reducible mass, or palpable defect, are often not found.

Table 42.1. Preoperative symptoms in 192 cases of nonpalpable inguinal hernias (from Spangen and Smedberg [1], with kind permission from Springer Science+Business Media).

Type of inguinal pain	**No.**
Dull, gnawing pain	190
Neuralgic pain only	2
Combined dull and neuralgic pain	136
Pain, radiating from the groin to the ipsilateral	
Thigh	101
Flank	62
Lower abdomen	33
Pain accentuated by	
Physical exertion	176
Menstruation	19
Mental stress	3

Table 42.2. Clinical findings in 192 cases of occult inguinal hernia (from Spangen and Smedberg [1], with kind permission from Springer Science+Business Media).

Finding	**No.**
Tenderness corresponding to the deep inguinal ring upon palpation during a Valsalva maneuver	192
Hyperalgesia of the skin corresponding to the distribution of the ilioinguinal nerve	121

Imaging is often required to confirm the diagnosis prior to committing the patient to surgical exploration and repair. Ultrasound and magnetic resonance imaging (MRI) are found to be good options, and CT scan is considered to be a poor option for the pelvis. In our study, ultrasound has a 100 % positive predictive value and 0 % negative predictive value [3]. Thus, if an ultrasound is negative, and symptoms and physical examination are suggestive of inguinal hernia, then MRI (not CT scan) is recommended as the next modality [3, 4]. MRI has a 95 % positive predictive value and 85 % negative predictive value. MRI is 91 % sensitive and 92 % specific for findings of occult inguinal hernia. On the other hand, among patients who underwent CT scan with negative findings, 91 % had occult hernias notable on MRI.

On operative exploration, the patient has a fat-containing hernia defect. This is typically preperitoneal fat only, without peritoneal involvement, i.e., no hernia sac. Thus, intraperitoneal examination by laparoscopy, without takedown of the peritoneum and fat to visualize the fascia itself, may provide a false-negative result.

Conclusion

Inguinal hernia can cause chronic pelvic pain. The absence of a hernia on examination should not rule out inguinal hernia as the cause of pain. A complete history and physical examination, followed by imaging (ultrasound or MRI), are necessary to rule out inguinal hernia as the cause of pain. Surgical treatment may provide immediate cure.

References

1. Spangen L, Smedberg SG. Nonpalpable inguinal hernia in women. In: Bendavid R, Abrahamson J, Arregui ME, Flament JB, Phillips EH, editors. Abdominal wall hernias: principles and management. New York: Springer; 2001. p. 625–9.
2. Saad CA, Kim DS, Towfigh S, Solnik MJ. Inguinal hernia as a cause of chronic pelvic pain: a key sign to make the diagnosis (abstract). J Min Invasive Gynecol. 2014;21 Suppl 6:S76.
3. Miller J, Cho J, Michael MJ, Saouaf R, Towfigh S. Role of imaging in the diagnosis of occult hernias. JAMA Surg. 2014;149(10):1077–80.
4. Robinson A, Light D, Kasim A, Nice C. A systematic review and meta-analysis of the role of radiology in the diagnosis of occult inguinal hernia. Surg Endosc. 2013; 27(1):11–8.

43. Thoracolumbar Syndrome

James A. Rydlewicz and Dean J. Mikami

Editor's Comment (BPJ)

Thoracolumbar syndrome (TLS) is a very rare but real etiology of lower back pain and referred chronic groin pain. The diagnosis is often by exclusion, and based primarily on history and physical examination alone, TLS is a challenge to diagnose and treat, but should be part of the groin pain differential diagnosis. Distal sensory nerves that originate from the posterior primary rami of the thoracolumbar spinal nerves T12–L2 can be irritated without any obvious magnetic resonance imaging (MRI) or X-ray findings, thus increasing the challenge of making the diagnosis. Treatment should initially be aimed at physical therapy and rehabilitation modalities. Pain management specialists can then inject the trigger points. Outcomes after surgical intervention have not been well documented.

Introduction

Patients who present with groin pain may be experiencing referred pain from a spinal pathology. Robert Maigne first described thoracolumbar syndrome in 1974; it is sometimes called Maigne's Syndrome [1]. The thoracolumbar junction is comprised of the T10–11, T11–12, and T12–L1 vertebrae. The dermatomes T10–L2 are responsible for the referred pain that patients experience [2]. Patients usually complain of low back pain, but can also have ipsilateral gluteal and groin pain.

Thoracolumbar syndrome is defined by a dysfunction of the thoracolumbar junction referring pain in the corresponding dermatomes of T10–L2. In particular, T12 and L1 are specifically located in the groin region, and they emerge at the level of the thoracolumbar junction.

B.P. Jacob et al. (eds.), *The SAGES Manual of Groin Pain*,
DOI 10.1007/978-3-319-21587-7_43

Low back pain may also be involved with groin pain, or groin pain can be an isolated complaint [2]. T12 is the transitional vertebra of the spine where the thoracic facet joint meets the lumbar facet joint. It is believed that the thoracolumbar facet joint irritation is the cause of the pain. This irritation causes unilateral pain to the distribution of the posterior primary rami of the lower thoracic and upper lumbar nerve roots [3].

Clinical Manifestations

Low back pain is the most common complaint of individuals with thoracolumbar syndrome and usually starts with a rotational twisting motion. The pain is usually unilateral, located in the sacroiliac or low lumbar region and may radiate to the lateral thigh. Pain is often made worse with extension and certain positions. Patients may also complain of lower abdominal, groin, pubic, or testicular pain. Patients describe the pain as a deep aching sensation, which is commonly mistaken for intestinal, urologic, or gynecologic disorders [4]. These clinical signs correlate with T12–L1 spinal nerve root innervations (Fig. 43.1).

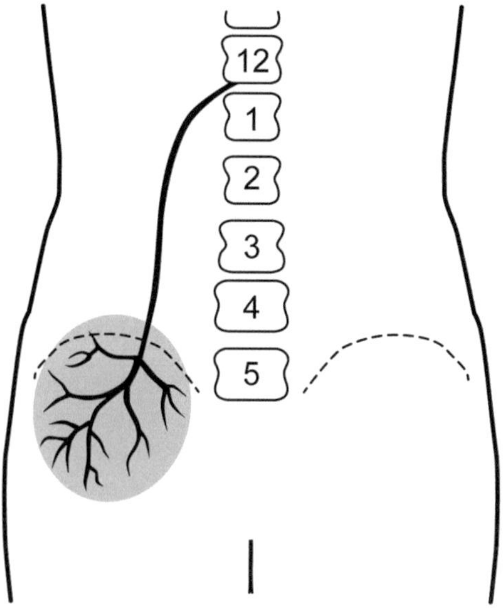

Fig. 43.1. T12–L1 thoracolumbar nerve root compression with referred pain.

The posterior ramus supplies the subcutaneous tissue of the lower waste and buttocks. The anterior ramus supplies the lower abdomen and groin [5]. Patients with TLS may also have other etiologies causing groin pain, in addition to the TLS, thus adding complexity to pinpointing TLS as one of the sources of discomfort.

Physical Examination

Physical exam begins with examination of the spine with the patient in a prone position. Lateral pressure on the spinous processes of T9–L3 should elicit unilateral pain. The compression should be performed in both a right and left movement. Direct compression over the affected facet will elicit the same tenderness (Fig. 43.2). The posterior iliac crest should then be palpated to identify point tenderness. Rubbing the crest in an up-and-down motion should elicit pain at a point usually 7 cm from the midline. The pain should be sharp in nature. This point is called the posterior iliac crest point; it is where the irritated cutaneous branches of T11–L1 are compressed [3] (Fig. 43.3).

The pinch–roll test is then performed to test for hyperalgesia of the skin and subcutaneous tissues of the gluteal and iliac crest region. Referred pain accompanies hyperalgesia and thickening of the skin. The test is performed by grasping a fold of skin between the thumb and forefinger and rolling the tissue in a controlled manner. The involved side should elicit tenderness compared to the opposite side [4] (Fig. 43.4).

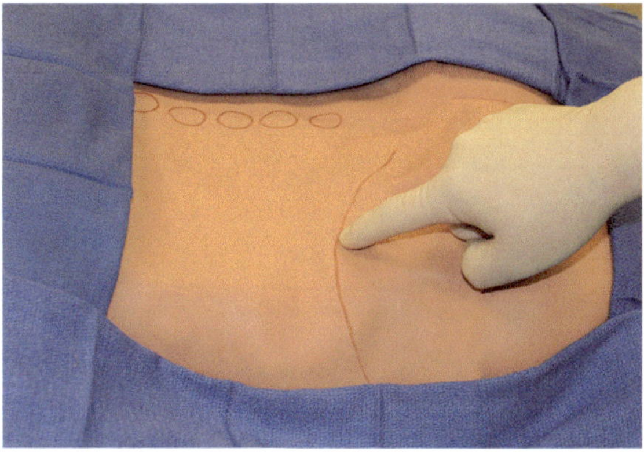

Fig. 43.2. Point pressure over iliac crest.

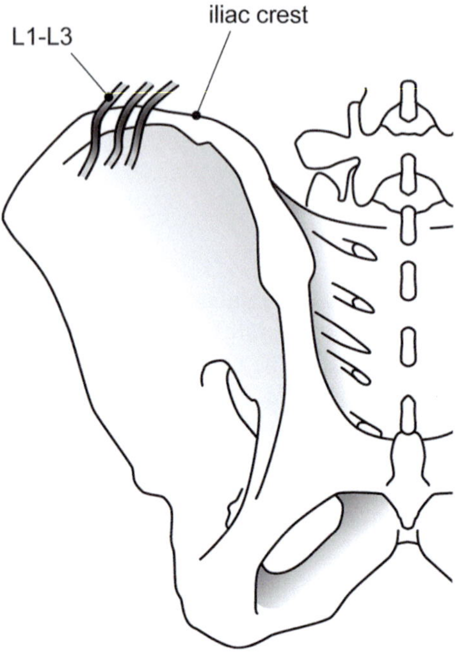

Fig. 43.3. Anatomical position of the L1–L3 as it transverses the iliac crest.

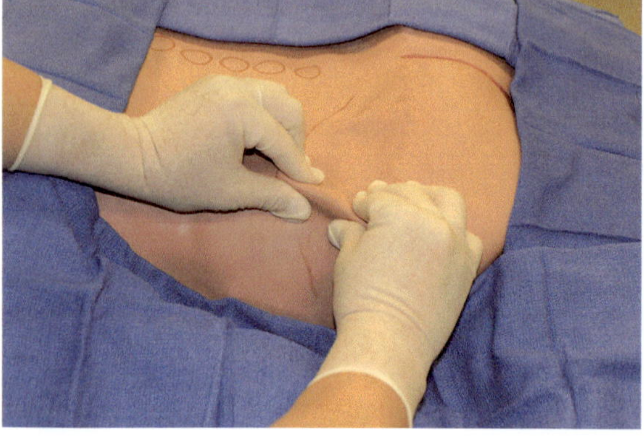

Fig. 43.4. Pinch–roll test.

Groin pain should be examined with standard inguinal examination to rule out inguinal hernia. Thoracolumbar syndrome has two clinical features that are characteristic: a positive pinch–roll test and tenderness on the superior aspect of the pubis. Palpation of the periosteum of the pubis should cause tenderness on the affected side. The pinch–roll test should be performed in a supine position in the inguinal area [2].

Workup and Diagnosis

Plain films of the thoracolumbar region should be obtained to rule out gross segmental instability. Computed tomography (CT) and MRI of the spine can evaluate for masses, disk herniation, spinal stenosis, and fractures. CT scan of the abdomen can be useful to rule out intra-abdominal pathology. Thoracolumbar syndrome is diagnosed clinically and most radiological studies will be normal [6].

Clinical criteria for diagnosis include (1) a positive pinch–roll test, (2) a positive posterior iliac crest point, (3) localized tenderness over the affected thoracolumbar segment, and (4) tenderness in the facet joints at the affected level [3]. A diagnostic nerve block can also be performed to confirm the diagnosis either at the bedside or with fluoroscopic guidance. The needle is inserted over the painful facet about 1 cm from the midline. Lidocaine 1 % is injected around the joint and laterally around the dorsal ramus. One can also inject lidocaine at the posterior iliac crest point. The pain should be gone within minutes of injection [2].

Treatment

Treatment of thoracolumbar syndrome is directed at the vertebral column. Spinal manipulative therapy directed at the correct thoracolumbar posterior joint is a first-line treatment. Manipulation is a forced movement applied to the joint and is contraindicated in severe osteoporosis [2]. If manipulative therapy does not cause relief of symptoms, then injections of corticosteroids around the painful facet joint and posterior iliac crest may help. Nonsteroidal anti-inflammatory agents, massage, and physical therapy have also been shown to be helpful. Surgical treatment is rarely indicated. Concern for nerve entrapment as the cause of the problem and failure of medical management would be an indication for surgery and nerve release [4].

Conclusion

Low back pain with associated groin pain can be caused by referred pain for irritation of the thoracolumbar facet joints. A detailed history and complete groin and spine exam are essential to diagnose thoracolumbar syndrome. Once diagnosed, spinal manipulative therapy or injection of trigger points is the best treatment option.

References

1. Maigne R. Origine dorso-lombaire de certaines lombalgies basses. Rôles des articulations interapophysaires et des branches postérieures des nerfs rachidiens. Rev Rhum. 1974;41(12):781–9 (article in French).
2. Maigne JY. Thoracolumbar junction and thoracolumbar spinal pain syndromes. Société Française de Médecine Manuelle, 1996. http://www.sofmmoo.com/english_section/4_thoracolumbar_junction/thoracolumbar_junction_australie.htm. Accessed 16 Mar 2015.
3. Proctor D, Dupuis P, Cassidy JD. Thoracolumbar syndrome as a cause of low-back pain: a report of two cases. J Can Chiropr Assoc. 1985;29(2):71–3.
4. Maigne R. Low back pain of thoracolumbar origin. Arch Phys Med Rehabil. 1980;61:389–95.
5. Kim SR, Lee MJ, Lee SJ, Suh YS, Kim DH, Hong JH. Thoracolumbar junction syndrome causing pain around posterior iliac crest: a case report. Korean J Fam Med. 2013;34(2):152–5.
6. Fortin JD. Thoracolumbar syndrome in athletes. Pain Physician. 2003;6(3):373–5.

44. Patient with Referred Hip Pain

Shirin Towfigh

Chief Complaint

Right groin pain, "like a tearing sensation at my groin crease."

History

The patient is a 41-year-old male with chronic right groin pain. He has had a full workup at major academic institutions, including evaluation by gastroenterologists and pain management specialists, and has undergone colonoscopy, endoscopy, and injections, with no improvement in his pain. His main complaint is a tearing sensation at the right groin crease. It radiates from the groin to his anterior superior iliac spine (ASIS). It also radiates down his leg and into his thigh. He denies pain radiating around to his back or any hip pain. He admits that the pain is "deep" and not at the surface of his groin. The pain is worse with activities. It is not better when lying flat. He has pain when stepping into the car on the driver side. He also notes a hip click on that side. When the pain is at its peak, he walks with a limp to protect himself from the pain and prefers not to bear weight on that right leg. It is painful to lie on his right side. He prefers to lie on his contralateral side in fetal position. He denies groin bulge. He has no testicular pain.

B.P. Jacob et al. (eds.), *The SAGES Manual of Groin Pain*,
DOI 10.1007/978-3-319-21587-7_44

Physical Exam

No visible bulge or palpable hernia defect in the right groin. Nonspecific tenderness 2+ at the right groin, at the internal ring region. Non-tender ASIS, hip area, pubic bone. Pain with passive flexion and internal rotation of the right hip.

Imaging

Magnetic resonance imaging (MRI) of the pelvis was ordered to evaluate for inguinal hernia. It showed suggestion of hip anterior acetabular labral tear, with increased signal at acetabulum (Fig. 44.1), as well as CAM-type femoroacetabular impingement (FAI).

Diagnosis

Right hip labral tear with FAI.

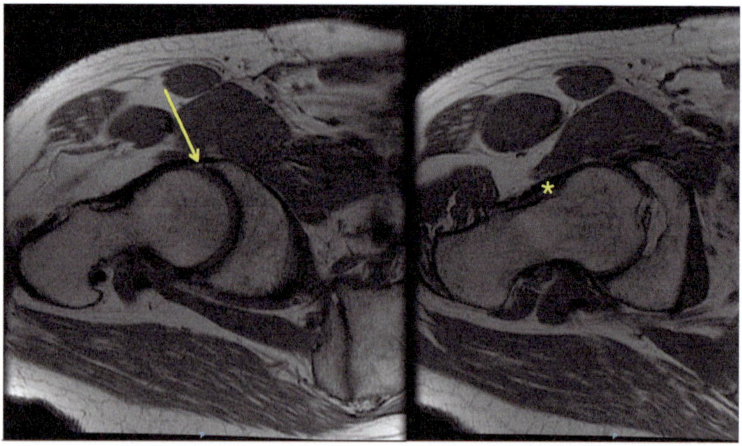

Fig. 44.1. MRI of the pelvis, non-contrast, T2 axial view. Right anterior acetabulum with intermediate linear increased intensity signal, suggestive of labral tear (*yellow arrow*). Also, concomitant osseous bump at femoral head-neck junction suggestive of CAM-type femoroacetabular impingement (*yellow asterisk*). MR arthrogram is indicated.

Treatment

The patient was referred to an orthopedic specialist for evaluation and treatment. MR arthrogram confirmed the diagnosis. He underwent joint injection, with resolution of his pain. He is scheduled for arthroscopic surgery.

Outcomes and Discussion

Hip disorders, especially labral tears, may present with groin pain. A detailed history and physical examination are key to help differentiate their diagnosis from other diagnoses, such as inguinal hernia. In fact, it is not uncommon for a patient to have an inguinal hernia at the same setting of a hip disorder. It is important to correctly diagnose the cause of the patient's groin pain and to provide treatment focused on that disorder. In my practice, I often see patients who undergo elective inguinal hernia repair, with no improvement in their preoperative symptoms. Workup proves that their preoperative symptoms were indeed due to another pathology, such as a hip disorder, and their inguinal hernia was asymptomatic. A poor outcome would be to have a complication from the elective inguinal hernia repair while at the same time the patient's main cause of preoperative pain was due to the hip disorder, thus altogether complicating the patient's plan of care, outcome, and recovery.

Key elements in the history can help differentiate hip disorders from groin pain. With hip disorders, the pain is often not relieved with lying flat; there may be a limp or difficulty of weight bearing on the ipsilateral leg. Many patients with hip disorders describe pain as wrapping around the ASIS, like the letter "C," cupping their thumb and fingers at their waist. The pain is often deeper than expected and lower than expected for an inguinal hernia. Also, a sensation of tearing or pulling is often described, as opposed to a dull or sharp pain. Activities that are remarkably painful include an adduction, such as stepping into or out of a car or sitting cross-legged on the floor. Many patients report a hip click or popping sensation. On physical exam, passive flexion and adduction, with internal rotation, elicit pain and limitation in movement.

Conclusion

Hip disorders such as labral tear, may present with groin pain. Their distribution of pain can be similar to that of an inguinal hernia, with groin pain radiating down the leg and around the back and worsening with activities. Key questions in the history can help differentiate a hip disorder from that of an inguinal hernia. Remember that patients with inguinal hernias may have a primary hip problem. A hip X-ray and MRI, preferably a dedicated hip MR arthrogram, are diagnostic. See Chap. 8, for a full review and discussion on this very important topic.

45. Value-Based Clinical Quality Improvement for Chronic Groin Pain After Inguinal Hernia Repair

Bruce Ramshaw

Introduction

Improving the value of patient care has become the challenge for healthcare in the twenty-first century. In healthcare, value should be defined by quality measures, patient safety and satisfaction, and the costs of care for a defined care process throughout the patient's entire cycle of care. Until recently, there have been no examples of patient care based on defined care processes and collected outcome measures that determine value. However, publications from business experts have proposed a model for patient care that would allow for defining, measuring, and improving value [1–3]. There have also been recent guidelines and a book chapter describing these concepts applied to healthcare [4, 5]. Continuing to provide patient care in a model designed in the nineteenth century, using the principles of reductionist science, evolved from the time of the Renaissance, is no longer adequate. Our current system structural design for "modern" patient care includes the hospital model with hierarchy, bureaucracy, and departmental silos, causing fragmentation in care that is becoming more inefficient as complexity increases [6, 7]. Another system structure for providing patient care is the individual physician model, which is also not sufficient in light of the exponential increase in medical knowledge [8]. Both core structures for providing patient care are inadequate given the increasing complexity of patient care and the increasing pace of change in our world in general. A complex systems science view of healthcare, which is based on principles that describe "complex phenomena demonstrated in systems characterized by nonlinear interactive components," allows us to simplify patient care by designing care around definable patient groups, diseases, and problems [9].

B.P. Jacob et al. (eds.), The SAGES Manual of Groin Pain,
DOI 10.1007/978-3-319-21587-7_45

The information generated by the care processes can be used to improve the outcomes of care over time. This continuous improvement of the patient's entire cycle of care has the potential to lead to improved quality, safety, and patient satisfaction at the same time that costs are lowered, resulting in improved value [1, 2].

The clinical quality improvement (CQI) program described in this chapter demonstrates an attempt at process improvement for patients with chronic groin pain after inguinal hernia repair. With chronic groin pain after inguinal hernia repair becoming a more common and sometimes devastating problem, there is an opportunity to significantly improve the outcomes for this group of hernia patients and decrease the costs of care for this patient group, thus increasing overall value.

Traditionally, improvements in patient care have been dependent on established clinical research tools such as prospective randomized controlled studies. However, using traditional research tools for a complex dynamic process—such as for a patient with chronic pain after hernia repair—with inherent uncontrollable variables can be inadequate to improve value for patients. Recently, principles of CQI have been introduced to improve clinical care. The value of applying these principles has already been established for a portion of a patient's cycle of care: reducing central line infection, for example [10]. Implementing the principles of CQI for the entire cycle of care for patients who develop chronic pain after abdominal wall (inguinal and ventral) hernia repair has not yet been demonstrated. A comparison between traditional clinical research and CQI is presented in Table 45.1.

Methods: Developing a CQI Program for Patients with Chronic Pain After Hernia Repair

Using principles of complex systems science and tools such as CQI programs and nonlinear data analytics (such as predictive analytics), we can define a variety of patient groups who had abdominal wall hernia disease and related complications, such as chronic pain after hernia repair. We have constructed a diverse hernia team to serve the needs of this patient group. Based on feedback from former patients and review of current literature, a dynamic care process is defined for the entire cycle of care, from the moment of first symptom or contact until full return to a maximum quality of life, with ongoing contact for long-term follow-up. One step in the process is to determine the factors involved in producing various outcomes. In reviewing the literature, two sources have produced

Table 45.1. Comparison between traditional research and CQI (with the kind permission of Surgical Momentum, LLC).

	Traditional clinical research		**Value-based clinical quality improvement**
Intent	Testing of hypotheses		Improving value
	Traditional clinical research		**Clinical quality improvement**
	Pro	**Con**	**Comparison**
Pre-market	FDA requirement	Costly and lengthy	Not currently acceptable for pre-market
Human subjects research (potential risk to patient)	Appropriate ethical protections	Costly, limited ability to interpret value	Not ethically appropriate in current environment
Post-market surveillance	Might help determine harmful devices	Costly and lengthy	Appropriate mechanism for post-market surveillance because information is gathered in real time in the real clinical world with no change in patient care
Clinical research	Traditionally known and accepted research mechanism	Costly, rarely answers clinical questions adequately	Ideal for real-world clinical research. Over time, can lead to improved value of care and opportunities to define unmet clinical needs
CQI project	Known research mechanism	Costly, wasteful, may be unethical, limited ability to interpret value	Ideal, as long as it is applied to the whole process of care, or if applied to a subprocess, the outcomes of the whole process are measured concurrently
Off label use/obtain additional indications	May be industry initiated (depending on risks to the patient)	Costly and lengthy	May be appropriate—clinician initiated, no increased risk to patient
	Traditional clinical research		**Clinical quality improvement**
Patient consent	Disclosed to patient that they are participating in a defined clinical research project		Disclosed to patient that CQI is a natural part of their care and that information collected will be used primarily to improve the patient care process for future patients

Table 45.2. Preoperative and postoperative factors that can contribute to the development of chronic pain after hernia repair.

Preoperative factors	Postoperative factors
Pain greater than 1 month	Pain
Repeat surgery	Post-op radiation
Psychological vulnerability	Neurotoxic chemotherapy
Anxiety	Depression
Females	Psychological vulnerability
Younger age	Anxiety
Worker's compensation	Neuroticism
Inefficient diffuse noxious inhibitory control	

Table 45.3. Factors that can contribute to the development of chronic pain after hernia repair and increase the vulnerability to pain.

Having English as a second language
Race and ethnicity
Income and education
Sex and gender
Age group
Geographic location
Military veteran status
Cognitive impairments
Surgical patient
Cancer patient
End of life

patient-related factors that contribute to the development of chronic pain after surgery. These factors are listed in Tables 45.2 and 45.3 [11, 12].

Patients are offered a variety of choices for treatment (including non-operative management) of their chronic pain after hernia repair. Most patients have already sought and received many nonsurgical treatments prior to seeking a surgical option for their chronic pain. The surgical treatment choices include a diagnostic laparoscopy with attention to the presence of intra-abdominal adhesions, the presence of interstitial and hidden hernias, and any foreign body such as mesh, tacks, sutures, etc. that may be contributing to the entrapment or irritation of nerves and potentially contributing to inflammation and chronic pain. A review of the current evidence helped to establish the dynamic care processes. In all cases, the patient and family are included in a shared decision process. Information is provided by the hernia team, including the director of patient care management, other patient care specialists, and a surgeon who is experienced with the surgical treatment for this problem. A set of patient education documents is given to the patient and family and includes a basic education summary for this problem, a group of frequently asked questions (FAQs) generated by former patients for the entire cycle of care, and a copy of a book chapter on this problem with a description of surgical treatments. Patients and family members are encouraged to do their own research, to talk with other patients, and to consider other opinions, including from other surgeons who are considered experts in this area.

In this model, we also define outcome measures that determine the value of care (quality, safety, satisfaction, etc.). These measures are obtained based on the subjective and objective input from a multidisciplinary hernia team, including the patient and family. As part of CQI, the hernia team enters into a data-sharing contract that allows the de-identified patient information to be shared with others who could add value to the process of interpreting the data and might contribute ideas for improved care. In addition to the core hernia team members, business operations specialists, engineers, and associates from the manufacturers of drugs and devices used in the care of this patient group can potentially contribute ideas and knowledge to improve the outcomes for this CQI program. As a part of the CQI program, a group of volunteer patient and family members, surgical residents, medical students, and other general surgeons may also participate at various times to add their perspective to the improvement process. This can occur through participation at regularly scheduled CQI meetings. The primary objective of this CQI program is for the improvement of value for the patient within the local care process. Secondary goals may include sharing this de-identified data and analysis with hospitals, other physicians, patients, medical device companies, regulatory bodies, and others within the healthcare value stream. Implementing CQI as a part of the actual patient care process allows for the coordination of care and quality improvement efforts to be exempt from the strict requirements of the Health Insurance Portability and Accountability Act (HIPAA), and the effort is not required to go through an Institutional Review Board (IRB) process [13].

The process of developing a CQI program for patients with chronic pain after hernia repair will initially be limited to those hernia practices that regularly treat this patient group. However, the data and analytics that are generated from these specialized hernia programs can be shared with any surgeon, hospital, or person with chronic pain after hernia repair that could benefit from access to this information. A sample worksheet that could be used to generate a CQI program for this condition is presented in Fig. 45.1.

Determining "What Matters"

Patient Care Process Definition Worksheet

Surgical Category	**Procedure**
Chronic pain after hernia repair	Laparoscopic and open operation to treat pain

Process Variables Worksheet

What Matters	**Specific Factors and Data Elements**	**Additional Comments**
Do patient demographics matter? If so, what patient variables?	Yes Age (DOB) Gender (M/F) BMI (Ht, Wt, Calc BMI) Smoker (Current/Former) Controlling personality (Y/N) Worker's Comp (Y/N) Active lawsuit (Y/N)	
Do symptoms matter? If so, what symptoms?	Yes Active wound infection (Y/N) Active mesh infection (Y/N) Pain level Ability to do normal activities Ability to work Functional assessment	Capture the duration of the symptoms and time of onset after hernia repair
Do pre-op tests matter? If so, what tests?	No	
Do medical history variables impact the disease process? If so, which medical history variables?	Yes Emotional complexity (low, medium, high) Medical complexity (low, medium, high) History of psychological disease (Y/N) Prior treatment for pain (list)	
Do surgical history variables impact the disease process? If so, which surgical history variables?	Yes # of recurrent hernias (for this hernia) # Prior abdominal surgeries (include all hernia repairs) History of mesh infection (Y/N) History of wound infection (Y/N)	
Are specific meds pertinent? If so, which meds?	Yes Current opioid use (drug, dose, frequency)	
For history of present illness, what questions are important to answer?	Effect on quality of life: - ability to work - ability to exercise - ability to do leisure activities	
What intra-op specific variables matter? What questions are important?	Date of surgery (age at time of surgery) Procedure: Intraoperative nerve block: R/L Bilateral Type Who performed surgery? Who performed TAP block?	
What intra-op process deviations could occur?	i.e., inadvertent bladder injury	

Fig. 45.1. A sample worksheet to generate a CQI program for chronic pain after hernia repair.

Post-op hospital stay/ Discharge variables	Length of stay (days) ICU stay Y/N # Days Pain medication (convert to morphine equivalent): IV/IM: Dil Morph Dem PO Dil 2 mg Perc 5 Loritab 7.5 Roxicet 10 PACU information: time (min), opioid use (morphine equivalents)	
Post-op – List possible complications	Respiratory Treatment Transfer to ICU Intubation (# days intubated) Bleeding Wound complication Minor Moderate Major Pneumonia Pulmonary effusion Treatment PE Ileus NG Tube Delayed D/C Cardiac Arrythmia MI Other List	
Post-op clinic visit Short term (0-90 days) recovery variables?	Pain at rest Pain while active Functional assessment Complications: Wound (minor, moderate, major) Re-Hosp Reason Patient satisfaction: Happy with cycle of care (Y/N) Improvement suggestions Any patient safety issues (list) Ability to return to work (Y/N, Y-full, part) Ability to return to leisure (Y/N, Y-full, part) Any other issues?	
Post-op clinic visit Long term (90+ days) recovery variables?	Pain at rest Pain while active Functional assessment Complications: Re-Hosp Reason Recurrence Chronic pain (Degree of improvement) Patient satisfaction: Happy with cycle of care (Y/N) Improvement suggestions Any patient safety issues (list) Ability to return to work (Y/N, Y-full, part) Ability to return to leisure (Y/N, Y-full, part) Any other issues? Costs for entire cycle of care	

Fig. 45.1. (continued)

Discussion

This chapter attempts to describe the principles of CQI and nonlinear statistical analytics applied to the entire cycle of care for patients with chronic groin pain after hernia repair. The point of CQI is not to prove or disprove a direct cause and effect with various process improvement interventions but to define, measure, and improve the value of care for patients. Implementation of ideas for process improvement is one way to attempt to improve outcomes that define value.

Another way to attempt to improve value is to analyze the data that is generated from real patient care to attempt to predict outcomes of treatment, termed predictive analytics [14, 15]. Predictive analytics is the practice of extracting information from existing data sets in order to determine patterns and to predict future outcomes and trends. Other nonlinear statistical methods such as factor analysis can produce weighted correlations (positive and negative). This analytical tool can help determine what factors contribute the most to outcomes. By identifying the factors that are important to producing the outcomes, ideas for process improvement can be generated. Another important concept to foster improvement will be the opportunity to develop multiple collaborations across organizations. Each team that is applying CQI for a patient process, such as management of patients with chronic pain after hernia repair, will develop different process improvement ideas and generate a pool of data. By pooling data and sharing ideas, there will be the opportunity to prevent overlearning, the tendency for a single team functioning in isolation to stop improving.

Traditional research methods such as prospective randomized controlled trials are producing diminishing returns in a world that is changing faster and faster. As with Newtonian principles applied to physics, traditional linear research and statistical methodologies are incomplete when applied to the real world of patient care. With a robust understanding of complex systems science, it is appropriate and necessary to apply more complete nonlinear scientific tools, such as CQI and nonlinear statistical methods, to our patient care. Instead of attempting to prove or disprove a hypothesis, value-based CQI is implemented to improve value to the patient. Traditional clinical research defines inclusion and exclusion criteria, primary outcomes, and length of the study. CQI has no inclusion or exclusion criteria, has many outcome measures, and never ends.

The use of CQI for improving patient care has been supported by healthcare law since the HIPAA law in 1996. These principles were again supported with the Patient Safety and Quality Improvement Act of 2005.

The need for human subjects research protections and the use of IRB processes have been challenged; when true CQI efforts are implemented, there is a clear distinction when compared to human subjects research that does require an IRB process. True CQI is focused on local process improvement and utilizes evidence-based medicine interpreted by the clinical team, ideally including the patient and family in a shared decision process. CQI is not appropriate for pre-market studies, for interventions that could clearly increase risks for patients, or for efforts that intend to produce generalizable knowledge as a priority, rather than local process improvement as a priority. The intent to publish is not sufficient to classify the effort as human subjects research. This information about the distinction between human subjects research and CQI is clearly presented in the FAQ format on the US Health and Human Services website [13]. It should be noted that the results of a CQI project in one local environment do not necessarily apply to another, different local environment. Local environmental variation can produce different patient results from the same process improvement intervention.

Summary

The use of CQI applied to the entire cycle of care for improving value-based outcomes is a complex systems solution for healthcare. The implementation of CQI is facilitated by implementing a multidisciplinary hernia team, by learning how to design dynamic clinical processes, by learning how to interpret data and data analyses, by learning how to generate and implement ideas for process improvement, and finally, by developing a patient and family committee to assist with the hernia team process improvement ideas. Future plans include adding a process activity-based cost model so that true value for the entire cycle of care can be measured and adding additional collaborative hernia teams in other locations so that knowledge can be shared and data can be pooled to define patterns and subpopulations from larger data sets, termed big data.

To our knowledge, this is the first publication demonstrating the use of CQI for patients with chronic pain after hernia repair. Additional prospective randomized controlled studies are not adequate or appropriate for this type of real-world attempt to improve patient value, because they are designed for hypothesis testing and generalizable knowledge, rather than for attempting to improve patient value in a local clinical environment. Continuing to refine processes, define value-based outcomes, and apply complex system data analytics has the potential to improve the value of care delivered in each local environment where these principles are implemented. Additional sites applying CQI and complex systems data analytics will be necessary to allow collaborations that will produce sustainable improvement of value for patients with chronic pain after hernia repair.

References

1. Bohmer RM. Designing care: aligning the nature and management of health care. Boston, MA: Harvard Business Press; 2009.
2. Porter ME, Lee TH. The strategy that will fix health care. Harv Bus Rev. 2013;91(10):50–70.
3. Kaplan RS, Porter ME. How to solve the health care cost crisis. Harv Bus Rev. 2011;89(9):46–52. 54, 56–61 passim.
4. Ramshaw B. Establishing a hernia program and follow-up regimen: a complex systems design for care and improvement. In: Jacob BP, Ramshaw B, editors. The SAGES manual of hernia repair. New York: Springer; 2013. p. 3–18.
5. Bittner R, Bingener-Casey J, Dietz U, Fabian M, Ferzli GS, Fortelny RH, et al. Guidelines for laparoscopic treatment of ventral and incisional abdominal wall hernias (International Endohernia Society–IEHS)-part 1. Surg Endosc. 2014;28(1):2–29.
6. Elhauge E. The fragmentation of the U.S. health care: causes and solutions. New York: Oxford University Press; 2010.
7. Cebul RD, Rebitzer JB, Taylor LJ, Vortruba ME. Organization fragmentation and care quality in the U.S. healthcare system. J Econ Perspect. 2008;22(4):93–113.
8. Hunt R, Newman R. Medical knowledge overload: a disturbing trend for physicians. Health Care Manag Rev. 1997;22(1):70–5.
9. Zimmerman B, Lindberg C, Plsek P. Edgeware: insights from complexity science for health care leaders. Irving, TX: VHA; 1998.
10. Bizzarro MJ, Sabo B, Noonan M, Bonfiglio MP, Northrup V, Diefenbach K, Central Venous Catheter Initiative Committee. A quality improvement initiative to reduce central line-associated bloodstream infections in a neonatal intensive care unit. Infect Control Hosp Epidemiol. 2010;13(3):241–8.
11. Institute of Medicine. Relieving pain in America: a blueprint for transforming prevention, care, education and research. Washington, DC: The National Academies Press; 2011.
12. Macintyre PE, Schug SA, Scott DA, Visser EJ, Walker SM, APMSE Working Group of the Australian and New Zealand College of Anaesthetists and Faculty of Pain Medicine. Acute pain management: scientific evidence. 3rd ed. Melbourne: ANZCA & FPM; 2010.
13. U.S. Department of Health & Human Services. Frequently asked questions about human research. 2015. http://answers.hhs.gov/ohrp/categories/1569. Accessed 20 Apr 2015
14. Mayer-Schönberger V, Cukier K. Big data: a revolution that will transform how we live, work and think. London: John Murray; 2013.
15. Siegel E. Predictive analytics: the power to predict who will click, buy, lie, or die. Hoboken, NJ: Wiley; 2013.

46. Patient Care Manager Perspective on Chronic Groin Pain After Hernia Repair

Brandie Forman and Bruce Ramshaw

Introduction

Where should one begin in defining "patient-centered care" for those suffering from chronic groin pain after hernia repair? Identifying and caring for the medical needs of a patient should certainly be at the center of the focus of the caregivers in any medical or surgical facility. State-of-the-art equipment and procedures, with highly trained physicians and nurses, are aimed at meeting the medical needs of every patient. To assume, however, that "patient-centered care" can and will result solely from the use of specialized equipment and the presence of dedicated physicians and nurses creates the potential for overlooking many important patient needs and fails to achieve true patient-centered care. The need for such care is especially important in dealing with more complex patients such as those who have chronic pain.

True "patient-centered care" begins with the broadest possible definition of a patient's needs. To limit that definition to a medical condition requiring specific medical or surgical procedures may overlook subordinate needs that can rise to overshadow the medical need. Patient-centered care must begin by understanding the total situation from the patient's perspective. This involves meticulous care in communicating with the patient as early as possible in the cycle of care and maintaining that communication far beyond the end of the clinic visit or the hospital stay. This communication must be aimed at establishing the highest level of understanding possible.

The communicator must be able to listen as intently as she speaks. What does a particular question or absence of a question suggest about

B.P. Jacob et al. (eds.), *The SAGES Manual of Groin Pain*,
DOI 10.1007/978-3-319-21587-7_46

the patient's understanding of the entire situation? Why was there such a long pause after your last explanation before the patient's next question? Did the patient really understand what was being explained? Instead of simply answering questions, the communicator must also ask appropriate questions to identify the level of the patient's understanding and to be sure that there are no gaps or holes in that understanding.

Such patient-centered communication must extend beyond the individual seeking and receiving care and must also include the family members who will be involved in the preparation for the care, who will be waiting as the procedure takes place, and who will be providing comfort and assistance for the patient following the clinic visit and/or surgical procedure. Communication with the patient's support system can also identify factors that could be of benefit or could be detrimental to the treatment of chronic pain. For example, our team has observed in some cases that a controlling female influence (mother and/or spouse) for an adult male suffering from chronic pain after hernia repair can predict a more challenging recovery and potentially a less successful outcome. This factor currently observed by our hernia team is potentially related to outcomes. Factors such as this will need to be evaluated in a factor analysis to determine the weighted correlation to various outcomes for a more objective evaluation.

The Role of the Patient Care Manager

The role of the patient care manager in the clinical team is to provide true patient-centered care by facilitating all levels of communication. The patient care manager must become the communicator who creates the environment that will identify the total needs of the patient—including the patient's family/support system—and who will coordinate meeting those needs in an appropriate manner with the entire clinical team. At the same time, the patient care manager and the clinical team must be able to transfer critical information from the medical team to the patient in a manner that will ensure that the ongoing needs of the patient will be met. Listening, explaining, and understanding are the key ingredients in facilitating the dynamics of this exchange.

As previously mentioned, this communication process must begin at a very early stage. In order to ensure proper patient-centered care in a program treating patients with chronic groin pain, the appropriate clearances must be obtained from primary care physicians or referring medical specialists in such areas as pulmonology, cardiology, and other specialized fields. The patient must be informed of the various alternatives for travel to and from the medical facility. Possible needs and alternatives must be explored for convenient accommodations for patient and family members before, during, and following the planned treatment, test, or procedure at the medical center. The patient must be aware of the specific paperwork required for employers, insurance providers, and others with whom the patient is connected. The patient care manager and the clinical team must be aware of and actively considering these needs in order to provide all of the information needed by the patient as soon as it is required. The viewpoint required for all of this is that of the patient. The clinical team should be thinking, "What does the patient need?"

Obviously, the situation of some patients suffering with chronic pain after hernia repair may be more complex than just one individual patient seeking help. Manipulation, opioid abuse, secondary gain, and a number of issues, both conscious and subconscious, may be present and can inhibit the relationship between a patient and the clinical team and inhibit a patient's potential to improve.

It is very important for the patient care manager and clinical team to confront any of these issues. But it is also important that the confrontation be done with empathy and love, not with judgment. Instead of going into a relationship with defenses up, our clinical team attempts to engage every patient with empathy. However, if this is abused, we do put up defenses as needed.

Shared Decision Process

As the medical team identifies the treatment options, there is likely to be a need for some interpretation for the patient and the patient's family. While the medical team may well deal on a daily basis with medical situations identical with or at least similar to those of a particular patient, in most cases the patient has never dealt with such a situation before and/or has difficulty understanding the options due to the presence of chronic pain. The patient care manager and the clinical team must be able to take the medical terminology and the explanation of symptoms, outcomes, and ongoing treatment options and break them down in such a way that the patient can understand their true significance. One cannot assume that the patient understands; there must be patient and family assurance based on the clear and concise communication process by the clinical team. The patient care manager and the clinical team must learn to think like the patient in order to provide the facilitation required to be certain that the patient's needs have been clearly identified and met as much as possible.

The Operation and Postoperative Care

During a surgical procedure, the patient care manager and the clinical team must become the coordinators who remove the vacuum of doubt that can surround those awaiting the outcome of the procedure. This is not a "hand-holding" process, but rather a clear effort to offer complete information regarding the processes taking place beyond the doors of the operating room. The questions must be addressed. The answers must be complete. The "next step" should be foreseen and explained. Never should there be the assumption that the less said the better. In most cases, more information is needed rather than less. This is one clear example of the patient's needs extending beyond the operating room and including those who are waiting nervously to see what happens next. The patient care manager and the clinical team must be available and responsive to that part of the patient that is in the waiting room while the surgery is taking place.

During the postoperative period, and even after discharge from the hospital, communication must continue if the patient's needs are to be met. The clinical recommendations must be transferred to the patient and to the family in such a way that they understand what is included in the postoperative protocol and what is expected of them. Communication between caregivers must also be completed to assure that the patient's needs are met. Once again it is the patient care manager and the clinical team who must facilitate all of this communication.

The role of the patient care manager and the clinical team during the transition from the surgical facility to the next level of care—whether to a rehabilitation center, another care facility, or to the patient's home—is particularly important. There must be assurance that all paperwork and other orders for medical follow-up are completed correctly and in a timely fashion. The patient and the medical team must be assured that care has been taken to ensure proper handling of this next level of care. Once again, communication among all who are involved in the care is critical. The patient care manager and the clinical team must be directly involved in all of this transfer of information. For patients with chronic groin pain, this often includes communication with not only families but also physicians in other locations. Developing relationships with the support and caregivers in the patient's hometown, often at a great distance, is another important role for the patient care manager.

Following discharge from the hospital, there must be no assumption that the care is complete or that all needs of the patient have been met. The patient care manager and the clinical team must make follow-up calls to be sure the recovery is taking the proper track. Additional clinic visits must be arranged and the patient made aware of the importance of those visits. If long-term care is required, the terms and basis for that care must be carefully set forth to assure that the ongoing needs of the patient are met. For patients on chronic opioid medications, the patient care manager often facilitates communication and pain management with a pain specialist near the patient's home.

Patient-Centered Care for the Entire Cycle of Care

"Patient-centered care" involves approaching every element of the entire process (medical, surgical, social, emotional, etc.) from before the initial clinic visit (first contact) to the final and ultimate end of the cycle of care from the patient's perspective. How the patient perceives the care received may very well determine whether or not the treatment was successful. Thinking like the patient, talking like the patient, and even feeling like the patient are essential elements in providing "patient-centered care."

The subject of "patient-centered care" is more critical in the mind of the patient and the patient's family than it can ever be to a caregiver or administrator at a medical facility. That is the foundation, in fact, of the entire concept of "patient-centered care." Viewing medical care and treatment from the perspective of the patient and the patient's family may well clarify the reasoning behind a method of care that focuses on more than the medical condition of the patient but also on the patient's emotions and psyche as he/she prepares for, undergoes, and follows up his/her own care. There are some specific behavioral patterns that are essential for the caregiver to follow in order to be sure the patient is truly at the center of the care given. This pattern from the perspective of the patient, and the spouse, sibling, parents, or offspring of the patient, involves seven specific and overlapping procedures—to listen, to understand, to question, to instruct, to listen, to reinforce, and to reassure.

Obviously, as you read these seven items, you notice that listening is included twice, indicating that it is the central and most important element to ensure that the entire care process is successful in the mind of the patient. Listening is the first element in the care cycle for the care team. As the patient comes forward and attempts to explain his/her condition, listening involves more than just hearing words and understanding symptoms. Listening is crucial to the process of recognizing the mind-set of the patient as he/she attempts to describe not only his/her condition but also what he/she has encountered prior to this phase of care in attempting to deal with the condition. The terminology that the patient uses will probably not be medically correct, so listening to the complete story from the patient involves getting into the patient's mind and understanding what he/she is trying to say and what he/she is feeling. By listening to hundreds of chronic patients suffering after hernia repair, we have heard of several patterns:

- The patient did not do research or think much about the decision to have the repair. The surgeon said it was "no big deal."
- The patient feels stupid for not doing more research.
- The patient expresses the wish that he/she had known more about what was going into their body (e.g., that a mesh was being placed and what type of mesh it would be).
- After surgery, the surgeon would not listen, refused to acknowledge that the patient was in pain, said that the cause was not the mesh or the operation and that the pain was "all in your head," and told the patient not to come back.

Understanding involves the process of leaving the caregiver's medical knowledge in the office and being able to put oneself into the position of the patient. There are a wide variety of emotional and psychological reactions on the part of most patients, as well as the physiological reactions to their medical condition. Fear, confusion, intimidation, uncertainty, and misinformation are some of the more obvious characteristics of many patients. Being placed in a medical institution—whether a doctor's office, a medical clinic, an emergency room, or hospital room—is very uncomfortable for many. For the caregiver, this environment is their normal workday setting. For the patient, however, it is like a foreign land. Not only is the setting uncomfortable, the language they hear—words that they cannot pronounce, with meanings that are either vague or completely unknown—further creates a sense of mystery and increases the likelihood of confusion and intimidation. In order for caregivers to be able to deal effectively with the patient and center their care on the total needs of the patient, they must understand where the patient is—in a strange place, surrounded by strange words, and attempting to deal with a situation they may not understand at all. The caregiver's understanding of the patient involves much more than recognizing and treating the medical symptoms; it involves relating to them as their equal and not their superior.

Understanding the patient in chronic pain after hernia repair includes various patterns with severe impact on the quality of life for the patient and their loved ones. These patterns include:

- Learned helplessness that can lead to a lack of hope, depression, and contemplation of suicide.
- Lack of validation of their pain and suffering. Many patients have stated that they wished that they had cancer so that friends, family, and others would show more empathy.
- Strain on relationships, especially a spouse.
- Significant posttraumatic stress syndrome and other emotional effects similar to those seen in torture victims.
- Confusion or difficulty understanding what has happened to them and how to explain it to others.
- Anger toward the surgeon who did the initial hernia repair or others who may have been involved in the process, and more recently anger toward the mesh company (which may in great part be due to potential financial gain through product liability lawsuits).

One of the best ways for the caregiver to grow from listening to understanding is to ask pertinent questions. These questions will, obviously, involve exploration of the medical condition of the patient, but they must also probe the understanding, the level of fear, the sense of intimidation, and the confusion that the patient is enduring. By recognizing what issues are present, there can be a foundation to care for the patient's total needs. As the caregiver asks questions that probe the areas of the patient's understanding, confidence, and uncertainty, there will develop a bond that will enable the patient to feel more comfortable about the care being offered. They will become more open both to asking the right questions without fear of being ridiculed by one who is wiser than they are and also to applying the instructions the caregiver offers about their condition, the care they are receiving, and the follow-up care upon leaving the medical facility.

As the listening, understanding, and questioning take place, a shared decision process about the best course of action for the patient often happens naturally. This is a uniquely individual process. Some go through the process quickly and others more slowly. The goal of the shared decision process is for the patient and family to feel that they have made a decision that is right for them so that they are confident moving forward with implementation of that shared plan.

The final part of the process of "patient-centered care" is reassurance. This may well take place after the treatment is concluded. The follow-up conversations or other contact with the patient will help the patient to have the assurance that the care received was necessary, timely, and as complete as possible. Even though the results of the procedure may be obvious to the medical team, one must never assume that those results are obvious to the patient. Going the extra mile in explaining after the fact what has happened, what is involved in the recovery and post-care period, and what the patient should expect as a long-term result of the treatment can prove invaluable to the patient as he or she looks back at what has occurred and forward to what is to come.

For patients with chronic inguinodynia after hernia repair who undergo another operation to attempt to relieve pain, it is common for them to need reassurance for several months after surgery. The acute surgical inflammation can lead to times when the patient feels the pain is as bad or worse than ever. This can even occur in patients who gain total pain relief eventually. However, this healing process can take months or even years.

Summary

The role of a patient care manager and the use of "patient-centered care" in the mind of the patient is the only care that should be offered. By involving the elements outlined here in the care process, the patient who has chronic pain after a hernia repair can feel throughout the care process and following the care process that they have been effectively and efficiently cared for.

Suggested Reading

1. Baker SK, Bank L. I'm sorry to hear that: real life responses to patients' 101 most common complaints about health care. Gulf Breeze FL: Fire Starter Publishing; 2008. ISBN 13: 978-0974998657.
2. Diering SL. Love your patients! Improving patient satisfaction with essential behaviors that enrich the lives of patients and professionals. Blue Dolphin: Nevada City, CA; 2004. ISBN 13: 978-1577331414.
3. Frampton SB, Charmel PA, editors. Putting patients first: best practices in patient-centered care. 2nd ed. San Francisco: Jossey-Bass; 2008. ISBN 13: 978-0470377024.
4. Gerteis M, Edgman-Levitan S, Daley J, Delbanco TL, editors. Through the patient's eyes: understanding and promoting patient-centered care. San Francisco: Jossey-Bass; 2003. ISBN 13: 978-0787962203.

47. Workers' Compensation: An Occupational Perspective on Groin Pain, Including Psychosocial Variables, Causality, and Return to Work

Joseph S. Pachman and Brian P. Jacob

Introduction

Groin pain and inguinal hernias are a frequent cause of lost work time [1]. Despite the fact that elective inguinal hernia repair is a commonly occurring surgery, there is surprisingly little evidence-based guidance available regarding return to work, causality determination, and psychosocial variables that impact post-herniorrhaphy functional recovery [2]. These issues are of particular relevance to disability insurance payers, such as workers' compensation carriers, which are contractually responsible for medical treatment, as well as indemnity payments for lost wages that are the result of a workplace injury. Especially since this financial responsibility may extend for years, there is an interest in addressing any potentially contributory comorbid conditions that might result in a more expeditious return to work. In addition to the fiduciary responsibility to accurately assess causation, there is an incentive to identify all of the factors that may have contributed to an injury, so that appropriate prevention practices can be applied as related to future claims.

B.P. Jacob et al. (eds.), *The SAGES Manual of Groin Pain*,
DOI 10.1007/978-3-319-21587-7_47

Evidence and Recommendations

There is significant variability in recommendations regarding return to work post-herniorrhaphy [2]. Survey data suggest that when the occupational job demands involve heavy lifting, return to work recommendations vary from a few days to as long as 3 months post-op [3]. There is evidence that post-herniorrhaphy recommendations for early return to work and unrestricted activity are more likely to result in functional recovery [4]. There is good evidence that return to full duty work, even with high physical demands, should generally not exceed 30 days and this time should generally be even less with laparoscopic surgery [2]. Even in the case of more conservative recommendations for return to work with physical demands that include frequent lifting of greater than 25 lbs., disability of more than 6–8 weeks is not supported by available evidence [5].

Return to Work

In most cases, return to work recommendations can include time-limited initial work restrictions (e.g., sedentary work). These recommendations should never be based on the patient report of job availability, but instead upon sound medical judgment regarding work capacity. Even if accommodated work is not available, this determination is occupational, not medical. Furthermore, there is a good deal of evidence that early return to work, even with appropriate time-limited restrictions, reduces long-term disability [6].

In general, workers' compensation carriers are motivated by expeditious return to work, quality outcomes, appropriately limited use of pre- and postoperative opiate analgesics, and the absence of recurrence. Regarding the latter, the available evidence suggests that there is no difference related to recurrence in the case of early return to work following elective inguinal repair [6]. Not surprisingly, self-employed post-herniorrhaphy patients have been found to return to work sooner than those patients who are receiving disability benefits [7]. There is also evidence that workers' compensation patients report a greater duration of pain and disability post-herniorrhaphy as compared to patients who are receiving group health benefits [8].

Psychosocial Variables

Much of the variation in disability following hernia repair appears to be a function of psychosocial variables. Jones et al. [2] found that apart from age, educational level, income level, occupation, symptoms of depression, and the expectation for return to work accounted for nearly two thirds of the variance in return to work. These authors found that depression significantly delayed return to work in this setting. Parés [9] emphasized the importance of preoperative expectations, as well as cultural and motivational issues related to return to work post-herniorrhaphy. The inflection point as related to likely prolonged disability in workers' compensation appears to be 3 months absence from work [10].

Pain as Basis for Disability Decision

If pain alone is considered to be an ambiguous indication for surgery, this subjective report is even more unclear when used as a basis for disability decisions. There is some evidence that at least 3–6 % of post-herniorrhaphy patients will report some degree of chronic pain and that this is more likely if there was a history of prior chronic pain [11]. It is reasonable to hypothesize that a history of prior workers' compensation claims would similarly be a predictor of chronic post-herniorrhaphy pain, and this may be worth considering in preoperative evaluation. It is also worth noting that opiate analgesics can be a particular concern when there is a claim of work-related pain, with regard to the potential for diversion, medication misuse, and prolonged disability [12].

Catastrophizing

Tripp and Nickel [13] have emphasized the role of "catastrophizing" as related to chronic groin pain and increased disability. Shaw et al. [14] have demonstrated that this psychosocial variable can significantly impact the duration of disability. In this case, catastrophizing refers to misattribution and exaggeration of physiological experiences of groin pain. There is emerging evidence that pain catastrophizing can be effectively mitigated [15]. For example, although a complete discussion of these issues is beyond the scope of this chapter, informing the patient that some time-limited postoperative pain is often evidence of tissue healing and repair incorporation, can reduce negative affect and improve outcome perception [16].

Causality

Physicians treating patients with groin pain will occasionally be asked by a workers' compensation carrier to render an opinion regarding causality. The issue here is whether the payer for the groin pain treatment is more appropriately the workers' compensation or the group health carrier. There is often a motivation by patients to shift this responsibility to the workers' compensation system, given the absence of a deductible or required co-pay. The rate of reimbursement to physicians and hospitals in workers' compensation can vary greatly from state to state.

In most jurisdictions, causality refers to the predominant cause of the symptoms and findings, in this case, groin pain. Usually the question relates to an estimation of causality as related to (greater than 50 %) medical probability (i.e., is the groin pain, more likely than not, the result of the workplace injury or the result of other unrelated processes?). However, it is worth noting that the causation threshold in different state workers' compensation systems can vary.

The question of causality, i.e., whether or not an inguinal hernia may be attributed to a single work-related strenuous lifting event, or even to recurrent strenuous activity, is often a source of litigation. However, despite the large number of claims in this area, there is little evidence to support work-related causality [1]. There is increasing support for the fact that inguinal hernias are more likely related to a congenital or acquired connective tissue weakness [17]. For example, the available evidence does not support an increased risk of developing an inguinal hernia among laborers [18]. It is also interesting to note that inguinal hernias are unusual occurrences in weight lifters [1]. As Hendry et al. further point out, in the work setting, it is likely that when a hernia diagnosis follows a specific lifting or strenuous event, "the event in question has merely brought forward the occurrence of a hernia, and it would most likely have occurred anyway around a similar time" (p. 362). In this respect, the hernia is analogous to a myocardial infarction that happened to occur at work, but was more likely than not the result of underlying atherosclerosis, not the occupational related event. There is also no evidence to support a relationship between a single or even recurrent strenuous work-related (or nonwork-related) event and subsequent hernia recurrence [1].

Maximal Medical Improvement

Physicians who are treating injured workers with groin pain might on occasion be asked to comment on maximal medical improvement (MMI). MMI is not necessarily related to a prescribed post-hernia repair time, for example, but rather to that point at which little or no further improvement is anticipated (more likely than not) and treatment gains appear to have plateaued. When an injured worker is determined to be at MMI, a physician might also be asked to provide a permanency rating. This rating of permanent impairment will help to determine the amount of benefits for the patient by the workers' compensation or disability insurance carrier. For example, usual successful hernia repair would be expected to result in a 0 % impairment rating. A patient who reports an absence of pain but for whom there is a recurring groin protrusion with increased abdominal pressure, and for whom there are some appropriate lifting restrictions (and surgery declination), would receive an impairment rating of 10 %. In the case of ongoing pain due to documented residual nerve entrapment, or in the case of recurrent inguinal hernias that are only partially reducible despite repeated surgical repair, it is advised that the American Medical Association's *Guides to the Evaluation of Permanent Impairment* be referenced [19], as these determinations are less straightforward.

Conclusion

In summary, there is increasing evidence that groin pain and inguinal hernias are not usually related to workplace injuries, that disability duration is often overestimated, and that chronic groin pain and herniorrhaphy recovery are in part related to psychosocial variables, some of which are modifiable. Disability and subjective reports of chronic pain are disproportionally higher among workers' compensation patients. Workers' compensation carriers are generally interested in value. If a hernia surgeon is able to document improved outcomes and decreased recurrence rate, there is often an opportunity for a preferred referral relationship (that is not necessarily related to reimbursement).

Worker's compensation can be arcane. Although it is described as a single system, in reality, it is a complex set of often challenging rules and regulations that vary from state to state. If case-specific questions arise, a treating physician can consider engaging the carrier medical director to help clarify.

Disclosure Dr. Pachman is the Regional Medical Director, Liberty Mutual Group; the views expressed here do not necessarily reflect the views of Liberty Mutual.

References

1. Hendry PO, Patterson-Brown S, de Beaux A. Work related aspects of inguinal hernia: a literature review. Surgeon. 2008;6(6):361–5.
2. Jones KR, Burney RE, Peterson M, Christy B. Return to work after inguinal hernia repair. Surgery. 2001;129(2):128–35.
3. Ismail W, Taylor SJ, Beddow E. Advice on driving after groin hernia surgery in the United Kingdom: questionnaire survey. Br Med J. 2000;321(7268):1056.
4. Callesen T, Klarskov V, Bech K, Kehlet H. Short convalescence after inguinal herniorrhaphy with standardized recommendations: duration and reasons for delayed return to work. Eur J Surg. 1999;165(3):236–41.
5. Forbes J, Fry N, Hwang H, Karimuddin AA. Timing of return to work after hernia repair: recommendations based on literature review. BC Med J. 2012;54(7):341–5.
6. Franche RL, Cullen K, Clarke J, Irvin E, Sinclair S, Frank J, Institute for Work & Health (IWH) Workplace-Based RTW Intervention Literature Review Research Team. Workplace-based return-to-work interventions: a systematic review of the quantitative literature. J Occup Rehabil. 2005;15(4):607–31.
7. Bourke JB, Lear PA, Taylor M. Effect of early return to work after elective repair of inguinal hernia: clinical and financial consequences at one year and three years. Lancet. 1981;2(8247):623–5.
8. Salcedo-Wasicek MC, Thirlby RC. Postoperative course after inguinal herniorrhaphy. A case-controlled comparison of patients receiving worker's compensation vs patients with commercial insurance. Arch Surg. 1995;130(1):29–32.
9. Parés D. Return to work after elective inguinal hernia repair. Cir Esp. 2013;91(8):473–5 (Article in Spanish).
10. Franklin GM, Wickizer TM, Coe NB, Fulton-Kehone D. Workers' compensation: poor quality health care and the growing disability problem in the United States. Am J Ind Med. 2014;58(3):245–51. doi:10.1002/ajim.22399.
11. Dennis R, O'Riordan D. Risk factors for chronic pain after inguinal hernia repair. Ann R Coll Surg Engl. 2007;89(3):218–20.
12. Franklin GM, Mai J, Turner J, Sullivan M, Wickizer T, Fulton-Kehoe D. Bending the prescription opioid dosing and mortality curve: impact of the Washington State opioid guideline. Am J Ind Med. 2012;55(4):225–31.
13. Tripp DA, Nickel JC. Psychosocial aspects of chronic pelvic pain. Pain Clinl Updates. 2013;21(1):1–6. Available from: http://www.iasp-pain.org/PublicationsNews/NewsletterIssue.aspx?ItemNumber=2063. Accessed 11 Feb 2015.
14. Shaw WS, Reme SE, Pransky G, Woiszillo MJ, Steenstra IA, Lindon SJ. The pain recovery inventory of concerns and expectations: a psychosocial screening instrument

to identify intervention needs among patients at elevated risk of back disability. J Occup Environ Med. 2013;55(8):885–94.
15. Darnall BD, Sturgeon JA, Kao MC, Hah JM, Mackey SC. From catastrophizing to recovery: a pilot study of single-session treatment of pain catastrophizing. J Pain Res. 2014;7:219–26.
16. Devine EC. Effects of psychoeducational care for adult surgical patients: a meta-analysis of 191 patients. Patient Educ Couns. 1992;19(2):129–42.
17. Szczesny W, Cerkaska K, Tretyn A, Dabrowiecki S. Etiology of inguinal hernia: ultrastructure of rectus sheath revisited. Hernia. 2006;10(3):266–71.
18. Abrahamson J. Etiology and pathophysiology of primary and recurrent groin hernia formation. Surg Clin North Am. 1998;78(6):953–72.
19. Rondinelli RD, Genovese E, Brigham CR, American Medical Association. Guides to the evaluation of permanent impairment. 6th ed. Chicago, IL: American Medical Association; 2008.

Index

B.P. Jacob et al. (eds.), *The SAGES Manual of Groin Pain*,
DOI 10.1007/978-3-319-21587-7

I

R

S

MIX
Papier aus verantwortungsvollen Quellen
Paper from responsible sources
FSC® C105338

If you have any concerns about our products,
you can contact us on
ProductSafety@springernature.com

In case Publisher is established outside the EU,
the EU authorized representative is:
Springer Nature Customer Service Center GmbH
Europaplatz 3, 69115 Heidelberg, Germany

Printed by Libri Plureos GmbH
in Hamburg, Germany